T0321719

Supernovae and Stellar Wind in the Interstellar Medium

Supernovae and Stellar Wind in the Interstellar Medium

Tatiana A. Lozinskaya

Translated by
Marc Damashek

Library of Congress Cataloging-in-Publication Data

Lozinskaĭa, T. A. (Tat'ĭana Aleksandrovana)
 [Sverkhnovye zvezdy i zvezdnyĭ veter. English]
 Supernovae and stellar wind in the interstellar medium /
Tatiana Lozinskaya.
     p.   cm.
 Translation of: Sverkhnovye zvezdy i zvezdnyĭ veter.
 Includes bibliographical references and index.
 ISBN 978-0-88318-659-6
 1. Supernovae.   2. Stellar winds.   3. Galaxies.   I. Title.
QB843.S95L6913   1991
523.8'446–dc20   91-46587
                                                        CIP

Translator's note

    With the kind permission of Dr. Lozinskaya, I dedicate this translation of her book to the memory of Dr. Phil Angerhofer, friend and colleague.

Contents

Chapter 4. The Effects of Supernovae and Stellar Wind on Gas and Dust in the Galaxy

Dedicated to the Memory of Iosif Samuilovich Shklovskii

❖ ◎ ❖

Preface to the First Edition

Supernovae and stellar wind have become pivotal problems of the day, linking such fields of research as stellar structure and evolution, the physics and chemistry of the interstellar medium, and star formation. The study of supernovae bears upon many of the central issues of modern astrophysics, such as neutron stars and black holes, nucleosynthesis and the origin of the primary cosmic rays, and neutrino and gravitational radiation. In this book, we consider just one aspect of a wide range of problems associated with supernovae and stellar wind, namely their multifaceted interaction with the interstellar medium

The burst of radiation and ejection of a supernova shell, as well as the slower mass outflow in the form of a stellar wind, combine to ionize, heat, and accelerate the surrounding gas, and they alter its density and chemical composition. This "disturbance" affects not just a star's immediate environment, but the Galaxy as a whole. On the other hand, it is precisely these properties of the interstellar medium that govern the complex of phenomena accompanying the expansion of the ejected stellar matter, and finally, the star formation process itself.

It would certainly be unthinkable to treat in a single book, however concisely, the full range of pertinent research bearing upon these subjects. Nevertheless, we have combined the two basic themes, somewhat to the detriment of both, although we are not simply catering to the biases of the author, who has spent more than 20 years experimentally investigating supernova remnants and the nebulae produced by stellar wind. It is now clear that both of these phenomena — supernovae and stellar wind — must be examined in concert with one another. Observations from space have shown that mass loss in the form of a stellar wind is present in all early-type stars, and is enhanced in the later stages of evolution, often being accompanied by the complete loss of the outer hydrogen layers. Mass loss from red supergiants, the immediate precursors of type II supernovae, typically takes place at a rate of $\dot{M} \approx 10^{-6} M_\odot$ /yr, and may be as high as $10^{-4} M_\odot$ /yr at a velocity $V_w \approx 10$ km/sec just before they explode. The wind from Wolf–Rayet stars, the possible progenitors of objects like Cassiopeia A, is even stronger: $\dot{M} \approx 10^{-5} M_\odot$ /yr, with $V_w \approx 2 \times 10^3$ km/sec. This means that the supernova shell is ejected into a medium already "disturbed" by the stellar wind, and only by taking into account the mass outflow that has occurred in previous stages of evolution can one correctly interpret observational results on supernovae and their remnants.

Stellar wind and supernovae are inextricably linked, as they constitute the agency that regulates the physical state of the gaseous medium in the Galaxy — more specifically, they control the heating of the gas, the formation of the giant gas and dust supershells surrounding OB associations, and the creation and destruction of cold gas clouds. Finally, supernovae and the wind from OB associations lead to gravitational instability of giant molecular clouds and trigger star formation; they are responsible for the recycling of matter and for enriching it with heavy elements.

The need for a joint study of stellar wind and supernova remnants has been recognized only fairly recently, and the two problems are still far from being completely understood. As before, there is no comprehensive answer to the question of how and why stars explode. Research on supernovae and stellar wind and their influence on the interstellar medium is being pursued on a broad front, and its tools consist of the entire arsenal of astrophysical techniques. The most copious sources of information have turned out to be at x-ray wavelengths, and these have only been properly exploited quite recently. Here is a vigorous branch of astrophysics that is undergoing continuous development, and the flow of new information is increasing rapidly. The results that we present here will therefore most assuredly reflect our present understanding, much of which is still controversial.

The author thanks N. G. Bochkarev, V. I. Slysh, and N. N. Chugai for reading this book in manuscript form and providing constructive criticism.

Preface to the Second Edition

The first Russian edition of this book was completed in late 1984 and early 1985. Having applied the finishing touches, the author was prepared to begin a new version of the manuscript immediately, reflecting up-to-the-minute developments. To a large extent, the present English edition represents that complete rewrite. But we have not attempted simply to examine the most important new results on individual objects, since in the two to three years intervening between the first and second editions there have been fundamentally important "global" developments in the subject matter dealt with here.

First of all, the classification of supernovae has changed, a manifestation of our new depth of understanding of the final evolutionary stages in stars of different masses, as well as of the supernova explosion phenomenon itself.

Secondly, the launch of the IRAS observatory has essentially opened a new infrared window, one that has turned out to be extremely promising for research on supernova remnants. This has stimulated a third major advance in the study of the nature and evolution of these objects (the first coming with the advent of radio astronomical observations, and the second with the exploitation of x rays as signaled by the launch of the Einstein Observatory). IRAS data seem to convey much information relevant to the study of wind-blown bubbles as well.

Thirdly, it has become perfectly clear that it is impossible to carry out a satisfactory analysis of supernovae or supernova remnants of arbitrary age without allowing for the effects of the radiation and stellar wind emanating from the progenitor into the surrounding interstellar medium.

Finally, the explosion of supernova 1987A in the Large Magellanic Cloud has shed new light on all of supernova science. It has altered our understanding of stellar explosions and the development of young supernova remnants, and will continue to do so for decades to come. Furthermore, having such a bright explosion so close by presents novel opportunities for the study of interstellar and intergalactic gas. We need only mention, for example, that observations of the light echo from SN 1987A have provided an unprecedented opportunity to perform three-dimensional tomography of the interstellar medium. The observational program for SN 1987A continues to grow, and every passing day, month, and year promises not just answers to long-held questions, but new puzzles as well.

These new topics have been included in the second edition, although it goes without saying that neither the old nor the new version of the book could possibly pretend to an exhaustive treatment of the problem. Not being able to keep up with the avalanche of new results seems inevitable in this rapidly developing branch of astronomy, but it does leave us hopeful — it is, after all, what makes our science so enthralling.

Chapter 1
Supernovae and Young Supernova Remnants

The term *supernova* was coined by Baade and Zwicky to distinguish from the more ordinary novae those anomalously bright stellar outbursts which at maximum often outshine their entire parent galaxy. Trimble (1982) has pinned down precise dates: the word supernova was first heard in lectures given in 1931, and became familiar to the astronomical community at the December, 1933 meeting of the American Physical Society.

It is striking that by 1934, Baade and Zwicky had brilliantly guessed that a supernova explosion is accompanied by the transformation of a "normal" star into a neutron star, and their notion of the energies involved is consistent with our present understanding. But as much as three decades later, in the preface to his fundamental treatise *Supernovae*, Shklovskii emphasized that "we cannot say with any assurance why it is that only certain stars explode, and what sets those stars apart from others." This was repeated in the second edition, ten years later — a decade marked by stupendous success in all aspects of supernova science — and we still have no comprehensive answer!

What we presently know of supernova explosions has been gleaned from two major sources: observations of extragalactic supernovae, and investigations of young supernova remnants within our own Galaxy, the remnants of the so-called "historical" supernovae. Manuscripts and historical chronicles have enabled us to identify supernova outbursts in the year 185, possibly in 393, in 1006, 1054, 1181, 1408(?), 1572, and 1604; to this list we may add the explosion in Cassiopeia, which has been quite reliably dated to 1680. A young supernova remnant with optical nebulosity, an extended radio and x-ray source, and in two instances a compact stellar remnant, have been identified with each of these. These objects have been closely scrutinized, and a great deal of observational data has been accumulated, for these are the few instances in which we are able to study nearby remnants in an early phase of development, when the matter ejected by the explosion is directly observable.

The panoply of phenomena involved in the explosion of a supernova is so complicated that it is impossible to analyze the observations without having some model in mind. We therefore anticipate our main exposition with a brief outline of the phenomenon, based on the totality of the current observational and theoretical data; see the diagram in Fig. 47.

The explosion of a supernova marks either the complete disruption of a star or the ejection of its outer layers as its core collapses, due to the loss of thermal and mechanical stability at the endpoint of stellar evolution. The matter thrown off expands into gas that left the surface of the progenitor at some previous stage of development, sweeping it up and communicating the energy of the explosion to the surrounding medium. This is accompanied by the appearance of two shock waves: the first — the blast wave — propagates outward into the surrounding gas, and the second, the "reverse" shock, moves back inward through the expanding ejecta. A convective layer forms at the contact discontinuity between the swept-up and ejected gas due to Rayleigh–Taylor instability.

The energy liberated by the series of reactions $^{56}Ni \rightarrow ^{56}Co \rightarrow ^{56}Fe$ and by the radiative cooling of the shell torn off by the shock wave becomes observable as the supernova outburst. The ejected gas and the swept-up gas of the stellar wind (or the interstellar medium), heated by the blast wave and reverse shock, radiate at ultraviolet and x-ray wavelengths; this radiation is visible immediately after the explosion and for centuries thereafter — i.e., in young supernova remnants. The enhancement of the magnetic field and the acceleration of relativistic particles in the convective layer at the boundary of the ejecta and at the front are responsible for the radio emission from supernovae and young shell-like remnants. Dense clumps of ejected matter and clumps of the wind or interstellar gas heated by the shock wave show up as bright optical filaments in young remnants. If a stellar remnant — a pulsar — is formed in the explosion, synchrotron emission from the relativistic electrons that it injects may possibly be observed as a supernova radio burst, and it will most certainly be observed as the radio, x-ray, and optical emission from the young remnant, a so-called *plerion*. The slow injection of energy provided by the pulsar governs the later behavior of the supernova light curve. The interaction of the pulsar wind with matter ejected in the explosion can perhaps alter the dynamics of the shell during the earliest stages of expansion, and completely governs the way in which the synchrotron radio and x-ray emission from the young Crab-like remnant changes in time.

The hot plasma responsible for thermal x-ray emission from the remnant is the source of collisional heating of the circumstellar dust (both interstellar dust and that associated with the progenitor stellar wind). The heated dust radiates in the infrared, and as has recently been revealed, cooling via radiation from dust is the dominant energy drain in young remnants (with a plasma temperature $> 10^6$ K).

With this scheme in mind for the interaction between the collapsed stellar remnant, expanding supernova shell, progenitor wind, and interstellar gas, we turn to an analysis of observations of extragalactic supernovae (Section 1), SN 1987A in the Large Magellanic Cloud (Section 2), and young remnants (Sections 3–5). Our present state of understanding is as follows. The outbursts of SN 1604 (Kepler), SN 1572 (Tycho), and

SN 1006 and their remnants fit well within the confines of our current ideas about type I supernovae (the latter two are definitely SN Ia). The Crab Nebula (SN 1054), 3C 58 (SN 1181), and the object 0540–69.3 in the Large Magellanic Cloud are young remnants of type II outbursts, but they differ considerably from one another. Cassiopeia A and similar objects, which have come to be known as oxygen-rich remnants, are neither SN Ia nor SN II, but instead form a separate class which is perhaps identifiable with SN Ib; this, however, remains an unsolved problem.

We attempt in Section 6 to give a coherent picture of how stars of different masses end their lives, based on observations of supernovae and supernova remnants. But we must relegate to a future time a lucid and incontrovertible comparison of the observational data on supernovae and young remnants with the theory of stellar interiors and evolution (including the complete disruption or collapse of the core). (We attempted such a comparison in the survey by Blinnikov et al. (1988), but the results were quite tentative, and not entirely unambiguous.)

1. Supernovae: classification, statistics, light curves, spectra, and radio emission

The last supernova to be observed in our Galaxy appeared nearly four centuries ago, and everything that we know about SN light curves, spectra, and rates of occurrence in galaxies of various morphological types comes from observations of extragalactic supernovae. The first extragalactic supernova ever discovered — S And — was found in M31 one hundred years ago. On the initiative of Zwicky, a systematic search for supernovae in nearby galaxies was begun in 1934. Our present databank of extragalactic supernovae numbers more than 600 objects, to which several dozen outbursts are added annually (Barbon et al., 1984, 1988).

The new classification of supernovae

Until recently, supernovae were conventionally classified into one of two families, type I or type II (SN I or SN II). This scheme, which was proposed by Minkowski in 1941, was in use up through 1985, although there were various attempts to improve upon it by relegating supernovae to as many as five or even eight types. The basis for classification consisted of the presence (SN II) or absence (SN I) of hydrogen lines in the spectrum near maximum light. This is a fundamental criterion, not only because one or two high-quality spectra suffice for a classification, but because it is of clear-cut physical import as well. We will argue below that stars that have shed their outer hydrogen layers in the course of evolution are pre-SN I stars, while those that have retained them up to the moment of the explosion are pre-SN II stars. Type I and II supernovae also display different light curves and luminosity at maximum light, but the introduction of more classification criteria by no means simplifies the task of classification. The spread in luminosity and variations in both light curves and spectra within a single type are large, so that there is effectively an "overlap" between the populations of SN I and SN II (for example, see Tammann, 1977; Trimble, 1982; Bartunov and Tsvetkov, 1986). Physical arguments have been put forth for dividing each of the types into two subtypes — rapidly and slowly decreasing brightness among the SN I, and linear and plateau light curves among the SN II — and then finding a correlation between the subtypes and the velocity of the matter ejected in the outburst, the color, the maximum brightness, and so on (Barbon et al., 1973, 1979; Branch, 1971, 1982; Trimble, 1982; Pskovskii, 1977a, b, 1984).

The separation of SN I into two subtypes suggested itself long ago (e.g., see Shklovskii (1984b), Oemler and Tinsley (1979)), as their occurrence throughout spiral and elliptical galaxies and their correlation with the rate of star formation attest to the existence of two populations of SN I precursors. Although type I supernovae were found to comprise a more homogeneous group than SN II, judging by their light curves, spectra, and maximum luminosity, a new system of classification was introduced in 1985–1986, with three types of supernova: SN Ia, SN Ib, and SN II. It was found that the so-called "peculiar" SN I (Bertola et al., 1965), which are characterized by the absence of Si II absorption at $\lambda \sim 6150$Å near maximum light, differ from "normal" SN I in terms of a great many parameters. This made it possible to identify the former "peculiar" SN I with the new SN Ib, in contrast to the "normal" SN Ia (Panagia, 1985; Panagia et al., 1986; Wheeler and Levreault, 1985; Elias et al., 1985; Porter and Filippenko, 1987). Thirteen supernovae are classed as SN Ib, including SN 1983N, SN 1984L, and SN 1985F. At maximum light, SN Ib tend to be somewhat weaker and redder than SN Ia, they typically show strong nonthermal radio emission near optical maximum, and their outbursts occur in S galaxies and regions of star formation. The most clear-cut differences show up in the later stages, at $t \gtrsim 250^d$: lines of [Fe II] and [Fe III] dominate the spectra of SN Ia, while lines of [O I] dominate the spectra of SN Ib (Gaskell et al., 1986). We shall demonstrate below that none of these distinctions is purely phenomenological — they all bear on the specific physics of the situation. Thus, the term *SN I* will henceforth be considered to be essentially meaningless unless it is further qualified as *SN Ia or SN Ib*.

As before, the new classification is based on spectral features. Within types Ia and Ib, there exist minor differences among the light curves within type II, the differences are particularly pronounced. Also as before, there remain anomalous and unclassified objects. The clearest example of the latter is SN 1961V in the galaxy NGC 1058, with a unique light curve, unusual spectrum, and anomalously low expansion velocity (Branch and Greenstein, 1971; Utrobin, 1984). The possibility also looms ever larger that we have not identified all the various types of supernovae: extragalactic objects that are faint at maximum light could easily escape detection at present.

Statistics of supernovae
The distribution of supernovae among galaxies of different morphological types and within their host galaxies is, as we shall see in Section 6, fundamental to our understanding of the nature of pre-supernovae. Related problems include the origin of cosmic rays, the physical and chemical state of the interstellar medium, the birth of pulsars, and so on. It would seem to be easy to find out what the supernova rate is by counting the number of supernovae in the various galaxies and knowing the time of observation. But here, in all its glory, we encounter the basic problem of all of observa-

tional astronomy: the need to take account of selection effects. The differences in outburst rates estimated by different authors seemingly arise through a multitude of effects that corrupt the observed distribution of supernovae. Apart from "general astronomical" factors affecting the completeness of the sample (we must be sure that all supernovae outbursts in a program galaxy are above the limit of detectability), such as interstellar extinction and distance, we need to be aware of observational biases peculiar to supernovae. For example, SN Ia, SN Ib, and SN II all have different maximum brightness, which can lead to an underestimate of weaker supernovae in distant galaxies. Supernovae concentrate toward the center of a galaxy, but the nuclear region may be overexposed on photographic plates, hindering identification. Types II and Ib concentrate within the spiral arms of galaxies, behavior not displayed by type Ia. The result may be an underestimate of the number of SN II and SN Ib, inasmuch as absorbing matter is also concentrated within the arms. Finally, the most difficult factor to take into account is the inclination of the host galaxy to the line of sight. Supernovae of all types (but especially SN II and SN Ib) form a flattened system; i.e., they are concentrated exactly where most of the gas and dust resides. This implies that all of the supernovae in an edge-on spiral may not be detectable, while this is not an important effect in ellipticals.

It is therefore necessary to correct observations for the inclination of spiral galaxies, or more ideally to count only those supernovae detected in face-on galaxies, which of course reduces the overall size of the sample and the statistical fidelity of the results. It also completely fails to provide any sort of reliable quantitative measure of absorption in irregular galaxies.

These selection effects and the most appropriate means of dealing with them have been discussed by Tammann (1977, 1982), van den Bergh and Maza (1976), Shklovskii (1976a), and others.

Supernova rates are quoted per unit galactic luminosity or per unit galactic mass, since the very first counts for late-type spiral systems showed that the number of outbursts is correlated with galactic mass and luminosity, and one would naturally expect such behavior in other types of galaxies as well.

The most detailed analyses of the occurrence of supernovae in galaxies of various morphological types were carried out by Tammann (1982), Tsvetkov (1983, 1987a), and van den Bergh et al. (1987) (see also van den Bergh (1987) and Cappellaro and Turatto (1988)). The samples employed for the counts were as follows.

Tammann: 400 galaxies from the Shapley–Ames catalog with radial velocities less than or equal to 1200 km/sec, including galaxies in the Virgo cluster, regardless of their velocity. For a Hubble constant $H = 50$ km/sec/Mpc, this sample is complete to a distance of 24 Mpc. In the period 1960–1976, a total of 77 supernovae were detected. A further sample of 2955 galaxies from the list of de Vaucouleurs et al. (1976) was used, comprising 173 supernovae; the latter list is not complete out to any particular distance, and was utilized to assess the relative frequency of supernovae.

Tsvetkov: photographic search for supernovae using the 40-cm astrograph of the Sternberg Astronomical Institute Crimean station. Sky coverage consisted of 32 areas, taking in approximately 1500 galaxies brighter than $m_{ph} = 15$. This sample is the least populous so far, with only 32 supernovae detected up to 1989, but it has the advantage of being one of the most homogeneous in observational terms. The effective observing time is determined by the frequency with which a given area is photographed and the length of time that the different types of supernovae spend above the limit of detectability in each individual galaxy, with due allowance for the SN light curve, the distance to the galaxy, its inclination, and absorption.

Cappellaro and Turatto (1988): photographic search for supernovae in 736 galaxies from the catalog of de Vaucouleurs (1976) at Asiago observatory. A total of 51 supernovae have been discovered since 1959. The value of this sample also lies in its observational homogeneity, enabling an accurate determination of the effective observing time.

van den Bergh et al. (1987): visual observations of 1017 galaxies by Evans from 1980 to 1985, using a 25-cm telescope. A homogeneous data set comes from observations of 748 Shapley–Ames galaxies. The sample is complete to $m_v \sim 14.5$. In five years, Evans was the discoverer of 11 supernovae; he also found an additional four, but was not the first to report them. The effective observation time is also different here for each galaxy, depending on the frequency of observations and the light curves of SN Ia, SN Ib, and SN II.

Visual observations have their advantages. Above all, they give one access to the bright nuclei of galaxies, which are often overexposed in photographs. Furthermore, to take maximum advantage of total available observing time in photographic searches, preference is given to rich clusters of galaxies, where early-type galaxies predominate.

Systematic errors in estimating supernova rates can stem from the fact that not all supernovae are detected in the allotted observing time (we shall see below that there are anomalously weak and "short-lived" supernovae), from overexposure of galactic nuclear regions in photographs, from the nonuniform distribution of dust in spiral galaxies, and from improper allowance for the inclination of a galaxy. Among the workers mentioned above, the overall error is probably less than 50%. Random errors are negligible for such supernova-rich morphological types as Sc, and may be as high as 30-50% in early- or very late-type galaxies.

The absolute supernova rates according to Tammann (1982) and Tsvetkov (1987a) are given in Table 1. There we use what has become the standard unit of measure, $1SNu \equiv 1SN/(10^{10} L_{B\odot} \cdot 100 \text{ yr})$, where $L_{B\odot}$ is the absolute blue magnitude of the sun, $L_{B\odot} = 5.^m 48$ (Tammann, 1982). The mutual consistency of the results is fairly good, considering the uncertainty in the relevant selection effects.

Table 1. Supernova Rates in SNu (1 SN per $10^{10} L_{B\odot}$ per 100 years).
Data of Tammann (1982) and Tsvetkov (1983, 1987a).

Type of galaxy	Tammann			Tsvetkov		
	All SN	SN I	SN II	All SN	SN I	SN II
E	0.22	0.22	0	0.1	0.1	0
S0	0.12	0.12	0	0.1	0.1	0
S0a, Sa	0.28	0.28	0	—	—	—
Sab, Sb	0.69	0.37	0.32	0.23	0.23	—
Sbc, Sc, Scd, Sd	1.38	0.77	0.61	1.03	0.47	0.56
Sdm, Sm, Im	1.02	0.88	0.19	1.3	1.3	—
I0	?	?	?	—	—	—

The supernova rate in S galaxies quoted in Table 1 may possibly be an overestimate, due to the very large correction for inclination required in such galaxies (van den Bergh, 1987; van den Bergh et al., 1987). Extrapolating the estimates of Tammann to the sample searched by Evans leads one to expect about 50 supernovae, but the number actually discovered is 15. The discrepancy between the statistics of Tsvetkov and the discoveries of Evans is somewhat smaller, but still significant.

The same holds true when one compares these average statistical estimates with observational results for M31. Tammann's statistics predict a mean time between supernova outbursts of $\tau = 21$ yr for M31 ($M_B = -21.67$, type Sb); according to Capellaro and Turatto, $\tau = 42$ yr. Now M31 was the host to SN 1885. The equivalent observing time for M31 in the ensuing century has been at least 72 years: since 1917, at least one photograph per year has been taken to a limiting magnitude $m_{ph} = 18$ or fainter (data kindly provided to the author by A. S. Sharov). With $\tau = 21$ yr, the lack of a supernova over the 72 years following SN 1885 may be viewed as a statistical fluctuation with an associated probability P(1 SN) = 3%. The probability of observing 2, 3, or 4 supernovae (including SN 1885) in this same period would be P(2 SN) = 11%, P(3 SN) = 19%, P(4 SN) = 22%, respectively, if the statistics are Poisson-distributed. The supernova rate quoted by Capellaro and Turatto with $\tau = 42$ yr yields corresponding probabilities P(1 SN) = 18%, P(2 SN) = 30%, P(3 SN) = 26%, P(4 SN) = 15%. (The lack of supernovae in other galaxies of the Local Group since observations began is fully consistent with the expected frequency. Tammann (1982) gives τ(M33) = 110 yr, τ(LMC) = 268 yr, and τ(SMC) = 1008 yr.) The supernova rate in our Galaxy is discussed further in Section 11.

The segregation of SN I into two new types requires a reconsideration of the statistics. The first steps in that direction have already been taken.

Table 2. Supernova Rate in an "Average" Shapley–Ames Galaxy. Data of van den Bergh et al., (1987) ($h \equiv H/100\ \mathrm{km}\cdot\mathrm{sec}^{-1}\cdot\mathrm{Mpc}^{-1}$).

Type of SN	Adopted M_V(max)	$\nu\,(SNu)$
SN Ia	-20.1^{m}	$0.28(0.25-0.33)h^2$
SN Ib	-18.6^{m}	$0.43(0.33-0.70)h^2$
SN II	-18.0^{m}	$1.09(0.66-2.10)h^2$

Table 2 lists the rates of SN Ia, SN Ib, and SN II in an "average" Shapley–Ames galaxy, based on the data of van den Bergh (1987) and van den Bergh et al. (1987). The values in parentheses correspond to limiting magnitudes $m_v(\mathrm{lim}) = 15$ and $m_v(\mathrm{lim}) = 14$, with an average value $m_v(\mathrm{lim}) = 14.5$; in Table 2, $h \equiv H/(100\ \mathrm{km/sec/Mpc})$.

In Fig. 1a, we have plotted the distribution of the absolute rate of occurrence of SN Ia, SN Ib, and SN II for galaxies of different morphological types, using the data of Tsvetkov (1989); Fig. 1b gives the results of counts by Cappellaro and Turatto (1988), which were carried out without dividing SN I into two classes, but using a larger sample. It can be seen that the relative rates of SN Ia and SN Ib in spiral galaxies are approximately the same (this was earlier pointed out by Branch (1986)).

We may summarize the basic, well-grounded results of supernova counts, which are important to an understanding of the origin of the different types of supernovae, as follows.

1. Only SN Ia occur in elliptical galaxies. This well-known fact provides evidence that the immediate precursors of SN Ia are low-mass, old-population stars, with some mechanism for prolonging the life of a SN Ia progenitor being required: see Section 6.

2. Spiral galaxies play host to supernovae of all types, with the SN II rate typically equal to the overall type I rate, SN Ia + SN Ib. Estimates of the SN Ib rate that take their lower luminosity into account indicate that they probably occur in S galaxies just as often as SN Ia.

3. Type Ia supernovae are evenly distributed throughout the disk of their host galaxy, and display no significant correlation with spiral arms. Type Ib and II outbursts occur primarily in spiral arms. Type II supernovae are correlated with H II regions; out of 13 type Ib supernovae, at least six or seven are located within such regions. The correlation of SN Ib and SN II with spiral arms and H II regions suggests that their progenitors ought to be short-lived stars of high initial mass $M_{MS} \gtrsim 6-10 M_\odot$.

4. The normalized supernova rate increases as one progresses from early elliptical galaxies through later types. The rate of SN II in Sd and Im galaxies is somewhat lower than in Sc and Scd, and this requires explanation, as these are galaxies with a plentiful supply of gas and young stars. The effect may possibly be the result of inadequate sample size.

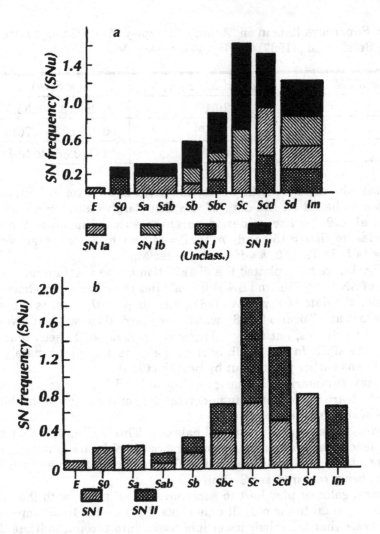

Fig. 1. a) Rate of occurrence of SN Ia, SN Ib, and SN II in various types of galaxies (Tsvetkov, 1988). **b)** Rate of occurrence of SN Ia+SN Ib, and SN II according to Capellaro and Turatto (1988).

5. Of the I0 galaxies, we can only state that SN Ia outbursts occur within them often. This is consistent with the basic implications of the statistics, since morphologically this type is contiguous with the ellipticals. Two outbursts have been seen in each of two I0 galaxies: SN 1895A and 1972E in NGC 5253, and SN 1965I and 1983G in NGC 4753. It is impossible, however, to properly assess the effects of absorption and orientation in such galaxies.

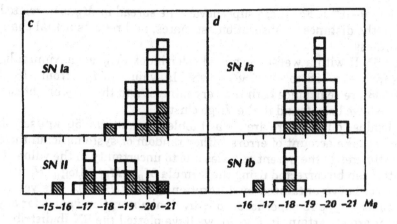

Fig. 1 (cont'd). c) Absolute magnitude at maximum light, M_B(max) for SN Ia and SN II with individual measurements of $E(B-V)$. **d)** The same quantity for SN Ia and SN Ib, uncorrected for absorption, from the data of Tsvetkov (for $H = 50$ km/sec/Mpc). Hatching indicates supernovae in the Virgo and Ursa Major clusters, as well as supernovae with a color excess $E(B-V) > 1$.

Absolute magnitude at maximum light

This quantity can be derived from the apparent magnitude for supernovae at well-determined distances if one makes allowance for absorption of light both in our own and in the host galaxy. Although the former is usually known more or less reliably, the latter correction in S and Im galaxies is rather uncertain. M_{max} can therefore only be found with any assurance for supernovae that have individually measured color indices. By assuming a normal color index at maximum, averaged over SN I and SN II, of $(B-V)_0 = -0.^m 15$ and $A_B = 4E(B-V)$, Tammann (1982) utilized 17 SN I in absorption-free E and S0 galaxies to find $M_{Bmax} = -19.^m 69 \pm 0.^m 14$, and from the nine most reliable objects he found $M_{Bmax} = -19.^m 73 \pm 0.^m 14$.

On the average, SN II are less luminous than SN Ia, and their maximum brightness spans a wide range, from $-15.^m 88$ to $-20.^m 88$. The large spread and fainter M_{max} are intrinsic properties of SN II, and not a consequence of improper allowance for local absorption in markedly and irregularly dust-laden spiral galaxies. This also follows from the histograms of the distribution in absolute magnitude compiled by Tsvetkov (1986, 1989) for SN Ia and SN II that have individually measured $E(B-V)$; see Fig. 1c. The shaded region identified by the legend in Fig. 1c corresponds to supernovae in the Virgo and Ursa Major clusters, for which the color excess is

$E(B-V) = 0-0.^m 1$. For these supernovae, the spread in M_B is unrelated to errors in the distance or absorption estimates, and reflects actual changes in brightness.

For SN II with a well-determined color index, M_B at maximum light ranges from $-15.^m 88$ to $-20.^m 88$. For SN Ia, values of M_B from $-18.^m 88$ to $-20.^m 88$ are real, since both the very faintest and the very brightest supernovae have been found in the Virgo cluster.

All values quoted here are for a Hubble constant $H = 50$ km/sec/Mpc, and do not take account of errors, either random or systematic, in the distance estimates to the parent galaxies due to uncertainty in the value of H. The latter can be corrected using the formula $M_H = M_{50} + 5\log_{10}(H/50)$.

A color curve untainted by absorption, giving $(B-V)_0$, has yet to be obtained for a type Ib supernova, so individual absorption values for any of the latter are uncertain. In Fig. 1d, we have plotted the SN Ib distribution of absolute magnitude M_B at maximum, uncorrected for absorption, and for comparison, the corresponding distribution for SN Ia.

The mean value for seven SN Ib has been found to be $\overline{M}_B(\text{SN Ib}) = -17.0$, which is $1.^m 5$ brighter than for SN Ia: $\overline{M}_B(\text{SN Ia}) = -18.5$. Noting, however, that SN Ib make their appearance in spiral arms and H II regions, while SN Ia are either uniformly distributed throughout the disk of spiral galaxies or erupt in gas- and dust-deficient ellipticals, one might expect interstellar absorption to be more prevalent among SN Ib. In fact, Rust (1974) has found the following mean values for supernovae: $\overline{A}_B = 0.^m 07$ in E and Im galaxies, $\overline{A}_B = 0.^m 47$ in spirals outside the arms, and $\overline{A}_B = 1.^m 57$ in spiral arms and H II regions. With this in mind, we may conclude that the difference in maximum light between SN Ib and SN Ia is less than approximately $1.^m 5$, and in any case the faintest SN Ia are dimmer than most SN Ib.

Light curves and spectra

Primitive light curves were in fact constructed by the very first eyewitnesses to the "historical" Galactic supernovae. It is remarkable how much astronomical information on historical supernovae is conveyed by ancient records like the following: "Second year of the Chung-p'ing reign period [of Emperor Hsiao-ling], tenth month, day kuei-hai, a guest star appeared within Nan-mên ['Southern Gate' ~ Centaurus]. It was as large as half a mat; it was multicolored [lit. 'It showed five colors'] and it scintillated. It gradually became smaller and disappeared in the sixth month of the year after next" (Clark and Stephenson, 1977). An engaging interpretation of a large number of ancient records of this type and their astrophysical consequences is to be found in Shklovskii (1976a), Pskovskii (1978a), Wang et al. (1986), and Clark and Stephenson (1982). Similar descriptions have stood the astronomical community in good stead, as they have made it possible to reconstruct the light curves of historical supernovae.

A detailed examination of supernova light curves and spectra lies beyond the scope of the present book. We shall merely identify the major features characterizing supernovae of types I and II, as well as those that are important to an understanding of the nature of supernova outbursts and their effects on the interstellar medium. (A detailed analysis of supernova light curves and spectra may be found in the book by Shklovskii (1976a), in a number of collections (Wheeler, 1980; Rees and Stoneham, 1982), in the reviews by Trimble (1982) and Blinnikov et al. (1988), in the forthcoming atlas by Filippenko (1989), and in cited papers on individual objects.)

Modern analysis of the outburst mechanism is based on composite light curves — that is, the superposition of light curves from several dozen supernovae. Quite informative composite light curves were constructed in the early 1970's, based primarily on data from the Asiago observatory (Barbon, 1980; Barbon et al., 1973, 1979). These indicated that in the main, all supernovae resemble each other, with a rapid rise in luminosity over 3–20 days, a relatively flat, domed maximum approximately 10 days wide, and a subsequent slow decline at $0.^m03-0.^m1$ per day. The rate of falloff in the brightness varies over the latter part of the light curve, forming a flat plateau in certain SN II light curves, and a characteristic abrupt transition from a rapid reduction in brightness to a slow exponential decrease in almost all SN Ia.

The changeover from photographic to photoelectric observational techniques improved the accuracy of brightness measurements significantly, but the bulk of the existing data still consists of old photographic light curves. The work of Tsvetkov (1986) is therefore noteworthy in this regard: he reduced old photographic observations taken at Asiago and at the Sternberg Astronomical Institute, as well as his own — begun in 1980 — to a uniform system based on new photoelectric measurements of comparison stars. The composite B and V light curves constructed by Tsvetkov are plotted in Fig. 2 (SN Ia) and Fig. 3 (SN II). The input data consisted of 15 SN II light curves and the 20 most reliable SN Ia light curves, all reduced to the new photoelectric system. The typical magnitude determination error is $0.^m1$ in the range $m = 14-16$, and $0.^m2-0.^m3$ for $m \geq 17$; what is new here is the way in which the light curves of individual supernovae are superposed. Barbon, Ciatti, and Rosino superposed light curves, as far as possible, over a full 80–100 days; this had the effect of smoothing out differences and emphasizing the similarity between the different outbursts. (It should come as no surprise that recent opinion has held that all SN I light curves are identical, with the apparent differences stemming from measurement errors.) The composite light curves of Figs. 2 and 3 were obtained by superimposing individual curves so as to obtain the best correspondence in the vicinity of the peak. The standards used for this purpose were the most thoroughly investigated outbursts near maximum light, namely SN 1981B, SN 1975N, and SN 1971I. This normalization to the maximum emphasizes the differences among the light curves.

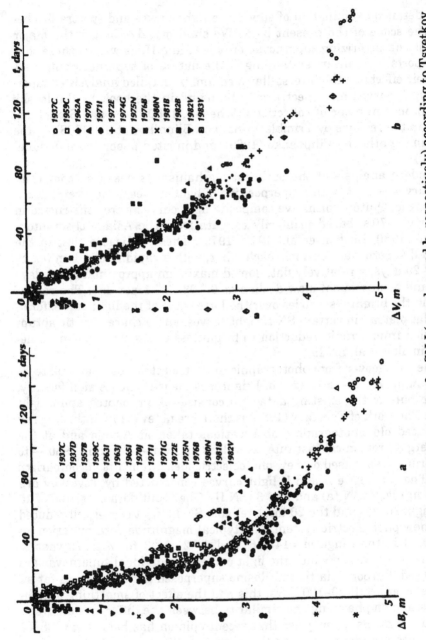

Fig. 2. Composite B and V light curves for SN I (panels a and b, respectively) according to Tsvetkov (1986). SN 1976B is a SN Ib; all others are SN Ia.

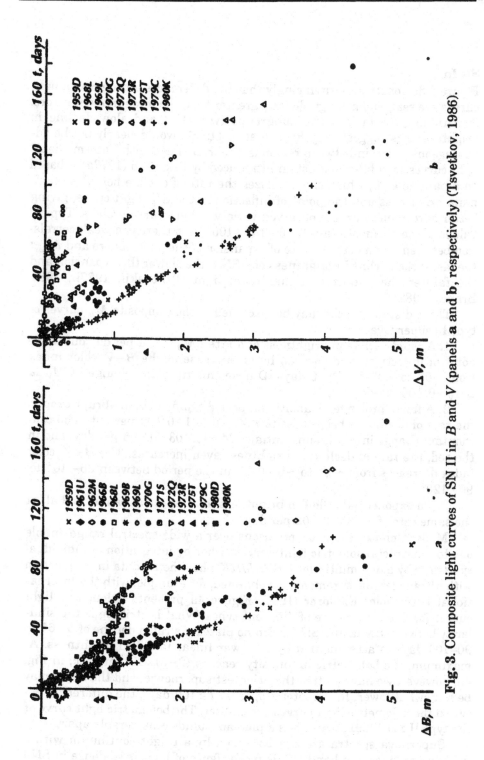

Fig. 3. Composite light curves of SN II in *B* and *V* (panels **a** and **b**, respectively) (Tsvetkov, 1986).

SN Ia

Figure 2 demonstrates convincingly that the differences among SN Ia light curves are real, and are significantly greater than the measurement errors. Partitioning SN Ia into two subgroups (with fast and slow decline in brightness), as suggested by Barbon et al. (1973), would clearly not be advantageous — the in-between region is uniformly filled with "intermediate" light curves. A subtler distinction introduced by Pskovskii (1977a) is based on a parameter β, which characterizes the rate of decline between maximum brightness and the point of inflection of the M_B light curve, giving better agreement with the observed variety of light curve profiles. Typical values of the parameter are $\beta = 6 - 14^m / 100^d$, and there is a weak correlation between β and both the rate of expansion of the photosphere and M_{max}: the faster the decline in brightness of a SN I, the slower the expansion and the fainter the outburst at maximum light (Pskovskii, 1977b, 1984; Branch, 1982).

Three distinct epochs may be discerned in the composite M_B curves of type Ia supernovae:

1. A rapid decline in brightness at a rate $\Delta M_B \approx 0.^m 1$ per day out to the point of inflection, accompanied by an increase in the $B - V$ color index from $\sim 0^m$ to $\sim 1^m$ in 30–40 days. During this time, the change in M_V is $\Delta M_V \approx 0.^m 06$ per day.

2. A fairly prolonged transitional period typified by an abrupt drop in the rate of decline in brightness to $\Delta M_B \approx 0.^m 01 - 0.^m 02$ per day. The concomitant change in M_V is substantial: $\Delta M_V \approx 0.^m 03 - 0.^m 05$ per day, and in the red, the rate of decline in brightness even increases. The $B - V$ color index decreases from $\sim 1^m$ to $\sim 0 - 0.^m 5$ in the period between $30 - 40^d$ to $90 - 120^d$.

3. An exponential falloff in brightness in all spectral bands, and all at the same rate of $0.^m 01 - 0.^m 02$ per day.

Modern photoelectric observations over a wide spectral range enable one to construct bolometric light curves either by integration of individual spectra or by using multicolor $UBVRIJHK$ photometric data in conjunction with ultraviolet measurements, as obtained, for example, with the International Ultraviolet Explorer (IUE). Figure 4a presents bolometric light curves for five supernovae of different types. What is striking is the similarity between SN Ia and SN II with no plateau: a rapid decline of $5 - 6^m$ in 90–100 days is accompanied by a slower linear falloff in brightness. At maximum, the bolometric luminosity reaches $(2 - 3) \times 10^{43}$ erg/sec; in Fig. 4a, however, we have plotted the brightest supernovae, and the mean may be somewhat lower. In the later stages, at $t \geq 100$ days, the differences between the bolometric light curves are greater. The bolometric light curve of the type II SN 1969L, which has a plateau, stands considerably apart.

Supernova spectra are characterized by a bright continuum with a great many lines and bands. This picket fence of lines is so dense in SN I

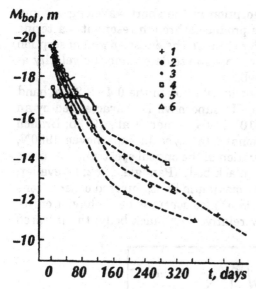

Fig. 4a. Bolometric light curves of SN Ia 1972E (curves 1 and 2; data from multicolor photometry and spectral energy distribution; SN Ia 1981B (curve 3); SN IIP 1969L (curve 4); SN IIL 1979C (curve 5); SN IIL 1980K (curve 6) (Bartunov and Tsvetkov, 1986).

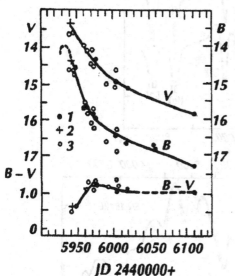

Fig. 4b. Light curves and color indices for SN Ib 1984L (Tsvetkov, 1987b).

that it took many years for them to be identified. The first step, laying the foundation for the modern interpretation of SN I spectra, was taken by Pskovskii (1968) and Mustel' (1971, 1972, 1973).

Without becoming immersed in details, let us describe the main points. The continuous spectrum of a type Ia supernova can be approximated by blackbody emission, and it provides one with a rough estimate of the radius of the photosphere in terms of the luminosity and temperature: $L = 4\pi R_{ph}^2 \sigma T^4$. Many of the lines exhibit P Cygni profiles, with emission in the long-wavelength wing and absorption in the short-wavelength wing, which suggests that these lines are produced through resonant scattering in an expanding atmosphere. Doppler shifts in the emission and absorption features of different spectral lines enable one to determine the velocity at different levels in the expanding shell.

Near maximum light, the energy distribution in the $0.4-2.2$ μm band of the continuous spectrum of a type Ia supernova is characterized by an effective temperature $T_{eff} \approx (1-2) \times 10^4$ K (Kirshner et al., 1973b; Branch et al., 1983). Ultraviolet observations of the type Ia supernovae 1980N, 1981B, and 1982B provide a continuation of the optical spectrum, but there is a marked deficiency relative to a black body (Panagia, 1982; Chevalier, 1984a, and references therein). Near maximum light, infrared observations of SN 1972E, 1980N, 1981B, and 1981D also continue the behavior of the optical spectrum, with a deficiency relative to a black body; the infrared

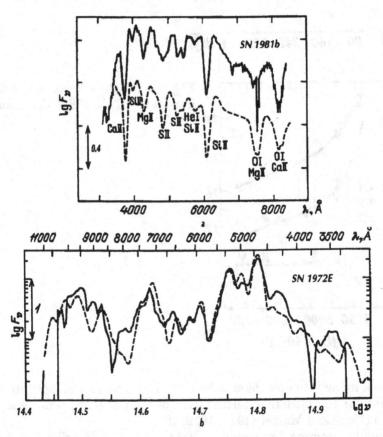

Fig. 5. Observed and synthesized spectra (solid and dashed curves, respectively) for SN Ia and SN II: **a)** SN I 1981B near maximum; **b)** SN Ia 1972E at $t = 260^d$ (cont'd).

flux has a secondary maximum after approximately 30 days (Trimble, 1982; Chevalier, 1984a).

At the "dome" stage of a SN Ia, the brightness behavior and spectral energy distribution can be satisfactorily described by a blackbody model of the expanding photosphere, with a temperature that drops from $\sim(1-2)\times10^4$ K at maximum light to 6×10^3 K at the breakpoint of the light curve (Kirshner et al., 1973b; Mustel' and Chugai, 1975). Near maximum, the radius of the photosphere is about 10^{15} cm, first increasing because of expansion, and then shrinking due to progressively greater transparency of the shell. The expansion velocity of the photosphere is of order 10^4 km/sec. It is presently believed that the optical emission from a type Ia supernova at all phases from maximum light onward derives from reprocessing of the energy provided by the radioactive decay $^{56}\text{Ni}\rightarrow^{56}\text{Co}\rightarrow^{56}\text{Fe}$ (see, for example, Colgate and McKee (1969), Trimble (1982), Blinnikov et al. (1988), Woosley and Weaver (1986), and references therein).

The multitude of broad maxima and minima in the spectrum of a SN Ia near maximum results from a succession of absorption bands produced by the broad overlapping lines of the Ca II, Si II, Fe II, and Mg II ions, and others as well. The line profiles are consistent with a model of a shell in free expansion, with $v=R/t$, where R is its radius and t is the time since the explosion (Nadezhin and Utrobin, 1977; Mustel' and Chugai, 1975; Utrobin and Chugai, 1979). To judge by the absorption lines, the chemical composition of the outer layers of a SN Ia shell is characterized by a major deficiency of hydrogen. The absence of $H\alpha$ lines at maximum

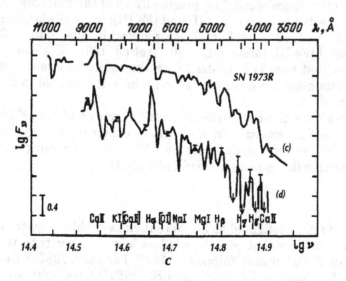

Fig. 5 (cont'd). c, d) SN II 1973R at $t=26^d$ (c) and $t=128^d$ (d) from data of Akselrod (1980), Branch (1982), and Kirshner and Kwan (1975).

light implies that $M_H < 0.01 M_\odot$ in the outer layers above the level at which $v \sim 10^4$ km/sec; see the paper by Blinnikov et al. (1988), and references therein. During the exponential decline in brightness, the flux in the continuum decreases, and strong bands appear; the latter are blends of large numbers of [Fe II] and [Fe III] lines. In Fig. 5a, we have plotted an example of the observed and synthetic spectra of the type Ia supernova SN 1981B near maximum. The synthetic spectrum was worked out by Branch (1982), and is quite consistent with the observations for absorption $A_v = 0.^m 3$, a color temperature at maximum of $T = 1.7 \times 10^3$ K, and a photospheric expansion velocity at maximum of $v = 1.2 \times 10^3$ km/sec. Fifteen to twenty days after maximum, the lines of Ca II, Si II, and possibly O I were still visible in the spectrum of SN 1981B, but the lines of Mg II and S II had disappeared, and the brighter bands of Fe II had appeared.

In Fig. 5b, we show the observed and synthetic spectra of SN 1972E, a type Ia supernova, at the later phase $t \approx 260^d$. The synthetic spectrum was worked out by Axelrod (1980), who calculated models for the formation of SN Ia spectra given a shell initially consisting entirely of $0.4 - 1 M_\odot$ of ^{56}Ni. The best agreement between theoretical and observational spectra of SN 1972E is obtained in a model with a two-layer shell, consisting of an inner shell expanding at somewhat less than 8000 km/sec and containing approximately $0.7 M_\odot$ of ^{56}Ni and its decay products, and an outer shell comprised of lighter elements like Ca, Si, C, O, and Mg. The relative hydrogen abundance in the shell of SN 1972E is negligible, and the Si/H ratio is at least 100 times higher than in a plasma with the solar composition (Mustel' and Chugai, 1975).

In the later stages of a SN Ia, practically all of the fundamental emission bands are due to lines of [Fe II] and [Fe III]; in particular, the strong maximum at $\lambda \sim 4700$ Å in Fig. 5b is produced by the 4648, 4701, and 4733 Å lines of [Fe III] (Meyerott, 1980; Axelrod, 1980). The dominance of iron lines at that time is consistent with the radioactive model of SN Ia which predicts some tenths of a solar mass in iron produced in the decay ^{56}Ni $\rightarrow ^{56}$Co $\rightarrow ^{56}$Fe.

Features have also been detected in the spectrum of SN 1972E which are associated with emission from Co III, and their temporal variation is reproduced by Axelrod's model quite well. We thus have indisputable evidence supporting the radioactive model of a SN Ia.

SN Ib

These were first investigated as a coherent class of objects quite recently, but it was soon clear that this class is less homogeneous than the SN Ia (Panagia, 1985; Porter and Filippenko, 1987; Tsvetkov, 1987b). Certain of the SN Ib (for example, SN 1964L and SN 1983I) have light curves that are no different than the "mean" SN Ia light curves. In Fig. 4b, we have plotted the light curves and color indices of SN 1984L derived by Tsvetkov (1987b); SN 1976B and 1985F have similar light curves. For these three objects, the fall in brightness from the maximum to the breakpoint and the rate of decline in the B band near maximum turns out to be less than for the average SN Ia (Tsvetkov and Chugai, 1987).

In the infrared ($J, H,$ and K bands), SN Ib light curves are markedly different from those of SN Ia, and resemble the light curves of the so-called linear SN II-L (Elias et al., 1985; Panagia, 1985; Tsvetkov, 1987b).

Near maximum light, the spectral distribution of a SN Ib in the 0.1–2.2 μm band is similar to that of a SN Ia, but it corresponds to a lower blackbody temperature, $T \sim 8300$ K (Panagia, 1985). The redder color of a SN Ib at maximum is related to this fact.

The most notable features of late SN Ib spectra ($t \gtrsim 200^d$) are the strongest lines, due to [O I] and [Ca II], the strong lines of Mg I], Na I, Ca II, and [C I], and the barely visible lines of N I, perhaps Si I, and others for which identification is uncertain, inasmuch as these lines are very broad, with $\Delta\lambda \gtrsim 100$ Å (Begelman and Sarazin, 1986). The Fe II and Fe III blends are weaker than at the corresponding phases of a SN Ia there are no hydrogen lines whatsoever. (Hα emission with a P Cygni profile has recently been detected in the early spectrum of SN 1987K, which later became a typical SN Ib. The early spectra of two other SN Ib — 1987M and 1988L — show a small local maximum near the expected position of Hα; see Filippenko (1988)).

The similar behavior of SN Ia and SN Ib light curves means that the energy source illuminating a SN Ib is also the radioactive decay ^{56}Ni $\to$ ^{56}Co $\to$ ^{56}Fe, but the mass of ^{56}Ni may be severalfold smaller than in a SN Ia. Wheeler and Levreault (1986) have found M(Ni, SN Ib)/M(Ni, SN Ia)≈ 0.25 by comparing the luminosities of types Ia and Ib supernovae at maximum light. In an analysis of the bolometric light curves of the classical type Ia SN 1972E and the type Ib 1983N, Chugai (1986) found

$$M(\text{Ni, SN 1983N}) = 0.14(D/5 \text{ Mpc})^2 M_\odot,$$

and

$$M(\text{Ni, SN 1972E}) = 0.64(D/2 \text{ Mpc})^2 M_\odot,$$

where D is the distance to the corresponding galaxy in Mpc. In other words, he found the same ^{56}Ni ratio between SN Ia and SN Ib. According to an estimate made by Graham et al. (1986), the mass of iron in the shell of SN 1983N is M(Fe)$\sim 0.3 M_\odot$, but this estimate was based on the intensity of the [Fe II] line, and depends crucially on the poorly determined state of ionization of the iron.

According to Begelman and Sarazin (1986), the mass of oxygen in SN 1985F is very large, M(O) $\gtrsim 5.6(D/10.6$ Mpc$)^2 M_\odot$, but it may turn out to be less if the inhomogeneity of the distribution of O throughout the shell is taken into account (Blinnikov et al., 1988). A comparative analysis of light curves and mean shell expansion velocities for SN Ia and SN Ib led Chugai (1986) to conclude that the total mass contained in the shells of SN Ib is somewhat larger than for SN Ia, but the differences do not amount to more than a factor of about 1.5 – 2. The masses of SN Ib shells appear to be less than $6 M_\odot$. The lower limit for the mass of a type Ib supernova is approximately equal to the mean mass for a SN Ia shell — that is, approximately $1 M_\odot$.

The expansion velocity of SN Ib shells has been found to be about 30% lower than for SN Ia at the same phase. This result derives from velocity measurements of the Ca II absorption line of SN 1984L at maximum light, the He I line of the same supernova a month later, the [O I] emission in the spectrum of SN 1985F some 280 days after the explosion, and others (see references in Blinnikov et al., 1988). In these measurements, it is found that the velocity at the boundary of the total incineration zone of a SN Ib is 2-3 times less than in a SN Ia: $v \sim 1500 – 3500$ km/sec instead of $5000 – 7000$ km/sec, as in a SN Ia (Chugai, 1986). This confirms the measurements of Graham et al. (1986), who found $v(\text{Fe}) \approx 2500 \pm 1000$ km/sec from iron lines in the spectrum of SN 1983N, while at the same time the iron in a SN Ia has a velocity of $(5-8) \times 10^3$ km/sec.

SN II

We shall examine the light curves and spectra of SN II a bit more closely, since SN 1987A in the Large Magellanic Cloud was of this type. The light curves of SN II are much more varied in form than those of SN Ia; see Fig. 3. The most clear-cut difference, enabling one to identify two groups within the SN II, is the presence or absence of a flat plateau at 50–100 days phase. The corresponding distinction is between SN II-P (plateau) and SN II-L (linear), although it is clear from Fig. 3 (see also Barbon et al., 1979; Schaefer, 1987) that there are also intermediate types of light curves in which the plateau is less well defined. The presence of a plateau correlates with spectral features of a SN II (Panagia et al., 1980). Type II supernovae are characterized, in their later stages, by a quasiexponential decline in brightness which is slower than in the SN Ia: $t_{1/2} \approx 93$ days in the former, and $t_{1/2} \approx 50$ days in the latter (Barbon et al., 1984). The $B-V$ color index of a SN II varies from about $-0.^m 15$ at maximum to about $+1^m$ after 30–40 days; the $U-B$ color index increases at the same time from about -1^m to about 0^m, after which reddening is reduced. In the U band, the light curves of SN II-P and SN II-L look the same, and no plateau is observed.

The continuous spectrum of a SN II is much more distinctive than that of either a SN Ia or SN Ib, and corresponds closely to a Planck distribution — over the full range from the UV to the near-infrared, that is from 0.2 to 2.2 μm, the spectrum conforms quite well to a single blackbody temperature. The temperature reaches $\sim(1-1.5) \times 10^4$ K at maximum, falls to $\sim(5-7) \times 10^3$ K after a number of weeks, and to $\sim 4 \times 10^3$ K in a few hundred days (Panagia, 1982; Kirshner et al., 1973a). In the infrared, there is

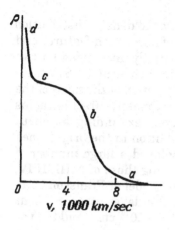

Fig. 6. A sketch of density as a function of velocity in the shell of SN II 1970G (Chugai, 1981).

excess emission at all stages which is associated with the radiation from dust (Dwek, 1983). An ultraviolet excess in the vicinity of 1500 Å which was observed in the spectrum of SN 1979C and 1980K was probably due to two-photon emission from hydrogen in the supernova atmosphere (Benvenuti et al., 1982).

The expansion velocity at the level of the photosphere decreases from $\sim(1-1.7)\times10^4$ km/sec at maximum to $\sim(5-6)\times10^3$ km/sec after a few weeks, and to $(1-2)\times10^3$ km/sec in a few hundred days, a result of the fact that as the expanding envelope becomes progressively more transparent, we observe radiation emerging from deeper and deeper layers of the ejecta.

At maximum light, the spectrum of a SN II is practically continuous, containing only the broad, strong Hα emission line. Other hydrogen lines in both emission and absorption appear subsequently, as do lines of Ca II, Na I, and Fe II, with typical P Cygni profiles. Strong emission lines become visible at $t > 100$ days, the most prominent being those of hydrogen, [O I], [Ca II], Mg I] , and Fe II blends (see, for example, Kirshner et al. (1973a), Kirshner and Kwan (1975), Pronik et al. (1976), Chugai (1980), and Uomoto and Kirshner, 1986).

Near maximum light, the spectra of SN II are saturated by overlapping lines and bands to a lesser extent than those of SN Ia, b. This enables one, albeit qualitatively, to reconstruct the mass distribution, velocity, and chemical composition along the line of sight through the expanding shell. A quantitative solution requires that one compare the observed spectral line profiles of many elements with those calculated for different shell models.

This is a different sort of investigation to undertake, in terms of both the necessary theoretical line modeling and the observations themselves, since it requires an extensive series of high-resolution spectra. So far, therefore, the analysis has only been carried through for several SN II. In spectra of SN 1969L, for example, the change in shape of Hβ is best fit by a model with an extended, freely expanding shell ($v = R/t$) whose density decreases with radius (Kirshner and Kwan, 1974). The evidence obtained from a number of lines in the spectrum of SN 1970G is that the shell is extended and spherically symmetric; its kinematics are consistent with free expansion, and the density falls off with increasing radius. Figure 6 gives a qualitative plot of the matter density as a function of velocity (Chugai, 1981). With a few possible exceptions (for instance, SN 1979C), this is probably representative of the shell structure of a SN II. SN 1979C is a

SN II-L, and is distinguished from other SN II by the density distribution and kinematics of its ejecta: the radial velocities of absorption features and the width of the Hα emission line remained essentially unchanged for two and one-half months following maximum light. Branch et al. (1981) interpreted these data as indicating that a significant fraction of the mass of the shell was plowed into a comparatively thin (approximately 10% the radius of the photosphere) spherically symmetric layer, expanding at about 8000 km/sec. Subsequently, at $t \gtrsim 70$ days, in addition to the bright lines of Hα and Na I, the spectrum of SN 1979C displayed a large number of lines in the UV, some of the more notable ones being those of N III, N IV, Si IV, and C IV (Panagia, 1982). According to the model of Fransson et al. (1984), these lines are formed in a geometrically thin layer with $\tau > 1$ near the photosphere, where the density approaches $10^9 - 10^{10}$ cm^{-3} and the velocity is about 8400 km/sec.

The principal characteristics of the light curves and spectra of type II-P supernovae are described by modeling the explosion of a red supergiant — a massive star with radius $R_0 \sim 10^3 R_\odot$ and a normal chemical composition (Grasberg et al., 1971; Litvinova and Nadezhin, 1985). In the plateau stage, the light curve results from the dissipation of the thermal energy stored in the shell by the passing shock wave. By comparing hydrodynamic models with observations of SN II-P, particularly with their shell expansion velocity, luminosity, and plateau-phase duration, Litvinova and Nadezhin obtain the following initial parameters for an "average" type II-P supernova: progenitor radius $R_0 \approx 500 R_\odot$, shell mass $M \sim 5 M_\odot$, supernova energy $E_0 = 10^{51}$ erg.

The duration of the plateau phase is proportional to the mass of the ejected shell — that is, the variety that we see in the shape of SN II light curves is due to shell masses that range from $\sim 1 M_\odot$ in SN II-L to $\sim 10 M_\odot$ (Litvinova and Nadezhin, 1985; Falk and Arnett, 1977).

The relatively high luminosity observed in the later stages of a SN II ($t > 100$ days) compels one to postulate an additional "long-lived" source injecting energy into the shell. The various candidates for such a source include a pulsar (Chevalier, 1977a), a shock wave propagating through the stellar wind of the presupernova (Renzini, 1978), and the energy derived from the radioactive decay ^{56}Co $\rightarrow ^{56}$Fe, given an initial mass of ^{56}Ni in the range $0.03 - 0.05 M_\odot$ (Weaver and Woosley, 1980; Trimble, 1982; Chugai, 1987; further references appear in all of the foregoing). There is as yet no consensus on a SN II-L model, but the low mass of the ejected shell, according to the model of Litvinova and Nadezhin, argues in favor of a pulsar.

By comparison with a SN Ia, the chemical composition of a SN II shell is close to normal. Absorption lines of Ca, Na, Mg, and Fe in SN II spectra fail to indicate any abundance anomalies for these elements (Kirshner and Kwan, 1975).

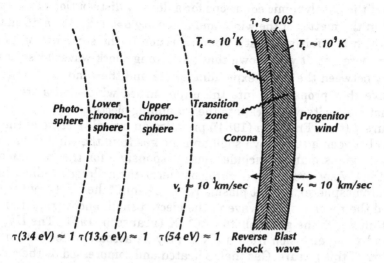

Fig. 7. Diagram of the expanding shell of a SN II (Fransson, 1984). Shown are the blast wave in the wind and the return shock in the ejecta. Ultraviolet spectral lines arise primarily in the upper chromosphere, and the Balmer lines of hydrogen and Mg II, in the lower chromosphere. A very hot corona is situated near the front; in the transition region, the temperature falls to $\sim 10^4$ K.

According to Chugai (1982), the relative line intensities in the spectrum of SN 1970G correspond to normal Ca, Fe, and Na abundances; the relative O abundance may be several times the solar value, but the mass of oxygen is less than $0.2 M_\odot$ for a total shell mass of $3-4 M_\odot$. The hydrogen and helium abundances are uncertain: the mass of ionized hydrogen, which is responsible for Hα recombination-line radiation at the later phase $t \sim 100$ days, is about $0.1 M_\odot$, and the mass of neutral gas should be substantially higher, $M_H \sim 2-3 M_\odot$ (Chugai, 1987).

At the end of the 1970's, supernova research was extended in both directions from its classical optical domain — toward both shorter (UV and x-ray) and longer (IR and radio) wavelengths. Even though such observations were obtained only for the most recent events, these developments opened a whole new page in the study of supernovae. It has become quite clear that the physics of supernovae, particularly of SN II, is largely governed by the interaction of the supernova shell with the gas furnished by the wind from the presupernova.

The theory of the braking of the supernova shell that has been ejected into the gas of the stellar wind was developed by Nadezhin (1981), Chevalier (1982b), and Grasberg and Nadezhin (1987). They found a self-similar

solution of the gas-dynamic equations for a density distribution of the form
$\rho_w \propto R^{-2}$ in the matter of the wind, and a density $\rho \propto (t/R)^k t^{-3}$, $k \geq 5$ in the
ejecta, which is a fairly realistic approximation to the structure of an ex-
ploding supergiant. It was shown that two strong shock waves arise at the
boundary between the freely expanding ejecta and the wind: an "outward"
blast wave that propagates into the unperturbed wind, and a "reverse"
shock that makes its way back into the expanding ejecta.

Figure 7 (from Fransson (1984)) provides an overall picture of the interaction between a supernova shell and the gas from the wind of the progenitor. Regions marked include those responsible for the formation of
lines in the UV spectrum (upper chromosphere), the hydrogen Balmer lines
(lower chromosphere), the hot plasma regions behind the blast front in the
wind and the reverse shock wave in the ejected shell, and a transition region within which the gas cools to $\sim 10^4$ K (Fransson, 1984). The UV, IR,
radio, and x ray emission from a SN II may be accounted for within the
framework of this picture. Gas that is heated and compressed by the outgoing and returning shock waves cools rapidly (in a matter of months) due to
radiative losses. The ejecta heated by the reverse shock radiate primarily
in x rays. X-ray photons may possibly be reradiated in UV lines corresponding to a high state of ionization; the layer of the wind material
heated by the blast wave can give rise to continuous UV emission. A certain fraction of the supernova's x-ray emission may come from the scattering of optical photons by the relativistic electrons responsible for radio
emission, as well as from synchrotron radiation by ultrarelativistic electrons and from the comptonization of photospheric radiation by hot electrons of the blast wave in the wind, but the final verdict is not yet in (see
Chevalier , 1984a; Fransson, 1984; Fransson et al., 1984; Blinnikov et al.,
1988, and references therein).

The unique Balmer profiles in the spectrum of SN 1984E, discovered
by Dopita et al. (1984), find their explanation in a model in which the supernova shell interacts with the very strong stellar wind of the progenitor:
there is a narrow P Cygni emission peak at a radial velocity close to that of
the parent galaxy, and that peak is superimposed on the broad component
of the P Cygni profile. These authors associate the narrow, unshifted component with super-wind from the progenitor at a mass-loss rate possibly
exceeding 10^{-4} to $10^{-5} M_\odot / yr$. Narrow absorption lines whose Doppler
shift increases with time have been discovered in the spectrum of
SN 1983K, probably indicative of acceleration of a dense wind by the expanding supernova shell (Niemela et al., 1985).

A narrow hydrogen absorption component was also observed in the optical spectrum of SN 1979C at maximum light, and was most likely connected with the stellar wind of the progenitor (Branch et al., 1981). Lines
of highly ionized N, C, and Si are readily visible in UV spectra of this supernova taken with IUE (Panagia, 1982; Fransson et al., 1984). The outer
layers of the shell, where these lines are formed, may be heated by soft x-

ray emission from the gas at the shock front; see Fig. 7. Highly anomalous relative abundances of the CNO elements derived from UV line measurements may also provide indirect evidence of a strong outflow from the progenitor of SN 1979C: $N/C \approx 8(N/C)_\odot$, $N/O \gtrsim 2(N/O)_\odot$. Such an anomaly may result from the loss of the outer hydrogen envelope of the progenitor, exposing layers that have been enriched by products of the CNO cycle (Fransson et al., 1984). Chevalier and Fransson (1985) have argued that late ($t \approx 1$ yr) Hα emission in SN 1979C is energized purely by shell–wind interaction; Chugai (1988) has reached a similar conclusion with respect to Hα emission from SN 1980K at $t \approx 600$ days.

The two type II supernovae 1979C and 1980K have been observed in the infrared (Dwek, 1983; Chevalier, 1984a, and references therein). Near maximum, the IR spectrum corresponded to $T_{eff} \sim 5000$ K, although the radiation was poorly described by a single temperature. Later, at days, excess emission was detected in the 10–30 μm range, due possibly to circumstellar dust heated by the optical radiation from the supernova. The dust is most likely to be localized at a distance of about 10^{17} cm, and is associated with the wind from the progenitor — the required wind-induced mass-loss rate is $10^{-5} - 10^{-4} M_\odot$ (Dwek, 1983).

Both of the foregoing supernovae were observed in x rays by the Einstein observatory. Thirty-five days after the optical maximum, the luminosity of SN 1980K was $L_{(0.2-4 \text{ keV})} = 2 \times 10^{39}$ erg/sec, and fell by a factor of two after another 50 days (no measurements were made at other times); see Canizares et al. (1982). In the context of the model described by Chevalier (1982a), this luminosity could be provided by emission from the ejecta, which were heated upon interaction with the wind from the progenitor, if the wind was characterized by $\dot{M} \sim 10^{-5} M_\odot/\text{yr}$, $V_w \sim 10$ km/sec, or in other words, by values typical of a red supergiant. No x-ray emission was detected from SN 1979C — for any reasonable assumptions about the mass-loss rate of the progenitor, it ought to have been (within the scope of the same model) close to the limiting sensitivity of the Einstein observatory.

Radio emission from supernovae

In 1971, a bright radio source was detected in the galaxy M101 at the position of SN 1970G, which had erupted a year previously (Allen et al., 1976, and references therein). This became the first "radio supernova" ever observed. Ten years later, radio emission was observed from SN 1979C (Weiler et al., 1981) and SN 1980K (Sramek et al., 1980); all three outbursts were of type II. One feature common to these supernovae is the delay of the radio maximum relative to the optical. Radio "light curves" are shown in Figs. 8a and 8c; in general outline, they resemble SN II optical light curves. An abrupt rise in radio flux density is accompanied by a smooth decline, ending in a plateau; at a wavelength of 20 cm, the maximum occurs approximately 100 days later than at 6 cm, and after about 500 days, there is a slight increase in flux density. The radio luminosity at maximum is $2 \times 10^{26} - 2 \times 10^{27}$ erg/sec/Hz, which is one to two orders of magnitude higher than that of young supernova remnants; linear polarization is less than 1%. The radio spectrum becomes flatter with time: the variation in the spectral index of SN 1979C and 1980K in the 1.5–5 GHz range is shown in Figs. 8b and 8d. The change in the slope of the spectrum clearly diminishes with time, and presumably the spectral index asymptotically approaches $\alpha \approx -0.2$ to -0.5, a value typical of the radio emission from supernova remnants; see Section 10.

The question posed immediately after the discovery of radio emission from SN II was whether such radio outbursts also accompanied SN I, the answer having fundamental importance in elucidating the nature of the radio emission from supernovae. Up through 1983, no radio emission was detected from any SN I. The situation was partially resolved when radio emission was discovered from the type I SN 1983N in M83 (Sramek et al., 1984). The radio light curve of SN 1983N, shown in Fig. 8e, differs markedly from the light curves of type II supernovae. The radio outburst was detected 11 days prior to optical maximum, and the decline in radio luminosity was given by $S_\nu \propto t^{-1.59}$ — that is, it was much more rapid than in the SN II. The 6–20 cm spectral index was $\alpha = -1.04 \pm 0.11$. At maximum, the radio luminosity of this supernova was approximately 10^{27} erg/sec/Hz at 6 cm, comparable to that of the brightest SN II.

We could now at least partially understand the negative results of searches for radio emission from previous type I supernovae: they all took place months after the optical outburst, when the radio luminosity had fallen by an order of magnitude, judging by SN 1983N. However, SN 1981B, which was observed fairly early on with the VLA, ought to have been detected, had it been similar to SN 1983N.

Here, then, was a fundamental question: was 1983N a typical type I supernova? Attempts to answer this question ultimately led to the segregation of the SN I into two classes, SN Ia and SN Ib.

We now know that radio outbursts accompany SN Ib, but not SN Ia (Panagia et al., 1986; Weiler et al., 1986). In addition to SN Ib 1983N, radio emission has been detected from SN Ib 1984L (Sramek et al., 1985). The maximum luminosity, the shape of the radio light curve, and the variation of the spectral index are all identical for these two SN Ib (Panagia et al., 1986).

Fig. 8. Radio emission from supernovae: radio light curves at 6 and 20 cm, and the spectral index $\alpha_{6\,cm}^{20\,cm}$ for SN IIL 1979C (**a, b**), SN IIL 1980K (**c, d**), and SN Ib 1983N (**e**) (Weiler et al., 1986).

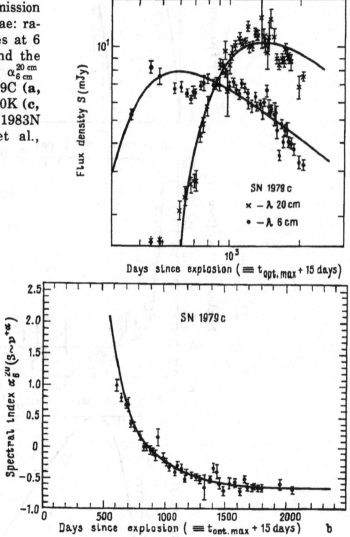

Fig. 8 (cont'd).

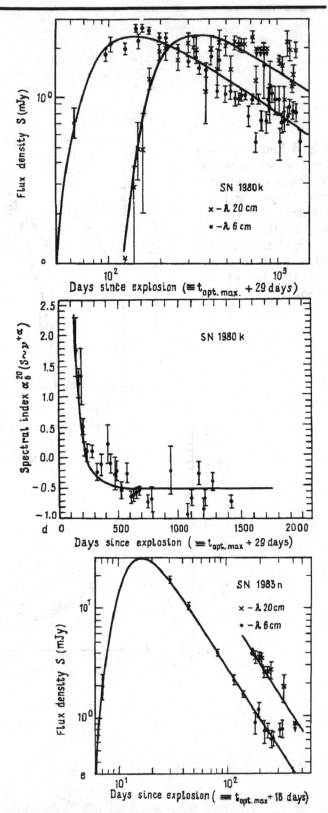

Radio emission has so far been detected from a total of a dozen supernovae. The basic data on these, as well as upper limits on the radio flux density for another seven, are presented in Table 3, where we have used the data of Cowan and Branch (1982, 1984, 1985), Ulmer et al. (1980), Weiler et al. (1983, 1986), Rupen et al. (1987), and Chevalier (1988). Radio light curves have been constructed for SN 1979C, 1980K, 1981K, 1983N, and 1984L; three of these are shown in Fig. 8, together with the behavior of the spectral index α (based on the comprehensive data of Weiler et al. (1986)). The light curves may be satisfactorily represented by curves of the form $F_\nu \propto \nu^{-\alpha}(t-t_0)^{-\beta}e^{-\tau}$, where $\tau_\nu \propto \nu^{-2.1}(t-t_0)^\delta$ is the optical depth with respect to free–free absorption, and t_0 is the time of the explosion. Corresponding values of α and β (Weiler et al., 1986) are given in Table 3, and $\delta = -3 + \alpha - \beta$ according to Chevalier (1984b).

Light curves with a very steep initial rise and a smooth decline can be explained in terms of nonthermal synchrotron radiation, in conjunction with free–free absorption in the outer layer of a thermal plasma; see Chevalier (1984c) and Weiler et al. (1986). The initial rise may be associated with the decrease in τ_{f-f} as the outer absorbing layer expands, while the subsequent decline following $\tau_{f-f} \approx 1$ may be connected with the falloff in relativistic particle density and magnetic field energy density as the shell expands. If the absorbing plasma is associated with the gas supplied by the wind of the progenitor, the latter having been ionized by the outburst (for example, by the burst of ultraviolet radiation accompanying the emergence of the shock wave from the surface of the supernova), the fact that the radio outburst of a SN II comes later than that of a SN Ib may be explained in a natural way by the higher-density stellar wind of the red supergiants, which are the precursors to the SN II.

Two mechanisms were immediately proposed to explain the radio emission from supernovae. According to Shklovskii (1981a), Pacini and Salvati (1981), and Bandiera et al. (1984), the radio emission is produced by relativistic particles moving in the magnetic field generated by a young, rapidly rotating pulsar ($P \lesssim 10$ msec). In the initial phase of ejection, the supernova envelope is completely opaque to the radio emission produced within the central cloud of relativistic plasma. For a typical ejected mass of $1-3M_\odot$ and an expansion velocity of 10^4 km/sec, the optical depth of the supernova envelope, which has not yet been heated by the reverse shock and is at a temperature of order 2×10^4 K, is $\tau_{f-f} \sim 10^6 - 10^8$. For the radio emission associated with the pulsar to be observable, it is necessary to assume that the ejected shell is quite patchy, consisting in fact of individual dense knots. The patchy structure might perhaps result from a Rayleigh–Taylor instability that occurs when the pulsar wind interacts with the ejecta (Bandiera et al., 1983). But even though the observational evidence is that young supernova remnants do indeed tend to be patchy (see Sections 3, 4, 5), the fragmentation of supernova shells at such an early stage (corresponding to the radio supernovae) is highly unlikely.

Table 3. Radio Emission from Supernovae.

Supernova	Type	Galaxy	Distance, Mpc	Age, yr*	S(6 cm), mJy	L (6 cm), erg/(sec·Hz)	α_6^{20} ($S \propto \nu^{-\alpha}$)	β ($S \propto t^{-\beta}$)	$\dot{M}/v_\infty$ ($M_\odot$/yr)/(km/sec)
SN 1950B	II?	M83	~7	30	0.5	3×10^{25}	0.4		
SN 1957D	II?	M83	~7	23	1.9	10^{26}	0.25		
SN 1970G	II	M101	7.2	1.4	~2.5	10^{26}	0.7		
SN 1979C	II	M100	17	1.2	8.3	2×10^{27}	0.72	0.7	10^{-5}
SN 1980K	II	NGC 6946	7	0.4	2.6	10^{26}	0.50	0.6	3×10^{-6}
SN 1981K	II	NGC 4258	6.6	0.5	~2	10^{26}	0.91	0.7	
SN 1983N	Ib	M83	~7	0.08	18.5	10^{27}	1.03	1.6	5×10^{-7}
SN 1984L	Ib	NGC 991	~24	0.14	0.7	4×10^{26}	1.0	1.5	
SN 1986J	II	NGC 891	8	4.4?	120	9×10^{27}	0.4		2×10^{-5}
SN 1939C	II	NGC 6946	10	40	≤0.6	$<7 \times 10^{25}$			
SN 1968L	II	M83	~7	13	≤0.3**				
SN 1968D	II	NGC 6946	10	11	≤0.6	$<7 \times 10^{25}$			
SN 1980N	Ia	NGC 1316	33	0.2	≤0.6	$<7 \times 10^{26}$			
SN 1981B	Ia	NGC 4536	33	0.05	≤0.3	$<3 \times 10^{26}$			
SN 1983G	Ia	NGC 4753	22	0.2	≤0.2	$<10^{26}$			
SN 1984A	Ia	NGC 4419	22	0.2	≤0.3	$<10^{26}$			
SN 1987A	II	LMC	0.055	2 days		4×10^{23}	1***		10^{-8}

*Age at the time of flux measurement; **λ = 20 cm; ***ν > 3 GHz

A fundamentally different mechanism for supernova radio emission was proposed by Chevalier (1981, 1982a, 1984a, b). The interaction between the supernova shell and the gas of the progenitor wind is subject to Rayleigh–Taylor instability. Turbulence arises at the boundary of the ejecta and the wind, entangling and thereby enhancing the magnetic field and leading to the acceleration of relativistic particles in the same way as was suggested by Gull (1973) for the beginning of the slowdown phase in young supernova remnants. Based on the observational data for young supernova remnants (see Section 10), this mechanism is efficient enough to account for the synchrotron radiation of radio supernovae.

The radio light curve for the initial phase enables one to estimate τ_{f-f}, and thus the density of the circumstellar ionized gas, which depends on the outflow from the pre-supernova. Corresponding values of $\dot{M}/V_w$ using data taken from Lundquist and Fransson (1988) and Chevalier (1988) are displayed in Table 3. At a stellar wind velocity $V_w = 10$ km / sec, the radio emission from a SN II gives values of the mass outflow rate that are typical of red supergiants, $\dot{M} \approx (2-5) \times 10^{-6}$ $M_\odot$/yr. It is important to note here that the mass-loss rate obtained in this way for SN 1979C is close to $\dot{M} \approx 5 \times 10^{-5}$ $M_\odot$/yr ($V_w = 10$ km / sec), the rate derived from UV spectra with the model of Fransson et al. (1984), and the mass loss from SN 1980K ($\dot{M} \approx 5 \times 10^{-5}$ $M_\odot$/yr and $V_w = 10$ km / sec) is consistent with its x-ray luminosity in the shock wave/stellar wind model of Chevalier (1982a).

Thus, the parameters of the progenitor wind based on radio measurements of SN II are consistent with the idea of the precursors of SN II being red supergiants. Notwithstanding the fact that the temporal variation ($S_\nu \propto t^{-1.6}$) and radio spectrum of the type Ib supernovae 1983N and 1984L are in perfect agreement with Chevalier's model for $\dot{M} \sim 10^{-6}$ $M_\odot$/yr, the existence of a wind about a pre-SN Ib requires explanation. According to Sramek et al. (1984), type Ib supernovae may be associated with the explosion of a white dwarf in a binary system with a supergiant. This makes it possible to combine the high density of circumstellar matter in the wind with the bolometric light curve of a SN Ib, which is inconsistent with the explosion of a supergiant. Another possibility is that progenitor itself loses matter at a rate

$$\dot{M} \sim 10^{-6} \left(\frac{V_w}{10 \text{ km/sec}} \right) M_\odot / \text{yr}.$$

Does the synchrotron radiation of radio supernovae represent an initial stage of radio emission from young supernova remnants like the Crab, Cassiopeia A, Tycho, and Kepler, or does it represent at least one of the three types to be discussed in Sections 3, 4, and 5? There is as yet no conclusive answer to this question. The theories of Bandiera et al. (1984) and Chevalier (1984c) provide for the "reprocessing" of radio emission from the supernova into radio emission from young remnants, either plerions (like the Crab) or shells (like Tycho and Cas A). But the theory of synchrotron-emission evolution from supernovae to young remnants, especially when a central pulsar is present, is model-dependent, and it also depends on many poorly known parameters, such as the density distribution in the ejected matter and in the circumstellar gas of the stellar wind (see Section 10). One is therefore better off to rely on a purely empirical comparison.

Figure 9, from Weiler et al. (1986), represents the luminosity of all radio supernovae, the most reliable radio light curves, and upper limits on the luminosity for supernovae with undetected radio emission. Here are also plotted the young Galactic remnants of the "historical" supernovae less than 1000 years old.

We see from Fig. 9 that radio SN II can develop into young radio remnants like the Crab if their radio luminosity varies in approximately the same way as their contemporaneous light curve. Radio remnants like Tycho and Kepler do not lie on the continuation of the radio light curves of SN 1983N and SN 1984L.

Note that radio supernovae are still quite rare, and radio light curves are even more so. There is no question that it is still premature to quote

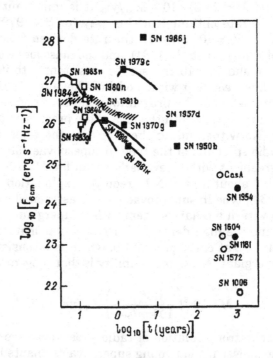

Fig. 9. Radio luminosity of supernovae and young supernova remnants (Weiler et al., 1986). Arrows indicate upper limits for (radio) undetected supernovae; smooth curves are light curves. SN Ia and SN Ib are denoted by open squares, and SN II by filled squares. Open circles are shell-like supernova remnants, filled circles are plerions.

statistics on radio supernovae. Nevertheless, Fig. 9 strongly suggests that the radio flux density from a SN Ia is several times weaker than that of a SN Ib or SN II. Figure 9 shows upper limits for the flux from the type Ia supernovae 1984A, 1983G, and 1981B and, as one can see, these supernovae would be detectable if only their radio emission were similar to that of a SN Ib at the corresponding phase.

To summarize, then, the major sources of information on the supernova outburst phenomenon are presently light curves, optical spectra, and the distribution of supernovae within their parent galaxies. Most recently, we have begun to exploit other highly productive regions of the spectrum, namely the ultraviolet, x-ray, infrared, and radio.

Supernovae may be cleanly divided into three classes: SN Ia, SN Ib, and SN II. In the latter case, we may further subdivide them into SN II-P and SN II-L.

Spectral and light-curve analysis of SN Ia lead to a model with an expanding shell having $M \sim 1-2\,M_\odot$, distributed over the velocity range $0 < v \lesssim 2 \times 10^4$ km/sec, and the shell kinematics imply free expansion, with $v = R/t$. The inner layers of the shell ($v < 8000$ km/sec) consist mainly of $0.4 - 1\,M_\odot$ of ^{56}Ni and its decay products (ultimately iron). The outer layers ($v > 8000$ km/sec) consist of intermediate elements like Ca and Si, or lighter ones, like C, O, and He; the mass of H in the shells is negligible. These parameters of SN Ia are in good agreement with models of the thermonuclear detonation of degenerate stars — white dwarfs with mass $M \sim 1.4\,M_\odot$ that have shed their external hydrogen layers in the course of evolution.

The SN Ib probably represent the explosion of more massive stars, also lacking a hydrogen envelope. Pre-supernova masses may reach $5-6\,M_\odot$, and approximately $1\,M_\odot$ of oxygen (several solar masses is a possibility) is ejected in the outburst. The explosion may take place in a binary system with a supergiant companion. It is still too soon to draw any firm conclusions about the total mass of the envelope and the mass of oxygen. It is conceivable that in this case we are observing the result of an explosion of the core of a fairly massive star ($M_{MS} \approx 20-30\,M_\odot$, possibly even $50\,M_\odot$) which has lost its hydrogen envelope, and probably a significant part of its helium envelope.

An analysis of the light curves and spectra of SN II suggests the explosion of a supergiant, the envelope of which consists mainly of hydrogen. The mass of the ejected envelope ranges between a few tenths and ten solar masses, and in most SN II is distributed monotonically over the velocity range from 10^3 to 10^4 km/sec. No pronounced chemical abundance anomalies are observed in the shells. The spectra of SN II indicate vigorous mass loss from the progenitor; this is suggested by observational data in the UV, x-ray, IR, and radio ranges. The typical mass-loss rate is $10^{-6} - 5 \times 10^{-5}\,M_\odot$/yr at a wind velocity of $V_w \sim 10$ km/sec, which is consistent with outflow from a red supergiant.

Recent results based on new high-quality optical spectra (Filippenko, 1988) have shown that supernova classification is not entirely definitive. Very interesting and quite complicated questions remain in the study of a wide range of late-phase supernova spectra in general, and SN Ib in particular. Distinctive Hα emission in the early spectra of SN Ib 1987K (as well as weak Hα in the early spectra of two other SN Ib) probably means that type II and type Ib supernovae may form a uniform sequence of objects that differ in the mass of the pre-supernova hydrogen envelopes. The most natural explanation would be physical continuity between the progenitors and explosion mechanisms of SN II and at least some SN Ib (Filippenko, 1988). All of these problems have most certainly become especially interesting in light of SN 1987A in the Large Magellanic Cloud.

2. Supernova 1987A in the Large Magellanic Cloud

The past decade has been marked by significant progress in our understanding of the physics of supernovae with regard to both the theory of the explosion and the wide range of observational manifestations of that explosion. What was pointedly lacking, though, was a test of competing ideas through direct observations of a bright, nearby supernova.

An astronomical event of enormous importance took place on February 23, 1987, when SN 1987A erupted in the Large Magellanic Cloud. For the first time in almost four hundred years, a supernova was visible to the naked eye; for the first time, we knew precisely which star had exploded and how it behaved prior to the explosion; for the first time, it became possible to observe a bright supernova at a qualitatively new level over the full electromagnetic spectrum from gamma rays and x rays through radio waves; and for the first time, neutrino radiation was detected, galvanizing the physical as well as the astronomical community.

The supernova was discovered by Ian Shelton, a Canadian astronomer working at Las Campanas observatory in Chile; it showed up when he developed his just-exposed plate of the LMC at the end of the night of 23–24 February. A reconstruction of the events leading up to the discovery is important here, since it enables us to ascertain the time interval between the neutrino burst (collapse of the core) and the optical outburst (emergence of the shock wave at the stellar surface). The optical outburst took place no earlier than February 23.39 Universal Time (UT) — at that time, the experienced New Zealand amateur A. Jones noticed no changes in the vicinity of 30 Dor, yielding an a posteriori brightness estimate of $m_{vis} \geq 7.5$. Slightly more than an hour later (February 23.444 UT), R. McNaught of Australia obtained a photograph of the LMC on which the supernova was already at sixth magnitude, but he missed out on the glory of being the discoverer by not developing his negative promptly! On the discovery plate taken by Shelton, the brightness had already reached $m_{ph} \sim 5$. The last "quiescent" photograph of the region was probably taken by Shelton, and covered February 23.06–23.10 UT.

A half hour later, at February 23.12 UT, the neutrino observatory at Mont Blanc registered the first pulse — five neutrino events in seven seconds. At 23.316 UT, three neutrino detectors in the USA (IMB), Japan (Kamiokande-II), and the USSR (Baksan neutrino station) recorded the main pulse: in all, 24 neutrino events in approximately one minute (see Aglietta et al.(1987), Koshiba (1987), Svoboda et al. (1987), and Alekseev et al. (1987), and references therein).

The coordinates of SN 1987A coincided with those of the star
Sk −69°202, a supergiant of spectral type B3 Ia and apparent magnitude
$m_v \sim 12.2$ (Sanduleak, 1969). An analysis of several dozen archival photographs of the region disclosed that there was a second blue star
($m_v \sim 14.5$) at a distance $\sim 3''$ to the northwest of Sk −69°202, and an even
fainter blue star ($m_v \sim 15.5$) located $\sim 1.''5$ to the southeast (West et al.
(1987), Walborn et al. (1987), and references therein).

It was not immediately clear which of these three stars was responsible for the outburst. Indeed, the first spectra of SN 1987A displayed strong
hydrogen lines, suggestive of a SN II. On the other hand, up to that time,
blue supergiants were not reckoned to be potential precursors of type II
supernovae. Moreover, the faint third star in the Sk −69°202 system was
not discerned immediately, and UV measurements of SN 1987A suggested
that there were two spatially unresolved sources; in addition, the UV spectrum was reminiscent of the spectra of blue stars. In the first few days,
then, it was suspected that the blue supergiant might still be intact.

The answer was not long in coming (see Sonneborn et al., 1987; Kirshner et al., 1987a; Gilmozzi et al., 1987; Gilmozzi, 1987). Ultraviolet observations by IUE began 14 hours after the outburst, and showed that the flux
from SN 1987A was already declining. In the first few days, the flux at
$\lambda < 1600$ Å fell by three orders of magnitude, and the radiation from the
supernova no longer swamped the radiation from the neighboring stars.
Analysis of the IUE data covering that period showed that the emission
was actually due to two point sources, but the distance between them, the
position angle, and the flux were all consistent with the two fainter blue
companions — the bright component, Sk −69°202, had disappeared. This
removed any vestige of doubt concerning the identity of the progenitor.

Astrometrically accurate coordinates for the supernova were published
April 29: according to Girard and van Altena (1987), the differences between the supernova coordinates and those of Sk −69°202 were
$\Delta\alpha = 0.^s003 \pm 0.^s012$ and $\Delta\delta = -0.''07 \pm 0.''09$. We thus knew for the first
time exactly which star had ended its evolution in a supernova explosion: it
was the blue supergiant Sk −69°202, of spectral type B3 Ia, with
$V = 12.^m24$, $B - V = 0.^m04$, and $U - B = -0.^m65$ (Rousseau, 1978). Estimates
of absorption have been inconsistent, with measurements by various investigators giving $A_V = 0.^m22 - 0.^m66$, the most probable mean value being
$A_V = 0.^m3$, with $E(B-V) = 0.^m11$ (see Dopita et al., 1987a, and references
therein. With $A_V = 0.^m3$, the parameters of Sk −69°202 are $M_{bol} = -7.^m71$,
$\log(L/L_\odot) = 4.98$, and $\log(T_{eff}) = 4.11$. The expected rate of mass loss in the
form of a stellar wind for such a star is $\dot{M} \approx (1-6) \times 10^{-6} \, M_\odot/\mathrm{yr}$ at a velocity of $V_w = 500 - 550$ km/sec.

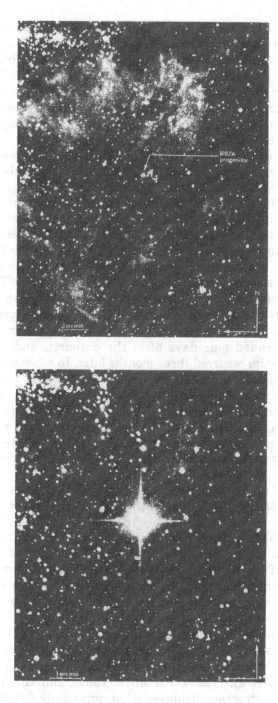

Fig. 10. Supernova 1987A in the Large Magellanic Cloud. Photographs of the field before **(a)** and after **(b)** the outburst obtained with the 3.6-m telescope of the European Southern Observatory.

The first spectra of SN 1987A, taken February 24, displayed hydrogen lines. The neutrino burst provided evidence for core collapse and the formation of a pulsar (and if we believe that there was a double neutrino burst, then possibly a black hole; see Hillebrandt et al., 1987a). It was therefore clear from the outset that we were dealing with the explosion of a SN II. However, our hopes that we might be able to study a "canonical" SN II from nearby in qualitatively more detail were not entirely justified: SN 1987A immediately behaved unconventionally.

The first surprise was that a SN II explosion had taken place in a blue supergiant (BSG), rather than a red supergiant (RSG). Strictly speaking, this was the chief difference from a "standard" SN II, since any other anomalies of SN 1987A could be accounted for by explosion models for a star of relatively small radius ($R \sim (1-5) \times 10^{12}$ cm instead of the usual RSG value of $R \sim 3 \times 10^{13}$ cm) and lacking substantial atmospheric outflow.

In its initial phases, the light curve of SN 1987A resembled neither a type II nor a type I; see Fig. 11. We may briefly summarize the results obtained by an international force of observers (Blanco et al., 1987; Menzies, 1987; Cristiani et al., 1987a, b). In the first few days following the explosion, the brightness and color variations of SN 1987A were several times faster than in other SN II. The optical light curve was characterized by two maxima: the first ended four days after the outburst, and the second smooth main maximum occurred three months later. In other words, it corresponded to the plateau phase in a conventional SN II-P; see Fig. 11a. After maximum light some time in the period 20–25 May, the distinctions gradually abated, and the light curve subsequently looked like a canonical SN II. Even at this late stage $t \gtrsim 100$ days, however, the color indices of SN 1987A remained anomalous (see Fig. 11): $B-V = 1^m 6 - 1^m 7$ and $U-B \sim 2.^m 6$, as compared with typical values for SN II, $B-V \sim 0.^m 7$ and $U-B \sim 0.^m 9$.

At the epoch of the first maximum (~ 4 days), the absolute magnitude of SN 1987A was $M_B = -14.5$, significantly fainter than other SN II; see Fig. 1b.

Variations in the bolometric luminosity of SN 1987A are plotted in Fig. 11c (taking $A_V = 0.^m 6$, and assuming a distance modulus to the Large Magellanic Cloud of $18.^m 5$). During the same period, the photometric data were quite consistent with blackbody radiation from a source at uniform temperature, facilitating a determination of the effective temperature and radius as a function of time. Utilizing the data of Bouchet et al. (1987a), we show the evolution of the latter two quantities in Fig. 12.

The spectrum of SN 1987A near maximum has been repeatedly described in the literature; for example, see Fosbury et al. (1987), Danziger et al. (1987), Menzies et al. (1987), Kirshner (1987), and Kirshner et al. (1987a). The first spectrogram displayed a hot, practically featureless continuum, with very broad, weak, blue-shifted Balmer lines and a very weak He I line. The further development of the spectrum, which was controlled by the optical clearing of the expanding shell, also took place rather more rapidly than in typical SN II. At first, it was dominated by strong, broad H lines exhibiting a P Cygni profile. From measurements of the blue edge of the Hα line, the maximum observed velocity of the ejected matter reached 31000 km/sec, and the maximum velocity of the absorption features in the first day or two was 17000–20000 km/sec, which is twice as high as the "normal" ejection velocity from a SN II (see Section 1). The highest velocity, 40000 km/sec, was measured on February 24 at the blue edge of the 2800 Å Mg II line.

In the first days and over the first 1–2 weeks, the velocity determined from the shifts in the absorption line centers quickly dropped, by approxi-

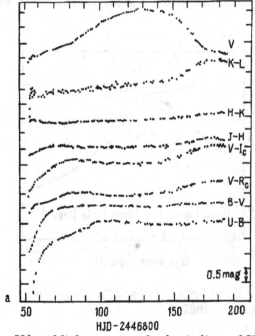

Fig. 11a. *V*-band light curve and color indices of SN 1987A (Menzies, 1987).

mately 800 km/sec per day. After a week, strong metal lines made their appearance, primarily those of singly-ionized elements in the iron group. The Hα line remained the most conspicuous, even after the other Balmer lines were no longer distinguishable because of blending with metal lines.

The quantitative analysis of the spectra of SN 1987A near maximum is made considerably more complicated by the large drop in velocity, temperature, and ionization in the line-forming region, as well as by the blending of a great many lines and bands.

Ultraviolet spectra obtained by the IUE and Astron satellites were particularly heavily blended (Kirshner, 1987; Kirshner et al., 1987a; Boyarchuk et al., 1987). A satisfactory analysis of the early results therefore requires that one calculate synthetic spectra incorporating a large number of elements in different stages of ionization. In Fig. 13, we show the observed and calculated spectra of SN 1987A at different phases, based on the data of Lucy (1987) and Fosbury et al. (1987). The "theoretical" spectra clearly provide a good approximation to the observed spectra, both near maximum light and after $\sim 100^d$. At this stage, the strongest lines and absorption bands in the visible and IR are due to blends of singly- and doubly-ionized Na, Mg, Si, Ca, Ti, Cr, Fe, and Ni. In the spectrum shown in Fig. 13 from March 15, the two strongest absorption bands in the wavelength ranges 400–450 nm and 500–540 nm are due to blends of H I, Ca I, Ti II, Cr I, Fe I, Fe II in the former and Mg I, Ti II, Cr II, Fe II in the latter.

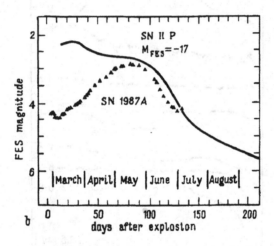

Fig. 11b. V-band light curve and "average" light curve of SN IIP (Panagia, 1987).

New lines appeared in the spectrum as the color temperature fell. In March, with $T_{phot} \leq 6000\,\mathrm{K}$ in the optical spectrum, lines probably attributable to Ba II and Sr II were identified for the first time in any supernova (Williams, 1988).

The Mg II line was the strongest in the early UV spectra, but a few days after the outburst it was practically lost due to blending, primarily with lines of Fe II. Lines of lighter elements, such as C II, Al II, Al III, Si II and Si IV dominated the short-wave UV (Kirshner et al., 1987a).

Figure 14 conveys some idea of the density distribution in the shell of SN 1987A as a function of velocity. This distribution was obtained by Kirshner (1987) from an analysis of the blue wing of the Hα line and of velocity variations during the first two weeks following the explosion. The slope of the curve gives $\rho \propto v^{-11}$, and the total mass lying above $v = 6000\,\mathrm{km/sec}$ is

1.8 $M_\odot$ (or 1 to 6 $M_\odot$ when measurement errors and model uncertainties are taken into account).

All of the aforementioned properties of SN 1987A that set it apart from other SN II — the lack of an initial dome-like maximum and the presence of a second maximum in the light curve, the low initial luminosity, the abrupt decline in UV brightness, the high initial velocities at the photospheric level and their rapid falloff with time, the rapid change in the spectrum — are accounted for quite well by hydrodynamic models of the explosion of a compact massive star. Independent modeling calculations (Woosley et al., 1987; Nomoto et al., 1987; Grasberg et al., 1987; Arnett, 1987, 1988; Schaefer et al., 1987; Shigeyama et al., 1988; see also Arnett and Fu (1989), and Imshennik and Nadezhin (1989) and references therein) have shown that the observed light curves, color variations, effec-

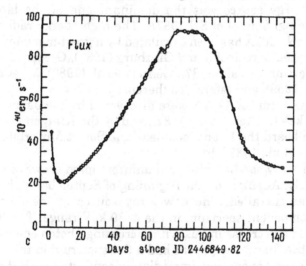

Fig. 11c. Bolometric light curve of SN 1987A (Menzies, 1987).

tive temperature, and velocity at the photosphere are best fit by the theory if the total stellar mass preceding the explosion lay in the range $M_{tot} \approx 10-20\,M_\odot$, the mass of the envelope is $M_{env} \approx 6-7$ to $15\,M_\odot$, the explosion kinetic energy was $E_0 \approx (0.5-1.5) \times 10^{51}$ erg, and the progenitor radius was $R \approx 30-50\,R_\odot$.

The observed neutrino emission, its energy characteristics, and the time delay between the neutrino and optical outburst of SN 1987A are also highly suggestive of the explosion of a high-mass star with a compact shell (Hillebrandt et al., 1987a; Shigeyama et al., 1987).

It is clear from Fig. 11 that SN 1987A only began to decline in brightness some 100 days after the outburst, twice as long as predicted by hydrodynamic models of the explosion of a compact star for radiation sustained by thermal and recombination energy. The evidence, then, is for an additional mechanism to heat the expanding shell. Because of the modest initial brightness of SN 1987A (associated with the compactness of the outer layers of the progenitor), the subsequent injection of energy into the shell is all the more evident: there is a second, clear-cut maximum of the light curve, rather than the plateau typical of other SN II-P (Fig. 11). The additional "long-lived" source could be either the energy deposited by the collapsed stellar remnant, or the energy imparted by the decay of approximately $0.1 M_\odot$ of ^{56}Ni. In either case, some of the energy goes into heating the shell, maintaining the optical luminosity of the supernova, while some fraction of the rest passes directly through the shell as x rays and gamma rays. The intensity of the latter ought to rise as the expanding shell becomes more and more transparent.

The gamma-ray and hard x-ray observations should have divulged exactly which energy source was the dominant one in the later stages $(t > 100^d)$ of the light curve of SN 1987A. The high-energy radiation to be expected from SN 1987A has been calculated by a great many investigators — for example, see Berezinskii and Ginzburg (1987), Grebenev and Syunyaev (1987), Bartunov et al. (1987), Gehrels et al. (1987), Xy et al. (1988), Fu and Arnett (1989), and references therein.

Hard x rays from SN 1987A were discovered by the Japanese Ginga satellite (4–30 keV) (Dotani et al., 1987), and by the Röntgen international observatory on board the Kvant module of the Soviet Mir orbiter (20–300 keV) (Syunyaev et al., 1987a, b).

The Ginga x-ray satellite observed uninterrupted growth in the x-ray flux through July, August, and the beginning of September; in September, the growth was curtailed. The new x-ray source in the LMC had an unusual two-component spectrum in the 4–30 keV range. The "soft" component exhibited a decrease in flux with increasing photon energy between 4 and 10 keV; the "hard" component had a flat spectrum in the 10–30 keV range. The soft component was immediately connected with the thermal radiation emanating from the supernova shell, which was interacting with the circumstellar gas, and the hard component was identified with Compton-scattered gamma rays originating in the decay of ^{56}Co (Itoh et al., 1987;

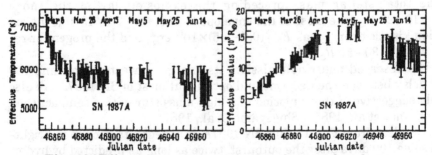

Fig. 12. Effective temperature and radius at the photospheric level in SN 1987A (Bouchet et al., 1987).

Masai et al., 1987).

The equipment aboard the Kvant module detected SN 1987A at 20–100 keV on August 10, and at 50–300 keV several days later (Syunyaev et al., 1987a; Syunyaev et al., 1987b). The spectral index in the latter range was $\alpha = -1.5$, and the spectrum flattened out in the 20–50 keV range. According to Syunyaev et al. (1987a, b), the 20–300 keV luminosity of SN 1987A was 2×10^{38} erg/sec. At first, there was no way, by comparing the observed spectrum and its temporal variations with the calculated behavior for the two alternative models powered by a young pulsar or gamma rays from ^{56}Co decay, to distinguish between these two sources of hard x rays within the opaque shell. In either case, it was possible to select the mass of the ejected envelope, its geometry (symmetric or toroidal; continuous or sparse), the ejection speed, and the chemical composition so as to provide a good fit to the observed spectrum. Distinctions would show up subsequently. Calculations carried out by Grebenev and Syunyaev (1987) for a young pulsar showed that the hard x-ray flux should increase over the first two years, as the expanding envelope clears, and then decline as a result of the spindown of the pulsar. For radioactive decay, on the other hand, the emitted flux ought to start dropping off after just one year.

By the spring of 1988, it had become clear that the evidence provided by the spectrum and temporal variation of the hard x-ray flux pointed to radioactive decay as the main energy source at that stage. The critical test, however, would be the detection of either the short-period variability associated with a central pulsar, or gamma-ray lines characteristic of the decay ^{56}Co $\rightarrow$ ^{56}Fe. These lines were searched for both in balloon experiments and by NASA's Solar Maximum Mission satellite (Solar Max).

Two balloons carrying spark chambers sensitive at 50–500 keV were launched in April, 1987 by scientists from Australia, England, Italy, the

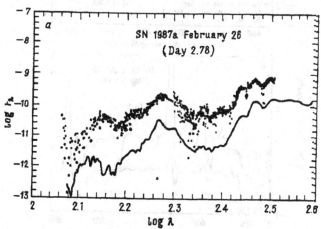

Fig. 13a. Observed and synthetic spectra of SN 1987A in the optical and near infrared; data of Lucy (1987) and Fosbury et al. (1987) (continued on next page).

United States, and West Germany. No positive results were obtained, but it was recognized that the supernova envelope should not yet have been transparent. However, an early detection of x rays was conceivable, and would have suggested patchiness of the ejecta, resulting in a reduction of the time needed for hard x rays to escape from the shell.

In the fall of 1987, soon after the discovery of hard x rays from SN 1987A, gamma-ray lines were detected at 847 and 1238 keV, corresponding to the first and second excited states of the ^{56}Fe nucleus due to

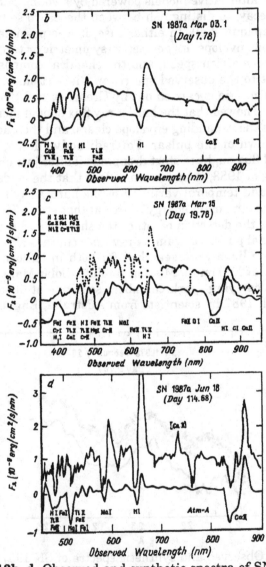

Fig. 13b–d. Observed and synthetic spectra of SN 1987A in the optical and near infrared.

the radioactive decay of ^{56}Co. This was the successful result of the gamma ray experiment aboard Solar Max (both lines have been observed almost continuously since August 1987; see Matz et al. (1987, 1988)) and of several balloon flights between October 1987 and January 1988 (Sandie et al., 1988a, b; Cook et al., 1988a, b; Rester et al., 1988; Mahoney et al., 1988).

The exponential decline of the light curve of SN 1987A after maximum has been shown to correspond to 0.07 $M_\odot$ of ^{56}Ni, which formed and then decayed to ^{56}Co (see Woosley (1988), Arnett and Fu (1989)). The gamma-ray flux at 847 keV indicated that a total of 2.3×10^{-4} $M_\odot$ of ^{56}Co was decaying in August–October 1987, or 1.3% of the total mass of ^{56}Co expected to have been in SN 1987A at that time (Matz et al., 1987, 1988).

The gamma-ray and x-ray emission of SN 1987A were detected much earlier than had been predicted theoretically by Xu et al. (1988); see also Fu and Arnett (1989), Imshennik and Nadezhin (1988), and references therein. This problem can be resolved if Co is extensively mixed into the supernova's outer layers and its core (Itoh et al., 1987; Kumagai et al., 1988; Pinto and Woosley, 1988; Ebisuzaki and Shibazaki, 1988; Grebenev and Syunyaev, 1987a).

Another piece of evidence that the principal energy source was the decay of ^{56}Co appeared at the end of March, 1988. Using the 2.2-m telescope at La Silla, Bouchet and Danziger (1988) obtained an IR spectrum of SN 1987A covering 8–13 μm. Their observations showed that the line at 10.53 μm was half as strong as it had been when they observed it at the end of November, 1987. This feature had been identified as a fine-structure transition line of Co II, and the mass of ^{56}Co existing in that ionization state was found to be $M(\text{Co II}) = 0.0023$ $M_\odot$ on March 30. The theory gives $M(\text{Co II}) = 0.0020$ $M_\odot$ at 400 days if the total mass of synthesized ^{56}Co is 0.07 $M_\odot$. Measurements at the end of November 1987, i.e., at 280 days, yielded $M(\text{Co II}) = 0.044$ $M_\odot$, also in good agreement with theory. These estimates, incidentally, incorporate the uncertainties in the model values

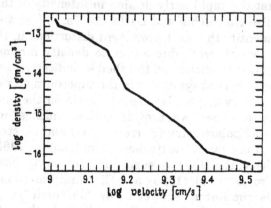

Fig. 14. Velocity dependence of shell density for SN 1987A (Kirshner, 1987).

adopted for the temperature and ionization state of the ^{56}Co.

The following facts, then, suggest that between ~100 days and ~2 years, the principal energy source of SN 1987A is radioactive decay. First, to high precision, the bolometric light curve is described by the function $F(t) = F_0 \exp(-t/\tau)$ with $\tau = 111.3^d$ during the period ~130–300^d, which is consistent with the half-life of ^{56}Co, 77 days (Whitelock et al., 1988); second, the nature of the spectrum and the temporal evolution of the x-ray emission are as expected; third, the 847 and 1238 keV gamma-ray lines of ^{56}Fe have been detected; fourth, the 10.53 μm fine-structure line of Co II has also been detected.

The main observational results obtained thus far for SN 1987A (light curve, optical, x-ray, and IR spectra) and compared with radiative hydrodynamic models and numerical modeling may be summarized as follows; see Woosley (1988), Arnett and Fu (1989), Fu and Arnett (1989), as well as Imshennik and Nadezhin (1989) and references therein. The key observational data are consistent with the explosion of a star with initial mass $M_{MS} \approx 19 M_\odot$. At the time of the explosion, the mass of the iron core was $1.45 M_\odot$, the mass of the helium core was $6 M_\odot$, and that of the hydrogen envelope was $M_{env} = 5 - 10 M_\odot$. A neutron star of mass $1.4 M_\odot$ was formed, releasing $(2-3) \times 10^{53}$ ergs of energy in a neutrino burst. The explosion's kinetic energy was $E_{exp} = 8 \times 10^{50} (M_{env}/5 M_\odot)$ ergs.

The late-stage light curve was powered by the radioactive decay of $0.07 M_\odot$ of freshly synthesized ^{56}Ni. The accurately exponential late-stage ($t > 100$ days) light curve constrains the possible pulsar's energy input to the expanding ejecta to $L < 2 \times 10^{39}$ erg/sec, and its period to $P > 1 \times 10^{-2}$ sec (Fu and Arnett, 1989).

SN 1987A may possibly have provided the first direct evidence of a nonspherical explosion, thanks to a fortunate confluence of two circumstances. The first is that the star's high apparent brightness made it possible to carry on uninterrupted spectrophotometry at a resolution of 10 Å. The second is that the rapid early decline in intensity of the Hα line, at $t \approx 30 - 40^d$, enabled Hanuschik and Dachs (1987a, b) (see also Hanuschik et al. (1988)) to identify the most prominent distortions in the absorption wing of the line, which were due either to density inhomogeneities or nonuniform excitation conditions in the shell — indicative, in other words, of a lack of perfect spherical symmetry in the supernova ejecta.

Polarimetric observations, also made possible for the first time by the high brightness of the supernova, provide further evidence for a nonspherical shell. Multicolor polarimetry (Barrett, 1987) and spectropolarimetry, which was carried out by an Australian team (Dopita, 1988) and by the Bolivian expedition of the USSR Academy of Sciences (Vid'machenko et al., 1988) (see also Cropper et al. (1988), and Clocchiatti and Marraco (1988)), convincingly discriminated between the "intrinsic" polarization of SN 1987A and that due to the interstellar medium. In absorption lines, the degree of polarization was found to be considerably higher (1–2%) than in the continuum (0.3–0.5%). The polarization depends on the wavelength in a very complicated way, the details of which are still not entirely clear.

There is still no universally accepted interpretation of the polarization data, but the possibility that the present picture results from toroidal asymmetry cannot be ruled out, leading to some interesting but highly speculative proposals (see below). Apart from this large-scale asymmetry, evidence soon appeared for small-scale inhomogeneity in the envelope. In 1988, profiles of Hα and [O I] 6300, 6364 Å were found to exhibit specific "saw-tooth" structure at a level far above the noise (see cover figure in Allen (1988)). Another piece of evidence comes from analysis of CO bands. Repin et al. (1988) found that intensities of the fundamental (4.6 μm) and second-harmonic (2.3 μm) lines of CO to $t \approx 200-250$ days suggest that CO is distributed in clumps, with a filling factor $f \approx 0.01-0.1$.

SN 1987A has been observed at radio wavelengths and, naturally, the radio outburst turned out to be anomalous as well. Emission was detected two days after the optical outburst at 0.84, 1.4, 2.29, and 8.41 GHz by Australian astronomers working with the MOST, FST, and PTI radio telescopes (Turtle et al., 1987).

Figure 15a (from Turtle et al. (1987)) presents radio flux density curves at four frequencies. It can be seen that the maximum occurred at 0.8 GHz on February 27, at 1.4 GHz on February 26, and at higher frequencies only the falloff in radio brightness was detected. The radio spectrum is shown in Fig. 15b; the steep falloff at $\nu > 3$ GHz corresponds to a spectral index $\alpha \approx -1$. The brightness temperature at 2.3 GHz was $T_{br} \sim 10^7$ K; that is, the radio emission was nonthermal. At maximum, the radio luminosity of SN 1987A was approximately 4×10^{23} erg/sec/Hz. Comparing the data with Table 3 and Figs. 8 and 9, we see that the radio outburst was probably a thousand times weaker and a hundred times shorter than in other radio supernovae. If other supernovae exploded with a radio burst that weak and that brief, they would be undetectable. At one time, therefore, the radio burst from SN 1987A was thought to be a precursor. Time passed, however, and there was no subsequent "normal" outburst from a type II radio supernova. According to current ideas about the nature of the radio emission from supernovae (see Section 1), such anomalously early, weak, and brief emission is also a natural consequence of the fact that the progenitor was a blue supergiant.

Estimates by Chevalier and Fransson (1987) indicate that the interaction of the supernova shock wave with the wind of a blue supergiant having a mass-loss rate $\dot{M} = 6 \times 10^{-6} M_{\odot}/\mathrm{yr}$ and velocity $V_w \sim 550$ km/sec provides a satisfactory explanation of the radio emission from SN 1987A. Relying on an earlier model (Chevalier, 1984c), they assume that the energy density in relativistic particles and the field equals a tiny fraction of the energy density of the gas behind the shock front, and that free–free absorption is associated with an exterior layer of ionized gas of the wind. The ionization is most likely to be due to the burst of ultraviolet radiation that ac-

companies the emergence of the shock wave from the surface of the supernova (see below).

There are other possibilities as well. Turtle et al. (1987) and Storey
and Manchester (1987) have proposed that the relativistic electrons responsible for synchrotron emission and the thermal electrons responsible
for free–free absorption coexist within the same thin layer at the outer
boundary of the supernova shell. Reasonable variations of the parameters
of this shell give good agreement between the calculated and observed light
curves, as can be seen from Fig. 15a.

Both schemes are highly simplified, of course. According to Lundqvist
and Fransson (1987a), the radio light curve is largely dictated by the detailed structure and evolution of the circumstellar gas shed by the progenitor. It is also conceivable that the magnetic field is not enhanced at the
shock front, but instead is "extracted" from the star by its stellar wind and
amplified via rotation, with synchrotron reabsorption playing an important
role (Fedorenko, 1987). Bisnovatyi-Kogan et al. (1988) have suggested that
the radio emission of SN 1987A may be due to an antineutrino flux during
the outburst which induced a stream of relativistic positrons, and the exci-

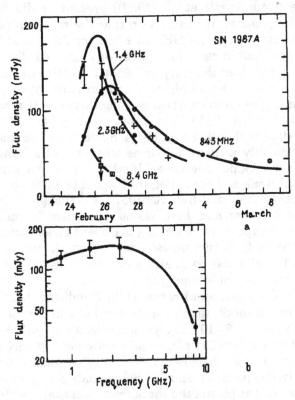

Fig. 15. Radio emission from SN 1987A (Turtle et al.,
1987): **a)** radio light curve; **b)** spectral index.

tation of plasma oscillations that are transformed into radio waves. The positrons may additionally be accelerated at the shock front.

Speckle interferometry of SN 1987A was carried out through March and April, 1987 in Chile and Australia. The objective of these observations was to measure the size of the expanding shell around the supernova. The results were astonishing: on March 25 and April 2, the speckle interferometer at the Cerro Tololo 4-m telescope divulged the existence of a "bright spot" 0."057 from SN 1987A (in position angle 194–196°). The spot was only 2.'"7 fainter than the supernova in Hα; that is, it was hundreds of times brighter than any other star in the LMC (Karovska et al., 1987; Nisenson et al., 1987). The reality of this "mystery spot" was quickly confirmed with the 4-m Anglo–Australian telescope (Marcher et al., 1987). Never before had a supernova posed such an enigma. Proposed solutions ran the gamut, including some that were quite exotic — a gravitational lens, the breakup of the stellar core. But the most reasonable explanation would seem to be that the spot was due to radiation from a dense clump of matter, probably ejected by the star prior to the supernova explosion, and ionized by the ultraviolet burst from the supernova itself (Dopita et al., 1987b; Hillebrandt et al., 1987c). According to Dopita et al. (1987b), the ultraviolet burst associated with the emergence of the shock wave is characterized by a temperature $T_{eff} \sim (1-2) \times 10^5$ K, luminosity $L_{UV} \sim (2-3) \times 10^8 L_\odot$, and duration $\sim 2-4$ hours. Being located 0."057 from the supernova, the bright spot corresponds to ionization taking place about March 23. The mass of ionized gas responsible for the radiation was $\sim 0.01-0.1 M_\odot$, and according to Dopita et al. (1987b), the clump was at least some 0.02 pc across, since the brightness of the spot remained unchanged for at least a month. Although there are reasons to doubt the reality of the companion spot (see Meikle (1988)), an interesting question arises: what might the origin be of the dense clumps at such a distance from the star? This question raises an even more fundamental one: why did the progenitor of SN 1987A explode as a blue, rather than a red, supergiant?

One possible reason was suggested by Shklovskii (1984b) and Arnett (1987). The low heavy-element abundance in the LMC reduces the opacity of the gas in the stellar atmosphere, and accordingly diminishes the radiation pressure that gives rise to an extended, outflowing atmosphere. The net result is that massive precursors in the LMC and other irregular galaxies may never reach the red supergiant stage.

However, the red supergiant region of the Hertzsprung–Russell diagram of the LMC is fully populated right up to $M \sim 50 M_\odot$ (Humphreys, 1987). An alternative was therefore proposed immediately, with the blue supergiant Sk –69°202 already having passed through the red supergiant stage (Kirshner, 1987). In order to clarify the evolutionary status of the progenitor of SN 1987A, Maeder (1987a, b) performed a numerical analysis of massive stars with mass loss. He demonstrated that the most likely evolutionary track for a star having the same photometric characteristics as Sk $-69°202$ and metallicity $Z(LMC) \sim 0.25\,Z_\odot$ in models with moderate mixing would correspond to an initial mass $M_{MS} \sim 17 - 18\,M_\odot$ and a presupernova mass $M_{fin} \sim 8 - 9\,M_\odot$. The star would pass through a final $BSG \to RSG \to BSG$ sequence, and lose approximately 50% of its mass in the form of a stellar wind. Maeder clearly obtained values of M_{MS} and M_{fin} close to those needed to reconcile the light curve of SN 1987A with hydrodynamic models of the explosion (Woosley et al., 1987; Arnett, 1988; Arnett and Fu, 1989).

Saio et al. (1988) have also conducted a numerical analysis of the evolution of a $20\,M_\odot$ star over a wide range of mass-loss rate, metallicity, and He abundance. They have found that the star evolves from blue to red when the mass of the hydrogen envelope decreases, and comes back from red to blue before carbon ignition if the He abundance is sufficiently enhanced by mass loss and mixing. For lower metallicity, as in the LMC, such evolution occurs over a wider range of parameters. Although mass loss is important, a progenitor can retain $5 - 10\,M_\odot$ of hydrogen-rich envelope.

There are a number of observational arguments suggesting that Sk $-69°202$ actually did pass through a red supergiant stage. Archival spectra of the star in the quiescent state have confirmed that it was a typical object of type B3 Ia, but with one peculiarity — an anomalously strong N II line at 3995 Å (Gonzales et al., 1987). A careful quantitative analysis of two other B3 Ia stars in the LMC has also revealed enhanced N and He abundances in the photosphere (Kudritzki et al., 1987a). The authors therefore conclude that at least these three B3 Ia stars are highly evolved post-RSG objects in which evolution was accompanied by copious mass loss.

Finally, the most compelling argument is based on UV spectroscopy of SN 1987A (Wamsteker et al., 1987; Kirshner, 1987; Cassatella, 1987). Narrow emission lines of N V (1240 Å), N IV] (1484 Å), He II (1640 Å), N III] (1750 Å), and other ions showed up in the spectrum in late May and early June of 1987. The strength of the lines indicated enhancement of the relative N/C and N/O ratios by a factor of 60 and 6, respectively, most probably due to nitrogen enrichment of the radiating gas in the course of the CNO cycle. The narrow linewidths (less than 1500 km/sec), the fact that the line centers are not shifted relative to the star, and the relatively low density ($\lesssim 10^5\ \mathrm{cm}^{-3}$, according to the intensity of the N IV doublet) suggest that these lines are not associated with the supernova shell.

At the same time, the anomalous chemical composition implies a stellar origin for the radiating matter. The most credible proposal concerning the nature of these lines (Cassatella, 1987; Kirshner, 1987; Fransson et al., 1987) is that they are produced by matter lost by the star during the red supergiant stage and ionized by the UV burst from SN 1987A. If we adopt a typical RSG stellar wind speed of $V_w \sim 10$ km/sec and a duration of $t \sim 10^5$ years for the RSG stage, the circumstellar envelope could reach ~ 1 pc in size. When the expanding envelope of the supernova arrives at that point (and judging by the expansion velocity, that ought to occur in about 10 years), we should witness an abrupt enhancement of x-ray and radio emission induced by the interaction of the shock wave with the circumstellar envelope of the red supergiant. It will, in fact, be precisely the "normal" radio burst that failed to materialize in the first few months following the explosion because of the fact that the slow, dense wind of the red supergiant was swept out of the immediate neighborhood of Sk –69°202 by the subsequent fast wind of the blue supergiant.

Returning now to the explanation of the "bright spot," one can speculate that it was due to radiation from a dense clump of matter previously ejected in the red supergiant stage (the ejection not having been spherically symmetric). Studies of ring nebulae around Wolf–Rayet and Of stars (see Chapter III) suggest that the evolution of massive stars is indeed accompanied by the ejection of dense, patchy, and often asymmetric shells. Evidence has also recently been found for the ejection of dense clumps of matter from the massive η Carinae system (Meaburn et al., 1987b).

Finally, it is possible to establish a link between the location of the bright spot and the position angle of the plane of polarization of the light from SN 1987A. According to Vid'michenko et al. (1988) and Cropper et al. (1988), the polarized component of the radiation may possibly stem from scattering in an equatorial toroidal formation. We emphasize that this is but one of the hypotheses that have been advanced, but if it is adopted, the bright spot then turns out to be localized in the polar region, and moves along the axis of the toroid, whose orientation may be determined from that of the plane of polarization.

We see, then, that SN 1987A has shed light not just on the phenomena involved in the explosion, but on the physics of the circumstellar gas as well, i.e., on mass loss from the progenitor in the early stages of its evolution. Furthermore, the bright supernova has also emerged as a probe of the interstellar gas at large distances from the star, using the "light echo" phenomenon. Optical light echoes from SN 1987A were originally discovered by Crotts (1988a) and Rosa (1988) at the end of August 1987; more detailed observations became available at later stages, in February–March 1988, using various ESO telescopes (see D'Odorico (1987), Crotts (1988b), Heathcote and Suntzeff (2988), and Gouiffes et al. (1988)).

On February 13, 1988, M. Rosa, using the 3.6-m ESO telescope, obtained a remarkable coronagraphic image of the environs of the supernova (that is, excluding the star itself) at a wavelength of 470 nm (Rosa, 1988). The photograph (Fig. 16a) shows two concentric rings, clearly representing a "light echo" from SN 1987A. The light echo phenomenon, the scattering of a spherical wavefront by interstellar dust, had been observed in conjunction with novae, and had long been predicted for supernovae (Zwicky, 1940; van den Bergh,1975, 1977), but this was the first time one had actually been detected. The light echo geometry is depicted in Fig. 16b, which gives the distance r to the scattering layer in terms of the radius R of the ring according to the relation $R^2 = 2ctr$. The two rings in Fig. 16, at radii of 32″

(7.8 pc) and 51″ (12.7 pc) correspond to reflecting layers at ~130 pc and ~340 pc from the supernova.

The rapid temporal evolution of the light echo promises to divulge the spatial geometry of the layers and of the individual clumps within. We have thus been provided with an opportunity to carry out tomography, as it were, on the interstellar medium in the vicinity of the supernova. For now, the image in Fig. 16 may be looked upon as an instantaneous cut through a structure consisting of two concentric shells, or of one shell in the line of sight. It would be reasonable to compare the formation of such shells near the Tarantula nebula, with typical sizes of 100–200 pc, to the effects of stellar wind and supernovae in the OB association NGC 2044 (LH 90) and the superassociation 30 Dor (such structures will be discussed in detail in Chapter IV). According to one localization scheme for the reflecting clouds (Crotts, 1988b), at least some of the light echo is reflected from the super-

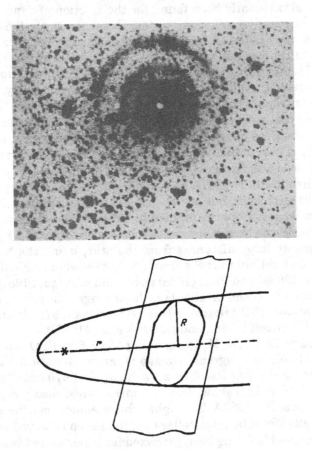

Fig. 16. Light echo from SN 1987A. **a)** Photograph obtained by M. Rosa (ESO); **b)** geometry of the light echo (see text).

bubble N 157C, and SN 1987A appears to be about 300 pc behind the LH 90/N 157C complex.

Months after the discovery of the light echo, Caldwell photographed the same region with the Cerro Tololo 1.5-m telescope, and Huchra and Olewin obtained low-dispersion spectra of the rings (Heathcote and Suntzeff, 1988). Comparison with earlier plates showed that the rings are expanding at approximately 1."8 per month, and a segment of a third ring (in position angle ~60°) had appeared. The spectrum of the rings resembled that of the supernova, but not at the time of observation — it corresponded to the stellar spectrum at an earlier time. The spectrum of the outer ring was found to be identical to the supernova spectrum in April, 1987, and that of the inner ring matched the spectrum of May, 1987. The slit of the spectrograph was narrower than the width of the ring, and the authors have not ruled out the possibility that different parts of the rings may scatter light emitted by the supernova at different epochs. The processing of related data obtained by direct photography and spectroscopy of the light echo thus provides an unprecedented opportunity to reconstruct the complete three-dimensional structure of the interstellar medium out to hundreds of parsecs from the supernova.

The echo phenomenon in SN 1987A was observed not just at visible wavelengths. Earlier, on days 113–119 and 163–164 following the explosion, Perrier (1987) and Chalabaev et al. (1987, 1988) carried out infrared speckle interferometry using the ESO 3.6-m telescope. They detected resolvable components at 2.2, 3.8, and 4.5 μm. Their power spectra on days 113–119 can be interpreted as resulting from either an axially symmetric disc (or ring) halo at about 120 milliarcsec, two spots at 62 milliarcsec, or one spot at 90 milliarcsec. The later observations correspond to displacements of 500 ± 60 or 350 ± 40 milliarcsec, respectively, for one or two spots. All cases have been treated in terms of heating of circumstellar dust by the initial UV flash, and might be considered an infrared echo from SN 1987A. The theory of the infrared echo from SN 1987A has been considered in full detail by Emmering and Chevalier (1989).

Finally, exciting new results have been reported by Gilmozzi (1988): an ultraviolet echo was detected by IUE on day 432 at about 40 arcsec from SN 1987A, resulting from the initial UV flash. The relevance of this observation lies in the fact that a UV echo would supply direct information about the spectrum of the supernova at the time of the shock breakout.

This section is a particularly difficult one to end. In any event, it should be clearly stated that the endpoint of the second Russian edition of the present book was reached in 1988. Since that time, a number of important new observational facts have come to light.

According to the ESO team, dust started to form in the supernova envelope on about day 530, resulting in a blue shift of the emission lines, a far-infrared excess in the spectrum of SN 1987A, and absorption of optical and near-infrared radiation.

Three more highly important results have been presented at the workshop *SN 1987A and other Supernovae* (Marciana Marina, Isola d'Elba, 17-22 September, 1990).

The most impressive — but still ambiguous — result concerns the behavior of the bolometric curve. If the data from the ESO team are confirmed, then we are seeing the effect, since December 1989, of some new and as-yet unidentified source of energy pumping infrared radiation out of SN 1987A with a luminosity of about 10^{38} erg/sec.

The expected radio emission due to interaction of the supernova envelope with the stellar wind of the red supergiant seems already to have appeared. In June–August 1990, an Australian team detected a rise in the radio flux of SN 1987A at 7 cm, 21 cm, and 30 cm at a level of several mJy.

In August 1990, the Hubble Space Telescope obtained a very interesting image of SN 1987A, showing ring-like structure of the circumstellar material in the [O III] λ 5007 line. The ring, with a radius of $\sim 5 \times 10^{17}$ cm and inclination of about 45°, was presumably formed by the interaction of the blue supergiant stellar wind of the presupernova with the wind of the red supergiant in the presence of significant deviations from spherical symmetry.

The explosion of SN 1987A in the Large Magellanic Cloud has presented us with a unique opportunity to study the supernova phenomenon. Observations will become more and more productive with the clearing of deeper and deeper layers of the shell — revealing heavy elements, the products of nuclear decay — with an examination of the cloud of relativistic plasma created by the pulsar wind, and so on. This is a singular opportunity, vouchsafed to us for the very first time, to trace the transformation of a supernova into a supernova remnant — a task that will, however, require the efforts of generations of astronomers yet to come.

For the time being, observing programs centered around SN 1987A continue to expand, and every passing day, month, and year promises not just answers to longstanding questions, but new mysteries as well.

3. The Remnants of Type I Supernovae: Tycho (A.D. 1572), Kepler (A.D. 1604), and SN 1006

Only two Galactic supernovae — SN 1572 and SN 1604 — have ever been observed by professional astronomers. Moreover, Kepler's supernova erupted in the same part of the sky where Mars and Jupiter happened to be in conjunction. Both events were therefore described in great detail, and their brightness was referenced to neighboring stars. The light curves reconstructed by Baade (1943) from eyewitness observations are quite similar to modern light curves of extragalactic supernovae. We begin our analysis of the remnants of historical supernovae with these two objects, as they have been reliably classified as type I supernovae on the basis of their light curves and color.

Tycho's Supernova Remnant (A.D. 1572)

The classic example of a type I supernova event is that of SN 1572. According to Clark and Stephenson (1982), the brightness at maximum was $m_V = -4.0$, at a distance of about $3\,kpc$. Pskovskii (1978a), on the other hand, gives $m_V = -4.5 \pm 0.2$, $E(B-V) = 0.^m64$, and $A_V = 2.^m11$ at a distance $r = 5.1\,kpc$. According to recent measurements by Albinson et al. (1986), with $A_V = 2.^m3 \pm 0.^m25$ and $m_{vis} \sim -4.0 \pm 0.3$ the brightness of SN 1572 at maximum light was $M_{vis} \sim -17.8^{+0.7}_{-1.2}$.

The first estimates of the kinematic distance based on 21-cm absorption yielded $r = 5 - 6\,kpc$. More recent observations (Schwarz et al., 1980) have demonstrated that absorption line features occur at velocities $+8 \leq v_{LSR} \leq -50\,km/sec$, and at the lowest velocities, the features are considerably weaker and may be associated with a local disturbance in the interstellar gas in the vicinity of the remnant[1] (see also Henbest (1980)). This then gives a kinematic distance of about $4 - 4.5\,kpc$, but such a distance is not to be relied upon in the direction of the Perseus arm, due to possible deviations from a purely circular rotation model of the Galaxy. The observations of Albinson and Gull (1982) yield a kinematic distance of $2 - 2.5\,kpc$.

Using the empirical relationship between the radio surface brightness and the linear size of a supernova remnant (the so-called Σ–D relation; see Section 10), the distance to Tycho is found to be $5\,kpc$ (Milne, 1979a). But if the suspicion of Tuohy et al. (1983b) that the radio remnants of this type of supernova are systematically fainter than previously thought turns out

---

[1] Throughout this book, v_{LSR} denotes the velocity relative to the Local Standard of Rest.

to be confirmed, then the Σ–D relation will tend to overestimate the true distance.

There is one further estimate, based on the angular size and rate of growth of the remnant, if one adopts as an upper limit on the expansion rate the velocity at the surface of the photosphere of a type I supernova, $V_0 = 10^4$ km/sec. This then gives a distance $r \leq 4$ kpc. This represents an upper limit on the distance, since it fails to take any possible deceleration of the shell into account.

These numbers typify the accuracy of distance estimates to a thoroughly studied supernova remnant, the explosion itself having been recorded. We shall adopt a distance of $r = 2.5 - 3$ kpc to Tycho, with the following justification. Another supernova remnant, Cassiopeia A, is also to be found in the Perseus arm at a distance $r = 2.8$ kpc, as measured by the only method known to be accurate — by comparing radial velocities and proper motions of two hundred filaments in the nebula (see Section 5). A multitude of absorption lines of H I, CO, formaldehyde, highly excited carbon, and other species are observed in the direction of Cas A (a survey of the data is given by Lozinskaya et al. (1986)). Absorption features are visible at the same velocities, $2 \leq v_{LSR} \leq -48$ km/sec, as in Tycho's SNR. The angular distance between the two remnant is approximately 8°, the deepest features in the absorption line profiles mimic each other, and it is not unreasonable to think that the absorption may be due to the same set of large-scale molecular clouds, i.e., that the distances to the two objects are approximately the same.

Based on new 21-cm observations, Albinson et al. (1986) came to the same conclusion regarding the location of Tycho on the near side of the Perseus arm at a distance of $2.2^{-1.5}_{-0.5}$ kpc. Kirshner et al. (1987b) have found $r = 2.0 - 2.8$ kpc by comparing the proper motion of filaments with the shock velocity, using observations of the broad and narrow components of the Hα line. However, the shock velocity determined in this way entails the uncertainty inherent in the theoretical model used — see below.

In 1949, using the 5-m Hale telescope at Mount Palomar, Baade obtained high-quality red plates of the region of the sky where SN 1572 was known to have appeared. A nebula was recorded on those plates, with long, thin, rather bright filaments, forming an incomplete shell about 8′ across. The optical spectrum of the nebula shows only the Balmer Hα and Hβ lines, and all attempts to find the [O II], [O III], [N II], [S II], and other lines typical of supernova remnants have thus far been unsuccessful (Kirshner and Chevalier, 1978).

Two components are distinguishable in the Hα and Hβ lines — a narrow component comparable in width with the instrumental profile, and a broad one, with $\Delta v = 1800 \pm 200$ km/sec (FWHM). The total intensities of the two components are approximately equal — $I(H\alpha) \approx 7 \times 10^5$ erg/cm^2/sec/sr — and the center of the line is shifted by less than 100 km/sec from zero radial velocity (Chevalier et al., 1980; Kirshner et al., 1987b).

Delicate optical filaments are located at the outer boundary of the extended nonthermal radio source, which exhibits a clearly defined shell structure (Fig. 17).

High-resolution radio observations (Henbest, 1980; Klein et al., 1979; Dickel and Jones, 1985, plus references in these sources) yield the following parameters for the radio remnant: outer radius of the shell, $218'' \pm 7''$ in the southwest and $257'' \pm 7''$ in the northeast sectors; thickness $\Delta R / R \approx 0.25$; the shell is fairly homogeneous and surrounded by a thin

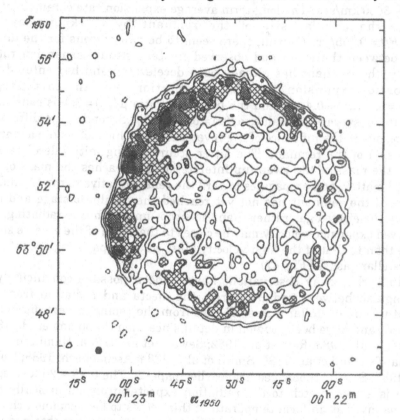

Fig. 17. Radio isophotes of Tycho's remnant at $\nu = 2.7$ GHz (Henbest, 1980).

rim. The radio spectrum is nonthermal, with $\alpha = -0.61 \pm 0.03$ (Vinyaikin et al., 1987), the spectral index remaining constant over both the outer and central regions of the remnant to within $\Delta\alpha = 0.1$ (suggestions in the literature that the periphery and central regions have different spectra have not been confirmed).

Prime-focus plates taken with the 5-m Hale telescope between 1949 and 1974 have revealed that the system of filaments is expanding away from the center at a mean angular rate of $\mu = 0.''20 \pm 0.''3$/yr (Kamper and van den Bergh, 1978). High-precision measurements of the bright radio knots made with the Westerbork radio telescope at epochs 1971 and 1979 yield a mean proper motion $\mu = 0.''256 \pm 0.''026/yr$ (Strom et al., 1982). The results are clearly consistent, with the radio observations being an order of magnitude more accurate (the author remembers well how enthusiastically the first radio observations with an angular resolution of $10°$ were greeted!).

For a distance of $2.5 - 3\,kpc$, this gives a linear expansion velocity of $3000 - 3600\,km/sec$. The long-term average expansion rate obtained by dividing the angular size of the remnant by its known age is $\langle\mu\rangle = R/t = 0.''55/yr$. Overall, there seem to be two reasons for the difference between the mean and observed "instantaneous" expansion rates. Firstly, the remnant has already been decelerated and has entered the adiabatic expansion stage (see Section 9). In this stage, $\dot{R} \equiv \mu = 0.4(R/t) \equiv 0.4\langle\mu\rangle$, and the predicted ratio $\mu/\langle\mu\rangle = 0.4$ is consistent with the observational value, $\mu/\langle\mu\rangle = 0.47 \pm 0.05$. Secondly, the difference may be due to the influence of the reverse shock, which slows the apparent expansion of the remnant. The latter occurs during initial deceleration, when the mass of the swept-up interstellar gas reaches the mass of the ejected matter, and propagates toward the center relative to the expanding ejecta. If the remnant has not yet reached the adiabatic stage and it is mainly the ejected gas rather than the swept-up gas that is radiating, the observed expansion velocity may be closer to the speed of the reverse shock wave than to that of the high-velocity blast wave propagating out into the interstellar gas (see Fig. 47).

In Tycho, x-ray observations have made it possible convincingly to distinguish between radiation from the ejecta and radiation from the swept-up circumstellar matter. X rays from the remnant were discovered in 1967, and have been studied in detail since then (Fabbiano et al., 1980; Becker et al., 1983; Reid et al., 1982; Seward et al., 1983a; Hamilton et al., 1986a; Tsunemi et al., 1986; Smith et al., 1988a; Aschenbach, 1988; Itoh et al., 1988; Smith, 1988a; see also earlier papers). The $0.15 - 25$ keV spectrum is a poor match to the radiation expected from an optically thin plasma layer at uniform temperature: this seems to be a common characteristic of young remnants. A better fit to the observations is obtained with a two-temperature plasma model having typical temperatures $(7-8)\times10^6$ K and $(7-8)\times10^7$ K (Pravdo et al., 1980). Interstellar absorption provides the explanation for the low-energy falloff. Taking absorption into account and assuming an atomic hydrogen column density of $N_H = 3\times10^{21}$ cm^{-2}, the luminosity of the remnant becomes $L_{0.15-4.5\ keV} = (6-7)\times10^{35}$ erg/sec (Reid et al., 1982).

The spectrum contains many lines of highly-ionized heavy elements; the brightest are the helium-like lines of Si and S, and the lines of Ar, Ca, and Fe as well (see Fig. 44). We shall see in Section 8 that the quantitative interpretation of the x-ray spectra of young supernova remnants is ambiguous, notwithstanding the large amount of information arriving in the x-ray band. Behind the high-speed shock front in a young remnant, the ionization temperature "lags behind" the electron temperature, which in turn lags behind the kinetic temperature of the ions. As a result, the impression produced is of a two-temperature spectrum with a deficiency of both high-temperature lines and a low-temperature continuum. This can then lead to an overestimate of the heavy-element abundance, which is determined using the intensity of x-ray lines. When allowance is made for the departure of the x-ray emitting plasma from ionization equilibrium, the Si, Ar, and Fe abundances in Tycho are 6–7 times higher, and the S and Ca abundances are 10–15 times higher, than their normal cosmic values (Tsunemi et al., 1986); see also Table 12. At the same time, it should be borne in mind that these figures depend on a great many model parameters, and the principal uncertainty in the latter derives from the poorly understood distribution of the ionization state behind both the outgoing blast wave and reverse shock fronts. Having analyzed the EXOSAT spectrum of Tycho, Smith (1988a) explained the observed intensity of the Fe line near 6.5 keV by radiation from circumstellar gas with a normal Fe abundance that was heated by the outgoing blast wave.

Observations by the Einstein observatory taken with 2″ resolution in the 0.2 – 4 keV band provided comprehensive data on the large-scale structure of the remnant, as depicted in Fig. 18 (see Seward et al., 1983a). Most of the radiation comes from a bright, highly circular shell of radius $R = 216″$ and thickness $\Delta R/R = 0.2$, whose brightness increases toward its outer edge. The shell is nonuniform: against a diffuse background, one can pick out approximately 400 bright, compact clumps. The latter are typically some 24″ across (0.34 pc at a distance of 3 kpc), and they jointly emit about 70% of the radiation in the 0.2 – 4 keV band. A fainter thin shell can be detected beyond the bright diffuse one, its outer radius being $R = 240″$, and its thickness, $\Delta R/R = 0.1$.

How is one to reconcile this three-component structure with modern ideas about the interaction between matter ejected in the explosion and the interstellar gas? In all probability, the outer thin shell consists of circumstellar gas that has been compressed and heated by the outgoing blast wave. The inner bright shell was formed by matter from the star that was ejected in the explosion and heated by the reverse shock; see Fig. 47.

Compact areas of enhanced brightness which are observable against the background of the diffuse shell are clumps of ejected matter. The explosion of the supernova was most likely accompanied by the immediate ejection of matter in the form of a diffuse shell, along with denser "debris." On the other hand, clumps might also have been formed not at the instant of the explosion, but as a result of instabilities that developed at the boundary between the ejected and swept-up matter at the beginning of deceleration. Future research on the kinematics and lifetime of x-ray clumps will assure a basis for deciding between these two possibilities.

The mass of the hot plasma responsible for the three components of the x-ray emission has been estimated by Seward et al. (1983a) (Table 4), who utilized calculations of the volume x-ray emissivity of nonequilibrium plasma (Shull, 1982). Seward et al. choose the electron temperature in the swept-up or ejected shells to be consistent with the radiation spectrum and temperature in the dense clumps, assuming that they are at the same

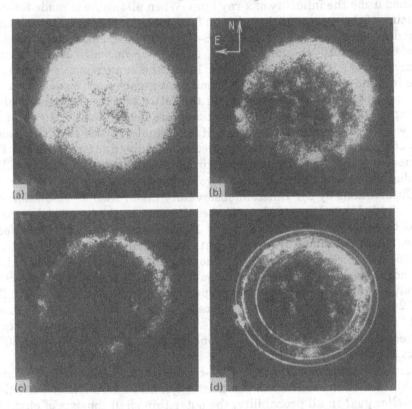

Fig. 18. X-ray image of Tycho's supernova remnant (Seward et al., 1983a). The four panels depict different brightness ranges, including **(a)** a faint outer shell, **(b)** discrete compact knots, **(c)** the brightest part of the shell, and **(d)** the overall double-layer structure of the shell.

pressure. The chemical composition of the swept-up gas (the outer shell) is assumed to have the normal cosmic makeup; the heavy-element abundance in the ejected matter has been estimated from the x-ray spectrum using the model of Shull (1982); see also Table 12. We see that the total mass of ejected diffuse matter and dense clumps is equal to the mass of swept-up gas, which is consistent with the idea that Tycho is in transition between

Table 4. Density and Mass of Three Components of Tycho's Remnant Based on X-ray Data (Seward et al., 1983a).

Model	Region	Chemical abundances	Adopted T_e, keV	n_e, cm^{-3}	$M, M_\odot$
Nonequilibrium	Swept-up gas	Solar	7	1.44	2.2
	Diffuse ejecta	From spectrum	7	0.61	1.2
	Ejected clumps	From spectrum	2	2.5	0.7
Ionization equilibrium	Swept-up gas	Solar	5	3.2	5
	Diffuse ejecta	From spectrum	5	1.15	2.3
	Ejected clumps	From spectrum	0.5	4.6	1.3
Ionization equilibrium + extreme heavy element enhancement	Swept-up gas	Solar	5	3.2	5
	Diffuse ejecta	$10^3 \times$ Solar	5	0.36	1.1
	Ejected clumps	$10^3 \times$ Solar	0.5	0.68	0.3

the free expansion stage and the adiabatic expansion stage. The two lower parts of the table correspond to plasma in ionization equilibrium and with anomalously high heavy-element abundances (1000 times the solar values). Even under these extreme conditions, the estimates differ by a factor of two or three at most.

The estimated mass of the x-ray plasma depends not only on the degree of departure from ionization equilibrium, but on the density distribution in the ejecta and swept-up gas as well. After the reverse shock has passed, the inner layers of what was initially a mass of homogeneous expanding ejecta can turn out to be less dense than the outer layers.

Taking account of the density distribution, temperature, and degree of ionization nonequilibrium in the ejecta behind the reverse shock front, Hamilton et al. (1985, 1986a) found a total mass of about 1.4 $M_\odot$ for the ejecta in Tycho. The theoretical model best agreeing with the observed x-ray spectrum involved a three-layer structure for the ejecta, the innermost layer containing $\sim 0.6\,M_\odot$ of iron (of which $\sim 0.4\,M_\odot$ is as yet unheated by the reverse shock), the middle layer consisting of approximately $0.8\,M_\odot$ of a mixture of O, Mg, Al, Si, S, Ar, Ca, Fe, and Ni, and the outermost layer containing about $0.4\,M_\odot$ of carbon (Hamilton et al., 1986a). The mass of swept-up circumstellar gas is approximately $1.3\,M_\odot$.

The stellar remnant of the explosion of Tycho's supernova has never been found. X-ray observations place an upper limit on the surface temperature of any remaining neutron star of $T_{NS} \le (1.1 - 1.8) \times 10^6$ K.

Kepler's supernova remnant (A.D. 1604)

The star that flared up in the Milky Way in 1604 has become the best-studied of any of the historical supernovae. Nonetheless, the photometric distance to the supernova has been reestimated several times, and has varied between 3 and 12 kpc! At maximum light, the supernova reached $m_V = -2.4$ according to van den Bergh and Kamper (1977), $m_V = -2.5 \pm 0.3$ according to the data of Clark and Stephenson (1982), and $m_V = -3.5 \pm 0.2$ according to Pskovskii (1978a). The most serious issue here is the discrepancy in the absorption as estimated by the various authors, as that is what makes the distance to the supernova vary over such a wide range. The most reliable estimates seem to be those of Danziger and Goss (1980) and Dennefeld (1982), who find $A_V = 3.^m47$ from the intensity ratio Hα/Hβ in the spectra of several dozen filaments (the Balmer decrement calculated by Raymond (1979) was adopted as the "theoretical" value). With $m_V = -3.0 \pm 0.3$, $A_V = 3.^m47$, and $M_V = -19$ for a type I supernova, the distance to Kepler's supernova becomes 3.2 ± 0.7 kpc, and its height above the galactic plane is $z = 380$ pc. This is in accord with the distance determined from the radio brightness and angular size of the remnant. If we correct for height above the galactic plane (and this correction is substantial for Kepler's supernova), we find $r = 5$ kpc from the Σ–D relation of Caswell and Lerche (1979), and $r = 3$ kpc from the Σ–D relation of Milne (1979a).

The optical nebula associated with SN 1604 consists of a system of bright, patchy filaments and condensations located primarily in the northern part of the radio shell. The overall structure of the remnant — a narrow spherical shell of radius $R = 80 - 90''$ (1.3 pc) and thickness $\Delta R/R = 0.2$ — is best traced out at radio and x-ray wavelengths (Strom and Sutton, 1975; Gull, 1975; White and Long, 1983). Optical filaments coincide with a zone of enhanced radio brightness. The radio spectrum is nonthermal, $\alpha = -0.58$ (Milne, 1979a), and the polarization reaches $p \approx 10\%$ near the periphery, falling off toward the center (Strom and Sutton, 1975). The direction of polarization suggests that there is a substantial radial component to the magnetic field. Matsui et al. (1985) estimate that the strength of the radial component is approximately 1.4×10^{-5} G, and that total magnetic field is approximately 7×10^{-5} G. New high-resolution radio maps of

Kepler obtained with the VLA by Dickel et al. (1988) show the complete structure of the radio shell. They discovered evolution of the thin radio shell by comparing two radio maps made four years apart with 2″ angular resolution. These observations at 6 and 20 cm give a mean expansion law $R \propto t^{0.5}$ for the shell, with $R \propto t^{0.35}$ at the bright northern rim, and $R \propto t^{0.65}$ in the eastern part. Such differences suggest that the remnant is expanding

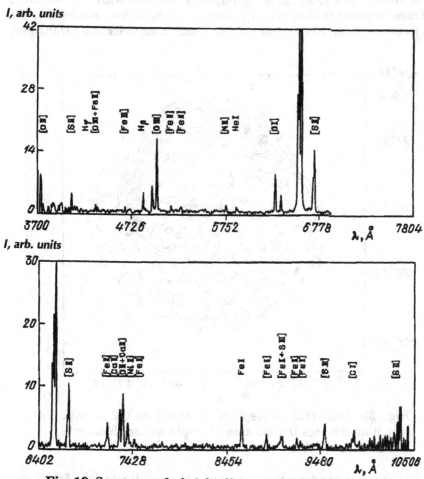

Fig. 19. Spectrum of a bright filament in Kepler's remnant (Dennefeld, 1982).

in the inhomogeneous ambient gas.

Spectra of the optical filaments contain a wealth of lines (see Fig. 19), the brightest being Hα, [N II], [O II], [O III], and [S II]. Particularly noteworthy are the many lines of iron in various stages of ionization (van den Bergh, 1980a; Dennefeld, 1982; Leibowitz and Danziger, 1983). Spectra in the far red were first investigated in this very remnant, yielding the first identification of the [N II] line, with detections of the lines of [S II], [S III], [Fe II], and [C I] as well.

There is as yet no rigorous quantitative interpretation available for the spectra; a comparison of the results obtained by Raymond (1979) and Shull and McKee (1979) suggests that the relative line intensities are best described by a shock wave propagation velocity of 90–170 km/sec, with a mean of about 110 km/sec, in a high-density medium with $n_0 \approx 300$ cm^{-3}; the nitrogen content is elevated by a factor of perhaps 3-4 compared to the normal cosmic abundance (Dennefeld, 1982; Leibowitz and Danziger,

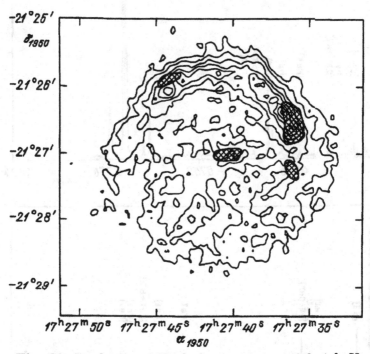

Fig. 20. Isophotes of Kepler's remnant at 0.2–4 keV. Hatching denotes the location of bright optical filaments (Matsui et al., 1985).

1983). These dense condensations occupy only a small fraction of the remnant (Fig. 20), and the medium between them consists of hot, tenuous gas radiating x rays.

The foregoing indirect estimate of the shock-wave velocity in the dense condensations is in agreement with direct measurements of filament velocities. Such measurements were made by Minkowsky (1959), and yielded radial velocities in the range $-275 \le v_{LSR} \le -140$ km/sec, with $\bar{v}_{LSR} = -220$ km/sec. The total linewidth in the filaments, measured far down in the wings, reaches $\Delta v \le 600$ km/sec, corresponding to a gas velocity $v \leq 300$ km/sec (Danziger et al., 1978). Since 1942, more than twenty high-quality plates of the nebulosity have been obtained, making it possible to detect the proper motion of the bright filaments (van den Bergh and Kamper, 1977). The results, spanning the period 1942–1976, are plotted in Fig. 21, superimposed on a radio map of the remnant. This figure provides a impression of the overall spatial structure of the shell and the localization of the bright filaments. The proper motions of the filaments, like their radial velocities, clearly do not reveal a systematic expansion of the shell. The mean angular expansion rate, $\mu \approx 0.''005$/yr, is less than the mean proper motion for the system of filaments as a whole, which is $\mu_{pec} \approx 0.''013 \pm 0.''003$/yr. Accordingly, at a distance of 3.2 kpc, this gives $v_{exp} \approx 77$ km/sec and $v_{pec} \approx 200$ km/sec; the latter value is in agreement with the measured mean radial velocity of the filaments. The mean rate of expansion of the remnant determined by its size and age is $\langle \mu \rangle = R/t = 0.''23$/yr, with $\langle v \rangle \approx 3500$ km/sec if deceleration has not yet begun.

The filaments and condensations vary in brightness with a characteristic time of 10 years. Certain of them have become brighter, and new clumps have appeared (van den Bergh and Kamper, 1977).

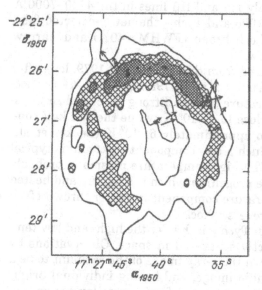

Fig. 21. Radio isophotes of Kepler's remnant. Arrows indicate velocities of proper motion from measurements by van den Bergh and Kamper (1977).

Based on the measurements, what can one conclude about the nature of the optical emission from the remnant? First of all, and this follows from the velocity estimates, the bright filaments and condensations could not have been ejected in the supernova explosion. Their characteristic expansion time (10^4 years) is much greater than the age of the remnant, and the high electron density in the region (determined from the intensity ratio of the [S II] lines at 6717 and 6731 Å, assuming $T_e = 10^4$ K), $n_e \gtrsim 3 \times 10^3$ cm^{-3}, suggests that they could not be decelerated within the lifetime of the remnant. In all likelihood, the filaments were formed during the pre-supernova phase out of matter ejected by the star as a stellar wind or a slow-moving envelope (planetary nebula), rather than from the interstellar gas. This is implied, firstly, by the enhanced nitrogen abundance in the filaments, and secondly, by the fact that SN 1604 exploded far from the plane of the Galaxy, at $z = 380$ pc, where the mean density of the interstellar medium is low, and one would not expect dense, slow-moving condensations to form from swept-up interstellar gas. The expansion of the condensations away from the center may then be due both to their original motion and that acquired through shock-wave acceleration. The motion of the overall system of filaments at approximately 200 km/sec may possibly be associated with the original motion of the progenitor.

Fesen et al. (1989) have recently obtained new CCD images and long-slit spectra of the filaments in Kepler that indicate for the first time the existence of faint Balmer-dominated high-velocity shock emission along much of the remnant's northern rim. Optical spectra of the faint emission seen along the northern limb show only Hα and Hβ lines in the 4700–7000 Å range. As in the remnant of SN 1006 and Tycho, the newly-detected faint Hα line in Kepler consists of both broad (FWHM = 40Å) and narrow (unresolved) components; see below.

X-ray emission was not detected from this object until 1979, long after the other historical supernovae (Tuohy et al., 1979a). The 0.15–4 keV spectrum is poorly fit by a one-temperature plasma. Strong Si, S, and Ar emission lines imply a temperature close to 6×10^6 K, while the 3–4 keV continuous spectrum corresponds to approximately 6×10^7 K (Becker et al., 1980a; White and Long, 1983). Such a two-component spectrum is typical of young supernova remnants. The high-temperature plasma is probably interstellar gas or gas shed by the progenitor which is swept up and heated at the blast front; the low-temperature component is matter thrown off in the explosion and heated by the reverse shock.

In contrast to the situation in Tycho, in Kepler the high- and low-temperature gas are not yet distinctly separated in space. Observations by White and Long (1983) have shown the x-ray image of the remnant to be a thin shell coinciding with the radio image, and having individual bright small-scale knots. The luminosity, corrected for interstellar absorption with $N_H = 2.8 \times 10^{21}$ cm^{-2} (this value, which is derived from the falloff in the low-energy x-ray spectrum, is consistent with the 21-cm observations of Danziger and Goss (1980)), is $L_{0.15-4.5\,keV} = 4 \times 10^{35}$ erg/sec at a distance of 3.2 kpc.

The Einstein x-ray data may be interpreted in the context of a number of models, such as one with free expansion or adiabatic expansion (with corresponding x-ray emission due mainly to ejecta heated by the reverse shock or interstellar gas heated by the outgoing blast wave); with a homogeneous or patchy interstellar medium; or with either equilibrium or nonequilibrium plasma. The existing observations do not enable one to choose between models, but in any event, we may draw the following conclusions (White and Long, 1983):

1. If the remnant is in the free expansion phase and its x-ray emission is mainly due to ejected matter, the mean density of the surrounding gas must be at least $n_0 \approx 0.1$ cm^{-3}.

2. If the remnant is already in the adiabatic expansion phase and the radiation comes mainly from swept-up gas, its density is $n_0 \approx 5\,\mathrm{cm}^{-3}$.

3. If the circumstellar gas is concentrated into small, dense cloudlets with large intervening separations, its mean density must be at least $n_0 \approx 0.1\,\mathrm{cm}^{-3}$.

4. The temperature of the radiating plasma is at most $T_e = 2\times 10^7\,\mathrm{K}$. This implies a lack of thermal equilibrium between the electron and ion gas ($T_e < T_i$) if the supernova remnant is in the adiabatic stage.

5. The total mass emitting x rays is at least $2-3\,M_\odot$, but the redistribution of density in the ejecta heated by the reverse shock reduces this estimate to $\sim 1.5\,M_\odot$ (Hamilton et al., 1985).

More recent EXOSAT observations up to energies in the 10-keV range have divulged the presence of the K and L lines of iron in the spectrum of Kepler, and these are stronger than in Tycho. Ballet et al. (1988a) and Smith (1988a) have analyzed the ionization conditions in the heated ejecta, and concluded that highly ionized lines of iron are most likely due to radiation from hot circumstellar gas with an iron abundance several times normal.

The remnant of SN 1006

The supernova outburst that took place in the constellation Lupus in the year 1006 was the brightest in the past thousand years, and was recorded in a multitude of Asiatic, Arabic, and European chronicles. The reconstructed light curve seems to indicate that the supernova was of type I, but estimates of the magnitude, and therefore the distance, are inconclusive. According to Pskovskii (1978a), $m_V \approx -6$, and the explosion occurred at a distance of approximately 4 kpc, while according to Stephenson et al. (1977), $m_V \approx -9$ and the distance was about 1 kpc.

The latter estimate is consistent with the distance of 1.3 kpc derived from the radio surface brightness, as corrected for height above the galactic plane (the z-correction for a high-latitude object is substantial). Absorption lines of iron have been detected in the spectrum of the blue subdwarf observed in projection against the remnant, and the line profiles suggest that the absorption takes place in an expanding supernova shell (Wu et al., 1983). The photometric distance to that star is approximately 1.5 to 3.5 kpc; see Fesen et al. (1988a).

Fesen et al. (1988a) obtained a similar distance, $\sim 1.5-2.3\,\mathrm{kpc}$, by comparing the maximum radial velocity of the Si II and O I absorption lines in the spectrum of the subdwarf with the proper motion of the filaments of the remnant; see below.

The radio source identified with SN 1006 possesses a clear-cut shell structure (see Fig. 69a, Section 10), with a spectral index $\alpha = -0.6$ (Stephenson et al., 1977; Green, 1988a; Roger et al. (1988) and references therein). Observations with $16'' \times 20''$ resolution have disclosed fine filamentary structure in the outlying regions whose thickness is at most

0.17 pc, with the structure and radius of curvature accurately preserved up to 90° around the symmetric shell (Reynolds and Gilmore, 1986). The observed structure suggests that these delicate radio filaments likely trace out the shock front, rather than the turbulent surface between the ejected and swept-up gas (Reynolds, 1988a). Recent high-resolution and high-sensitivity radio maps of the remnant by Roger et al. (1988) demonstrate all features of so-called "barrel-like" axially symmetric morphology; see Section 10. New numerical simulations by Bisnovaty-Kogan, Lozinskaya, and Silich (1990) show the axially symmetric structure of this youngest barrel-shaped remnant most probably to be the result of an asymmetric supernova explosion.

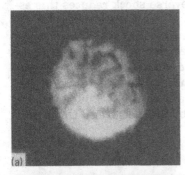

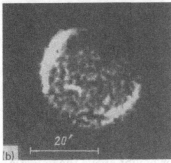

Fig. 22. X-ray images of the remnant of SN 1006 at **(a)** 0.1–0.3 keV and **(b)** 1.2–4 keV (Pye et al., 1981).

The structure of the remnant in x rays is similar to the radio map: it is a thin, symmetric shell 30′ in diameter (the linear diameter is 10 pc at a distance of 1.2 kpc) — see Fig. 22 and the data of Galas et al. (1982) and Vartanian et al. (1985).

Baade searched unsuccessfully for the optical nebulosity associated with the explosion, and it was finally found by van den Bergh (1976). The nebulosity is extremely delicate: a very thin (1–2″) and long (about 8–10′) perfectly regular filament, split here and there. The filament is localized at the periphery of the shell, but in areas where the x-ray and radio emission are weak, in contrast to other remnants. Such perfectly formed delicate features had never before been observed in young supernova remnants. Attempts to find other filaments and condensations associated with SN 1006 were unsuccessful (Long et al., 1988b).

The optical spectrum of the filament showed only the Balmer lines of hydrogen, with $I_{H\alpha} = 4 \times 10^{-6}$ erg/cm²/sec/sr; none of the other bright lines normally observed in supernova remnants, such as [N II], [S II], or [O III], were detected (Lasker, 1981). The optical nebula is clearly an analog of Tycho. A broad Hα component has also been detected in the remnant of SN 1006 (Kirshner et al., 1987b), with FWHM = 2600 ± 100 km/sec. The radial velocity of the filament is $v_{LSR} = -40.5 \pm 1.1$ km/sec, which remains unchanged along its length (Schweizer and Lasker, 1978).

New photographs of the region were obtained in 1980, and a comparison with the 1976 plates revealed proper motion of the filament from the center amounting to $\mu = 0."39 \pm 0."06/yr$, which corresponds to a velocity of 2300 km/sec (Hesser and van den Bergh, 1981). New improved proper motion measurements of the filament obtained by Long et al. (1988b) yield $\mu = 0."30 \pm 0."04/yr$. Based on an estimated blast wave velocity of 2800–3800 km/sec inferred from an analysis of the broad Hα component and the measured proper motion of the filament, Long et al. (1988b) derived a distance of 1.7–3.1 kpc, consistent with the distance for the subdwarf located behind the SNR (see also Kirshner et al., 1987b). The mean expansion rate obtained by dividing the size by the age is $\langle \mu \rangle = 0."83/yr$.

The x-ray spectrum is harder at the periphery of the shell than in the vicinity of its center (see Pye et al. (1981), Galas et al. (1982), Vartanian et al. (1985), and references therein). Assuming a plasma in ionization equilibrium, the spectrum corresponds to $T_e \approx 5 \times 10^7$ K at the edge and $T_e \approx 1.8 \times 10^6$ K within the shell; the plasma density is $0.3 \lesssim n_e \lesssim 0.6 \text{ cm}^{-3}$ near the periphery and $n_e \approx 1 \text{ cm}^{-3}$ in the interior. If ionization nonequilibrium is taken into account, the spectrum better corresponds to radiation from a one-temperature plasma at $T_e \approx 10^8$ K. The broad emission feature near 0.59 keV (see Fig. 44) is a blend of the O VII and O VIII lines, suggesting that the radiation is thermal. Some fleeting references in the literature to SN 1006 having a nonthermal spectrum were most likely due to the lack of plasma ionization equilibrium not being properly taken into account; see below.

The two-component structure of the spectrum and the x-ray map of the remnant are consistent with a transition between free expansion and adiabatic deceleration: the high-temperature, low-density shell is probably a manifestation of emission from swept-up circumstellar gas, while the low-temperature inner region consists of ejected gas that has been heated by the reverse shock. Pye et al. (1981) obtained an overestimate for the mass of hot plasma, up to $5-15\,M_\odot$, and Galas et al. (1982) found a mass $\sim 3\,M_\odot$; neither group allowed for nonequilibrium conditions in the plasma or density redistribution after passage of the reverse shock. New, more reliable estimates by Hamilton et al. (1985) and Hamilton et al. (1986b) yield $M_{ej} \sim 1.4\,M_\odot$ if the outermost ejected layer consists mostly of oxygen, and $M_{ej} \sim 2.5\,M_\odot$ if it is mostly helium. The x-ray spectrum, which is an unusual one for young shell remnants (apparently a power-law continuum with no lines), is consistent with the model of a nonequilibrium plasma heated by the reverse shock if the ejecta exhibit a stratified chemical composition. As in Tycho, the best agreement between the theoretical and observational spectra is obtained when the ejected matter consists of three layers: the innermost layer with about $0.6\,M_\odot$ of Fe (of which $\sim 0.2\,M_\odot$ has yet to be heated by the shock wave), a middle layer, made up of a mixture of O, Si, S, and Fe (a total of approximately $0.4\,M_\odot$), and on the outside, about $0.4\,M_\odot$ of carbon, with a total mass of swept-up gas totaling some $3.4\,M_\odot$ (Hamilton et al., 1986b).

The ultraviolet spectrum of the blue subdwarf seen in projection against the center of the remnant shows strong, broad absorption lines of Fe II at 2385 and 2590 Å, as well as other weaker lines (Wu et al., 1983). The center of the lines is at zero radial velocity, and they are 70 Å wide, corresponding to 10^4 km/sec. Also seen are redshifted lines of Si II, Si III, and Si IV, with radial velocity $v \approx 5 \times 10^3$ km/sec. Such velocities constitute direct proof of the fact that the absorption is taking place in the expanding supernova shell.

Fesen et al. (1988a) have undertaken a new series of UV observations and a detailed analysis of absorption line profiles, and have obtained intriguing results. The subdwarf spectrum also exhibits absorption lines of S II and O I, and stratification of chemical composition by velocity has been clearly established. The inner layer of the shell consists of iron expanding at ±5000 km/sec, followed by individual clumps enriched in Si II and Si IV moving at 5200±400 km/sec, S II at 6000±1000 km/sec, and O I at 6500±300 km/sec. With regard to the observed chemical composition and stratification, the freely expanding iron-rich matter of the core and the high-speed O-, Si-, and S-enriched clumps confirm the identification of SN 1006 as the deflagration explosion of a white dwarf. The mass of Fe II corresponding to the depth of the absorption lines is about $0.01 M_\odot$. This is but a small fraction of the amount of iron predicted by the theory of a white dwarf explosion, but according to Fesen et al. (1988a) and Hamilton and Fesen (1988), the iron in the ejecta may have been photoionized primarily to Fe III or Fe IV by the UV radiation liberated by the passage of the reverse shock. We see, then, that these UV observations of absorption lines in the spectrum of the background star have provided direct evidence for the chemical stratification of the ejecta, as proposed by Hamilton et al. (1986a, b) in modeling the remnant's x-ray emission. The observations of Fesen et al. (1988a) also attest to the marked patchiness of the ejecta; that is, there are dense fragments that are dispersing with practically no retardation and that have not yet been heated by the reverse shock, which accounts for the lack of any highly-ionized x-ray lines in the spectrum of the remnant of SN 1006.

The stellar remnant of the explosion has never been found. The blue subdwarf was a candidate for a time, but the fact that the iron absorption lines in its spectrum are produced by gas on both sides of the expanding shell means that the star lies beyond the supernova remnant. X-ray searches for thermal radiation from a possible neutron star have placed an upper limit on the blackbody surface temperature: $T \leq 8 \times 10^5$ K (Pye et al., 1981).

To summarize, then, three young remnants of type I supernovae have been observed in our Galaxy. We have been able, more or less, to fit the physical conditions in each into the framework of contemporary ideas about the initial phase of shell expansion. Next, we shall examine them as a unified class of objects at a similar stage in their development, although such an approach perhaps raises more questions than it answers.

Inasmuch as we are able to determine the type of these supernovae quite reliably from their light curves, we are definitely in luck, since contrary to expectations (recall that the absence of hydrogen lines from the spectrum is one of the basic earmarks of type I supernovae), the Balmer lines of hydrogen are among the strongest in the spectrum of Kepler, while there are no other emission lines at all in Tycho or the remnant of SN 1006. Being convinced that we have classified these supernovae correctly as to type, we arrive at the only reasonable explanation for this apparent contradiction: at optical wavelengths, we are simply not seeing the matter ejected by the explosion! Indeed, Kirshner and Chevalier (1978), Bychkov and Lebedev (1979), and Chevalier et al. (1980) maintain that the purely hydrogen spectrum of the Tycho and SN 1006 remnants is due to collisional excitation prior to ionization of neutral atoms that traverse the shock front in the intercloud medium (we shall return to this point in Section 8). The broad component of the Balmer lines is a consequence of charge exchange between high-speed protons behind the front and slow, neutral atoms.

The spectrum emitted by the bright filaments and condensations in Kepler is more like that of old remnants, rather than young ones. In Section 7, we shall see that the line ratios observed in Kepler — $I_{[N III]}/I_{H\alpha} = 2.3$, $I_{[S III]}/I_{H\alpha} = 0.5$, $I_{[O III]}/I_{H\beta} = 4.3$ — are typical of old shells; the lines of [Fe II] and [Fe III] have comparable intensities in the old objects IC 443 and N 49. This may be accounted for by postulating anomalously high density in the compact clumps of stellar or interstellar matter, engendering the rapid radiative cooling of the gas behind the shock front. The bright condensations in Kepler therefore demonstrate a later evolutionary stage of interaction between the shock front and the surrounding gas.

The remnant of SN 1006 is a prominent x-ray source, apparently lacking any strong lines of heavy elements.

Inasmuch as only three young type I supernova remnants have been observed, it is hard to say which is the norm and which the exception. In this regard, Tuohy et al. (1983b) have presented very interesting results in which they have identified four remnants in the Large Magellanic Cloud with hydrogen Balmer lines dominating the spectrum. We list the parameters of these remnants in Table 5. Two of them, 0505–67.9 and 0548–70.4 are clearly old shells, as implied by their radii of 20–30 pc, masses of 50–100 $M_\odot$, and weak [O III] lines. The objects 0519–69.0 and 0509–67.5 are analogous to Tycho and the remnant of SN 1006. The only spectral lines present are Hα and Hβ; a broad component of Hα is detectable in the first, and may be detectable in the second. The shock velocity derived from Hα, ~2900 km/sec, corresponds to a remnant about 500 years old, assuming adiabatic expansion. We see, then, that the object that most resembles Tycho and the remnant of SN 1006 is probably the most advanced among the three in an evolutionary sense. One should then consider the optical spectrum of Kepler to be the deviant from the norm, its distinctiveness stemming from the anomalously high density of gas (of either stellar or interstellar origin) in the vicinity of the supernova (see also van den Bergh, 1988).

Is it possible to associate SN 1006, SN 1572, and SN 1604 with SN Ia and SN Ib outbursts, based on observations of the young remnants that we have discussed?

There is at least one argument that seems fairly convincing. We have seen in Section 1 that iron lines in SN Ia spectra near maximum light yield expansion velocities of $(5-8) \times 10^3$ km/sec, while the iron detected in the spectrum of SN 1983N (type Ib) is moving at only about 3000 km/sec. Iron lines in the spectrum of the blue dwarf located beyond the remnant of SN 1006 are produced by absorbing clumps moving at ± 5000 km/sec. The implication is thus that SN 1006 was a type Ia supernova (Blinnikov et al., 1988); and since we have already decided that Tycho and the remnant of SN 1006 are all but identical, it would also seem that SN 1572 must also have been a SN Ia. Kepler, on the one hand, is somewhat different from SN 1006 and Tycho, but on the other, it is located too high above the galactic plane to have been a massive SN Ib progenitor.

At the same time, bearing in mind the high speed at which the filamentary system of Kepler is moving perpendicular to the galactic plane, it is conceivable that the progenitor was a runaway star. Bandiera (1987) concluded that the progenitor of Kepler was indeed a massive star, with a high mass-loss rate — i.e., that the supernova was a type Ib — based on the detailed morphology and kinematics of the remnant. In his model, the bright optical clumps in Kepler might have been formed by the bow shock of a runaway star with a mass-loss rate of $\dot{M} \sim 5 \cdot 10^{-5} M_\odot$/yr, and $V_w \sim 10$ km/sec. If the motion of the clumps is representative of the velocity of the progenitor, then the latter must have begun its journey out of the plane $\sim 3 \times 10^6$ years ago, and may have lost up to $10 M_\odot$ in the form of a stellar wind; in other words, it was quite a massive star. These arguments may possibly favor a SN Ib, but they are doubtless just circumstantial and rather speculative suggestions, and are much less forceful than the arguments that identify SN 1006 and Tycho's supernova as type Ia.

Let us now summarize the common properties of young remnants of type Ia supernovae based on the data presented in this section.

1. A low-mass ($M \sim 1-2 M_\odot$) symmetric shell is thrown off by the explosion, containing dense, compact clumps of poorly mixed stellar matter.

2. The chemical composition of the ejecta and the amount of iron expelled may be consistent with a presumptive white dwarf pre-supernova.

3. It is likely that no neutron star is formed in the explosion. This follows from the lack of a compact x-ray source and the fact that the radio and thermal x-ray emission emanate strictly from shells; see Section 6.

Table 5. Galactic Remnants of SN I, and Balmer-line Dominated Remnants in the LMC.

	Kepler	Tycho	SN 1006	0519−69.0	0500−67.5	0505−67.9	0548−70.4
Radius, pc	1.3	3.3	5−6.5	7.5	6.7	20	27.5
Assumed distance, kpc	3.2	3	1.2−1.6	55	55	55	55
v_{meas}, km/sec	≤300	3600	6500	~2900			
$\langle v \rangle = R/t$, km/sec	3500	7600	7000				
$I_{H\alpha}$, erg·cm^{-2}·sec^{-1}·sr^{-1}	2×10^{-4}	7×10^{-5}	4×10^{-6}	1.1×10^{-5}	5×10^{-6}	4.5×10^{-6} weak [O III]	$<10^{-5}$ weak [O III]
$L(0.15-4.5$ keV$)$, erg/sec	4×10^{35}	$(6-7)\times10^{36}$	$(2-4)\times10^{35}$	1.1×10^{35}	3.4×10^{36}	10^{37}	10^{36}
M_z, $M_\odot$	1−2**	1−2**	1−2**	~24	~11	~100	~51

*The Hα brightness of Galactic remnants refers to the brightest filaments; the brightness in the LMC is taken from the measurements of Tuohy et al. (1983).

**Assuming a reverse shock in the ejecta (Hamilton et al., 1985).

4. The Tycho, Kepler, and SN 1006 remnants are all in transition from free expansion to adiabatic slowing. The evidence comes primarily from the near equality of the mass of the ejecta and the mass of swept-up circumstellar gas, and from the x-ray emission coming from behind the reverse shock front, which develops at about the same time that slowing begins. On the other hand, there is some indication that the reverse shock has not yet progressed past the level corresponding to a velocity of ~7000 km/sec, and a substantial fraction of the innermost layers of the shell are still in free expansion.

The evolutionary age of Tycho, as given by the ratio of the swept-up mass to ejected mass, is probably less than that for SN 1006 (assuming, of course, that they expelled the same amount of mass!). But the expansion velocity in the remnant of SN 1006 is higher (Kirshner et al., 1987b), and the state of ionization equilibrium is not as far advanced as in Tycho, a result of the lower density of circumstellar gas: n_0(SN 1006) = 0.055 cm^{-3}, and n_0(Tycho) = 0.28 cm^{-3} (Hamilton et al., 1986a, b).

5. Evidently the ejecta are not expanding into the interstellar medium, but into matter shed by the progenitor. This is suggested mainly by the anomalous chemical composition in conjunction with the low velocity of optical condensations in Kepler. The radio structure of Tycho (specifically, the narrow outer radio rim) and its x-ray structure and spectrum correspond rather poorly to numerical models, unless one considers the possible collision of the supernova shell with gas in a slow-moving shell (planetary nebula?) expelled by the progenitor (Dickel and Jones, 1985; Smith and Jones, 1988). The circumstellar gas, whatever its origin, is not fully ionized by the explosion of a SN Ia, in any event at distances greater than about 1–2 pc. This is suggested by the presence of delicate optical filaments with a pure Balmer spectrum and a broad component in the Hα line — as long as we have correctly interpreted the evidence; see Section 8.

4. The Remnants of Type II Supernovae: The Crab Nebula (A.D. 1054) and 3C 58 (A.D. 1181)

The Crab Nebula

The Crab Nebula was discovered in 1731 by the English physician and amateur astronomer John Bevis; it was rediscovered in 1758 by Charles Messier, and this "nebulous object" became the first entry in his famous catalog. Two hundred years later, when Hubble first recognized the connection between the Crab Nebula and the supernova of 1054, it again became "object number one" for all of contemporary astrophysics, and it continues still to puzzle both theorists and observers. The Crab Nebula and associated problems were the sole topics of IAU Symposium No. 46 and the symposium entitled *The Crab Nebula and Related Supernova Remnants* (1985). The nature of the remnant has also been discussed by Shklovskii (1976a) and in the reviews by Trimble (1983) and Davidson and Fesen (1985).

We shall not dwell upon facts that are already familiar, but consider instead only recent results that have clarified the relationship between the Crab and other supernova remnants.

The optical emission from the remnant consists of two fundamentally different components: these are the bright filaments and the amorphous nebula. Most of the mass resides in the filaments, which comprise a substantial shell $5' \times 7'$ in size (a radius of approximately 2 pc at a distance of 2 kpc), and which emit solely recombination radiation. The continuous radiation from the amorphous nebula is synchrotron radiation emitted by ultrarelativistic electrons.

This idea was advanced in 1953, and 15 years later a pulsar was discovered in the Crab Nebula — the source of its relativistic particles and magnetic field. The synchrotron origin of the radiation from the Crab has now been firmly established over the entire range from radio waves to x rays and gamma rays — i.e., from 10^7 to 10^{24} Hz. The synchrotron emission spectrum of the Crab Nebula is plotted in Fig. 24, based on the data of Marsden et al. (1984), Manchanda et al. (1982), Iudin et al. (1984), and Mezger et al. (1986). The break in the spectrum near 10^{13} Hz is attributable to synchrotron energy losses. Since we know the age of the nebula, we may use Eq. (10.7)[1] and the frequency at the turnover in the flux distribution to find the magnetic field strength, $H = 3 \times 10^{-4}$ G. The synchrotron spectrum steepens at energies greater than about 150 keV,

---

[1] We shall occasionally make use of relationships established in subsequent sections.

$$\alpha_{E<150\,keV} = 2.13 \pm 0.05; \quad \alpha_{E>150\,keV} = 2.7 \pm 0.3,$$

where $F(\text{photons}/\text{cm}^2/\text{sec}/\text{GeV}) \propto E^{-\alpha}(\text{GeV})$, and that slope is maintained at least up to $E = 500\,\text{MeV}$ (Clear et al., 1987). The total synchrotron power radiated by the Crab at all frequencies is approximately $(2-3) \times 10^{38}$ erg/sec. We may also estimate the rate at which the Crab pulsar, PSR 0531+21, loses rotational energy by using the measured period

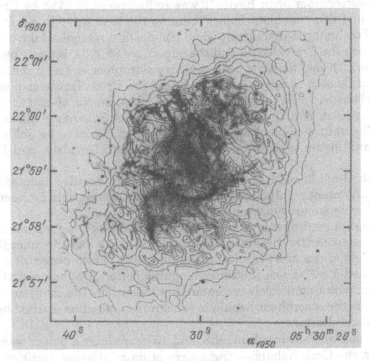

Fig. 23. The Crab Nebula. Bright filaments and amorphous glow with 2–3 GHz radio isophotes superimposed (Wright and Forster, 1980).

$P = 0.033\,\text{sec}$ and period derivative $\dot{P} = 423 \times 10^{-15}$ sec/sec (see Section 6): $L = 4\pi^2 I P^{-3} \dot{P}$, where I is the moment of inertia, which yields the rotational kinetic energy $\frac{1}{2} I \Omega^2$, and Ω is the angular velocity. For the Crab pulsar, we have $I \approx 10^{45}$ g·cm^2 and $L \approx (3-5) \times 10^{38}$ erg/sec, so the rotational energy loss of the pulsar is sufficient to support the synchrotron radiation of the nebula.

The x-ray emission from the Crab Nebula is concentrated in a region of size 1–2′, which is considerably smaller than the radio source and the amorphous nebula; the region is displaced toward one of the central wisps, which are evidently associated with the ongoing ejection of relativistic particles, raising the plasma pressure of the pulsar wind on the shell. This structure is consistent with the differing synchrotron loss rates of particles radiating at radio and x-ray wavelengths.

The x-ray emission is polarized: $p = 19-20\%$ (Weisskopf et al., 1978) between 2.6 and 5 keV, which affirms its synchrotron origin. Thermal radiation from hot plasma in the remnant (if any) is less than 2% of the synchrotron intensity level in the $4-50$ keV range, and less than 8% near 0.5 keV (Toor et al., 1976; Pravdo and Serlemitsos, 1981). No spectral lines of highly ionized elements have been found; an isolated, narrow feature near

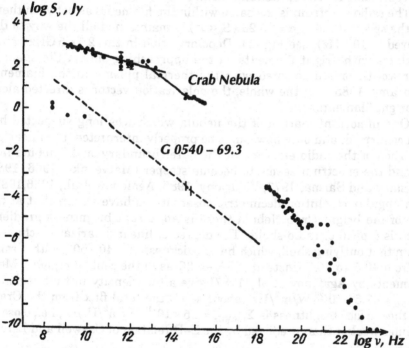

Fig. 24. Synchrotron spectra of the Crab Nebula and G 0540–69.3.

73 keV is probably due to cyclotron radiation in the pulsar magnetosphere (Manchanda et al., 1982). Hameury et al. (1983) detected neither the line at 73 keV nor one at 400 keV (also reported in the literature), so these features may be variable.

Novel results have lately been obtained at radio wavelengths by virtue of the high resolution afforded by modern radio telescopes (in this respect, too, the Crab Nebula has been unique, as lunar occultations provided astronomers with unprecedented resolution during the infancy of radio astronomy, far ahead of its time). Radio maps at 2.7, 5, and 23 GHz with arcsecond resolution are fully consistent with the delicate details observed at optical wavelengths (Swinbank and Pooley, 1979; Swinbank, 1980; Wright and Forster, 1980; McLean et al., 1983). But at higher resolutions, this consistency disappears: there are no radio features smaller than 0.1 pc, which abound when observed in optical lines, and possibly in the optical continuum as well (Wilson et al., 1985a). The difference between the fine-scale morphology of the optical and radio synchrotron emission, if it is confirmed, will be extremely important in analyzing the magnetic field of the nebula, since it should enable one to discriminate between the field into which old relativistic particles (which are responsible for the radio emission) are injected, and the field that is host to young particles (which are responsible for the optical emission).

The radio spectrum is the same within the filaments as between them, and the spectral index $\alpha = -0.26$ ($S_\nu \propto \nu^\alpha$) is maintained all the way to the infrared ($\sim 10^{13}$ Hz); see Fig. 24. Depolarization in the 2.7–5 GHz range correlates with bright filaments on the approaching side of the shell; in other words, it is associated with the thermal plasma in the filaments (Velusamy, 1985). On the whole, the polarization vector is directed along the bright filaments.

One important feature of the nebula which was long suspected has been confirmed, and only now can it be properly interpreted: the linear polarization of the radio emission at the outer boundary of the nebula falls off, and the spectrum seems to become steeper (Matveenko, 1966, 1984; Velusamy and Sarma, 1977; Velusamy, 1985; Agafonov et al., 1986,1987). High angular resolution decimetric observations have shown that at the edge of the bright Crab Nebula there is an abrupt brightness gradient, which is typical of radio shells. The degree of linear polarization changes within that outlying shell, which has a thickness of $\sim 40-50''$, with α ranging from -0.5 to -0.7 (instead of $\alpha = -0.26$ as in the central region). Measurements by Agafonov et al. (1987) give a flux density in that region of $S_{750\,\text{MHz}} = 47.5 \times 10^{-26}$ W/m^2/Hz (about 4–5% the total flux from the Crab), and the surface brightness is $\Sigma_{750\,\text{MHz}} = 1.5 \times 10^{-19}$ W/m^2/Hz/sr , i.e., close to 20 times weaker than predicted by the Σ–D relation of Caswell and Lerche 1979).

These facts, if confirmed, suggest that the Crab is essentially not Crab-like, but is instead an extremely young composite remnant: a plerion surrounded by a shell in the process of formation (see Section 10 and Fig. 47).

Surrounding the pulsar, there is a local minimum in the synchrotron brightness at x-ray, radio, and optical wavelengths, as well as a decrease in the degree of linear polarization. What has probably happened is that plasma with an entrained magnetic field has been swept clear of that region by the freely expanding pulsar wind (Brinkman et al., 1985). The boundary of the region of reduced brightness maps out the localized standing shock wave in the pulsar plasma which thermalizes the wind. Near the center, the magnetic field lines in projection look like concentric circles around the pulsar (Schmidt and Angel, 1979), supporting the suggestion of Kardashev (1964) that the field in the nebula may derive from the tangled field lines of a rapidly rotating collapsed central star.

The supernova of 1054 was originally classed as type I. Having reconstructed the light curve based on eyewitness records and using newer and more reliable estimates of interstellar absorption, Pskovskii (1978a) and Chevalier (1977a) concluded that the outburst was in fact of type II. But the Crab Nebula is different from the other historical remnants — it has an anomalously low expansion velocity, with the bright filaments moving outward at a mean velocity of 1400 km/sec, while the matter ejected in a SN II explosion, as determined by supernova spectra at maximum light, moves at $(5-10)\times10^3$ km/sec. Deceleration of the shell seems to be out of the question, as the system of bright filaments is actually being accelerated.

In order to account for the low expansion velocity, Shklovskii (1978) postulated that the bright filaments consist of matter from the stellar envelope thrown off prior to the explosion, much like a planetary nebula, and that they have been accelerated by the shock wave induced by the subsequent explosion. Chevalier (1977a), on the other hand, has proposed that the filaments are the innermost layers of the stellar mantle, which were ejected at low velocity. According to both hypotheses, the Crab Nebula should be surrounded by an outer weak envelope approximately $10-15$ pc in size, formed by a shock wave propagating outward at a speed typical of type II supernovae, $(5-6)\times10^3$ km/sec.

The expected parameters of this outer "high-velocity" shell have been examined in detail by Chevalier (1985) and Lundqvist et al. (1986); possible ways of detecting it have been widely discussed. Searches for weak emission beyond the bright Crab Nebula, as well as for high-temperature plasma or high-velocity motion corresponding to the putative fast shock wave, have already been under way for 10 years.

A series of spectrograms completely encompassing the field of the nebula was obtained in 1977–79 with the 4-m Anglo-Australian Telescope (velocities of more than 3000 filaments were measured in all), enabling investigators to construct a spatial model of the nebula that represented a significant qualitative advance (Clark et al., 1983). It was revealed that the thick shell is neither uniformly nor randomly filled with filaments, but instead has the clearly delineated structure shown in Fig. 25.

The shell consists of a bright inner layer and a faint outer layer surrounded by an extended halo. The thick, two-layer shell is bounded by two concentric filamentary surfaces, the inner one expanding at an average velocity of 720 km/sec, the outer one at 1800 km/sec. The north–south diameter of the inner surface is 135″, while that of the outer is 340″. The filaments in the shell are primarily arrayed tangentially, but there are a few radial "spokes" as well. The radial velocity distribution is symmetric along the major and minor axes, implying that the elliptical nebula is only slightly inclined to the plane of the sky. The filaments show obvious stratification inside the shell, not just in brightness and velocity, but also

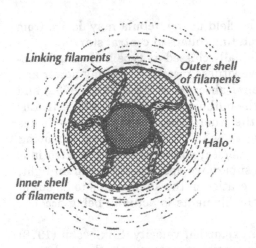

Linking filaments

Outer shell of filaments

Halo

Inner shell of filaments

Fig. 25. Sketch of the Crab Nebula, based on data of Clark et al. (1983). The denser cross-hatched area corresponds to bright synchrotron emission from inside the inner filamentary shell; the outer cross-hatching is weak synchrotron emission from between the inner and outer filamentary shells. The outermost emission comes from a faint halo.

spectroscopically. This double shell with spokes is easily discerned in the Balmer lines, [O II], and [O III], while lines of He are emitted only by the inner part of the shell. The observed stratification most likely reflects the structure of the ejecta, with different layers of the star, differing in chemical composition, flying off at different velocities.

The brightness of the amorphous nebula changes dramatically at the inner boundary of the filamentary envelope: bright synchrotron radiation is concentrated near its center, and weak, amorphous emission can be detected as far as 5.'4 from the center; i.e., it reaches the boundary of the outer envelope. This tends to confirm the proposals advanced by Pikel'ner (1961) to the effect that the network of bright filaments "confines" the magnetic field and relativistic particles responsible for the synchrotron radiation of the bright amorphous nebula. The particles and fields penetrating the interior bright filaments give rise to the weaker synchrotron "plateau" between the inner and outer filamentary shells.

Clark et al. (1983) and Henry et al. (1984) have measured maximum velocities of faint features in the spectrum of the Crab reaching +3600 to −2400 km/sec. Dennefeld and Pequignot (1983) have reported velocities of +3700 km/sec and +5200 km/sec. At one time, even higher-velocity weak spectral-line features were discussed, but those suspicions have yet to be confirmed. Searches for high-velocity motion in the Crab specially undertaken by Fesen and Ketelsen (1985) have not detected any velocities higher than 2000 km/sec at a flux greater than 3×10^{-15} erg/cm^2/sec, corresponding to an emission measure $E_m \geq 10$ cm^{-6} · pc. Those authors suspected only a single weak high-velocity feature, but even the latter corresponded to a velocity of no more than 3000 km/sec.

Murdin and Clark (1981) have reported on a faint extended Hα halo surrounding the Crab Nebula, 12′×28′ in size, whose brightness yields $E_m \sim 7$ cm^{-6} · pc, suggesting photoionization. But first of all, subsequent observations have yet to confirm its existence, and secondly, the presence of a halo, in and of itself, still does not imply the existence of a fast-moving shell, since the halo may be formed by the wind of the progenitor and ionized in the explosion, or by synchrotron UV from the bright nebula.

The hypothesized fast shock wave ought to have been able to heat the surrounding gas to a temperature $\gtrsim 10^7$ K. X-ray observations by the Einstein observatory were therefore fervently awaited, and in fact an x-ray halo was detected. But the faint halo-like emission surrounding the brighter Crab Nebula may have come from scattering by interstellar dust and/or the telescope mirror (Harnden, 1983). Mauche and Gorenstein (1985) have calculated that these effects could completely account for the observed brightness of the halo. At the same time, it remains possible that emission from the putative shell may make some contribution, but no more than $L_{sh}(0.5-4\,\text{keV}) \leq 2 \times 10^{35}$ erg/sec. This is a very weak constraint: recall that the remnant of SN 1006 is about that bright (Table 5).

Radio searches for a large, faint outer shell have also proved fruitless. Wilson and Weiler (1982) estimate that outside the nebula, at 5 to 10 pc from the center, the radio brightness at 610 MHz is at most 2×10^{-20} W/m^2/Hz/sr, and it is no more than 0.6×10^{-20} W/m^2/Hz/sr at 1400 MHz according to Velusamy (1984) and Matveenko (1984). These limits, however, are still higher than the brightness of the similarly aged shell remnant of SN 1006.

The most stringent limit on the radio brightness of the putative fast outer shell was recently obtained by Trushkin (1986a): he found $\Sigma_{7.6\,\text{cm}} < 10^{-22}$ W/cm^2/Hz/sr for a shell less than $30'$ in size, and $\Sigma_{7.6\,\text{cm}} < 3 \times 10^{-23}$ W/cm^2/Hz/sr for sizes between $30'$ and $60'$.

Thus, no outer fast-moving shell around the Crab, delineating a shock front propagating at a velocity typical of a SN II, has yet been detected. Furthermore, the fact alluded to above, that the Crab is probably not a plerion, but a young composite remnant, may render such searches for a distant high-velocity shell moot. In fact, if a radio shell forms at the periphery of a bright nebula as a consequence of enhancement of the magnetic field and acceleration of relativistic particles due to instability of the contact discontinuity between the ejecta and interstellar gas swept up by the shock wave (see Fig. 47 and Section 10), it would be pointless to search for that shock wave far from the boundary of the Crab Nebula.

But there is another possibility. Field enhancement could, in principle, also be related to instability of the contact surface between the pulsar plasma cloud and the ejecta (Chevalier, 1984b). In that event, radiation from the gas behind the blast wave front may simply not yet have been observed, and there remains some hope that that radiation might be detectable far from the boundary of the bright nebula.

Systematic spectral observations of the filaments in the Crab Nebula have been going on for at least 40 years (see Pronik, 1963; Davidson, 1979; Golovatyi and Pronik, 1977, 1986a,1986b; Fesen and Kirshner, 1982; Henry and MacAlpine, 1982; Pequignot and Dennefeld, 1983; Davidson, 1985; Golovatyi and Novosyadlyi, 1986; the foregoing contain further references). The brightest spectral lines are those of [O II], [O III], [N II], [S II],

He II, He I, C III], C IV, and H I, and there are a number of weaker lines of [Ne III], [Ar III], [Fe II], [Fe III], [Fe V], [Fe VII], C I, and [O I]. A simple enumeration of the most prominent spectral lines has much to say about the significant stratification of ionization and excitation conditions in the filaments.

Although detailed analysis is badly complicated by the blending of lines from different filaments in the line of sight that are moving at different radial velocities, some fairly definite inferences can still be drawn about the physical conditions in the filamentary envelope. In [O III] regions, the gas temperature determined by measuring the $I_{5007+4959}/I_{4363}$ line ratio ranges between 11000 and 16000 K; in [O II] regions, the range is 7700 – 16000 K; likewise, it is 7000 – 13000 K for [S II] and approximately 9600 K for [N II]. The electron density in the filaments falls in the range 500 – 3000 cm^{-3} (as given by the relative intensity of the lines in the [S II] doublet). There is a systematic variation in the [O III]/[O II] intensity ratio as a function of distance from the center; the [O III]/Hβ ratio varies from 3 to 45, with a mean value of 15; the [N II]/Hα ratio ranges from 0.44 to 19. In addition, the He I(λ5876)/Hα and He II(λ4686)/Hβ ratios vary widely.

A knowledge of the chemical composition of the filaments in young remnants, in which one may directly observe the matter expelled by the explosion, is vitally important to an understanding of the nature of supernovae. Many investigators have therefore constructed detailed photoionization models of the filaments of the Crab Nebula. The major conclusions, now more than 10 years old, may be summarized as follows.

1) In contrast to other supernova remnants, the light from the filaments in the Crab Nebula comes entirely from photoionization — collisional excitation at the shock front can be neglected. The source of the ionizing radiation is the ultraviolet synchrotron emission from the central amorphous nebula.

2) The relative He/H abundance is higher than normal.

3) The heavy-element abundance is close to the norm, but is poorly determined.

The most accurate analysis of the light from the nebula's filaments appears in the work of Henry and MacAlpine (1982) and Pequignot and Dennefeld (1983), who made use of up-to-date cross sections for all elementary processes and space-based measurements of the flux of ionizing radiation. According to Henry and MacAlpine, variations in the line intensities of helium, the most abundant element, are due to differences in chemical composition, not excitation conditions. The differences observed can be either systematic or random — the He content on the inner surface of the bright filamentary shell, as an example of the former, is several times its value on the outer surface.

One lower bound on relative abundance is $N(\text{He})/N(\text{H}) \leq 40$. This yields the mass of ionized gas in the filaments, $M_{ion} = 0.52 M_\odot$, and the mass of neutral gas, $M_{neutr} = 0.65 M_\odot$. Earlier estimates based on analyses of the radiation spectrum and the brightness of the filaments came up with a similar value of the total mass of the system of filaments, $M = 0.5 - 2.5 M_\odot$. It is worth recalling that Pikel'ner (1956) (see also Shklovskii, 1976a) obtained a nebular mass $M = 0.1 - 0.5 M_\odot$ based on independent considerations, having assumed that the filaments are being accelerated by magnetic field pressure.

The heavy-element abundance in the nebular filaments is less well-determined. The data of Henry and MacAlpine indicate that the C abundance is similar to the solar value, the N and O content is a factor of two to three below the norm, and the S and Ne abundance is perhaps somewhat below the norm. Pequignot and Dennefeld maintain, however, that the heavy-element abundance is higher than normal, and relative to oxygen, the nitrogen content is probably low, the neon content high, and the carbon content, close to the norm. These latter abundance estimates are more consistent with the idea of a massive progenitor. Golovatyi and Pronik (1986a, b) and Golovatyi and Novosyadlyi (1986) recently performed a systematic analysis of all available spectrophotometric observations of the filaments in the Crab Nebula. They identified typical observable regularities of spectral line intensity ratios, and utilized the results to construct three synthetic spectra for three groups of filaments. Basically, they concluded that observed variations in the intensity ratios could not result solely from temperature differences, different distances from the center, or different geometric thicknesses of the optically thick filaments; that stratification of the chemical makeup is required to explain the observations of a large complex of filaments; that on the average, the He abundance in the nebula is several times higher than normal, although it varies from filament to filament in the range $0.2 < \text{He/H} < 1.4$; and that heavy-element abundances are uncertain.

The classical optical range for spectrophotometric investigations of the nebula has now been extended into the ultraviolet (Davidson et al., 1982) and near infrared (Dennefeld and Pequignot, 1982; Henry et al., 1984); the observations of at least nine filaments span the range from 3700 to 10000 Å. The ultraviolet region of the spectrum, containing the strong lines of C IV (1549 Å), C III (1908 Å), and He II (1640 Å) seems to be the most promising avenue for determining the carbon abundance in the remnant. As yet there has been no rigorous analysis comparable to those in the optical, but a rough guess gives $0.5 \leq N(\text{C})/N(\text{O}) \leq 1.5$ for the carbon-to-oxygen ratio. In addition, the amount of He present is greater than usual.

The strongest iron lines in the spectrum show up in the infrared: the [Fe II] line at 8617 Å is fully 25% as strong as Hβ. Brightness variations in the [Fe II], [S II] 6717–31 Å, and [Ni II] 7378 Å lines are correlated; to judge by their ionization potential (7.9 eV, 10.4 eV, and 7.6 eV respectively), these three lines ought to be radiated from the same filamentary regions hosting neutral hydrogen. The intensity ratio of the ultraviolet and infrared lines of [O II] at 3727 Å and 7325 Å yields an electron temperature in the filaments of $T_e = (6 - 10) \times 10^3$ K, in agreement with estimates based on lines in the optical spectrum. Henry et al. (1984) have convincingly demonstrated that at that temperature and a density $n_e \approx 10^3 \text{ cm}^{-3}$, the nickel-to-iron ratio is tens of times greater than in a plasma with the solar abundances: $N(\text{Ni})/N(\text{Fe}) = 43 \pm 18$ on the average in the spectra of 14 filaments. At the same time, the nickel-to-sulfur ratio is 3–23 times higher than in the sun, and the iron-to-sulfur ratio is several times lower (the sulfur abundance is similar to the solar value).

The fact that the nickel content of the filaments of the Crab Nebula is elevated is exceedingly important, since it may be indicative of an enhanced abundance of iron-group elements in the ejecta. To test this proposition, a similar spectral analysis of the Orion Nebula was carried out, but it turned out in that classic H II region that Ni/Fe was an order of magnitude higher, Fe/S somewhat lower, and Ni/S about the same as the solar plasma (Henry, 1984). An examination of the conditions under which Ni and Fe are excited and adhere to dust grains is clearly called for, as well as further measurements of appropriate lines in H II regions, where there is no reason to expect any enrichment in the products of nucleosynthesis, and in supernova remnants; see Henry and Fesen (1988a). Henry and Fesen (1988b) provide a more detailed discussion of recent spectral data and of the Ni abundance problem.

In most of the Crab Nebula filaments that were investigated, the [C I] 9850 Å line was several times stronger than predicted purely on the basis of photoionization. It is possible that the emission from [C I] is partly due to excitation resulting from the collision of a shock wave with neutral hydrogen atoms (Henry et al., 1984).

This has been far from a comprehensive review of the fundamental spectrophotometric results for the Crab Nebula; a more complete summary of old and new observations and, most importantly, a discussion of those questions as yet unanswered because of the difficulties inherent in properly analyzing the observational data, has been given by Davidson (1985). Discrepancies among recent theoretical calculations of the spectrum of the Crab Nebula are substantial, especially as these pertain to heavy-element abundances. And although observational selection effects may be important (thus far, only some dozens of the brightest filaments have been observed over a wide spectral range; the various filaments and their spectra have been analyzed using a variety of apertures, an important consideration when physical conditions within the filaments are highly stratified), the differences more than likely reflect the imperfections of the theoretical models.

The latest deep narrow-band images and long-slit spectra (MacAlpine et al., 1989) divulge large-scale structure in the Crab's filaments: a nearly pure helium torus (~95% by mass), and two polar helium-rich lobes. The Ni-overabundant filaments are found to be located mainly in regions of relatively low He abundance, in particular near the base of the northern jet; see below.

The Crab Nebula continues to pose new questions, and examples come readily to hand. While attempting to find a faint, outer shell, Gull and Fe-

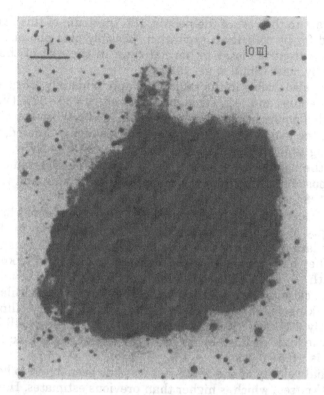

Fig. 26. Northern jet of the Crab Nebula. [O III] photograph by Gull and Fesen (1982).

sen (1982) obtained a number of deep plates of the area (using special high-contrast emulsions and long exposures taken through narrow-band filters in order to record faint emission features). Deep images in [O III] and Hα+[N II] revealed the amazing structure shown in Fig. 26, a well-defined straight-walled jet approximately 45″ (0.4 pc) wide, extending about 80″ from the northern edge of the bright nebula. This feature had in fact been detected by van den Bergh in 1970, but only now was its structure clear. Judging by the photographic image, it is a hollow, thin-walled cylinder, and its straight, collinear boundaries are remarkable when viewed in the context of the chaotic structure of the other filaments. The jet has no connection with the filaments of the nebula, and its axis misses the pulsar: it is displaced approximately 30–35″ to the east of the pulsar, and 10–15″ away from the center of expansion of the filaments.

Spectral line splitting has been observed (Shull et al., 1984), corresponding to expansion of the cylinder at 360 km/sec. Proper motion of individual knots in the cylinder walls (Fesen, 1985) suggests that the cylinder is lengthening at no more than perhaps 4000 km/sec. Both estimates imply a kinematic age of about 600 years — i.e., they show that the jet might have formed long after the explosion of the supernova.

In contrast to the rest of the nebula, the spectrum of this feature can be accounted for by collisional excitation (possibly plus photoionization). The helium content is closer to normal than in the rest of the nebula, roughly 50% by mass (Shull et al., 1984).

The jet is also visible in the radio. It is approximately 1% as bright as the central nebula, and it is strongly polarized: $p = 30-50\%$ (Velusamy, 1984). Detailed polarization measurements made by Wilson et al. (1985b) suggest that the magnetic field in the cylinder is highly ordered, that it is directed along the cylinder axis, and that it is only slightly weaker than the field in the nebula itself, with $H = (2-4) \times 10^{-4}$ G. The radio spectral index may possibly differ from the mean for the nebula, and is $\alpha = -0.8$ (Velusamy, 1984).

Recent observations by Fesen and Gull (1986) have somewhat diluted the peculiarity of the phenomenon. Very deep plates taken in the [O III] line reveal a large number of fine-scale knots and condensations 0.03–0.05 pc in size. They have refined the proper motion of one of these knots, coming up with $0.''22 \pm 0.''05$/yr, which corresponds to a velocity of 2100 km/sec, close to the expansion velocity of the outermost filaments of the Crab. Taking the mean speed from the "radial velocity ellipse" to be approximately 275 km/sec (Shull et al., 1984), Fesen and Gull have deduced that the jet is inclined by about 8° to the plane of the sky, and that its true size is $\sim 0.5 \times 0.9$ pc.

The radial velocity of one of the features has been found to be approximately 500 km/sec, which is higher than previous estimates. Having analyzed the morphology and kinematics in some detail, Fesen and Gull have concluded that despite the apparent displacement of the jet relative to the center of the nebula, it seems on the whole to share the common center of expansion of the Crab's system of filaments (this was previously also noted by Morrison and Roberts (1985)).

Woltjer and Véron-Cetty (1987) have detected weak continuum emission from the jet in the vicinity of $\lambda = 5354$ Å. A comparison of the optical and radio brightness suggests that optical emission in the continuum derives from a synchrotron source having the same spectral index as the center of the nebula.

We stress as well that the spectral index of the jet in the radio is not that much different from the index of the Crab Nebula itself, bearing in mind the aforementioned "turnover" in the spectrum toward the rim of the nebula. The relative underabundance of He mentioned by Fesen and Gull (1986) is also unremarkable, as a similar value is observed in a number of filaments on the outskirts of the Crab.

Several ideas have advanced as to the nature of this jet, but so far they all remain speculative. The first thing that springs to mind is that it is a "shaftway" through which the pulsar beams a fresh supply of relativistic particles (Gull and Fesen, 1982). But it is not clear why the shaftway then fails to pass through the pulsar, and why there is no sign of a connection with the pulsar, either at optical or radio wavelengths. Furthermore, we

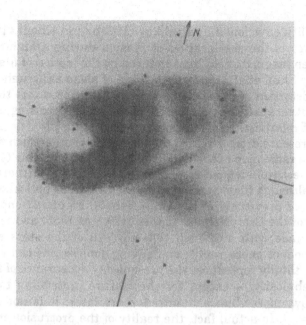

Fig. 27. Drawing of the Crab Nebula dated 1855, based on
visual observations with a 72-inch reflector.

already know the direction in which beaming is taking place — it is well
defined by delicate wisps in the amorphous nebulosity near the pulsar, and
these are 6–7″ to the west of the pulsar and moving west.

Other hypotheses considered include the wake that might have been
left behind by the progenitor as it shed matter in the form of a stellar wind
(Blandford et al., 1983), and yet another "shaftway," this one formed by the
interaction of the fast outer envelope with a very dense clump of interstellar gas (Morrison and Roberts, 1985). The first of these was refuted by the
high radial and transverse velocities of the filaments, and by the synchrotron radiation emanating from the jet. The second, in general, has
been bolstered by an array of indirect evidence stemming from the morphology and kinematics of this unique formation (this question has been
discussed in depth by Fesen and Gull (1986)). The expected size and density of a hypothetical clump should be ~0.5 pc and ~10^2 cm^{-3}; the clump
ought to be located within the present-day shell of the Crab, and should
have a typical velocity of about 100 km/sec.

In principle, such a clump could be easily accounted for by expulsion
from the progenitor, as seen in Kepler's remnant and Cas A (see also
Chapter 3). The problem is that there is as yet no observational evidence
for a halo or fast-moving shell around the Crab with which the clump could
interact.

Most likely, we are seeing the expulsion of relativistic plasma resulting
from an instability at the boundary between the pulsar wind and the enve-

lope, a possibility envisioned by Bychkov (1974b) and Kundt (1983). Another possibility is the acceleration of plasma during a high-speed encounter between magnetic field lines induced by the motion of filaments, or as discussed by Shull et al. (1984), the escape of plasma through a hole resulting from distortion of the shell's magnetic field frozen into the moving filaments. In either case, the uniqueness of the jet is surprising, so long as the eruption of relativistic plasma from one part of the shell does not lead to an abrupt pressure drop or retard the formation of other such features.

However, a radio map of the nebula obtained by Velusamy (1984) does reveal several less well-defined structures similar to the northern jet. It is also possible, although highly speculative, that a similar optical entity may also have existed previously in the Crab Nebula. Figure 27 shows a well known sketch of the Crab Nebula. It was drawn in 1855, and is based on observations made with a 72-inch reflector. All of the stars are easily identified on modern photographs, and having done so, one can verify that the drawing faithfully reproduces the present-day appearance of the bright amorphous nebulosity — except for the feature protruding toward the southwest. Now, more than a hundred years later, it is not nearly so clearly delineated. In actual fact, the reality of the protrusion in the century-old drawing can scarcely be doubted, given the fidelity of the rest of the picture to modern photographs. Could it be that a structure resembling the collinear jet on the northern part of the shell did indeed exist, and that this drawing records a late stage in its resorption into the interstellar medium?

3C 58 (SN 1181)

The explosion of SN 1181 has been identified with the radio source 3C 58, which is a complete analog of the Crab Nebula at radio wavelengths. It has played an important role in enhancing our understanding, since with its discovery it became clear that the Crab Nebula is not an exception, but the exemplar of an entire class of remnants which have come to be known as plerions (or Crab-like) (see Section 10).

Data on the brightness of SN 1181 at maximum are meager; unlike the other historical supernovae, its celestial position had not been accurately determined, and some doubt as to the link between SN 1181 and the remnant 3C 58 had crept into the literature. In 1982, however, new information was gleaned from Chinese chronicles that yielded a more accurate position, and high-speed motion of the gas in the remnant was detected, thereby dispelling any doubts about the identification.

The supernova was probably of type II, with a rapid decline in brightness after the plateau phase; the brightness at maximum light was $m_V = 0$ (Clark and Stephenson, 1982; Pskovskii, 1978a). The distance to 3C 58 has been determined through 21-cm absorption measurements, and up until recently, it was taken to be 8 kpc (Goss et al. (1973)). At that distance, the linear size and mean expansion velocity would be five times greater than

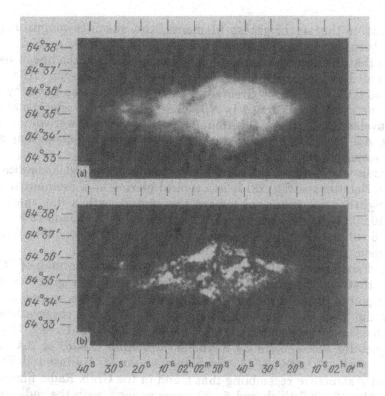

Fig. 28. Brightness distribution of (a) total and (b) linearly polarized radio emission from 3C 58 at λ = 6 cm (Aller and Reynolds, 1985).

for the Crab Nebula, although they are of the same age, and it would be greater than for any of the other historical supernova remnants. Furthermore, the first 21-cm measurements suggested that absorption features with $v_{\mathrm{LSR}} < -30$ km/sec were very weak. Green and Gull (1982) therefore undertook new, detailed investigations of the 21-cm absorption in the direction of 3C 58 with an angular resolution of $7' \times 9'$, excluding errors due to the fine-scale structure of the absorbing clouds. They reduced the distance to the remnant by a factor of three, resulting in a reassessment of the basic parameters of the object. In particular, SN 1181 then became anomalously faint. At the same time, it must be borne in mind that the kinematic distance in the vicinity of the Perseus arm is highly unreliable, owing to possible large-scale deviations from circular rotation of the Galaxy.

The distance to 3C 58 was recently determined by the most reliable method available: by comparing radial velocities and proper motions of a large number of optical filaments, Fesen et al. (1988b) obtained $r = 3-4.5$ kpc. At that distance, and with a color excess $E(B-V) = 0.^{m}6 - 0.^{m}7$ $(A_V = 2.^{m}0 \pm 0.^{m}25)$ as determined from the Balmer decrement of the six brightest filaments, the absolute magnitude of SN 1181 turns out to have been between $M_V = -14.5$ and $M_V = -15.5$.

Note that the number of absorbing hydrogen atoms in the line of sight, as given by absorption measured in the 21-cm line, is $N_H = (2.55 \pm 0.3) \times 10^{21}$ cm^{-2} (Green and Gull, 1982), and as given by the drop in the x-ray spectrum, it is $N_H = (1.8 - 2) \times 10^{21}$ cm^{-2} (Becker et al., 1982; Davelaar et al., 1986); this column density yields a lower value of absorption, $A_V \sim 1.^m3$. The brightness at maximum light cited above may possibly be somewhat high, but not by more than $0.^m5 - 0.^m7$.

At radio wavelengths, the source corresponding to 3C 58 is identical to the Crab Nebula; see Fig. 28. It is a typical plerion — a remnant whose radio brightness is enhanced toward the center, with a flat spectrum ($\alpha = -0.10 \pm 0.02$), and a high degree of polarization — 2, 15, and 25% at 50, 21, and 6 cm wavelength respectively (Wilson and Weiler, 1976; Weiler, 1980; Green, 1986a).

The distribution of linear polarization over the radio map of 3C 58 suggests that the structure of the magnetic field is quite regular; overall, the field points in the direction of elongation of the radio source, and is nearly radial near the minor axis. High angular resolution observations — $\sim 7''$ at 5 GHz (Wilson and Weiler, 1976), $4''$ at 2.7 GHz (Green, 1986a), and especially those made by Reynolds and Aller (1988) at 1.45 and 4.9 GHz with $2''$ and $2.''45$ resolution — have disclosed delicate filamentary structure resembling that found in the Crab. Radio filaments are typically $10 - 30''$ thick and $5 - 20$ times as long, with the radio spectrum being the same within the filaments as it is in the space between them. The data of Reynolds and Aller (obtained with sensitivity two orders of magnitude higher than ever before) indicate that the radio source is larger than previously believed. The bright central region of 3C 58 is enveloped in a faint shell $10.'3 \times 6.'3$ in size. At a distance of $3 - 4.5$ kpc, this yields $9 - 13$ and $5.5 - 8.2$ pc, which is approximately 50% larger than the Crab Nebula, and corresponds to a mean expansion velocity $\langle v \rangle = R/t \approx 6000$ km/sec, which is a typical velocity for the shell of a SN II; see Section 1.

Very faint filamentary optical nebulosity has been identified with the radio source (van den Bergh, 1978a). The brightest of the filaments, on the north side of the remnant, has $E_m \sim 50$ cm$^{-6} \cdot$ pc; the remainder are even weaker. Spectra of about 60 faint filaments obtained recently by Fesen et al. (1988b) show radial velocities near the center of 3C 58 ranging from 1050 to $-1060 (\pm 75)$ km/sec. Proper motions of the best-defined filaments at the rim reach $0.''05 - 0.''07$/yr.

The spectrum of the filaments shows lines of H I, [N II], and [S II]; in the bright northern filament, [O II] and [O III] are also visible. The intensity ratio $I_{[N II]}/I_{H\alpha} \gtrsim 1.5 - 2$ (Fesen, 1983; Fesen et al., 1988b), which is close to the value near the center of the Crab Nebula. The intensity of the [S II] lines yields typical values of $n_e \sim 10^3$ cm^{-3} for the density in the filaments, again similar to the Crab (Kirshner and Fesen, 1978; Kirshner, 1982; Fesen et al., 1988b).

The optical filaments of 3C 58 coincide with brightening of the synchrotron radio emission, and also, as a rule, with local minima in the degree of linear polarization (Weiler, 1980). The latter makes sense if, as in the Crab, thermal plasma in the filaments of 3C 58 is responsible for depolarization of the radio emission. The most obvious region of minimum polarization in Fig. 28 is the "shaftway" extending through the compact central x-ray source (see below) along the minor axis of 3C 58, in the direction of the bright northern filament and the northern radio protruberance. The "shaftway" is devoid of bright filaments, and is situated in a region of enhanced radio brightness, so it is probably associated with a local fine-scale perturbation of the regular structure of the magnetic field (such a structure, if it passed through the pulsar, would help to explain the northern jet in the Crab Nebula).

According to Reynolds (1988b), the formation of narrow radio filaments (synchrotron plasma) and optical filaments (thermal plasma) in the Crab Nebula, 3C 58, and like objects results from instabilities in the interaction process between the pulsar wind and the supernova ejecta at an early stage in the development of a plerion. It is therefore to be expected that just as in the Crab, the synchrotron filaments in 3C 58 ought to contain thermal plasma. The fact that the optical filaments of 3C 58 are much fainter than those of the Crab may be due to the absence from the former of synchrotron ultraviolet emission to ionize the gas in the filaments.

Like the Crab Nebula, 3C 58 possesses a featureless nonthermal x-ray spectrum (Becker et al., 1982). The spectral index corresponding to the slope of the spectrum in the energy range $\Delta E = 2 - 10 \, \text{keV}$, derived from EXOSAT observations, is $\alpha = 1.04 - 1.56$ ($F \propto E^{-\alpha}$) (Davelaar et al., 1986); that is, it is twice the value given by measurements with the Einstein observatory. Observations made with a resolution of $4''$ show that the x-ray emission comes from two components: a compact central source of size $< 5''$, and an extended source of size $\sim 1.'5$ (Becker et al., 1982). Approximately 5% of the emission from 3C 58 in the $0.4 - 4 \, \text{keV}$ band is concentrated in the central source, and at a distance of 3 kpc, its luminosity is $L_{0.1-4 \, \text{keV}} = 5 \times 10^{32} \, \text{erg/sec}$. The point source is located near the center, in a region of enhanced radio brightness, but it is displaced $26''$ eastward of the central radio maximum (Green, 1986a; Reynolds and Aller, 1988). The luminosity of the extended source is $L_{0.1-4 \, \text{keV}} = 2 \times 10^{34} \, \text{erg/sec}$, and as in the Crab, it is much smaller than the radio source 3C 58, a consequence of the short synchrotron-loss lifetime at x-ray wavelengths.

The thermal x-ray component of 3C 58 has yet to be identified. X-ray emission in the $0.1 - 10 \, \text{keV}$ range is considered to result strictly from the synchrotron mechanism. In all likelihood, the central point source is a manifestation of radiation from the hot surface of a pulsar (neutron star) whose beam is unfavorably oriented for terrestrial observers.

Thus 3C 58, the remnant of SN 1181, resembles the Crab Nebula in x rays and radio waves. Optically, though, there is a difference: 3C 58 lacks an optical synchrotron nebula, and the filamentary nebula is much fainter. The latter circumstance is due to the lack of an ionizing ultraviolet synchrotron source. SN 1054 and SN 1181 have been identified as type II supernovae, although the evidence, coming as it does from light curves, is not conclusive; see, for example, Utrobin (1978). In order to classify the Crab as an indisputable canonical SN II remnant, it would be necessary to find the aforementioned fast outer shell or high-velocity gas within the nebula.

In both objects, however, one sees radiation coming from ejected gas with normal or near-normal hydrogen abundance. This implies that the envelope of the supernova contained hydrogen in large quantities — in other words, the outburst was due to a SN II. Moreover, the relative abundance of CNO elements in the Crab is practically identical to that in the ejected envelopes of type II supernovae.

But the main argument that favors the identification of SN 1054 and SN 1181 as SN II is the presence of a collapsed stellar remnant, a rapidly rotating neutron star with a strong magnetic field. In 3C 58, its existence has been confirmed by the synchrotron x-ray spectrum and the x-ray point source at the center of the nebula. Also, we now know, on the one hand, that SN II occur in galaxies like our own just as often as SN I, so that there ought to be 2–3 SN II remnants among the historical supernovae. On the other hand, we have demonstrated in Section 3 that the remnant of SN 1006, Tycho's remnant, and Kepler's remnant are all of type I, and that Cas A cannot be a SN II remnant by virtue of the lack of hydrogen in the ejecta (see Section 5).

We are thus compelled to treat the Crab Nebula, 3C 58, and consequently other plerions as well (see Section 10) as remnants of type II supernova explosions. In addition, the peculiarity of the Crab Nebula and the outburst of SN 1181, like the variety of SN II luminosities and light curves, should be explicable in terms of recent proposals dealing with type II progenitors, namely supergiants with vigorous atmospheric outflows. In point of fact, the interaction between the shock wave, the ejecta, and the wind from the progenitor, which in concert determine both the light curve of the SN II and the properties of the very young remnant, depend on the density distribution in the supergiant atmosphere, which can vary markedly from one star to the next.

The principal parameters of the two young SN II remnants are listed in Table 6. The most important feature of these remnants is the fact that the magnetic field and relativistic particles responsible for the synchrotron radiation over the entire electromagnetic spectrum (in the Crab Nebula), or at least at radio and x-ray wavelengths (in 3C 58), are generated by a central pulsar (see also Section 10).

It is important to point out that in contrast to the historical SN Ia remnants (Section 3) and the young remnant Cas A (Section 5), radiation from the gas behind the shock front induced by expansion of the shell in

Table 6. Young Galactic Remnants of SN II.

Observational data	Crab Nebula	3C 58	CTB 80??
Age, yr	930	800	560??
Angular size	$5 \times 7'$	$\sim 6 \times 10'$	$1'; 10'; (40''?)$
Distance, kpc	2.0	~ 3–4.5	~ 2
Linear size, pc	3×4	$\sim 11 \times 7$	$\sim 30; \sim 6; 0.6$
$\langle v_{exp} \rangle \equiv R/t$, km/sec	2100	5000	
v_{exp} (obs.), km/sec	1500–3000	1100	1200?
z, pc	200	~ 150	~ 100
α (radio)	0.26	0.09	$\sim 0^*, 0.4$
Σ_{1GHz}, $W \cdot m^{-2} \cdot Hz^{-1} \cdot sr^{-1}$	1.2×10^{-17}	2.6×10^{-19}	
$L(10^7 - 10^{11}\,Hz)$, erg/sec	1.8×10^{35}	$(2.5 - 5.7) \times 10^{34}$	$5 \times 10^{31};^* \; 10^{33}$
$L(0.1 - 4\,keV)$, erg/sec	2.5×10^{37}	$(2 - 4) \times 10^{34}$	$\sim 10^{34}; \; 0.7 \times 10^{33}$

*Refers to the core of CTB 80.

both the Crab Nebula and 3C 58 is negligible. Optical emission from the Crab Nebula is synchrotron radiation in the amorphous part and is due to photoionization in the filaments. The faint filaments in 3C 58 are the sole clear-cut visible manifestation of shock-wave action on the circumstellar gas, inasmuch as there are no other sources of ionizing radiation within this object. As we have shown, the x-ray emission from both remnants is synchrotron radiation, except for a weak thermal component in the Crab at the 2–8% level. At radio wavelengths, the interaction of the blast wave with the surrounding gas at this stage of development can show up in the guise of a shell-type radio source, a consequence of field enhancement and particle acceleration in the turbulent layer at the boundary between the ejecta and the swept-up gas. Only a rudimentary shell of this sort seems to be visible in the Crab Nebula (see above).

Thus, not only do the Crab Nebula and 3C 58 differ from the other historical remnants in having a central source that continues to inject relativistic plasma carrying a magnetic field — they also lack any signs of a strong blast wave interacting with the interstellar gas. The latter may result from the surrounding gas being of low density, most likely as a consequence of the progenitor's stellar wind. IRAS and 21-cm observations of the interstellar medium around the Crab Nebula (Romani et al., 1990) do indeed show evidence of an extensive bubble surrounded by a shell ~ 180 pc in diameter. The large bubble around the Crab was most probably swept up by stellar winds from the progenitor of SN 1054 and nearby O stars.

Observational manifestations of the interaction between the shock wave produced by the outburst and the gas of the interstellar medium will evidently become more noticeable in the future, as more and more interstellar gas is swept up. Because of the low kinetic energy of the explosion (with high-speed motions in the halo as yet unconfirmed, we have $M_0 v_0^2 / 2 \approx 4 \times 10^{49}$ erg — that is, an order of magnitude lower than in other SN II and SN I), the Crab Nebula is likely to retain its peculiarities through its later stages as well.

0540–69.3 in the Large Magellanic Cloud

It had become clear by the end of 1984 that the object 0540–69.3 is not just similar to the Crab Nebula — to a much greater extent, it satisfies our current suppositions about the remnants of type II supernovae. It was originally classified as "oxygen-rich" (see Section 5). Optically, it appears with a bright compact central nebula 8″ (2 pc) in size, with a faint outer filament 30″ from the center. The bright nebula shows lines of [O I], [O II], [O III], and [S II], the lines of neon are either absent or very weak, and the ratio $I_{[O\,III]}/I_{H\beta} > 60$ (Mathewson et al., 1980, 1984; Dopita and Tuohy, 1984). The [O III] line consists of a narrow, bright component and a weak, broad one ($\Delta v = 2500 - 3000$ km/sec) shifted by +600 km/sec relative to the maximum. The age of the bright central shell inferred from the size and expansion rate, and assuming free expansion, is about 800 years.

Besides these characteristics typical of oxygen-rich remnants, the x-ray spectrum is devoid of lines and conforms to a power law (Clark et al., 1982), with the brightness increasing toward the center (Mathewson et al., 1984), suggesting a synchrotron radiation source and raising the suspicion of a central stellar remnant (a pulsar). The suspicion was well-founded: an x-ray and optical pulsar was indeed found to be at the center of the compact bright source, with period $P = 0.0502$ sec and period derivative $\dot{P} = 4.8 \times 10^{-13}$ sec/sec (Seward and Harnden, 1984; Harnden and Seward, 1984; Middleditch and Pennypacker, 1985, 1987). According to Manchester et al. (1985), the pulsar is not a member of a multiple system and is young, with a characteristic age $t \approx P/2\dot{P} = 1.7 \times 10^3$ yr, and if $I = 10^{45}$ erg·cm^2, the rotational energy loss is $L = 1.5 \times 10^{38}$ erg/sec, only a factor of two or three less than that of the Crab Nebula (see also the discussion in Section 10).

Right after the discovery of the pulsar, an attempt was made to detect the synchrotron radiation from relativistic particles that it must be injecting into the nebula. The experiment was a resounding success: not only were radio and x-ray emission detected, but so was optical synchrotron radiation, hitherto observed only in the Crab. Chanan, Helfand, and Reynolds (1984) demonstrated that the central nebulosity, which is bright in the [O III] line, is itself a shell around a more compact nucleus 4″ (1 pc) in size, which radiates a continuous spectrum in the blue. After correction for absorption, the continuous optical emission from the nucleus falls on a straight line connecting the radio and x-ray synchrotron emission of 0540–69.3 (see Fig. 24). Overall, the spectrum is described by the power law $S_\nu \propto \nu^{-0.8}$ over the frequency range from $5 \times 10^8 - 10^{18}$ Hz; absorption is taken to be $A_V = 0.^m 8 - 1.^m 0$. The optical spectrum then corresponds to $\alpha = -0.96$ to -0.8, and $\alpha = -0.4$ at radio wavelengths. Finally, the synchrotron origin of the continuous optical radiation from the nucleus ought to be confirmed by polarization measurements, which will probably not be long in coming.

The overall x-ray luminosity of the remnant is $L_{0.2\text{-}4\,\text{keV}} = 10^{37}$ erg/sec, twice as high as the Crab Nebula; as in the latter, the synchrotron luminosity over the entire energy span can be comfortably accounted for by rotational energy losses of the pulsar (Chanan et al., 1984; Reynolds, 1985). Most of the x-ray flux is associated with the central unresolved spot, which is less than 2" in size. Approximately 10-20% of the x-ray flux is due to thermal emission from the outer shell, which radiates strongly in the [O III] line.

The existence of this outer "thermal" shell surrounding a "synchrotron nebula," in contrast to the situation in the Crab Nebula, makes 0540–69.3 a more typical exemplar of the complex of phenomena presently thought to accompany the explosion of a type II supernova A detailed investigation of this remnant and searches for other similar ones would make for a most interesting and promising observational project.

Recently obtained spectra of 0540–69.3 covering 3500 to 9800 Å (Kirshner et al., 1989) display synchrotron emission over the whole range, along with strong [O I], [O II], [O III], [S II], [S III], [Ar III], [Ni II], [Fe II], [Fe III], and Hα; lines of [Fe V] and [Fe VII] may also be present. The average linewidth, 2735 ± 200 km/sec (full width at zero (baseline) intensity), suggests a kinematic age of about 760 yr. The lines are shifted by +370 km/sec relative to the rest velocity of the LMC. Although the pulsar and optical synchrotron emission of 0540–69.3 are indicative of an overall resemblance to the Crab Nebula, these authors find strong O and S lines with properties similar to those found in Cas A (see Section 5), as well as Fe and Ni mixed into the oxygen-rich zone, as in SN 1987A.

To conclude this section, we note that the three young remnants containing pulsars that we have discussed here are unquestionably objects of the same class, and in all likelihood are associated with SN II. But along with the remarkable similarities that these three remnants of approximately the same age share, we also see definite differences. The basic one is that the Crab and 3C 58 are plerions, while 0540–69.3 is a composite, with a clearly identifiable outer shell. This poses an interesting problem, which may have a twofold solution. On the one hand, it might be presumed that there are two kinds of SN II, one forming plerions and the other forming plerions with a shell. The evidence consists of the subdivision of SN II into SN II-L and SN II-P — see Section 1. On the other hand, as suggested by Lozinskaya (1980b) (see also Lozinskaya, 1986), it may be that plerions and composite remnants represent two phases in the evolution of a single kind of object. The observation that a shell has begun to form around the plerion in the Crab, if confirmed, probably favors this suggestion. (This problem relates to the interaction conditions between the pulsar plasma, the supernova ejecta, and the circumstellar gas; it is taken up in detail in Section 10.)

5. Cassiopeia A and the Oxygen-Rich Supernova Remnants

Cassiopeia A

In 1951, Baade and Minkowski discovered the optical nebulosity associated with the brightest radio source in the sky, Cas A. The very first photographs and spectrograms revealed the complex morphology and kinematics of the nebula, as well as the anomalous chemical composition of the filaments. Radio observations had shown that this was the most "dynamic" of the remnants in the Galaxy, both its brightness and fine-scale structure being rapidly variable.

Nowadays, classical optical methods are not the only means by which Cas A may be studied: proper motions down to hundredths of an arcsecond have been measured in the radio, and for the first time, radial velocities of radiating plasma have been detected through the x-ray lines of highly ionized elements. We should also state here that in the 1960's, when the two-component structure of the nebula was revealed and the "fast-moving knots" radiating in the lines of oxygen were identified with ejected matter, while the "stationary flocculi" were identified with condensations in the interstellar gas accelerated by the shock wave, the remnant seemed much more well-understood than it does today.

The optical nebula, consisting of patchy clumps and filaments about $2-5''$ in size, lies within the radio and x-ray shell (Fig. 29). There are no lines of H, He, C, or N in the fast-moving knots, which are dominated by the lines of oxygen and its nuclear combustion products: the strongest are the lines of [O I], [O II], [O III]; the lines of [S II] and [S III] are bright, those of [Ar III], [Ar IV], and [Ar V] are visible, and there are weak traces of [Fe II] and perhaps [Ni II] (Chevalier and Kirshner, 1979, and references therein). The line ratios vary markedly from filament to filament; there are clumps whose spectrum shows only oxygen, and these lie closer to the center than the sulfur-rich filaments. The stationary flocculi display lines of H, N I, [N II], [O I], [O II], and [O III], with strong lines of [Fe II] and He I (Chevalier and Kirshner, 1978). In chemical composition, the fast-moving knots are quite anomalous (see Table 7), with typical densities $n_e = 10^3 - 3 \times 10^4$ cm^{-3} and temperatures $T_e = (2-8) \times 10^4$ K. The stationary flocculi have a chemical composition closer to the norm, but they also show enhanced ratios of He/H and N/O, suggesting that the matter is of stellar origin (Peimbert and van den Bergh, 1971; Chevalier and Kirshner, 1978). The gas density in the stationary flocculi is $(5-7) \times 10^3$ cm^{-3} and the temperature about 7×10^3 K.

More than likely, the fast-moving knots and stationary flocculi are the cooling regions behind a shock wave propagating through dense clumps of matter (ejecta and stellar wind, respectively), due to the high pressure of the surrounding hot gas (Bychkov, 1973, 1974a; Chevalier and Kirshner, 1978; Contini, 1987). The total mass in all of the visible filaments and condensations is approximately $0.05 - 0.1 M_\odot$, which as we shall demonstrate below is only a negligible fraction of the mass of the remnant. A faint Hα glow with a radius of approximately 6' surrounds the bright shell of Cas A (Peimbert, 1971). The same kind of weak, plateau-like halo has also been observed in radio waves and x rays (Stewart et al., 1983).

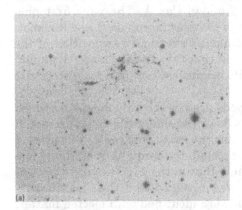
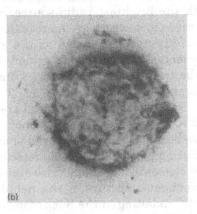

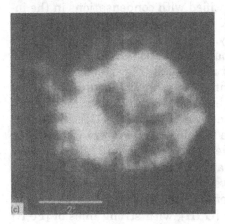

Fig. 29. Cassiopeia A in **(a)** visible light, **(b)** radio, and **(c)** x rays. All images are on the same scale. Observational data of Bell (1977) and Fabian et al. (1980).

The kinematics of Cas A have been investigated in detail, with the greatest debt being owed to van den Bergh and Kamper (Kamper and van den Bergh, 1976; van den Bergh and Kamper, 1983, 1985). They obtained an extensive collection of plates with the Hale 5-m telescope covering more than 35 years, the foundation having been laid by Baade in 1949. In all, the proper motion and radial velocity of approximately two hundred filaments and condensations were measured. Proper motions of fast-moving knots range from 0."2 to 0."5 per year; radial velocities go from 4000 to 9500 km/sec, with the mean at 5500 km/sec. Filament velocities are proportional to their distance from the center of the shell, indicating that they were ejected from a common center and are moving with practically no slowdown. On the other hand, changes in the brightness and geometry of filaments take place on a typical time scale of ten to twenty years; new filaments appear, brighten, sometimes change in shape, and gradually disappear.

The stationary flocculi develop in much the same way, changing shape and brightness on a time scale of more than 25 years. Several of the stationary condensations located outside the radio shell have been observed to change. Proper motions of the stationary condensations must be less than 0."2/yr, and radial velocities range from −80 to −430 km/sec. The kinematic age of the stationary condensations, as determined by their velocity and distance from the center, is 11000 ± 2000 years.

Table 7. Chemical Composition of Cassiopeia A: Content by Mass Relative to Oxygen (Chevalier and Kirshner, 1978). Solar Data Provided for Comparison.

Element (X)	X/O Cas A	X/O Sun
H	0.02	95
He	0.42	32
C	0.003	0.38
N	6×10^{-6}	0.12
Ne	0.03	0.16
Mg	0.06?	0.06
S	0.13	0.05
Ar	0.01	0.02
Ca	0.003	0.008
Fe	0.01?	0.21

The velocity, chemical composition, and morphology of the fast-moving knots have been found to be correlated (van den Bergh and Kamper, 1985). Filaments radiating in the [O III] line are more "ephemeral" than compact clumps that are bright in [S II]; at the same time, the fastest-moving filaments are only visible in the [S II] line, and do not radiate in the [O III] line. This probably means that the ejected matter from the exploded star is poorly mixed.

A completely new class of optical feature was recently discovered in Cas A (Fesen et al., 1987, 1988c). The features, combining the properties of fast-moving knots and stationary flocculi, have been designated *fast-moving flocculi*. These filaments are the farthest from the center of the remnant, and are located at distances of 144 – 192″ (i.e., outside the clear-cut radio and x-ray shell; see below). The spectrum of the fast-moving flocculi resembles that of the stationary condensations: the only lines it exhibits are those of Hα and [N II], and their relative intensity suggests that the nitrogen is probably somewhat overabundant. In contrast to the stationary condensations, however, the fast-moving flocculi have proper motions in the range 0."50 – 0."65/yr, which corresponds to an expansion velocity of 7600 – 8600 km/sec; the radial velocities range from +400 to –2200 km/sec.

The fast-moving flocculi are thus the clumps in the outermost layers, near the photosphere of the progenitor, that are moving at the highest velocity. The implication is that the outer layers of the supernova contained hydrogen, but only a small quantity, which is still only poorly determined.

The proper motions and radial distances of these flocculi have enabled Fesen et al. to date the explosion of the supernova associated with Cas A to 1680. This is later than the previous estimate of 1658 ± 3, and is consistent with the suggestion by Brecher and Wasserman (1980) that Flamsteed may perhaps have observed the supernova in Cassiopeia in August of 1680.

Comparison of the radial velocities and proper motions of a large number of filaments yields a distance of 2.8 kpc to the remnant (van den Bergh, 1971). Accordingly, the radius of the shell, which is quite well-defined in the radio and x-ray regions of the spectrum (see Fig. 29) is 1.3 pc in the southeast and 1.7 pc in the northwest.

The radio remnant is a diffuse shell in which more than 400 bright, compact knots are embedded. There is no one-to-one correspondence between the optical filaments and the radio knots, although they all lie within the bounds of the diffuse radio shell. As do the optical filaments, the radio knots vary in brightness on a time scale of decades. A close comparison of "radio photographs" taken nine years apart has revealed that most of the condensations became weaker, the average rate being 3–4% per year (although some increased in brightness by just as much) (Dickel and Greisen, 1979; Tuffs, 1983). The general impression is that the secular decrease in radio flux density from Cas A, about which we shall have more to say in Section 10, is at least a feature of the diffuse shell, but it also appears to be associated with very fine structure — perhaps with weak, unresolved radio condensations.

The first proper motion measurements of 30 compact radio knots were made by Bell in 1977; by now, the number of radio features with measured velocity has grown by more than an order of magnitude. Four long series of observations have been carried out: by Bell (1977) and Tuffs (1983) using the Cambridge 5-km radio telescope at 5 GHz, by Dickel and Greisen (1979) at 2.7 GHz using the NRAO three-element interferometer, and by Angerhofer and Perley (see Tuffs (1983)) using the VLA. In the main, these authors agree: the radio knots are in random motion. One occasionally sees transverse or inward motion, and if there is indeed any systematic outward expansion of the radio filaments, it must in any case be slower than the expansion of the optical filaments.

On the last point, however, opinions differ. Superimposed on a chaotic background, Bell and Tuffs found a system of radio knots expanding with a characteristic time scale of 950 years — i.e., with a velocity of about 1500 – 2000 km/sec. Two teams of American investigators, on the other hand, found no systematic expansion. Negative results obtained with the

NRAO three-element interferometer may have been related to the difficulty of properly allowing for the instrumental response in observations taken with a limited number of baselines. The VLA observations covered only a short time span: preliminary results discussed at IAU Symposium No. 101 (1983) cover only 21 months. The results obtained by Tuffs that are shown in Fig. 30 seem so far to be the most reliable, all the more so since the expansion that they imply is shared not just by the compact knots, but by a number of extended features as well.

More recently, exhaustive processing of the data (Tuffs, 1986) has confirmed these findings. Proper-motion measurements of 342 radio knots to astrometric accuracy has shown that the systematic expansion rate of this system of radio knots is only one-third that of the optical filaments. Random motion of the radio knots at the 0."1/yr level has also been confirmed, significantly increasing the measurement errors.

But a comparison of relatively low-resolution (1') radio maps of Cas A at 151 MHz taken in 1984 and 1986 reveals changes in the large-scale features of the shell on a characteristic time scale commensurate with the expansion time of the system of optical filaments (Green, 1988b). The overall impression is that the radio shell as a whole is expanding more rapidly than might be thought based on the random motion of the small-scale radio condensations within it, which would make sense if the expansion of the radio shell reflected the propagation of the blast wave.

Random motion of the radio knots might result from the interaction of dense, fast-moving clumps in the inner layers of the ejecta with the diffuse ejected shell, which has already begun to slow down. This is just the picture painted by Braun et al. (1987), who analyzed three radio maps of Cas A obtained with the VLA in 1981, 1983, and 1985. Several of the bright radio knots seemingly thrust into the diffuse radio shell from within, providing a beautiful illustration of the fragmentation of condensations, as well as the change in direction of fragment motion by approximately 45° from the local radius once they have passed the inner boundary of the shell.

X-ray observations in the 0.5–4.5 keV range suggest that emission is confined to two thin concentric shells: the inner, brighter one has a radius $R = 120''$ and thickness $\Delta R = 17''$ (1.5 and 0.25 pc at 2.8 kpc), while the outer one is less well-defined, with a radius of about 150", and $\Delta R \approx 20''$ (Murray et al., 1979; Fabian et al., 1980). This bright, two-layer shell is surrounded by a faint x-ray halo that progressively fades in the outward radial direction, and can be detected out to $R = 6'$ — in other words, it coincides with the optical halo (Stewart et al., 1983). The luminosity of the halo, $L_{0.5-3 \, keV} = 5 \times 10^{34}$ erg/sec, comes to about 2% of the luminosity of the bright nebula. The mass of the x-ray emitting plasma in the two-layer shell, if derived assuming ionization equilibrium and normal chemical abundances, amounts to $15-20 \, M_\odot$ (Fabian et al., 1980).

As in other young remnants, however (see Section 3), the x-ray emitting plasma in Cas A has not yet attained ionization equilibrium. The evidence, specifically, comes from recent observations made with EXOSAT (Smith, 1988a; Jansen et al., 1988) and Tenma (Tsunemi et al., 1986). Taking the ionization nonequilibrium of the radiating plasma into account, Tsunemi et al. (1986) obtained a mass $M_{x\text{-}ray} = 2.4 M_\odot$, an estimate that seems more correct. Interpretation of the x-ray data in an ionization equilibrium model leads one to a dual-temperature spectrum with $T_1 \sim 10^7$ K and $T_2 \sim (5-7) \times 10^7$ K. The chemical composition of the low-temperature plasma, which is deemed responsible for the line radiation, turns out to be quite anomalous: the Si abundance is about 50% higher than the norm, and relative to the solar abundance, estimates for the other elements are $S/Si \sim 2(S/Si)_\odot$, $Ar/Si \sim 4(Ar/Si)_\odot$, $Ca/Si \sim 2(Ca/Si)_\odot$, and $Mg/Si \sim 0.1(Mg/Si)_\odot$ (Holt, 1983). In addition, iron has been found to be underabundant relative to the sun (Becker et al., 1979). Taking plasma nonequilibrium into consideration lowers the heavy-element abundance, but the chemical composition still remains anomalous (Jansen et al. (1988), Aschenbach (1988), and references therein).

The EXOSAT and Tenma observations differ with regard to spectral interpretation. The latter may be approximated by a one-temperature spectrum with $T_e \sim 4.35 \times 10^7$ K, and the chemical composition of the plasma turns out to be close to normal. But the analysis of the EXOSAT data carried out by Jansen et al. (1988) (see also Aschenbach (1988) and references therein) suggests that even when the departure from ionization equilibrium is taken into account, the spectrum from 0.5 to 20 keV is best represented by radiation from a dual-temperature plasma. The high-temperature plasma to which Jansen et al. (1988) fitted a temperature $T_s \sim 3.3 \times 10^7$ K is most probably circumstellar matter shocked by the blast wave that has not attained ionization equilibrium. The low-temperature plasma ($T_s \sim 7.5 \times 10^6$ K) is most likely reverse-shocked ejecta, a hypothesis which they independently supported with the radial temperature distribution of emission in the $0.5-2.1$ keV band for several position angles. No evidence is found by Jansen et al. (1988) for a high-temperature (energy > 20 keV) component.

The temperature implied by the x-ray spectrum depends on position in the map of the remnant; see Murray et al. (1979); Fabian et al. (1980); Aschenbach (1985, 1988); Jansen et al. (1988). At a distance of 3' from the center, in the outer shell, one sees primarily concentrations of high-temperature plasma; in the inner shell, low-temperature plasma prevails. Dense, bright regions are characterized by low-temperature, and weak, diffuse regions by high. There is also some stratification of the regions radiating in highly-ionized lines of S and Ne (Markert et al., 1988).

Even the most conservative estimate of the mass of hot plasma suggests that most of the matter in Cas A radiates not in the optical spectrum, but at x rays. It would therefore be particularly interesting to study hot-plasma kinematics using lines in the x-ray spectrum.

Observing the brightest and best-resolved lines of Si XIII ($1.84-1.86$ keV), S XV ($2.43-2.46$ keV), and S XVI (2.62 keV) with high spectral resolution ($E / \Delta E \geq 100$), Markert et al. (1983) detected Doppler shifts in the lines emitted in the northwest and southeast sectors of the shell, unambiguously implying a relative velocity of 1800 ± 300 km/sec. The total linewidth in each sector is ~5000 km/sec. The velocity and brightness asymmetry indicate that what we are observing is a broad ring, inclined to the line of sight, and expanding at ~5500 km/sec.

This has provided the first direct observational proof that most of the hot plasma in the remnant is moving at the same speed as the expanding optical filaments, and as such, it would be difficult to overestimate its importance. The velocity asymmetry of the optical filaments (the receding filaments in the northwest have a mean velocity of 4000–5000 km/sec, and the approaching ones, 2000 km/sec) is consistent with this toroidal geometry of the ejecta. Radio observations at 86 GHz have also revealed a certain asymmetry in the distribution of rotation measure, probably due to the toroidal geometry of Cas A (Kenney and Dent, 1985).

Let us attempt now to construct a general model of the remnant based on the aggregate observational evidence that has been set out. The two-layer structure of the x-ray remnant yields to the same natural interpretation as in the case of Tycho's remnant: the inner, bright shell is made up of ejected matter that has been heated by the reverse shock wave, and the weak outer shell consists of swept-up circumstellar gas. The observed Doppler shifts of sulfur and silicon x-ray lines characterize the low-temperature plasma of the inner shell, as it is precisely that plasma that is responsible for the line radiation.

The high-temperature plasma is comprised of circumstellar gas that has been heated by the outgoing blast wave. Fast-moving optical knots are dense clumps of ejected matter, as witness their velocity and anomalous chemical composition. The appearance, development, and gradual disappearance of fast-moving knots may be accounted for by their being "switched on" at such time as ejected fragments impact the hot, high-pressure gas behind the shock front. The fact that both components of the ejecta — the stellar fragments and the diffuse gas — are moving at similar velocities indicates that the shell is not yet being effectively slowed. The chemical composition of the fast-moving knots varies quite widely, and it differs from that of the x-ray emitting plasma. This makes sense if the individual clumps are fragments of unmixed matter from fairly deep within the star.

The stationary flocculi are clumps from the outer layers of the star that were expelled early on, most likely as a slow-moving shell or a powerful wind from the progenitor. Their appearance and development, as well as their localization in the vicinity of the x-ray shell, is explained by the same turn-on process as they impact surrounding hot gas behind the blast wave and subsequently expand when the pressure of the hot gas subsides. A "cold" stationary condensation is also seen at 21 cm (Goss et al., 1988).

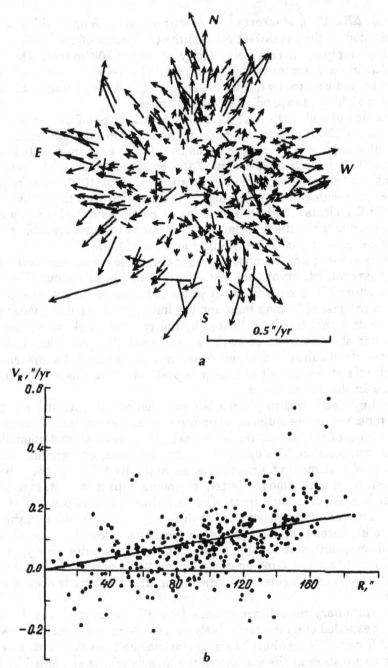

Fig. 30. Proper motion of the radio knots in Cas A (Tuffs, 1983). **a)** Velocity and direction are shown by arrows originating at each corresponding knot; **b)** radial projection of the velocity of proper motion as a function of distance from the center of the shell.

The matter expelled by the progenitor also has a characteristic two-component structure, with stationary flocculi consisting of dense clumps heated by the shock wave, and a diffuse component that constitutes the outer x-ray emitting shell. The weaker outer halo is probably made up of gas from the stellar wind of the progenitor which has not yet been disrupted by the supernova blast wave — further observations will be necessary to clarify the x-ray emission mechanism.

The chaotic motions exhibited by the radio knots are probably not directly indicative of the motion of matter, but rather of small-scale fluctuations of the magnetic field.

The toroidal structure of the x-ray shell may derive from the asymmetry of the ejecta or the asymmetry of the density distribution in the ambient medium — in other words, from inhomogeneous mass loss by the progenitor star (Markert et al., 1983). If the density distribution of the surrounding gas is toroidal as a result of a nonuniform wind (with a stronger outflow at the equator, for example), then the reverse shock will appear much sooner in this dense region, and will heat the ejected matter there more vigorously, even if the latter is symmetric.

It would seem simpler to suppose that the matter ejected in the supernova explosion was thrown off toroidally, as might result in the case of rapid rotation of a massive progenitor. Calculations performed by Bodenheimer and Woosley (1983) suggest that the matter cast off in the equatorial plane would then be enriched in oxygen and the products of oxygen burning (see also Ardelyan et al. (1979) and Chechetkin et al. (1988)).

The supernova explosion in Cassiopeia went unobserved (it has been purported that Flamsteed may possibly have observed the supernova in 1680, but this has never been reliably confirmed). Bearing in mind the advanced state of European astronomy at that time, and the complete lack of any record of the event in Chinese chronicles, it stands to reason that the supernova may have been several magnitudes fainter than a "normal" SN Ia or SN II, with $M_V \approx -16.5$ at maximum light, allowing for interstellar absorption.

An anomalously faint outburst may be associated with a massive star that sheds its outer hydrogen-rich layers as it evolves prior to the explosion, and fails to go through a supergiant stage with an extensive outflowing atmosphere. In that event, according to calculations by Imshennik and Nadezhin (1970) and Chevalier (1976), the supernova may be $5-6^m$ fainter at maximum light than the usual value. The proposition that a massive progenitor lost matter prior to the explosion is consistent with the presumed stellar origin of the gas in the vicinity of the remnant, deduced on the basis of observations of the nebula.

The entire corpus of observations of Cas A thus leads one to a model of the remnant in which two-component matter ejected by the supernova (both compact fragments and a diffuse shell, probably toroidal) interacts with two-component circumstellar matter furnished by a progenitor (dense clumps plus the tenuous, homogeneous gas of the wind). Effective deceleration of the ejected shell, much less the dense fragments, has not yet begun. On the main sequence, the progenitor had an initial mass of at least $10 M_\odot$, most likely $20-25 M_\odot$; the star lost its outer hydrogen shell prior to the explosion. One likely candidate would be the compact helium core of a massive, evolved star — i.e., a Wolf–Rayet star. According to Fesen et al. (1987), the presence of hydrogen and marked overabundance of nitrogen in the fast-moving flocculi, or in other words in the outer layers of the super-nova, mean that the progenitor was probably a WN7, WN8, or WN9. The stationary flocculi most likely represent material of an "ejecta"-type Wolf–Rayet ring nebula (see Sections 14, 15), which are frequently associated with WN7–WN9 stars.

G 292.0+1.8

This galactic supernova remnant is an analog of Cas A: its optical spectrum is dominated by the lines of oxygen and neon, and the x-ray image suggests that the ejected matter has toroidal geometry. The radio remnant was originally tagged as a composite type, consisting as it does of a bright central source approximately 2′ across, and a weaker plateau about 9–10′ across (Lockhart et al., 1977). But the spectral index distribution — $\alpha = -0.2$ near the periphery and $\alpha = -0.4$ at the center — the thermal x-ray spectrum, and the lack of a central compact x-ray source favor a different interpretation.

The oxygen- and neon-bright filaments are nonuniformly distributed across the radio remnant. The spectral profiles of these lines consist of a bright, narrow peak ($\Delta v \le 90$ km/sec) and a broad, squared-off pedestal ($\Delta v = 2000-2500$ km/sec) (Goss et al., 1979; Murdin and Clark, 1979). Narrow lines, including Hβ, appear throughout the remnant and even out-side its boundaries, and are most likely associated with the nearby H II re-gion. The broad O and Ne lines are visible only within the dense, compact knots lying inside the radio source, and there is no corresponding broad hydrogen component. The temperature inside the filaments is $(2-3) \times 10^4$ K and the density is close to 2500 cm^{-3} (Dopita and Tuohy, 1984). Studies of the velocity field have shown that the filaments constitute part of a shell that is expanding with a mean velocity of some 2200 km/sec (Braun et al., 1983).

X-ray observations of this object revealed a unique structure (Tuohy et al., 1982). It was on the heels of these observations, in fact, that the idea of toroidal geometry of the ejecta in oxygen-rich remnants was advanced. The structure exhibited by the x-ray source is that of an elliptical disk 8′×6.′5 across, crossed by a bright double bridge along its minor axis. The x-ray disk virtually duplicates the radio plateau. The bridge is a rather delicate one, about 6′ long, its components being separated by about 1′. This sort of geometry may result from the toroidal distribution of radiating matter (for example, in the equatorial plane of the star), the system being viewed edge-on.

The brightness and velocity distribution of the optical filaments does not display the ring structure (Braun et al., 1983), but even in Cas A, the ring of ejecta is plainly visible only in x rays. The distance to the remnant deduced from 21-cm measurements is $r \sim 3.7 \pm 2$ kpc; $r \sim 4 \pm 4$ kpc from optical absorption, and $r = 3.6 \pm 1.5$ kpc from the density of dust in the line of sight given by IRAS (Braun et al., 1986). Taking a mean distance $r \sim 4$ kpc, the radius of the weak radio plateau is about 6 pc, and the radius of the bright radio core and the bright, patchy optical nebulosity is about 3 pc. If we assume that the expansion velocity of the torus is $v = 2200$ km/sec (a reasonable supposition, since the size of the bridge is roughly equal to the radius of the remnant), we find an age of about 1500–2000 years for the ejecta if slowing has not yet begun. Infrared observations suggest that the explosion took place at the boundary of the dense cloud (Braun et al., 1986).

As in Cas A, the Ne/O ratio in the filaments seems to indicate that the progenitor of the supernova was a fairly massive star, with $M_{init} \gtrsim 25 \, M_\odot$ (Tuohy et al., 1982).

N 132D in the Large Magellanic Cloud

This extragalactic object is a member of the same class of supernova remnants. As in Cas A, one sees here a system of fast-moving filaments and slow-moving condensations. The former typically exhibit radial velocities in the range 1000–3000 km/sec, and radiate only in the lines of [O II], [O III], [Ne III], and [Ar III], with a complete absence of hydrogen: $I_{[O III]}/I_{H\beta} > 1000$. The spectrum of the stationary condensations is closer to normal, with radial velocities of less than 600 km/sec and densities of $(2-8) \times 10^3$ cm^{-3} (Lasker et al., 1978, 1980). The bright optical filaments and condensations are enclosed within a spherically symmetric region 6 pc in diameter, surrounded by a weak, diffuse disk 32 pc across, which in addition to oxygen lines also radiates in Hα, Hβ, and [N II]. A bright radio synchrotron source coincides with the central filamentary nebulosity. The x-ray source is associated with a diffuse, extended disk about 30 pc across, and displays a two-component thermal spectrum, with $T = 6.6 \times 10^6$ K and $T = 4.6 \times 10^7$ K, showing bright lines of O, Mg, Si, and S. The heavy-element abundance is elevated (Clark et al., 1982).

According to more recent data obtained by Hughes (1987), there are two components to the x-ray emission. The extended disk has a clearly defined shell structure attributable to radiation from plasma trailing the shock front, propagating in a cavity swept out by the wind of the progenitor. Radiation from the central region, which coincides with nebulosity that is bright in the lines of [O III], is due to plasma in the ejecta. A strong line of O VIII in the spectrum may mean that the oxygen abundance is enhanced.

The brightness and radial velocity distributions of the oxygen-rich optical filaments correspond to a ring oriented at 45°, expanding at a mean velocity of 2250 km/sec (Lasker, 1980). The age indicated by the size and expansion velocity of the ring of ejecta is 1300 years, assuming free expansion. If the boundary of the x-ray source and diffuse optical disk define the shock front, then the blast wave must be propagating at approximately 10^4 km/sec. Then again, by analogy with Cas A, it may be that the real supernova remnant is mapped out by the bright nebulosity and 6-pc radio synchrotron source, while the faint, disk-shaped halo around the remnant can be ascribed to the effects of the stellar wind of the progenitor.

IE 0102.2–7219

This object is the brightest x-ray source in the Small Magellanic Cloud; its optical filaments radiate principally in the lines of oxygen. A bright nebula ~24″ (7 pc) across is surrounded by a faint outer halo about 3.′5 in diameter, whose spectrum is typical of a high-excitation H II region (Dopita et al., 1981). An annular region between the bright filaments and halo is practically devoid of optical emission, but x-ray emission is observed out to ~20″, where the halo begins. The velocity of the filaments ranges from –2500 to +4000 km/sec, and their distribution in the plane of the sky corresponds to a severely warped torus expanding at 3300 km/sec (Tuohy and Dopita, 1983). This is the only object among the oxygen-rich remnants in which the torus is distinctly nonplanar, perhaps suggesting some strong perturbation at the time of expulsion. The age indicated by the size and velocity is about 1000 years.

This was the first remnant to show lines of He II (λ4686) in the halo, indicating that rather than being due to an early-type star, the ionization was in fact due to the supernova at the moment of its outburst or to hard radiation from the shock wave. The spectrum of the light coming from the fast-moving filaments is similar to that of G 292.0+1.8 and N 132D — only the lines of [O I], [O II], [O III], [Ne III], and [Ne V] are visible. Typical temperatures and densities are $T_e = 23000$ K and $n_e \approx 100$ cm^{-3} (Dopita and Tuohy, 1984). Lines detected in the UV include those of [Ne IV], [O II], Mg II, C III], and C IV, the Si IV–O IV blend at 1400 Å, and the strong O I] line at 1356 Å (Blair et al., 1988).

The x-ray image of the remnant is not consistent with the model of a spherical shell. Hughes (1988) has shown that the x-ray emission is concentrated in a narrow ring approximately 38″ (12 pc) in size located almost in the plane of the sky, and encircling the optical filaments. The soft x-ray spectrum is unusual, and requires assumptions of plasma departing significantly from equilibrium and having an anomalous chemical makeup (perhaps ejecta consisting almost entirely of Ne?).

The supernova remnant in NGC 4449

In 1978, a nonthermal radio source 25 times brighter than Cas A, the brightest Galactic source, was discovered on the outskirts of the irregular galaxy NGC 4449 (see Blair et al. (1983, 1984a); Bignell and Seaquist (1983), and references therein). The object was classified on the basis of its radio spectrum ($\alpha = -0.6$) as a young supernova remnant. Two systems of lines were present in its optical spectrum: the narrow lines associated with a neighboring H II region, and broad lines of [O I], [O II], [O III], [Ne III], and [S II], whose width equates to radial velocities of 7000 km/sec (Blair et al., 1983). No broad hydrogen lines have been detected; the spectrum, clearly, resembles that of the fast-moving knots in Cas A. The temperature in the [O III] emitting region is $(4-5)\times 10^4$ K, and the density in the bright filaments is $n_e = 10^5 - 10^6$ cm^{-3} — that is, it is considerably higher than in the filaments of Cas A. In all, the remnant contains about $0.01 M_\odot$ of oxygen, 50 times more than in the fast filaments of Cas A (Blair et al., 1983). The angular size of the radio remnant is at most $0.''04 - 0.''06$ (de Bruyn, 1983), implying a radius of less than 0.7 pc. An expansion velocity of ~ 3500 km/sec yields an age limit $t \leq 200$ yr, assuming that the shell is still expanding unimpeded. The remnant is associated with a strong x-ray source, $L_{0.2\text{-}4\,\text{keV}} = 8 \times 10^{38}$ erg/sec (Blair et al., 1983). The spectrum of the latter is unknown, but assuming by analogy with Cas A that the temperature in the remnant is close to 10^7 K, these authors very roughly estimate the mass of hot plasma to be $M_X \sim 25 M_\odot$. Heavy elements are 5–50 times more abundant than in the solar plasma. A gross estimate of the neon and sulfur abundance relative to oxygen, based on the intensity ratios [Ne III]/[O III] and [Si II]/[O II] is consistent with the explosion of a massive star with $M_{init} \sim 15 - 20 M_\odot$, within the framework of the model of Woosley and Weaver (1982).

We see, then, that the bright object on the periphery of NGC 4449 represents an early stage in the development of an oxygen-rich remnant. The upper limits on the size and age bespeak the extreme youth of the remnant, which would account for its high radio and x-ray luminosity.

Puppis A

It is conceivable that the well-known old supernova remnant Puppis A is also of this type (see Section 7). Fairly recently, Winkler and Kirshner (1984) discovered new, fast-moving filaments in Pup A, with very strong lines of oxygen and weak lines of hydrogen. These filaments, expanding away from the center at approximately 1600 km/sec, may possibly be analogous to the fast-moving knots of Cas A; the bright, slow-moving filaments moving at less than 300 km/sec, on the other hand, may resemble the stationary flocculi. Proper motions of the fast-moving oxygen filaments, as measured by Winkler et al. (1988), reach $0.''1 - 0.''2$/yr, which yields a kinematic age of 3500 years. This, then, is probably the oldest of the oxygen-rich remnants that we are considering.

0540–69.3 in the Large Magellanic Cloud

This is an extremely interesting object, combining as it does the earmarks of oxygen-rich remnants and plerions (see Section 4). Its identification as an oxygen-rich remnant derives from the spectrum of the shell surrounding the central synchrotron nebulosity. A slight anomaly in the spectrum, however — the absence of [Ne III] and [Ne V] lines, and very strong [S II] emission — distinguishes 0540–69.3 from the other objects discussed here; see also the new data of Kirshner et al. (1989). The task in future detailed investigations covering the whole energy spectrum will be to decide whether these differences are meaningful, or whether the association of plerions with oxygen-rich shells is a natural culmination of the evolution of certain stars, and how these stars might differ from others. So far, as we remarked in Section 4, 0540–69.3 seems to be the best candidate that we have for the remnant of a canonical type II supernova.

Let us summarize the observational data. We can state that there exists a unified and fairly well populated class of supernova remnants, typified by Cas A. This class, the *oxygen-rich remnants*, exhibits the following general properties.

1. The main identifier may be the spectrum of the filaments, which are dominated by lines of oxygen; fairly strong lines of neon or sulfur are evident, and there are no lines at all of hydrogen or nitrogen (the latter is only true of observations made with high angular resolution, enabling one to discriminate between the fast-moving filaments, the diffuse shell, and the halo).

2. Young objects in this class typically display toroidal structure, particularly in soft x rays. The toroid most likely consists of oxygen-rich ejected matter that has been heated by the reverse shock wave. It has an expansion velocity that ranges between 2000 and 6000 km/sec. Along with the torus, a faint, spherical shell of about the same size may also be seen; i.e., both spherically symmetric and toroidal ejection of matter takes place in the explosion.

3. Remnants in this class are surrounded by a faint extended halo whose optical spectrum typically is that of a high-excitation H II region (He II $\lambda 4686$ is visible); the halo around Cas A emits soft x rays.

4. The radio and x-ray luminosity of oxygen-rich remnants is dozens of times higher than SN Ia and SN II remnants of comparable age. Of course, this could explain the abundance of extragalactic examples of the class. But if we bear in mind that the basic identifying earmark — oxygen-line dominance — can only be observed at the beginning of a remnant's evolution, when it is no more than 1000–2000 years old and the mass of swept-up interstellar gas is not much greater than that of the ejecta, we may conclude that explosions of this type are frequent. We will demonstrate in Section 10 that those SN II that form plerions can also be easily identified during their first several thousand years. Since the number of plerions is

comparable to the number of oxygen-rich remnants (of the six historical Galactic supernovae, two are plerions and one is oxygen-rich; among the 25 reliably identified remnants in the Large Magellanic Cloud, three are plerions and two are oxygen-rich), it is plausible that the two types of explosions occur equally often.

5. Oxygen-rich remnants are formed in the explosion of massive stars. This follows, firstly, from estimates of the ejected mass based on x-ray luminosity, and secondly, from a comparison of oxygen and neon abundances in the filaments with modern theoretical predictions for nucleosynthesis in massive stars.

6. A stellar remnant — a pulsar — has been found in only one "atypical" object, 0540–69.3. In the other oxygen-rich remnants, the lack of a central pulsar is inferred both from direct evidence — the absence of any compact x-ray sources — and indirect — the thermal x-ray spectrum, the shell structure of the radio image, and the steep radio spectrum, $\alpha \approx -0.5$; see Section (10).

7. The massive precursor of a supernova exhibits vigorous mass outflow. Firstly, there is almost no hydrogen to be seen in the spectrum of the ejecta, implying that the star has lost its outer hydrogen layers prior to the explosion. Secondly, the chemical composition of the stationary condensations in Cas A suggests that they are of stellar origin, and their low velocity, that their constituent gas was not expelled in the explosion. Thirdly, the faint halo around remnants of this type is, in all likelihood, matter swept up by the stellar wind and ionized by the explosion.

8. It is conceivable that supernovae of this type have anomalously low luminosity at maximum light, being perhaps $3-5^m$ fainter than the average SN Ia and SN II, but this is surmised solely on the basis of the estimated upper limit of brightness for the unobserved SN in Cas A. (It must also be borne in mind, however, that absorption in the direction of Cas A is highly irregular: $A_V \sim 4-8^m$ per square arcminute (Troland, 1985), and even more in compact molecular cloudlets.)

Based strictly on observational data, then, we may draw some fundamental conclusions about explosions of this type. The progenitor must be a high-mass star that loses copious amounts of matter as it evolves. Inasmuch as the ejecta are devoid of hydrogen, we may safely assume that the precursor was the compact helium core of a massive, evolved star, the most likely candidate being a Wolf–Rayet star. The toroidal geometry of the oxygen-rich matter is indicative of either asymmetric supernova ejection or asymmetric mass outflow from the progenitor. Such asymmetry immediately suggests rapid stellar rotation.

It is also quite possible that the Wolf–Rayet star might be the secondary component of a binary system — that is, that the WR star has a compact, relativistic companion, such as are observed at the center of ring nebulae (see Chapter III). In that event, the toroidal geometry of the ejected matter or equatorial outflow preceding the outburst could result from orbital motion of the compact companion through the atmosphere of the other star, possibly terminating in the formation of a binary core.

What type supernovae formed Cas A and the oxygen-rich remnants?

The lack of hydrogen lines in the optical spectrum of the ejecta probably means that the oxygen-rich remnants are not associated with SN II, but one cannot rule out the possibility that the composition of the hot gas in the ejecta is significantly different from that of the dense oxygen filaments. Furthermore, fast-moving ejected flocculi showing hydrogen lines in their spectrum have recently been detected in Cas A, so the whole question of the hydrogen abundance in Cas A remains partly unsettled. We can say with assurance, however, that the oxygen-rich remnants differ from the Crab and 3C 58, both of which are associated with the explosion of a SN II and with the formation of a pulsar. In addition, Cas A and like objects do not resemble the remnant of SN 1006 or Tycho's remnant, thereby disassociating the oxygen-rich remnants from supernovae of type Ia.

In principle, Cas A might be the remnant of a SN Ib, the latter being oxygen-rich. We argued in Section 1 that a SN Ib — SN 1985F, for example — expels several solar masses of oxygen when it explodes. The one thing that presently prevents us from conclusively classing the oxygen-rich remnants with the explosion of a SN Ib is the toroidal geometry of the ejecta observed in all objects of this class that have been inspected with sufficient resolution, which unambiguously distinguishes them from the remnants of SN Ia and SN II (as we have discussed above). Thus far not one SN Ib has shown any observational evidence whatever of an asymmetric explosion. On the other hand, the identification of a SN Ib with an explosion in a binary system would make it possible to account for toroidal ejecta or outflow in the orbital plane.

So far, then, it remains unclear to what extent the conclusion of Fesen et al. (1987, 1988c) that hydrogen and nitrogen were present in the outer layers of the progenitor of Cas A is inconsistent with the possible identification of that remnant with a SN Ib.

6. Progenitors and compact stellar remnants

A supernova outburst is accompanied by the release of energy — both mechanical and radiative — in amounts exceeding 10^{51} erg. This estimate is derived both from supernova observations and from observations of supernova remnants that have undergone deceleration. On a stellar scale, the energetics of the event may be associated with the gravitational collapse of the core (the binding energy released upon compression to the size of a neutron star, $R = 10$ km, is $GM/R^2 \sim 10^{53}$ erg), or with the thermonuclear detonation of the core of a sufficiently massive star (the fusion energy available from a $1 M_\odot$ carbon core is $\sim 2 \times 10^{51}$ erg).

There is as yet no comprehensive theory to describe the final stages of evolution of a star: the formation of a white dwarf, the collapse to a neutron star or black hole, or the total disruption of a star by thermonuclear detonation. In fact, the mechanism of the supernova outburst — the culmination of certain paths of stellar evolution — is not entirely clear. On the one hand, there have been a great many studies of stellar evolution, producing models of internal structure and evolution for stars of practically any initial mass. On the other, a theory of types I and II supernovae has been developed that accounts rather well for observed supernova spectra and light curves. It is the marriage between these two lines of theoretical research that proves to be the weak link: there is no clear and consistent transition from the equilibrium stages of evolution, as described by a stellar evolutionary track on the Hertzsprung–Russell diagram, to the final thermonuclear detonation or collapse of the core, which transfers the star's binding energy to the outer layers that are blown off. Certain steps in the complex chain of events leading to the supernova explosion, however, have already been mapped out with a fair degree of assurance.

Here we consider the purely observational facts, for when all is said and done, it is the observations and not the theory that will provide the answer to the fundamental questions: What is the initial stellar mass that leads to a supernova explosion? Will a given star explode or collapse? Will there be a stellar remnant — either a neutron star or black hole?

The initial masses of supernova progenitors can be estimated on the basis of the way in which supernovae are distributed among galaxies of different morphological types, and within individual galaxies. The clustering of SN II and SN Ib within spiral arms and the lack of any concentration of SN Ia toward the arms comprises one of the most reliable methods for estimating lifetimes — i.e., the initial mass of the progenitor. But the amount of time spent by a star of given mass in an arm can only be determined very roughly, since it depends on the trajectory of the gas complex and the star within the arm, as well as the speed at which the spiral pattern moves. In addition, the mass estimate naturally entails uncertainty in a strictly theoretical parameter, the lifetime of a star of given mass.

We argued in Section 1 that type II supernovae only flare up in spiral galaxies, never in ellipticals. The rate at which SN II appear correlates with the color of the galaxy, which is governed by amount of star-formation going on. Type II supernovae are distributed inhomogeneously within spiral galaxies, and they show a strong tendency to be concentrated toward the spiral arms. Shklovskii (1960a) and Tinsley (1975, 1977) concluded from this that the precursors of SN II should be young, type O or early B high-mass stars on the main sequence, with $M_{MS} \geq 6 - 8\,M_\odot$, whose lifetime is at most $\sim 3 \times 10^7$ yr.

A linear dependence has been found between the rate of SN II and the Hα flux emanating from S galaxies, the latter being determined by the number of OB stars and their ultraviolet luminosity, integrated over the lifetimes of the corresponding stars (Kennicutt, 1984). This enables one to place a lower bound on the initial stellar mass resulting in a SN II explosion. Allowing for errors associated with possible underestimates of the SN II rate, the uncertainty of the initial mass function, the crude way in which the ionizing flux has been computed, based on model stellar atmospheres and the evolutionary tracks of stars of various masses, errors in the estimated galactic distances, etc., Kennicutt (1984) puts the lower bound on M_{MS} (SN II) between $5 - 6$ and $12\,M_\odot$, with the most probable value being $M_{MS}(\mathrm{SN\,II}) > 8\,M_\odot$.

The initial mass of the type II progenitor of the outburst that took place in A.D. 1054 is consistent with estimates derived from the statistics of extragalactic supernovae. According to Nomoto (1985) and Chevalier (1985), the lack of any oxygen or carbon enrichment of the ejected matter in the Crab suggests that the star that exploded had an initial mass $M_{MS} \lesssim 13\,M_\odot$, while the existence of the pulsar implies that $M_{MS} \gtrsim 8M_\odot$. Since the helium and hydrogen layers are efficiently mixed in stars with $M_{MS} = 8 - 13\,M_\odot$, the presence of filaments with enhanced He/H in the Crab means that a progenitor in that mass range ought to display strong outflow. For $M_{MS} \sim 11 - 13\,M_\odot$, mixing is less important, and that range of values of M_{MS} would apply if there existed a "fast-moving" outer hydrogen shell around the Crab; so far there is absolutely no observational evidence for such a shell (see Section 4). Thus, the chemical composition and the presence of a pulsar in the Crab imply that the most reasonable value for the initial mass of the progenitor is $M_{MS} \sim 8 - 10\,M_\odot$; see also Davidson et al. (1982), Henry (1986), and Chevalier (1985). Note also that according to new spectral data and deep imaging by MacAlpine et al. (1989), the combined mass of the neutron star and the gas in the Crab's filaments may be as high as $6 - 9\,M_\odot$, and the precursor's initial mass may have been about $20 - 30\,M_\odot$.

One can estimate the upper limit on the initial mass of a star that ends its life as a supernova with rather less confidence.

First of all, the mass of the matter ejected has been estimated directly from its x-ray luminosity, and as demonstrated by observations of young galactic remnants, the mass of the ejecta range from 1–2 solar masses up to $\sim 3-4\,M_\odot$ in Cas A. When the slow mass loss prior to the outburst is taken into consideration, the initial mass is accordingly much higher.

Secondly, the heavy-element abundance observed in the filaments of Cas A and similar objects can only be explained in terms of nucleosynthesis in massive stars ($M_{MS} \sim 20-30\,M_\odot$), and by assuming that a considerable amount of matter, perhaps $\sim 3-4\,M_\odot$, has been cast off.

Meanwhile, observations have been made that probably suggest that massive stars with $M_{MS} \geq 15-25\,M_\odot$ can collapse without exploding as a supernova. CO observations of Galactic open clusters by Bash et al. (1977) have shown that emission is present in young clusters containing stars earlier than type B0, but not in clusters lacking such stars. The statistics are quite convincing: CO emission has been found in 24 out of 28 clusters with young O stars, and has not been found in 35 out of 38 clusters in which the earliest stars are of type B1. To explain these results, Wheeler and Bash (1977) have proposed that high-mass O stars with $M_{MS} \geq 15-25\,M_\odot$ do not explode, and that gas is swept out of these clusters only after less massive stars that end their lives as supernovae begin to leave the main sequence. Since this seems to be the sole observational evidence so far that places an upper limit on the mass of a supernova progenitor, careful consideration must certainly be given to future observations and the analysis of possible selection effects.

Type Ia supernovae make their appearance in elliptical galaxies, where they are often observed on the outskirts (van den Bergh and Maza, 1976). The implication is that the progenitors are old halo stars, 10^{10} years old, with a mass close to $1\,M_\odot$. These are the only stars that have a lifetime approaching the age of an elliptical galaxy; if star formation is presently taking place in the latter, the rate is very low, due to the paucity of cold gas.

Recent estimates of the mass of the ejecta from SN Ia based on x-ray emission from young remnants yield $M_{ej} \sim 1.4\,M_\odot$, consistent with the explosion of a low-mass star — see Section 3.

The suggestion that the precursors of SN I might in fact be massive stars was made long before that class was split into SN Ia and SN Ib. In particular, Shklovskii (1984) deduced that there must be two kinds of SN I progenitor, given that SN I occur more often (per unit mass) in spirals than ellipticals, challenging the identification of SN I in spirals with the explosion of stars belonging to the old halo population. Further evidence, as noted by Tsvetkov (1987a), is provided by the lack of any significant difference in the rates, radial distributions, or z-coordinates of SN I and SN II in spirals.

Yet another argument favoring a younger population of type I supernova progenitors than might be inferred from their frequency in ellipticals is the high rate of SN I in galaxies harboring unusually vigorous star-forming activity. The large number of outbursts in these star systems is naturally associated with the multitude of massive stars, rather than with long-lived, low-mass stars. These are the I0 galaxies; morphologically, they are similar to early ellipticals, but they exhibit all the earmarks of ongoing star-formation. Tinsley (1979) and Oemler and Tinsley (1979) have therefore concluded that the initial mass of type I progenitors is typically $4-6\,M_\odot$.

We now know that SN Ib are associated with massive progenitors. At the present time, there is absolutely no evidence that the initial mass of a pre-SN Ib is less than that of a pre-SN II. Both types occur in the arms of spiral galaxies, and are often associated with H II regions. Furthermore, if the identification of SN Ib with oxygen-rich remnants holds up, it will provide additional evidence favoring the higher mass for SN Ib progenitors, $M_{MS} \sim 20-30\,M_\odot$, as implied by the chemical composition of Cas A and similar remnants; see Section 5.

It is conceivable, meanwhile, that star formation goes on in certain E and S0 galaxies, feeding on the gas lost by other stars. This, in any event, is precisely how Caldwell and Oemler (1981) interpret the correlation detected between the supernova rate in a given galaxy and its position within a cluster, i.e., with the rate at which gas is swept up by virtue of the dynamic pressure of the hot intergalactic medium. Guseinov et al. (1980) also remark that more than half the E and S0 galaxies in which SN I have been detected display anomalies that probably attest to continued star formation. A noteworthy recent event was the direct detection of x-ray emission from hot gas in elliptical galaxies, the quantity of gas involved being practically the same as in spiral systems (Fabian, 1985). It is still premature to speculate on how seriously this fact will affect our ideas about star formation in E galaxies, and therefore about the initial masses of SN Ia progenitors. The weight of the evidence seems to indicate that star-formation continues unabated in these stellar systems, but that it is low-mass stars that are being produced.

To conclude, let us discuss the lower limit on the mass of a star that gives rise to a supernova explosion at the end of its evolution. What is the mass of those stars that end their lives "peacefully," turning into white dwarfs after expelling a planetary nebula and slowly cooling? These estimates can be made by taking advantage of observations of open clusters (Wheeler, 1978; Tinsley, 1977). The number of white dwarfs to be found in the Hyades implies that they had initial masses of $2-6\,M_\odot$ at most, with the most likely value being $M_{MS} \leq 3-4\,M_\odot$ (van den Heuvel, 1975). Observations of other clusters yield $M_{MS} \leq 5-6\,M_\odot$. At the same time, there has been a tendency to ascribe greater mass to the precursors of planetary nebulae; in particular, the value of M_{MS} for the nucleus of NGC 7027, which was determined with allowance for the mass of the halo based on CO-line observations, is $\sim 6\,M_\odot$ (Knapp et al., 1982). The overlap between the initial mass ranges of planetary nebulae and SN Ia is not discomfiting, since only an insignificant fraction of all white dwarfs can subsequently engender a SN Ia. This follows from a simple estimate: one to three planetary nebulae are formed annually in the Galaxy, while SN Ia explosions occur no more often than every 50 years.

The foregoing observational facts yield the following crude classification for stars in spiral galaxies (see also Wheeler (1978), Tinsley (1979), and Shklovskii (1978, 1983)): white dwarfs result from the evolution of stars whose initial mass is less than $4-6\,M_{\odot}$, with a lifetime of $\gtrsim 10^8$ years; some stars with a mass between 4 and $6-7\,M_{\odot}$ engender type Ia supernovae after evolving for $(3-9)\times 10^7$ yr; type II supernovae derive from stars whose initial mass lies between 6 and $15-20\,M_{\odot}$, which live to be $(2-3)\times 10^7$ yr; stars of the same or greater mass that have shed their hydrogen envelope by the time they explode form SN Ib; more massive stars ($M_{MS} \gtrsim 20-30\,M_{\odot}$) may possibly not explode as supernovae, instead forming black holes, but that is presently a matter of speculation. The progenitors of type Ia and Ib supernovae differ only in mass, but in both cases we are dealing with highly evolved stars that have completely lost their outer hydrogen layers, just like white dwarfs, helium cores of massive stars, and Wolf–Rayet stars. The loss of the outer layers is probably connected with mass overflow in a close binary system. On the other hand, type Ib and type II supernovae may have similar initial masses on the main sequence and differ only in the mass of hydrogen in the pre-SN outer layer just before the explosion; see Section 1.

Now let us examine the purely observational data bearing on what happens to the core of star in a supernova explosion — specifically, how such explosions produce a neutron star or black hole, and under what circumstances the core may be completely disrupted.

The Crab and Vela pulsars represent a stunning confirmation of the proposal advanced by Baade and Zwicky that supernovae are somehow related to the transformation of a normal star into a neutron star. In the 1970's, it was thought that the explosion of a supernova was always accompanied by the formation of a neutron star, and that the lack of pulsars in other supernova remnants was merely the result of inopportune orientation, that is, of the fact that the terrestrial observer was not illuminated by the narrow beam emitted by the pulsar. This situation changed fundamentally with the launch of the Einstein observatory, one of the most important tasks of which was to search for neutron stars in young supernova remnants. Rather than expand on the search results and corresponding conclusions, let us list the observational evidence that would enable one to detect the collapsed core of an exploded star — a neutron star or black hole.

1. The one indisputable piece of evidence would be a pulsar near the center of the remnant, observable at radio, x-ray, optical, and gamma-ray wavelengths. Allowing for the sizable velocity acquired by the stellar remnant of an explosion in a binary system, or by virtue of an asymmetric explosion ($100-300$ km/sec, to judge by pulsar observations), the pulsar might end up off-center — or even outside the supernova remnant for the

very oldest ones. The common origin of a pulsar and an extended remnant would remain to be proved in each case.

2. The hot surface of a neutron star would be detectable as a compact x-ray source with a thermal spectrum. In contrast to that of a pulsar, the radiation could be observed at any orientation, which is precisely why such great expectations were pinned to the Einstein observatory. The surface temperature at the poles of a neutron star can differ from the temperature at the equator — in other words, the thermal radiation might also be modulated at the period of neutron star rotation.

3. A neutron star or black hole in a close binary system might be observable as an x-ray binary: x-ray emission is associated with the accretion of matter overflowing the Roche lobe of the companion, while periodic variability results from orbital motion and precession.

Apart from the direct evidence, the existence of a stellar remnant injecting relativistic particles can be demonstrated indirectly:

4. One might detect an extended synchrotron x-ray source whose brightness increases toward the center (the source would be smaller than the thermal plasma source behind the shock front, and would have a harder spectrum); see Section 10.

5. The radio remnant might be either a plerion or a composite (plerion plus shell); see Section 10.

Ideally, of course, one would observe all of the above.

There are presently five pulsars that have reliably been identified with supernova explosions: apart from the well-known ones in the Crab Nebula (Section 4) and Vela supernova remnant (Section 7), pulsars have recently been found in the remnants MSH 15–52 (Section 7), CTB 80 (Section 7), and 0540–69.3 in the Large Magellanic Cloud (Section 4). Information about these five is presented in Table 8, as follows. The first row identifies the remnant, and the second, the pulsar; row 3 gives the spectral ranges in which pulsed radiation has been detected; rows 4 and 5 give the pulsar period and its derivative; row 6 gives the characteristic age of the pulsar, $t = \frac{1}{2} P/\dot{P}$; row 7 gives the age of the supernova remnant (kinematic age, except in the case of SN 1054); row 8 gives the rate at which the pulsar is losing rotational energy; row 9 gives the temperature of the neutron star as derived from the thermal x-ray flux emanating from a compact source, assuming blackbody radiation and a radius of 15 km; rows 10 and 11 give the luminosity and size of the extended synchrotron x-ray source; row 12 lists the type of radio supernova remnant. These data have been compiled from the following sources: Seward (1983), Helfand (1983), Seward and Harnden (1984), Harnden and Seward (1984), Manchester et al. (1985), Middleditch and Pennypacker (1985), Wong and Seward (1984), Clifton et al. (1987), and Braun et al. (1989).

Table 8. Pulsars Associated with Supernova Remnants.

	Remnant				
	Crab Nebula	Vela XYZ	MSH 15–52	0540–69.3	CTB 80
Pulsar	PSR 0531+21	PSR 0833–45	PSR 1509–58	PSR 0540–69	PSR 1951+32
Spectral band	R, O, x, γ	R, O, x, γ	R, x	O, x	R
P, sec	0.033	0.089	0.150	0.0502	0.039
$\dot{P}$, 10^{-13} sec·sec^{-1}	4.23	1.25	14.9	4.8	0.06
$t = \frac{1}{2}P/\dot{P}$, yr	1230	11,200	1690	1.7×10^3	10^5
t_{rem}, yr	930	15×10^3	2×10^4	$\sim 800 - 110$	10^5
L, 10^{36} erg/sec	500	7	20	100–200	3–4
T_{ns}, 10^6 K	2.0–2.5	0.9–1.5	2.5		
L_x (synch), 10^{36} erg/sec	20	0.0004 (Vela A) 0.005 (Vela B)	0.16	8	0.02
D_x (synch), pc	0.8×0.5	~ 0.1 (Vela A) 4×8 (Vela B)	4.7×8.5	≤ 2	~ 0.6 3×6
Remnant type	Plerion	Composite	Shell	Composite	Composite

Notes.

Spectral band: R = radio; O = optical; x = x-ray; γ = gamma-ray.

L_x, D_x: Vela A is an x-ray source $\sim 1'$ in size; Vela B is the nebula, which is approximately $30' \times 60'$.

The neutron star temperature estimates listed here are very much in the nature of educated guesses, since even in the best-studied of the pulsars — PSR 0531+21 — they come from measurements of the emission in between pulses, at ~1% of maximum intensity, and do not allow for any possible variability in the thermal emission. In PSR 0833–45, the x-ray emission is constant and has been assumed thermal; in PSR 1509–58, the temperature was obtained by analyzing the pulse shape, and it assumes a sizable gradient between the poles and equator.

The pulsar in the Crab Nebula is clearly the youngest and most energetic. The age of PSR 1509–58 is ~1700 yr, which is quite different from the age of the nebula, ~10^4 yr (see Section 7); the age of PSR 1951+32 is consistent with the age of CTB 80 implied by the size of the infrared shell recently discovered by Fesen et al. (1988d).

The pulsars are surrounded by extended synchrotron x-ray sources that are smaller than the corresponding radio remnants. In addition, the nearest pulsar, PSR 0833–45, lies within a compact x-ray nebula about 0.1 pc in diameter; such a nebula would be below the angular resolution limit of the Einstein observatory for the other remnants.

In the radio, the Crab Nebula is a pure plerion (or one with a nascent shell), Vela XYZ and CTB 80 are composites, and so is 0540–69.3, but with enhanced oxygen abundance in the filaments.

In Table 9, we summarize what is known about supernova remnants containing compact x-ray sources, which may be associated with thermal radiation from a neutron star. We give the neutron star temperature, assuming blackbody radiation and a radius of 7 – 15 km, based on the data of Nomoto and Tsuruta (1983) and Helfand et al. (1980), as well as information on any x-ray or radio synchrotron remnant.

The remnant G 27.4+0.0 is very far away, so that the assumed absorption and estimated temperature are highly unreliable. We see that among the remnants with compact x-ray sources there is a plerion (3C 58) and a classic shell source (RCW 103). If in addition we can state with some assurance that a pulsar has been created in the plerion 3C 58, but is inopportunely oriented for a terrestrial observer (see Section 4), we can also say that in the shell remnant RCW 103, which exhibits none of the aforementioned hallmarks of continuing pulsar activity, a neutron star was most probably formed that had a weak magnetic field or rotated slowly. Naturally, it is also possible that the remnants containing compact sources but none of the other signs of pulsar activity simply represent positional coincidences with galactic or extragalactic objects of different kinds. In this regard, the most dubious case is that of the compact source in G 127.1+0.5 (see Goss and van Gorkom, 1984).

By considering the total number of x-ray point sources and the area occupied by an old remnant on the celestial sphere, Helfand and Becker (1984) have estimated the probability of a chance coincidence between the remnant and an x-ray source with a flux above some lower limit. They find $P = 35\%$ for the remnants W 28 and G 127.1+0.5, and $P = 4\%$ for

PKS 1209–52. The probability of a chance alignment is negligible for the other objects in the table.

Close binary systems with a relativistic companion — a collapsed stellar core — have been found in the two old supernova remnants W 50 and CTB 109. We shall have more to say about the unique source SS 433 at the center of W 50 in Section 7. The current picture is that this massive close system contains a "normal" component, a star of type B2–B5 with a mass of 20 $M_\odot$ that is overflowing its Roche lobe, and a relativistic companion — a black hole with a mass of 5–6 $M_\odot$ (Margon, 1982, 1984; Goncharskii et al., 1984; Antokhina and Cherepashchuk, 1985). The binary system loses matter in two ways: first, in the form of two relativistic jets streaming away at 80000 km/sec, with a mass loss rate $\dot{M} = 10^{-7} M_\odot$ /yr, and second, in a spherically symmetric manner at approximately 1000 km/sec, with $\dot{M} = 10^{-4} M_\odot$ /yr. The interaction of the jets with the shell of W 50 substantially determines the nature of the supernova remnant.

The x-ray binary system IE 2259+586 appears at the center of the radio shell remnant CTB 109: the pulsar radiates in x rays, radio, and infrared with a pulse period $P = 7$ sec, and it has an orbital period $P_{orb} = 2300$ sec (Fahlman and Gregory, 1983; Middleditch et al., 1983). A faint star with $m_B = 23.5$ appears in a photograph at the position of the x-ray pulsar.

A model of a binary x-ray pulsar based on these properties has been proposed by Lipunov and Postnov (1985), in which a 1.4 $M_\odot$ pulsar has a low-mass helium-rich companion star. But recent x-ray data from the EXOSAT and Ginga satellites have produced no evidence for a binary with such characteristics (Koyama et al., 1987; Morini et al., 1988). Observations spanning seven years have revealed an unusually stable (among the known x-ray binary pulsars) spin-down rate of 7×10^{-13} sec/sec. This poses serious difficulties for a conventional x-ray pulsar model, and two new interesting possibilities have recently been suggested. According to Carlini and Treves (1989), IE 2259+59 may represent a rapidly rotating neutron star ($P \approx 4$ msec) in free precession ($P = 7$ sec); Morini et al. (1988) have discussed the possibility of a rapidly rotating magnetized white dwarf slowing down via dipole electromagnetic radiation. In any event, the enigmatic properties of CTB 109's stellar remnant remain to be explained.

No symptoms of energy injection by a stellar remnant are detectable in the comparatively slow pulsar in CTB 109. It is a typical steep-spectrum ($\alpha = -0.5$) shell remnant; at 2.7 GHz, the degree of polarization is at most 5% (Downes, 1983; Gregory et al., 1983; Sofue et al., 1983; Hughes et al., 1984b). The age implied by the 18-pc linear radius (at a distance of 4 kpc) is $\sim 1.5 \times 10^4$ yr if the shell is expanding adiabatically. The pulsar is displaced relative to the center of the symmetric shell; this displacement corresponds to motion perpendicular to the line of sight at approximately 200 km/sec.

Table 9. Compact X-Ray Sources in Supernova Remnants: Search Results.

Remnant	Age, yr	T_w, 10^6 K	Assumed d, kpc	L_s (synchr), erg/sec	Type of radio remnant)	Reference (recent)
3C 58	800	2–2.4	2.6	1.8×10^{34}	Plerion	Becker et al. (1982)
RCW 103	2×10^3	1.7–2.2	2	None	Shell	Tuohy et al. (1983a)
W 28	6×10^4	1.8	2.3	Present??	Composite?	Matsui and Long (1985)
G 127.1+0.5		Poorly known	3.8	None	Analog of W 50	Geldzahler and Shaffer (1982); Helfand (1983)
G 296.5+9.7 (PKS 1209–52)		1.6	2	None		Matsui et al. (1988); Kellett et al. (1987)
G 27.4+00		~9?	26	Present	Shell	Kriss et al. (1985)
Cas A	300	≤1.5	2.8	None	Shell	Murray et al. (1979)
Tycho	380	≤1.8	3	None	Shell	Helfand et al. (1980)
Kepler	400	≤1.7	1	None	Shell	Helfand et al. (1980)
SN 1006	980	≤0.8	1	None	Shell	Pye et al. (1981)
RCW 86	2×10^3	≤1.8	2.5	None	Shell	Helfand et al. (1980)
G 350.0–1.8	8×10^3	≤1.5	4	None	Shell	Helfand et al. (1980)
G 22.7–0.2	10^4	≤2	4.8	None	Shell	Helfand et al. (1980)

The most detailed investigations of CTB 109 and its interaction with its nearby interstellar cloud have been carried out by Tatematsu et al. (1987, 1988).

The extended SNR PKS 1209–52 looks like a typical old shell-like object at radio wavelengths, with optical filaments located near the edge of the shell. Optical spectra obtained by Ruiz (1983) show line ratios typical of evolved supernova remnants. X-ray images recently obtained by Kellett et al. (1987) using the EXOSAT Observatory and Matsui et al. (1988) with the Einstein Observatory are similar to the radio map. The x-ray spectra correspond to a characteristic plasma temperature around $T_e \sim 1.7 \times 10^6$ K, and an ambient density around $0.1 \, \text{cm}^{-3}$. The "Sedov age" of the SNR is $t \approx 2 \times 10^4$ yr (Matsui et al., 1988), so a 6' displacement of the point x-ray source from the center of the shell corresponds to a transverse velocity of ~ 170 km/sec, which is typical of pulsars. Although blackbody radiation from the surface of a hot neutron star appears to be the most likely explanation, the true nature of the x-ray point source is still uncertain.

There have been attempts reported in the recent literature to identify certain pulsars with old supernova remnants on the basis of spatial coincidence, but they have not been successful: having measured the proper motion of 26 pulsars, Lyne et al. (1982) have demonstrated that none of the pairs proposed as being inherently related actually are, since the pulsars in question could not have originated in the corresponding remnants. On the other hand, a very fruitful idea has been developed by Fesen et al (1988d) (see Section 10): a high-velocity energetic pulsar may "rejuvenate" an old supernova remnant, where the pulsar may be located on the periphery of the shell (as in the case of CTB 80) or even outside it. This may make it possible to identify some previously unrecognized pairings of pulsars and supernova remnants.

Of all the objects we have listed that have compact stellar remnants, the type of supernova involved is only known for two, the Crab Nebula and 3C 58, and those with all the stipulations mentioned in Section 4. Apart from those two historical SN II, we haven't a single direct means of comparing the collapse of the core and formation of a stellar remnant with the type of supernova. Membership in an OB association might provide indirect evidence favoring a massive progenitor for the Vela XYZ and W 28 remnants (Lozinskaya, 1980b), but in other remnants identified with OB associations, such as IC 443 and the Monoceros Loop, no compact sources have been found. As we have already remarked, there are no compact sources in any of the historical SN Ia remnants or Cas A, nor are there any other indications of ongoing pulsar activity. In Table 9, we have listed upper limits on the surface temperature of a possible neutron star, as set by the receiver sensitivity of the Einstein observatory, assuming blackbody radiation and a radius of 7 – 15 km.

The lack of a compact x-ray source means one of two things: either there is no neutron star, or it has already cooled down. There have been numerous calculations of the neutrino and photon luminosity of a neutron star, which determine its cooling rate, but these estimates are not on the firmest of foundations, inasmuch as they depend on the properties of superdense matter near the center of a neutron star, which are poorly understood. Attempts to understand whether there are indeed cooled-down neutron stars in those young remnants where no compact x-ray sources have been detected have encouraged new calculations of the cooling rate that allow, among other things, for a different composition at the center (neutrons, pions, or quarks), superconductivity, and a strong magnetic field in the surface layers of the star, where the temperature varies the most rapidly (see Yakovlev and Urpin, 1981; Nomoto and Tsuruta, 1983, 1986, 1987; van Riper, 1983; references in these sources).

These calculations have shown that by allowing for rapid cooling of a neutron star due to pion or quark matter, one can account for the absence of compact sources in Tycho's remnant, Cas A, SN 1006, and Kepler's remnant. This comes into conflict, however, with calculations on the observed neutron stars in old remnants like RCW 103 and G 296.5+9.7 (PKS 1209–52).

It would therefore be more natural to assume that the historical type Ia outbursts of Tycho's and Kepler's supernovae and SN 1006, as well as that accompanying the formation of the oxygen-rich remnant Cas A, did not result in the formation of a neutron star. Accordingly, out of six explosions in the past 1000 years in our Galaxy, only in two did the core collapse to a neutron star — a pulsar.

A rough calculation of the total number of remnants containing compact sources yields approximately the same percentage of all explosions ending in core collapse. During the lifetime of the Einstein observatory, 65 galactic remnants and 31 objects in the Magellanic Clouds were searched for compact x-ray sources. Einstein was capable of detecting any neutron star with a surface temperature of $(1-2) \times 10^6$ K out to a distance $r \leq 5$ kpc (Helfand and Becker, 1984). There are 33 known supernova remnants within the corresponding volume of the Galaxy, and these constitute a more or less complete sample: the catalogs of nonthermal radio sources — supernova remnants — are complete to $r \leq 5$ kpc, and contain 25–28 objects no more than 30 pc in diameter within this region. Compact sources, the possible stellar remnants of supernovae, have been detected in 11 of these 33 objects. In a medium of density $n_0 = 1\,\mathrm{cm}^{-3}$, the typical lifetime of a 30-pc nebular remnant is $\sim 1.5 \times 10^4$ yr. Likewise, the typical active lifetime of a neutron star, as dictated by spindown and cooling, is also about 10^4 yr — in any event, neutron stars have indeed been observed in such remnants, including Vela XYZ, RCW 103, and MSH 15–52. These admittedly crude estimates are thus not inconsistent with the idea that perhaps one-third of all supernova explosions are accompanied by the formation of a neutron star, either isolated or in a binary system.

The theory of the final stages of stellar evolution and the climactic explosion of a supernova is an enormous problem unto itself, one that outreaches the scope of this book. Recent accomplishments in that realm have been reviewed by Trimble (1982, 1983), Imshennik and Nadezhin (1985), and Woosley and Weaver (1986). Blinnikov et al. (1988) have recently attempted to simultaneously relate the supernova explosion mechanism to the spectrum and light curve over a wide energy range, the type of remnant and chemical composition of the matter expelled by the explosion, and the theory of stellar evolution. Without reiterating their results, some of which are highly controversial, we note here the most reliable of their conclusions, which have been confirmed by observations of supernovae (Sections 1, 2) and young remnants (Sections 3, 4, 5).

Detailed gas-dynamic calculations suggest that in a model compact star of radius $R = 1-10 R_\odot$, and with instantaneous liberation of the explosion energy (without being specific about the mechanism involved), the "theoretical" light curves diverge from the observed ones: they display a lower luminosity, without the broad maximum typical of both SN I and SN II. The theory supplies two escape routes: one can either assume that the stellar precursor is not a compact object, but is instead enveloped by an extended atmosphere, or that the radiation from the supernova is derived from an ongoing injection of energy into the dispersing shell. Accordingly, the thermonuclear detonation of a compact degenerate star that lacks hydrogen, leading to complete disruption and no stellar remnant, would nowadays be classed as a SN Ia, while the core collapse of a massive star with a hydrogen atmosphere would be a SN II.

The slow injection of energy in the explosion of a type Ia supernova results from nuclear decay, coming from the sequence $^{56}\text{Ni} \to ^{56}\text{Co} \to ^{56}\text{Fe}$. The decay of ^{56}Ni, of which $\sim 0.5 M_\odot$ is synthesized in the explosion of a carbon–oxygen degenerate core with a mass of approximately $1.4 M_\odot$, fuels the SN Ia. The confluence of SN Ia light curves mentioned in Section 1 is then a consequence of the broadly similar initial conditions dictated by the Chandrasekhar limit.

Confirmation of the radioactive model of the SN Ia light curve comes from spectra exhibiting strong iron emission lines early on, the amount of iron being consistent with the reaction $^{56}\text{Ni} \to ^{56}\text{Co} \to ^{56}\text{Fe}$.

As we have seen in Section 1, synthetic spectra for a SN Ia as given by the radioactive model for carbon deflagration are in good agreement with the observational evidence. The theoretically predicted lines of ^{56}Co, which weaken with time, are detectable. Nucleosynthesis in deflagatory waves of burning can also produce such "intermediate" elements as ^{40}Ca, ^{28}Si, ^{32}S, and ^{36}Ar, which is also consistent with the observations. As indicated in Section 1, the light curves and spectra of SN Ia yield, in reasonable models of the shell, an ejected mass close to $1 M_\odot$. The ejecta consist mainly of iron and lighter elements; hydrogen is entirely absent or almost so, while a certain amount of helium may be present. The mass of iron in the ejecta is only a crude estimate, but it might come to $0.5 M_\odot$.

Observations of young remnants of SN Ia are also consistent with this model. In the remnant of SN 1006, one sees an interior layer derived from a freely expanding iron core, while at successively higher velocities, there are clumps with enhanced Si, S, and O abundance (see Section 3).

The hydrodynamic theory of SN II with a plateau has been worked out most thoroughly; see the series of papers by Litvinova and Nadezhin (1982, 1985). The best fit to the observed light curves near maximum and in the plateau region is provided by an ejected mass $M = 1 - 10 \, M_\odot$, radius $\sim 500 \, R_\odot$, and energy $E_0 \approx 10^{51}$ erg. At $t > 100$ days, steady injection of energy at a rate $10^{41} - 10^{42}$ erg/sec is necessary to compensate for the rapid dissipation of thermal energy by the ejected shell. This injection of energy is required even sooner after maximum in a SN II with no plateau. Possible energy sources that have been considered include magnetic spindown of a newborn pulsar, the interaction of the ejecta with the gas of the progenitor's stellar wind (Renzini, 1978), and the radioactive decay of ^{56}Co. The recent observations of SN 1987A provide definitive evidence that the principal energy source at $t \sim 1 - 1.5$ yr is radioactive decay; see Section 2.

It is possible that just before the eruption of a SN II there is a brief "superwind" phase, with velocities of some 3000 km/sec. This was the interpretation applied to the narrow P Cygni features observed in the near-maximum spectrum of SN 1984E (Dopita et al., 1984) and SN 1983K (Niemela et al., 1985). The phenomenon of narrow lines near maximum light can, however, also be explained by a SN II model with two explosions in succession (Grasberg and Nadezhin, 1986).

The rapid decline in brightness associated with a model explosion in which energy is liberated instantaneously and there is no extended atmosphere may account for those supernovae that are optically faint at maximum light, such as Cas A. The toroidally disposed ejecta typical of the oxygen-rich remnants might result from a magnetorotational supernova explosion mechanism or the thermonuclear explosion of a rapidly rotating pre-supernova; see Bisnovatyi-Kogan (1970), Bodenheimer and Woosley (1983), Ardelyan et al. (1979), and Chechetkin et al. (1988). Bodenheimer and Woosley (1983) have shown that the rebound of infalling matter in the equatorial plane induced by rotation, coupled with nuclear detonation heating, ensures that enough matter is ejected to account for the energy liberated in the supernova explosion. They considered the interior of a massive ($M_{MS} = 25 \, M_\odot$) star, $1.5 \leq M(R) \leq 8 \, M_\odot$ in Lagrangian coordinates, starting at the instant of formation of a collapsed core with $M = 1.5 \, M_\odot$, the infall velocity being $\sim 10^3$ km/sec and the initial angular momentum $J = 4.5 \times 10^{51}$ erg·sec, which corresponds to a rigid-body rotation speed of about 200 km/sec at the surface of a type O star. Two-dimensional numerical modeling incorporating centrifugal forces has demonstrated that while most of the mass in collapsed matter remains within 60° of the rotation axis, there is an outflow of matter in the equatorial plane that is restricted to a width of approximately 10°. Bodenheimer and Woosley traced the development of this system to the point at which the mass in the collapsed core reached $3.8 \, M_\odot$, and about $0.5 \, M_\odot$ of matter consisting of the products of oxygen burning had been expelled along the equator at some

7000 km/sec. This mechanism can thus provide for the toroidal geometry of the ejecta, a kinetic energy of 5×10^{50} erg in the ejecta, and enrichment of the ejected matter with oxygen and the products of oxygen burning. Although it is presently unclear just how realistic the accepted value of initial angular momentum is, and the interaction of the ejecta with overlying layers of the stellar mantle and shell have yet to be investigated, this model very probably provides the best explanation for the objects discussed in Section 5. The high mass ($3.8\,M_\odot$) of the collapsed core suggests that a black hole can indeed be left behind as the stellar remnant of the explosion of a massive, rapidly rotating star (Shklovskii (1979) proposed that Cas A harbors a black hole; see also Kundt (1980), however).

The final verdict is not yet in on the possible identification of oxygen-rich remnants with SN Ib. Despite an array of arguments favoring such an identification, the SN Ib problem is so new that speculation is still rife. There are, however, certain indisputable facts.

First, a SN Ib signals the explosion of a massive star (see above). Second, it is the explosion of a relatively compact star, lacking outer hydrogen layers, given the evidence of the light curves and spectra. Third, the precursors of SN Ib (or their companions in close binary systems) shed matter copiously, and the supernova shell interacts with the dense gas supplied by this wind, as evinced by the radio emission from SN Ib. Fourth, a significant amount of oxygen (perhaps $2-5\,M_\odot$) is expelled in the explosion of a SN Ib, pointing to a possible connection with oxygen-rich remnants. In particular, the chemical composition of the ejecta from SN 1985F implied by the spectrum of the supernova is in good agreement with the composition of the fast-moving filaments in Cas A. This list of properties suggests that Wolf–Rayet stars and the helium cores of massive stars are likely candidates for the role of SN Ib progenitors (for example, see Wheeler et al., 1987; Lander and Eid, 1986; Begelman and Sarazin, 1986).

A comparison of these properties of SN Ib (including a WR star as a likely progenitor) with the parameters of oxygen-rich remnants (see Section 5) points to a possible relationship between them. As we noted above, however, all of the oxygen remnants with thoroughly studied kinematics exhibit toroidal geometry, while no sign of such asymmetry has yet been revealed by any SN Ib. Meanwhile, evidence of an asymmetric explosion has thus far only been found in the very bright SN 1987A (see Section 2), and these observations imply that the corresponding effect would lie below the threshold of detectability in any of the other extragalactic supernovae. We therefore have what seems to be a likely chain of events, namely {massive star (in a binary system?)} $\rightarrow$ WR star $\rightarrow$ SN Ib $\rightarrow$ {Cas A and oxygen-rich remnants}.

Nevertheless, there are other versions of this story, in which Kepler's remnant is the debris left by a SN Ib (see Section 3). Among the other suggested progenitors for a SN Ib are a helium supergiant with a hot main-sequence companion (Uomoto, 1986), an accreting white dwarf (in a WD + RG system) (Branch and Nomoto, 1986; Khokhlov and Érgma, 1986), and

Table 10. Supernovae and Supernova Remnants.

	Type		
	SN Ia	*SN II*	*SN Ib*
Initial mass on MS, $M_\odot$	~4 to 6–7	6–8 to 12–(20?)	≥6
Presupernova	WD (in a binary system?)	RSG (BSG)	WD + SG? WR?
Type of instability, explosion mechanism	Thermal, thermonuclear	Collapse, magnetorotational	Collapse? Thermonuclear? Magnetorotational?
Stellar remnant	None	Neutron star	Black hole? Neutron star?
M_{ejecta}, $M_\odot$	1.4 (1–2)	1–10	(1–6?)
$M_{(Fe)}$, $M_\odot$	0.4–1	0.04	0.2–0.3
$M_{(O)}$, $M_\odot$	?	<0.2	1–5
Young remnant	Tycho SN 1006 Kepler?	Crab, 3C 58, 0540–69.3 CTB 80	Cas A? Oxygen-rich? (Kepler??)
Remarks	Why is Kepler different from Tycho and SN 1006?	Where is the fast-moving shell of the Crab? Are CTB 80 and SN 1408 related?	Was the supernova that produced Cas A anomalously weak? What are the earmarks of an asymmetrical SN explosion?

white-dwarf mergers in a common envelope (Iben, 1986; Cameron and Iben, 1986). The explosion of a compact star paired with a giant adequately accounts for the radio emission of a SN Ib, on the one hand, and makes it possible to relate the toroidal ejecta or outflow to the orbital plane of the system on the other.

It is quite clear that duplicity in supernova progenitors is of fundamental importance. Duplicity is probably associated with the segregation of massive stars into two classes: solitary stars turn into SN II, and binaries lose their outer hydrogen layers through mass transfer in a close system and become SN Ib. It also seems most likely that SN Ia occur in binary systems (Iben and Tutukov, 1984a, b; Iben, 1986; Cameron and Iben, 1986). Roche-lobe overflow in a close binary provides a mechanism that "conserves" or extends the lifetime of a pre-SN Ia. In particular, this is the only way in which one can account for the SN Ia seen in elliptical galaxies, where there should simply be no stars heavier than $1.4\,M_\odot$ to explode as supernovae.

Thus, at least three types of supernovae occur in spiral galaxies: SN Ia, SN Ib, and SN II. Young remnants may be segregated into three morphological classes. This situation is summarized in Table 10, which presents our most comprehensive picture of the causative chain {main-sequence star} → {type of SN} → {class of supernova remnant}. Naturally, the suggested relationships reflect our current level of understanding of the problem. The remarks in Table 10 single out some of the most important questions; clearly, there are more questions than well-founded evidence here (see also van den Bergh, 1988).

Chapter 2
The Evolution of
Supernova Remnants

The evolution of supernova remnants is completely governed by their interaction with the gas that surrounds them — at first, with the gas influenced by the radiation and wind of the progenitor, and subsequently, with the interstellar medium. Approximately 1000 years after the explosion, the individual properties of a supernova, which are so important to an understanding of the physics of young remnants, lose their significance. Conditions in old remnants, which comprise the vast majority of all such objects, are determined principally by the characteristics of the interstellar medium, primarily the density distribution of interstellar gas. Of all the parameters of an explosion, only two are important — the initial kinetic energy E_0 of the ejecta, and the presence or absence of a pulsar, which will inject relativistic particles and a magnetic field. A pulsar, if present, will exert a significant influence for something like the first $10^3 - 10^4$ years.

The propagation in the interstellar medium of the shock waves induced by the expulsion of a shell by a supernova is responsible for the gamut of observational effects seen in old remnants — hot plasma, radiating in x rays and coronal lines; wisps of optical nebulosity, representing cooling regions heated by the passage of shock waves through dense clouds; infrared emission due to shock-heated dust; extended radio and x-ray synchrotron sources due to relativistic electrons in a magnetic field. The relativistic particles and field can be injected by a pulsar or particles can be accelerated and the field amplified within the turbulent layer near the shock front, but even in the former case, the properties of the radio remnant are determined by the dynamics of the relativistic plasma cloud — in other words, by its interaction with the interstellar medium.

7. Optical nebulae: old supernova remnants

About 150 nonthermal radio sources that are supernova remnants have been identified in the Galaxy, but only 40 to 45 of these have been associated with optical nebulae (see van den Bergh (1978b, 1983)). Nonetheless, the foundations of supernova science were laid using the tools of optical astronomy. Optical observations were historically the first; they provided the first evidence for the peculiarity of these objects; they set the distance scale for supernova remnants; and they made it possible to study the deceleration of supernova shells in the interstellar medium.

The first identifications of old remnants were based on their characteristic morphology, consisting of delicate filaments filling a more or less regular shell, with no exciting star. This was how the class of filamentary nebulae exemplified by the Cygnus Loop, Simeiz 147, Vela XYZ, and IC 443 was distinguished. The second morphological type of old supernova remnant, a diffuse shell (e.g., W 28, HB 9, HB 21), was identified on the basis of its radio spectrum. Currently, searches for faint optical supernova remnants are based on deep photography with narrow-band filters centered on the lines of [N II], [S II], and [O III].

The list of optical supernova remnants in nearby galaxies is growing rapidly (see d'Odorico et al., 1980). Optical remnants have been discovered in the Magellanic Clouds using deep plates taken at the 4-m Anglo-Australian Telescope in the lines of [S II] and [N II]; see Lasker (1979) and references therein. (A complete list of optical, radio, and x-ray supernova remnants in the Magellanic Clouds — some 40 objects — may be found in the paper by Mathewson et al. (1985).) The spectra and morphology of these nebulae are not so different from those in our own galaxy, and detailed investigations of the kinematics, coronal-line, and x-ray emission suggest that interaction conditions between the shock wave and interstellar gas, which determine the evolution of a remnant, are also identical in the various galaxies (Dopita and Mathewson, 1979; Mathewson et al., 1983, 1984; Rosado et al., 1983).

More than 20 remnants have been found in M31 and M33 — galaxies belonging to the Local Group — and these have been minutely studied spectroscopically (see Sabbadin, 1979; Danziger et al., 1979; Blair and Kirshner, 1985; Long et al., 1988; these papers all contain further references).

We shall dwell here on observational results obtained from several old Galactic remnants, using these as a vehicle to examine the various interaction conditions between a supernova shell and the interstellar medium, including explosions in dense clouds, in the tenuous intercloud medium, and in a medium with a substantial density gradient. We shall also consider those objects whose properties are governed by ongoing pulsar activity.

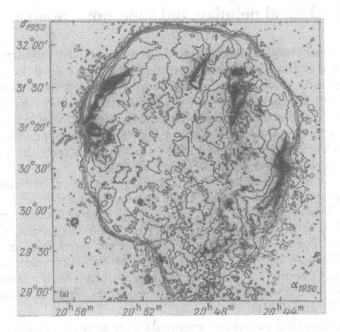

Fig. 31a. The Cygnus Loop: 0.1–4 keV x-ray isophotes superimposed on an optical photograph (Ku et al., 1984).

Data on the remainder of the optical nebulae that are old Galactic supernova remnants are concisely summarized in Table 11.

The Cygnus Loop (NGC 6960, NGC 6992-5)

This is one of the best-known supernova remnants; Fig. 31 shows optical, radio, infrared, and x-ray images. The fact that the nebula is close by — and therefore large and bright, facilitating observations with high spectral and angular resolution — is not the only reason why it attracted the attention of researchers. As early as 1937, Hubble had discovered that the two opposite sides of the shell are expanding away from the center at 0."03/yr; subsequently, Fesenkov et al. (1954) measured the proper motion of some dozens of bright filaments, and Minkowski (1958) determined radial velocities for hundreds of filaments, showing that they were moving away from the center at velocities ranging from 46 to 116 km/sec. By comparing proper motions with radial velocities, one could then determine the distance to the nebula to be 770 pc (yielding a linear size of 40 pc), thereby revealing the grand scale on which this phenomenon was being played out.

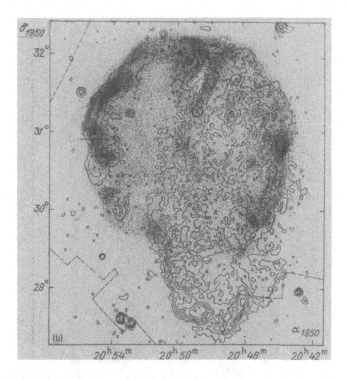

Fig. 31b. The Cygnus Loop: radio isophotes at 2.7 GHz superimposed on an x-ray image (Ku et al., 1984).

The thin bright filaments of the nebula are observed against a weaker diffuse background. The velocity dispersion of the gas in the filaments is less than that in the diffuse interfilament medium: in the bright filaments, the full width of the Hα line at half-maximum intensity corresponds to 10–30 km/sec, and in the diffuse medium, it is 40–85 km/sec (Doroshenko, 1970; Shull et al., 1982). At less than 10% of the maximum intensity, the weak, broad line wings are observable out to velocities of ±300 km/sec (Kirshner and Taylor, 1976; Doroshenko and Lozinskaya, 1977). According to Doroshenko and Lozinskaya (1977), these faint high-velocity wings are emitted not by the filaments but by the diffuse medium in which the bright filaments are immersed. The results of studies of the kinematics of the nebula are plotted in Fig. 32, which gives radial velocities of individual filaments and velocity intervals for the faint wings of Hα as a function of distance from the center of the nebula. When projection effects are taken into account, assuming a thin spherical shell, the observations are consistent with expansion of the system of bright filaments at

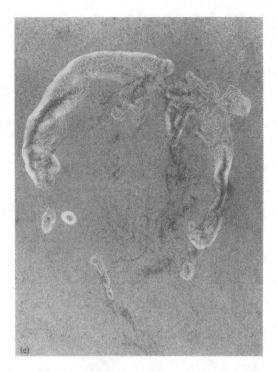

Fig. 31c. IRAS infrared isophotes superimposed on an optical photograph (Braun and Strom, 1986a).

~100 km/sec and motion of the diffuse gas at velocities ranging up to 300 km/sec.

Spectral investigations of the Cygnus Loop have been going on for decades (see Fesen et al. (1982) and Raymond et al. (1988); both contain further references). Some dozens of lines of various elements show up in the spectrum of the bright filaments; relative intensities vary widely within the confines of the nebula. The average gas density in the filaments, as determined from the lines of [S II] and [O II], is $100-300$ cm^{-3}; the temperature in regions emitting [O II], [N II], and [S II] is $(1-2) \times 10^4$ K, and in [O III] regions it is $(2-6) \times 10^4$ K. The nebular emission spectrum is poorly modeled by an optically thin plasma layer at uniform temperature, a conclusion reached by Pikel'ner (1954) as long as three decades ago. The currently accepted explanation for the spectra of filaments in old remnants is that they originate in gas behind a shock front that is propagating into dense regions of the interstellar medium. The gas in dense clouds behind

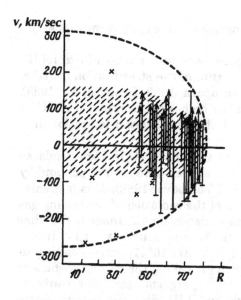

Fig. 32. Radial velocity of the filaments and interfilamentary gas of the Cygnus Loop as a function of distance from the center of the shell. The shaded region corresponds to the bright filaments measured by Minkowski (1958) and Doroshenko (1970); crosses denote the results obtained by Kirshner and Taylor (1976); vertical arrows show the high-velocity Hα wings in the measurements of Doroshenko and Lozinskaya (1977); the dashed curve takes into account geometrical projection for a thin spherical shell model.

the front is quickly cooled by radiative losses and collapses into a dense, cool layer several hundredths of a parsec thick, within which there is an abrupt drop in temperature from $\sim 10^6$ K to $\sim 10^4$ K and an abrupt change in density by a factor ranging from tens to hundreds. The observed relative line intensities in the filament spectra can be compared with the emission spectrum calculated for the hot gas behind the front. These calculations have been carried out many times, and span a wide range of shock front velocities, densities, and chemical compositions for the interstellar gas (see Cox, 1972a, b; Raymond et al., 1976; Raymond, 1979; Shull and McKee, 1979; Kaplan and Pikel'ner, 1979; Contini et al., 1980; Balinskaya and Bychkov, 1979, 1981; see also Section 8). For purposes of comparison, the most informative lines are those of the ionization states O I, O II, and O III, since they are among the brightest in the spectrum and reflect upon the state of the gas in the temperature range $10^4 - 2 \times 10^5$ K. The relative line intensities [O I]:[O II]:[O III] and others in the filaments in the Cygnus Loop best agree with values calculated for a shock wave propagating at 70 to 100 km/sec through a medium with unperturbed density $n_0 = 5 - 10 \, \text{cm}^{-3}$, and with a virtually normal chemical composition, which is entirely consistent with direct measurements of the velocities of the filaments.

The result of inhomogeneity of the interstellar medium is that radiative cooling does not begin simultaneously everywhere on a spherical shock front, and it occurs at a variety of rates. The regular shell structure breaks down due to thermal instability, and inhomogeneities in the interstellar medium lead to the reflection, focusing, and intersection of shock waves; several layers of emitting gas are visible in the line of sight. All of the foregoing may account for differences among the spectra of filaments appearing within a single remnant.

Monochromatic plates of the Cygnus Loop in the lines of Hα, [N II], [S II], and [O III] provide a clear-cut picture of the stratification of radiation from the cooling gas (Sitnik and Toropova, 1982; Hester et al., 1983). As can be seen in Fig. 33, radiation in the high-ionization line of [O III] is observed closer to the shock front than that from the lower-ionization lines of [N II] and [S II].

Condensations that are brightest in the line of [O III] are displaced relative to [N II] by $12-24''$ ($0.05-0.1$ pc at a distance of 770 pc), and by $0.05-0.5$ pc relative to [S II] (Sitnik and Toropova, 1982). This is greater than can be obtained from stratification of the radiation of the cooling gas behind the shock front within a single dense cloud. Indeed, the time needed to cool the gas from its temperature at the front, $T_s \approx 14 v_s^2 [\text{km/sec}] = (2-5) \times 10^5$ K, to $T_e \approx (5-8) \times 10^4$ K, corresponding to the strong [O III] emission, is $t_{cool} \approx 500/n$ yr, taking the cooling coefficient in the temperature range to be $\Lambda \approx 5 \times 10^{-27} T_e$ erg·cm^3·sec^{-1} in accordance with Fig. 52. Filaments that are bright in [O III] dim in a characteristic time

$$t_{rec} \approx (k_{O^{++}} n_e)^{-1} \approx 2 \times 10^3 n_e^{-1} \text{ yr},$$

where $k_{O^{++}}$ is the recombination coefficient for oxygen going to the state O^{++}. In that time, a wave in a dense cloud with[1] $n_{0cl} \approx 10$ cm^{-3}, propagating at about $v_{cl} \approx 100$ km/sec, will traverse a distance of $0.02-0.03$ pc, i.e., several times less than the observed displacement. The large-scale dynamics of a supernova remnant is governed by the propagation of the shock wave in the tenuous medium between dense clouds (see Sections 8 and 9). The speed at which the shock front of this fast wave is moving can be determined from the x-ray spectrum or from the faint, high-velocity wings of Hα; for the Cygnus Loop, it is $v_s = 400$ km/sec. The stratification of radiation observed in the remnant in all likelihood represents a cross-sectional snapshot of this fast-moving shock wave in a highly inhomogeneous medium: we see those emitting regions in different dense clouds at various distances from the front that correspond to different stages of cooling and recombination.

The Cygnus Loop is the best-studied remnant at ultraviolet wavelengths. The UV brightness of the filaments correlates with their visual brightness; the spectrum contains strong lines of C III, N III, and possibly O VI at 1031 and 1038 Å, although the latter identification is somewhat uncertain (see Raymond (1984) and Raymond et al. (1988), plus references

---

[1]For the sake of consistency, hereafter we use the notation to be introduced in Section 8: the subscript "0 cl" refers to the unperturbed gas within a cloud, and the subscript "0 i" to the unperturbed intercloud gas.

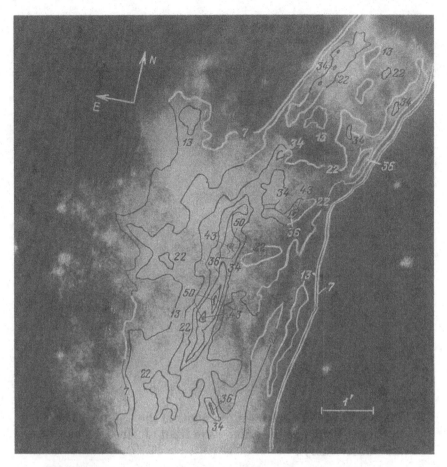

Fig. 33. Bright filament on the western side of the Cygnus Loop: [O III] isophotes superimposed on a [N II] image (Sitnik and Toropova, 1982; see also Bychkov et al. (1986)).

therein). Bright lines of C IV, N V, and O V correspond to higher temperatures and shock wave velocities than the lines in the visible spectrum. The combination of oxygen and nitrogen lines in the UV and visible ranges yields the relative abundance of the elements in five successive stages of ionization. The observations correlate poorly with the calculated emission spectrum of the shock wave, probably due to a superposition of several layers of gas behind a number of shock fronts lying in the line of sight.

Radio emission from the Cygnus Loop has been studied in the range from 10 MHz to 5 GHz; the radio spectrum is nonthermal, with a break at about 1 GHz. At higher frequencies, $\alpha = -0.84 \pm 0.4$, and at lower frequencies, $\alpha = -0.38 \pm 0.4$. Different spectra are observed within the confines of the bright shell: within NGC 6992–5, the spectrum is flatter, while within NGC 6974 it is steep (Abranin et al., 1977; Udal'tsov et al., 1978; Sastry et

Fig 34. The faint outermost filaments of the Cygnus Loop in Hα, which define the blast-wave front (J. Hester, Palomar Observatory).

al., 1981). Observations made with high angular resolution have demonstrated good agreement between the bright radio features and the optical filaments (see Green, 1984a).

Figures 31a and 31b show that some of the radio emission from the Cygnus Loop comes from outside the bright optical nebula, but that it coincides with the x-ray emission region. X-ray and coronal line emission from the remnant have been studied in great detail (see Section 8; see also Ku et al. (1984), Charles et al. (1985), Teske and Kirshner (1985), Ballet et al. (1988a, 1989), and references therein). At the outer boundary of the radio and x-ray images of the remnant, one finds very faint, delicate filaments radiating primarily in the hydrogen Balmer lines (Raymond et al., 1980, 1983; Treffers, 1981; Fesen and Itoh, 1985; Hester and Cox, 1986; Hester et al., 1986). These faint outermost filaments and the boundary of the x-ray image define the location of the blast-wave front; see Fig. 34

IRAS observations have shown that infrared emission from warm dust in the Cygnus Loop comes from the same locations as thermal x rays (Braun and Strom, 1986a). This is easily seen in Fig. 31c, which shows the correlation between the IR and x-ray emission, as well as local brightening of the IR emission in the vicinity of optical filaments.

A scheme describing the interaction between shock waves induced by the expansion of the supernova shell and the interstellar gas, which provides a reasonable description of all available observational data on the Cygnus Loop at visible, x-ray, and infrared wavelengths, is set out in Section 8 (the Cygnus Loop is our principal laboratory for studying the nature of old supernova remnants). We therefore limit ourselves here to stating the basic observational facts, which lead to the following conclusions.

The hot plasma behind the blast shock front radiates x rays with a spectrum corresponding to $T_e \sim (2-3) \times 10^6$ K (velocity $v_s \sim 350-400$ km/sec), and the initial density of the intercloud gas is $n_0 \sim 0.1-1$ cm^{-3}. Weak variations in the spectrum and brightening in the vicinity of optical filaments have been detected. Observations in the [Fe X] and [Fe XIV] coronal lines (Teske and Kirshner, 1985; Ballet et al., 1988a, 1989) show that the plasma temperature is lower there than in the x-ray emitting region, and corresponds to a shock velocity $v_s \sim 250-300$ km/sec; the density is $n_0 \sim 0.5-1$ cm^{-3}. The coronal line brightness is enhanced near filaments, most likely due to evaporation. Fabry–Perot observations obtained by Ballet et al. (1989) in the lines of [Fe X] and [Fe XIV] have established the bright filaments as having velocities up to 100 km/sec.

Optical emission is represented by two populations of thin filaments, each of which is seen against a background of fainter diffuse emission. A system of filaments that is bright in the lines of hydrogen and in forbidden lines "lags behind" the front, and most likely identifies a cooling region of strongly emitting gas that has been heated by the shock wave. In front of these bright "radiative" filaments, near the blast wave front identified with the boundary of the x-ray and radio images, is a system of very faint filaments radiating primarily in Hα. Figure 34 shows a photograph of these filaments (graciously provided by J. Hester of Palomar Observatory); both they and the diffuse Hα emission in between are observed just behind the "nonradiative" blast wave front. We shall have more to say, at the end of the present section and in Section 8, about the delicate filamentary structure and the emission mechanism of old remnants. Here we note that Hester et al. (1986) have measured proper motions of the faint outermost "nonradiative" filaments to be $\mu = 0.''06$/yr, i.e., twice the displacement of the bright filaments in the Cygnus Loop. A comparison of the proper motions of "nonradiative" filaments at the front with the velocity of the shock wave determined from the x-ray spectrum and the radial velocity of the faint wings of Hα, $v_s \sim 350$ km/sec, yields a distance to the remnant of ~ 1000 pc. This is close to the distance obtained by Minkowski from radial velocities and proper motions of the system of bright filaments. The discrepancies are understandable, since in both instances what was measured was either mean proper motions of the system of filaments or, indirectly, the mean velocity of the shock wave. A more accurate assessment of the

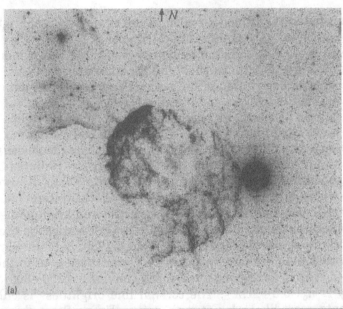

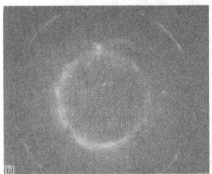
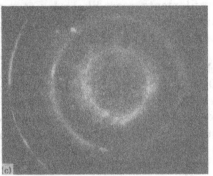

Fig. 35. IC 443. **a)** Appearance of the remnant on the red print of the Palomar Observatory Sky Survey. **b, c)** Fabry–Perot images of the nebula in Hα obtained by the author.

distance would come from measurements of the velocity and angular displacement of individual filaments.

The spherical symmetry of the Cygnus Loop is disrupted by an extended jet to the south, which is quite noticeable at x-ray, radio, and optical wavelengths. The radio emission from this jet has enhanced linear polarization, with $p = 25\%$ at $3 - 10$ GHz as compared with $p = 4 - 5\%$ in the rest of the nebula (Moffat, 1971). The magnetic field there is also more regular. The protruberance is probably associated with the large-scale structure of the magnetic field of the Galaxy, which is compressed by the expanding shell. It is possible for relativistic plasma to pour out along galactic magnetic field lines; a thermal instability at the epoch of formation

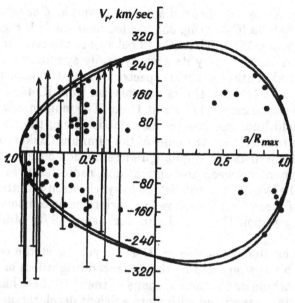

Fig. 36. Radial velocity of the filaments of IC 443 as a function of normalized distance from the center (author's measurements). Vertical line segments represent the high-velocity wings of lines emitted in the interfilament diffuse medium (see text). The curve shows the projection effect in the thin-shell model, assuming the validity of Eq. (9.12) for an explosion in a medium with a smooth density gradient.

of a cold, dense shell would assist in disrupting the shell's regular structure in the field direction (Moffat, 1975a; see also Section 9).

The age of a remnant can be determined from the observed shock wave velocity in the intercloud medium and the remnant's linear size, if one assumes a particular time dependence for the radius. We shall show in Section 9 that the set of equations (9.2) which describes the adiabatic expansion of a shell — the Sedov solution (Sedov, 1957) — provides a rather good approximation for the Cygnus Loop; the corresponding age is 2×10^4 yr.

Note that although small-scale, dense cloudlets are immersed in a tenuous medium, the large-scale density distribution of the unperturbed gas in the vicinity of the remnant is fairly uniform. Evidence for this comes both from the spherical symmetry of the shell and from direct 21-cm observations in the neighborhood of the remnant.

IC 443

We now consider the appearance of an old supernova remnant that has erupted into a medium with small-scale fluctuations superimposed on a pronounced large-scale density gradient. One example of such an object is the filamentary nebula IC 443 (fig. 35), located near the H II regions S 249 and S 247 (Sharpless, 1959). That IC 443 belongs to the class of old supernova remnants is suggested by its archetypically symmetric filamentary shell structure and nonthermal radio spectrum ($\alpha = -0.36$ according to Erickson and Mahoney (1985)). The radio emission from IC 443 is polarized at the 6 – 8% level at 3 cm, and at 2% at 11 cm. The magnetic field is fairly regular, with field lines parallel to the galactic plane; the field becomes somewhat more tangled near the edge of the remnant (Baker et al., 1973; Velusamy and Kundu, 1974). High angular resolution observations reveal complete agreement between the optical and radio features; the radio emission from the bright filaments is due to synchrotron radiation, and the spectral index of those filaments shows no variations as compared with the interfilamentary region (Duin and van der Laan, 1975; Mufson et al., 1986).

The explosion that gave rise to IC 443 occurred at the boundary of dense clouds. This is suggested by the enhanced brightness of the northeast side of the nebula and its actual shape — the shell there has a smaller radius of curvature, associated with more efficient deceleration and radiation in a medium with higher density. The cloud is directly observable at 21 cm: it has a mean gas density of $n_{0\,cl} = 10 - 20\,\mathrm{cm}^{-3}$, with individual denser knots ($n_{0\,cl} = 100 - 200\,\mathrm{cm}^{-3}$) that overlay the radio filaments; the total mass of the cloud is approximately $2 \times 10^3\,M_\odot$ (DeNoyer, 1978). In the same location there is also a dense molecular CO cloud which is almost certainly directly related to IC 443 — CO emission is concentrated about the bright nebulosity, and decreases right in the regions occupied by the bright filaments. The mean density within the molecular cloud is $n_{H_2} = 100\,\mathrm{cm}^{-3}$ (Cornett et al., 1977; Scoville et al., 1977). This massive ($M \approx 10^4\,M_\odot$) and thin (only about 3 pc) molecular cloud covers the southern and western parts of IC 443, and is responsible for strong extinction in the central regions of the remnant. The fact that the expanding supernova remnant is physically interacting with the molecular cloud is confirmed by the detection of a dozen different shocked molecular cloudlets in H_2, CO, and OH lines, all with velocity and velocity dispersion similar to that in the remnant (see DeNoyer (1979), DeNoyer and Frerking (1981), Treffers (1979), and Giovanelli and Haynes (1979)).

The kinematics of this nebula were first studied by the present author in 1967 – 1968. Observing the bright region in the northeast with a Fabry–Perot etalon and an image intensifier, we noticed a systematic broadening of the Hα line from the periphery of the shell to its center which could be interpreted as an expansion of the system of bright filaments at a mean velocity of 65 km/sec. It then became clear that certain of the fainter filaments were moving away from the center at 120 – 150 km/sec. We subsequently continued our investigation of the kinematics of IC 443, including the faint filaments in the southwest sector (Lozinskaya, 1975a, 1979b), with the following results.

The gas motions in IC 443 are typical of old supernova remnants. The lines of Hα, [N II], and other species emitted by the shell display a complicated structure with multiple peaks (see Fig. 35b). Throughout the nebula and in nearby H II regions, one sees "unshifted" line components at radial velocities $|v_{LSR}| < 20$ km/sec, representing the combined radiation of the galactic background and low-velocity dense gas clouds within the remnant. Also visible within the shell are "shifted" features in the line profiles at velocities from −200 to +200 km/sec, as well as faint, broad wings extending out to −350 to +240 km/sec, forming a more or less flat, elevated background pedestal. In Fig. 36 we have plotted the radial velocity of the "shifted" components (points) and diffuse wings (lines) as a function of the normalized distance to the shell's center of symmetry. The lines end in arrowheads in those cases constrained by instrumental sensitivity, resulting from overlapping orders in the Fabry–Perot etalon. As in the Cygnus Loop, the high-velocity wings are radiated by the diffuse interfilamentary gas residing in the bright sector of the nebula, rather than by the thin filaments themselves.

Assuming the validity of the Schmidt galactic rotation law (Schmidt, 1965) with $r_0 = 10$ kpc, the mean radial velocity $v_{LSR} = +3 \pm 3$ km/sec of the outermost filaments yields a kinematic distance to IC 443 of $r = 0.7 - 1.5$ kpc. This estimate is consistent with the commonly accepted distance of 1.5 − 2 kpc based on the assumed connection of the remnant with the H II region S 249, which is excited by the stars of the Gem OB1 association; the photometric distance to the latter lies in the range 0.85 − 2.5 kpc (Humphreys, 1978). Observations at 21 cm yield a kinematic distance of 2.2 − 3 kpc, but the line profile in the direction of IC 443 is quite complicated, and the situation is further confused by the cloud near the remnant; this estimate would seem to be somewhat less reliable. The suggestion by van den Bergh (1973) that the remnant is 500 pc away is based solely on the possibility that the star exciting the cloud to the northeast of the shell is HD 43836. A relatively recent method for estimating the distance on the basis of the low-frequency cutoff in the x-ray spectrum combined with presumed interstellar reddening yields, for the empirical value $N_H/E_{B-V} = (6.8 \pm 1.6) \times 10^{21}$ atoms/cm^2 per stellar magnitude, a lower limit of 1 kpc (Malina et al., 1976). The aggregate data give a most probable distance of 1.5 kpc, and a mean shell radius of 9 pc. (We shall dwell in some detail on distance estimation methods in order to emphasize that the fundamental parameter — the linear size of a supernova remnant — is often subject to 50 − 100% error. IC 443 is one of the best studied objects, and has been used to calibrate the Σ–D relation (see Section 10). If the distance to a remnant is in fact determined in some other way, especially via the Σ–D relation, it must be borne in mind that a sizable error is quite possible!)

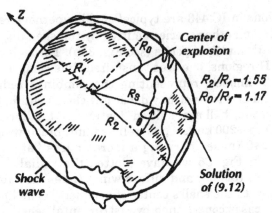

Fig. 37. The solution obtained by Kompaneets (1960), which is given in Eq. (9.12) and best represents the shape of IC 443 in the plane of the sky.

The asymmetric shape of IC 443 and the dense cloud to its northeast attest to the pronounced large-scale inhomogeneity of the interstellar gas, a circumstance that rules out consideration of the standard adiabatic Sedov solution (9.2) for the interpretation of the observational results. The problem of a point explosion in a medium with a plane exponential density gradient was solved by Kompaneets (1960). We shall make use of this self-similar solution, as given by Eqs. (9.12), since the nebula actually has approximately the appropriate form (see Gulliford (1974) and Fig. 37). The observed distribution of the filaments' radial velocity v_r and the shape of the nebula in projection on the plane of the sky enable one to determine, in terms of (9.12), the characteristic scale height of the density distribution ($H = 0.8R_0 = 7$ pc), the orientation of the surface — the angle by which the major axis is inclined to the plane of the sky ($\alpha = 0 \pm 15°$), the ratio of major to minor axes ($R_2/R_1 = 1.5 - 1.6$), and the mean expansion velocity ($v_0 = 280$ km/sec) (Lozinskaya, 1979b). With these parameters, the calculated dependence of v_r on R/R_s is shown in Fig. 36. The two curves correspond to limiting polar angles in the range $-45° \le \theta \le +45°$ and $135° \le \theta \le 230°$, for which velocity measurements have been carried out. As in the spherically symmetric Cygnus Loop, the observed points (v_r, R/R_s) randomly fill the interior of the theoretical "velocity ellipse." The velocity of the adiabatic shock front ($v_s = \frac{4}{3}v_{gas}$) corresponding to the high-speed filaments is $v_s = 370 \pm 50$ km/sec, ranging from 200 to 370 km/sec in the northeast and from 370 to 530 km/sec in the southwest sectors. With geometric projection taken into account, the faint high-velocity Hα emission from the diffuse interfilament medium corresponds to a shock front velocity of up to $v_s = 600 - 700$ km/sec.

X-ray emission from IC 443 is enhanced in the central and northern reaches of the optical shell (Lewin et al., 1979; Watson et al., 1983a et seq.), which is quite compatible with the assumed model. For an exponential density distribution having the aforementioned scale height $H = 7$ pc, the hot plasma density ($n_X \approx 4n_0$ behind a strong adiabatic shock front) at the southwestern part of the shell is a factor of 2 to 3 lower than at the center, and the x-ray luminosity $L_X \propto n_X^2$ (see Eq. (8.1)) is accordingly 5-10 times lower. Emission at the northeast boundary is associated with a lower shock wave velocity and temperature ($T_s \propto v_s^2$) due to the collision with the dense cloud. Further evidence for the temperature actually being lower there than at the center is provided by observations of the coronal line of [Fe X] (Woodgate et al., 1979). Emission from [Fe X] is concentrated near the bright eastern filaments, and corresponds to $T_e = 1.2 \times 10^6$ K.

A crescent-shaped H I cloud is part of the shell bordering the bright eastern filaments (Giovanelli and Haynes, 1979); it spans the radial velocity range −100 to +70 km/sec. The crescent shape and the velocity spread, which matches the expansion velocity of the system of bright filaments, suggest that this gas has been swept up by the expanding shell.

The spectrum of the bright filaments in the nebula displays more than 50 lines of different elements in various states of ionization. The relative intensities of the brightest lines are best fit by the theoretical spectrum of gas behind a shock front propagating through a dense medium with $n_{0d} = 10-20$ cm^{-3} at 65-90 km/sec (Fesen and Kirshner, 1980). The gas temperature in the filaments is $T_e \approx 2.4 \times 10^4$ K in the [O III] emission region and $(8-12) \times 10^3$ K in the [N II] and [S II] regions; the density of $n_e = 100-500$ cm^{-3} in the bright filaments is based on measurements in [S II].

The age of the nebula implied by its size and the blast-wave velocity, which in turn has been obtained either from the x-ray spectrum or the highest-velocity features in Hα, is approximately 5000 years.

In IC 443, as in the Cygnus Loop, we see a break in the regular shell structure and an outward flow of hot plasma. Beyond the bright tangential filaments in the northeast, there is a radial filament about 15′ long (see Fig. 35). Its optical spectrum resembles the spectrum of the remnant, and differs from that of the nearby H II region S 249, suggesting collisional excitation at the shock front (Fesen, 1984). In this same region there is weak x-ray emission coming from 10-20′ beyond the bright nebula (Watson et al., 1983a), as well as broad Hα emission (Lozinskaya, 1979b). It is conceivable that due to a local decrease in the density of the unperturbed medium, the shock front has advanced considerably farther here.

Most recently, research on the nature of the remnant IC 443 has been carried to new heights. It has become possible to make detailed comparisons of the nebular structure across the radio spectrum (using the VLA, at a resolution of ~3″), in the 21-cm line (20×56″ resolution), in the visible lines of Hα, [N II], [S II], and [Fe X] (1-2″ resolution), in the infrared (IRAS data with 25-100″ resolution), in the ultraviolet (IUE observations with 3″ resolution), and at x rays using the Einstein observatory (16″ resolution). These analyses were conducted by Mufson et al. (1986), McCullough and Mufson (1988), Braun and Strom (1986c), Petre et al. (1988), and Brown et al. (1988), and they confirm the basic conclusions drawn by previous investigators. Over a very wide energy range, the radiation from IC 443 is accounted for by the propagation of a shock wave through a medium with marked density inhomogeneities. The radio filaments overlay the optical; close to 80% of the emission at $\lambda \approx 20$ cm is associated with diffuse regions $\sim 20'$ in size.

The x-ray image of IC 443 differs significantly from its optical and radio appearance. The rather unusual x-ray properties (atypical morphology, low surface brightness, pronounced spectral hardening in the fainter southern region) are explained by the interaction of the expanding remnant with the large-scale northeast H I cloud and the molecular cloud in front of the shell. Petre et al. (1988) report that most of the remnant has a temperature of about 1.2×10^7 K and a relatively uniform interior density distribution, with absorption-induced hardening of the spectrum due to the higher column density in the foreground molecular cloud. In the bright northeast x-ray spot, the spectrum can be fitted well by either a two-temperature (2.2×10^6 K and 1.1×10^7 K) plasma in ionization equilibrium or a nonequilibrium plasma at 1.8×10^7 K. Observations in the vicinity of the bright x-ray spot in the [Fe X] line (Brown et al., 1988) have revealed a knotty structure of hot [Fe X]-emitting plasma with a mean density $n_e = 60$ cm^{-3} and gas pressure 7×10^7 cm$^{-3} \cdot$K. No direct correlation between x-ray emission and [Fe X] knots has been detected. An assumption of pressure equilibrium between [Fe X] and x-ray emitting plasma leads Petre et al. (1988) to conclude that despite pronounced brightening within the spot, the x-ray emission in IC 443 may be confined to a thin shell about 0.1 pc thick.

Three components of the IR radiation have been identified in the nebula: emission from dust heated by collisions with hot plasma, emission from dust heated by ionizing radiation, and emission in the lines of [Ne II], [Fe II], [Si II], and [O I]. The line radiation has been shown to make a large contribution to the flux observed in the IRAS bands, and with that contribution taken into consideration, the dust temperature is typically $T_d \sim 42 - 45$ K.

The strong line of [Fe II] at $1.64 \mu m$ has actually been detected in the spectrum of IC 443 (Graham et al., 1987a); it is 500 times as strong as in conventional H II regions, most likely as a consequence of shock-induced heating.

Convincing new evidence has been found for the interaction of the remnant with a dense molecular complex. Clouds containing CO, HCO, and HCN have been detected near the bright border of IC 443; these are characterized by emission lines with similar profiles and large velocity dispersion, $\Delta v \sim 90$ km/sec. This may be taken as proof that the shock wave has shocked the cloud, and confirms our measurements of the velocity of the system of bright optical filaments.

The density within the nearest molecular complex is nonuniform. Having analyzed the brightness and velocity distributions in the lines of H_2,

CO, HCN, HCO, and CS, Burton (1988) discovered a broad ring of "shocked" molecular clouds, bounding the region over which the expanding remnant has collided with a plane molecular layer. Braun and Strom (1986c) have interpreted the IR brightness distribution to be that of a triple shell structure in IC 443, associated with the fact that the supernova repeatedly heated three voids swept out by the stellar wind of the Gem OB1 association.

Finally, although the foregoing model must unquestionably be confirmed by a study of the kinematics of the optical filaments of IC 443, one thing is certain: the remnant is expanding into a highly inhomogeneous dense molecular complex that is subject to the disruptive influence of the radiation and winds from the stellar association.

G 78.2+2.1 and the nebula near γ Cygni

A bright symmetric nebula approximately 4' in size near the star γ Cygni was initially counted as a supernova remnant (van den Bergh, 1973), since it has been identified with the nonthermal radio source DR 4, the brightest one in the Cygnus X complex. The nonthermal nature of the radio emission is supported both by its spectrum ($\alpha = -0.7$) and its linear polarization ($p = 5\%$) (Johnson, 1974), thereby providing a basis for identifying the nebula with the supernova remnant.

Investigations of the nebula's spectrum and kinematics, however, have demonstrated that it is not a supernova remnant after all, but a conventional H II region (Lozinskaya, 1975). A candidate exciting star has also been found at the center (Arkhipova and Lozinskaya, 1978a). To account for the nonthermal radio spectrum of the H II region, we have proposed that an expanding shell — a supernova remnant — is colliding with the nebula; the supernova presumably exploded nearby, but did not have any inherent connection with the nebula (Lozinskaya, 1977).

Deformation of the hypothetical supernova shell in the vicinity of the collision and enhancement of the gas density due to the intense radiative cooling in the dense cloud leads to local compression of the magnetic field lines of the remnant, with a corresponding increase in the synchrotron emissivity. (A quantitative estimate can be obtained using Eq. (10.2); the basic equations relating to the theory of synchrotron radiation are derived in Section 10.) By taking into account the leakage of relativistic particles from the compression region and their isotropization behind the shock front due to scattering from inhomogeneities in the magnetic field (Bychkov, 1978a), it can be shown that the emissivity can be enhanced by a factor of 3 to 4.

An observer for whom the dense cloud is projected onto the putative supernova remnant will perceive an H II region that coincides with a source of synchrotron radiation. The gas velocity inside the cloud will then be given by

$$v_{cl} \approx v_s (n_{0\,i}/n_{0\,cl})^{0.5}$$

(see Section 8), where v_{cl} is the shock velocity in the dense cloud, and v_s is

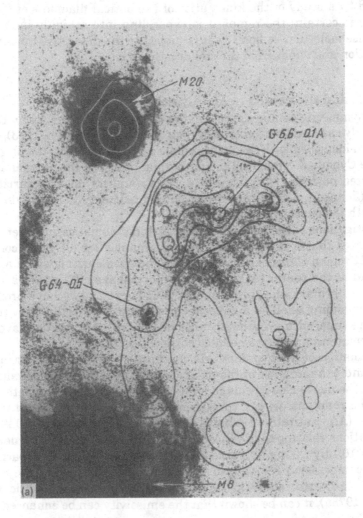

Fig. 38a Radio isophotes of W 28 superimposed on a photograph in taken in the red. The bright nebulae M8 and M20 are apparent; in addition, we see two flat-spectrum radio-bright regions, G 6.4–0.5 and a possible plerion, G 6.6–0.1 (see text).

the shock velocity in the intercloud medium; see Fig. 47. If the density contrast (n_{0i}/n_{0cl}) is large enough, gas motions within the cloud may be subsonic.

This model predicts that in the vicinity of the bright nebula near γ Cygni, there ought to exist a weak synchrotron radio source — an extended supernova remnant — and that prediction has in fact been confirmed. Higgs et al. (1977) and Baars et al. (1978) have actually found a shell approximately 1° across with a nonthermal radio spectrum ($\alpha = -0.65$), the genuine supernova remnant G 78.2+2.1. The object DR 4 and the bright nebula near γ Cygni comprise but a small part of this shell. The surface brightness of G 78.2+2.1 yields a remnant distance of ~1.8 kpc, which is close to the kinematic distance to the H II region; the linear size of the shell is then ~33 pc. Faint optical filaments have also been found to fill the radio shell; strong sulfur lines in the spectrum ($I_{[SII]}/I_{H\alpha} \approx 1$) suggest that this is indeed the optical supernova remnant and not just background radiation, which is quite bright in Cygnus (van den Bergh, 1978b). The supernova remnant G 78.2+2.1 is surrounded by an H I envelope that is expanding at a mean velocity of about 25 km/sec.

Dense, compact H I and CO clouds shocked by the expanding supernova shell have also been identified (Landecker et al., 1980; Braun and Strom, 1986c; Fukui and Tatematsu, 1988). It is conceivable that the gamma rays detected by COS-B and emanating from the same general direction result from the collision of the remnant with a dense cloud (Pollock, 1985). X rays have also been found to be coming from the extended shell;

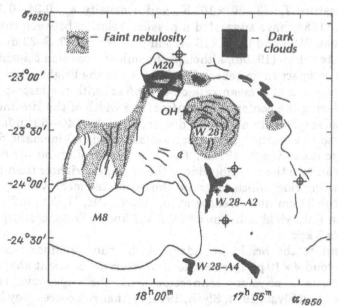

Fig. 38b. Large-scale structure of W 28 according to Zealey et al. (1983), with the brightest stars marked.

the temperature implied by the x-ray spectrum is 1.5×10^7 K, and the x-ray luminosity yields an ambient gas density in the range $0.1-0.2\,\text{cm}^{-3}$ (Higgs et al., 1983).

The nonthermal radio source W 28

This object is the result of a supernova explosion in a dense cloud. Photographs of the region taken in red light (see Fig. 38a) reveal an extensive gas and dust complex, taking in the bright H II regions M8 and M20, as well as the supernova remnant itself, comprised of faint, diffuse filaments that coincide with the nonthermal radio source. The remnant's radio spectral index is $\alpha = -0.42$ (Milne and Wilson, 1971; Green, 1987a); isolated bright knots with a flatter spectrum, $\alpha = -0.21$, are probably due to ionized gas radiating thermally, a presumption supported by their radio recombination-line emission. Linear polarization reaches $15-25\%$ at 5 GHz, and the magnetic field is regular (Angerhofer et al., 1977).

One of the compact, bright knots, G 6.6–0.1A, may possibly be a plerion, its flat radio spectrum being associated with relativistic particle injection by the stellar remnant of the explosion (see Section 6).

In x rays, there is both a bright, extended source and a second, more compact one; the latter is displaced from the center of the former (Matsui and Long, 1985). The extended source is brighter toward the center, rather than toward the periphery, but there is as yet no reliable way to choose between thermal radiation and synchrotron emission due to particles injected by the putative stellar remnant. If the radiation is thermal, it corresponds to a temperature $T_e \sim (3-50) \times 10^6$ K, and a density $n_e \sim 0.34 - 0.14\,\text{cm}^{-3}$. Shull et al. (1989) have suggested a possible relation between the supernova remnant W 28 and the 0.415-second pulsar PSR 1758–23 discovered by Manchester et al. (1985b), although the pulsar's position coincides with neither the compact radio source G 6.6–0.1A nor the bright x-ray nebula. W 28 is expanding at a mean rate of 40 km/sec, with the fastest-moving filaments fleeing the center at 80 km/sec; the width of the Hα line in the filaments corresponds to a velocity dispersion of $\Delta v = 40-50$ km/sec , and the mean radial velocity $v_{LSR} = +18 \pm 5$ km/sec yields a kinematic distance of $3.5-5$ kpc (Lozinskaya, 1973a). This is an upper limit on the distance, as the filaments on the receding side of the shell are brighter than those on the approaching side, which can raise the estimated mean velocity. Observations of the 21-cm line (Wenger et al., 1982), OH, H_2CO, and radio recombination lines yield a distance of $2.8-3.9$ kpc to W 28; the largest reliable value is 3 kpc.

Adjacent to the bright boundary of the radio remnant is an OH molecular cloud 4×10 pc in size; the characteristic crescent shape of the latter suggests a physical connection with the supernova remnant (Pashchenko and Slysh, 1974; Slysh, 1975; further references may be found in both of these). In addition, the particular structure of the outer boundary of the radio remnant betokens a collision with the cloud (Velusamy,

1988). The cloud's velocity dispersion, $\Delta v = 50$ km/sec, is consistent with the expansion velocity of W 28, and the OH abundance corresponds to a hydrogen density $n_H = 10^3$ cm^{-3}. The latter figure gives the gas density in the collisionally compressed region of the cloud; in the unshocked gas, it is much lower. In this "disturbed" region near the boundary of W 28, one finds a class IIa OH maser source, which is typical of supernova remnants. Here also is an extensive and dense CO cloud, whose radial velocity $v_{LSR} = +22$ km/sec is close to the velocity of the supernova remnant (Wilson et al., 1974; DeNoyer, 1983). Observations of the formaldehyde line also suggest a high cloud density associated with the supernova remnant: $n_H = 50$ cm^{-3} in the part adjoining M20, and $n_H = 20-35$ cm^{-3} near the bright boundary of W 28 (Slysh et al., 1979). A number of compact infrared sources are to be found near M20 and the supernova remnant (Wright et al., 1976). The synchrotron radio source in W 28 and its associated optical nebula are enveloped by a neutral gas shell (Wenger et al., 1976). The inner radius of the H I shell is 24 pc, i.e., it is practically the same as the synchrotron radio source; the outer radius is 40 pc. The mean hydrogen density is $n_H = 12$ cm^{-3}, the mass of the neutral shell is $M_{HI} = 7 \times 10^4$ $M_\odot$, and the expansion velocity is 20 km/sec.

An extended shell about 1.5° in size has also been identified which consists of emission nebulae and absorbing clouds; it surrounds the radio sources W 28 and W 28–A2, and passes through M20 and M8 (see Fig. 38b). This may possibly be a second, older supernova remnant (Zealey et al., 1983) or a shell swept up by the stellar wind from the OB association.

The old supernova remnant W 50

W 50 became one of the most intriguing astronomical objects ever when the centrally located peculiar object SS 433 was discovered (see Section 6). SS 433 was in fact so unusual that initially it was even thought not to be associated with W 50, despite its being precisely at the center of the shell. Two systems of bright H and He emission lines were detected in that fairly faint ($m_V = 14$) star which were shifted with respect to a third stationary component by 3×10^4 to 5×10^4 km/sec. The Doppler shifts were accompanied by abrupt changes in line profile, and were modulated with a 164^d period. A narrow feature near the maximum of the stationary component was shifted at about 100 km/sec, with a period of 13.1 days.

Our present interpretation of these facts is as follows (see Shklovskii (1981b), Margon (1982, 1984), Goncharskii et al. (1984), and references therein). SS 433 is a massive eclipsing x-ray binary system consisting of a B star ($M = 10-20\,M_\odot$) with a relativistic companion, probably a black hole, with a mass of $5-6\,M_\odot$. The B star is overflowing its Roche lobe, forcing the system into a supercritical accretion regime; this results in the formation of a powerful accretion disk and two collinear relativistic jets perpendicular to it. The accretion disk and the jets precess with a period of 164^d, in synchrony with the precession of the B star's rotation axis; the

13.1-day period corresponds to the orbital motion of the system. The rapid displacements of the high-velocity spectral components observed optically may be ascribed to emission from the precessing jets, which are oriented approximately 80° to the line of sight.

Detailed photometric investigations of SS 433 are still under way (for example, see Kemp et al. (1986) and Gladyshev et al (1987)); the mechanism whereby the relativistic jets are formed is under discussion, and novel evolutionary scenarios have been proposed for the system (see Kundt (1985), Bodo et al. (1985), Collins (1985), Antokhina and Cherepashchuk (1985), Helter and Savedoff (1986), Fabian et al. (1986)). But to digress for a moment from the unique and still largely unfathomable nature of the central stellar object, it is important to state that it is rapidly shedding mass, and that, to a large extent, is what governs the properties of the extended remnant W 50. As we have already noted, the rate of outflow in the jets is $\dot{M} = 10^{-7}$ $M_\odot$/yr at $V_\infty = 8 \times 10^4$ km/sec, while in the spherically symmetric flow, $\dot{M} = 10^{-4}$ $M_\odot$/yr and $V_\infty = 10^3$ km/sec.

There is no longer any question of the intimate connection between W 50 and SS 433, since the interaction between the jets and the matter in the shell has been observed directly. What would otherwise be the spherical shape of the radio shell is distorted by the presence of two symmetric lobes pointing in the same direction as the precessing jets (fig. 39). Recent observations have revealed a central radio source several arcseconds across, extended in the same direction, and varying in shape with a period of 164 days; here, then, we are directly observing synchrotron radio emission from the plasma in the jets (Fejes (1986) and references therein).

The compact central source and the two plasma jets are also visible in x rays, and account for approximately 90% of the emitted flux. Apart from the central source, there are two elongated x-ray spots at distances between 27 and 70 pc on either side of SS 433 (assuming a most probable distance to the remnant of 5.5 kpc, as given by Margon (1982)); see Fig. 39. The brightest region of each x-ray spot lie near the base of its respective "radio lobe"; the temperature of the radiating plasma varies from $kT \sim 4$ keV to $kT \sim 1.5$ keV as one moves progressively outward along the spots. X-ray synchrotron emission may also possibly contribute to the radiation (see Watson et al. (1983b) and references therein). The luminosity of each of the spots is about 6×10^{34} erg/sec, and the total thermal energy of the radiating plasma is $E_t \sim 0.6 \times 10^{51}$ erg. The brightness distribution in the x-ray spots implies a plasma density of $n_e = 0.2$ cm^{-3}, which rises to $n_e = 1$ cm^{-3} in the bright areas near the points of attachment of the radio lobes. Here one also finds delicate optical filaments that are more or less contiguous with the spherical part of the shell at those same points of attachment; see Fig. 39 (van den Bergh, 1980b; Zealey et al., 1980).

The influence of the relativistic jets upon the formation of the shell is so obvious that for a while there was some doubt as to whether W 50 as a whole is in fact a supernova remnant, rather than simply being a cavity blown out by the wind from SS 433 (see Königl (1983), for example). But

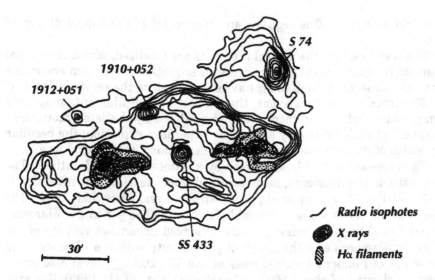

Fig. 39. Radio isophotes of W 50, with superimposed x-ray isophotes (Watson et al., 1983b). The compact stellar remnant SS 433 shows up, along with the bright optical filaments. The objects 1912+051, 1910+052, and S 74 are unrelated to the supernova remnant.

although theory does not (in principle) rule out a synchrotron radio spectrum when the magnetic field frozen into circumstellar gas swept up by a wind is compressed, we have not a single instance of such behavior in the ring nebulae surrounding known strong sources of stellar wind (see Chapter 3). The localization of optical filaments within the shell, rather than near the outer rim, is also atypical of the ring nebulae produced by stellar winds. Both problems are disposed of by the model of Shklovskii (1981b) and Geldzahler (1980), in which W 50 is an old composite supernova remnant, consisting of a shell plus a central source of relativistic particles. Here we are apparently viewing a limiting case, a plerion that completely fills the interior of a shell. The radio shell structure is actually quite well-defined, but in contrast to classical radio shell remnants, the brightness at the center is elevated (Geldzahler et al., 1980), and there is pronounced linear polarization: $p = 10\%$ at 1.7 GHz and $p = 40\%$ at 2.7 GHz (Downes et al., 1981).

Even in the model with a plerion filling the shell, W 50 is still anomalous: the radio spectral index varies from -0.4 to -0.7 (with a mean of -0.66) over the entire remnant (Downes et al., 1986), which differs from the behavior of typical plerion spectra; see Section 10. Likewise, even where relativistic particles from the jets of SS 433 ought to be impinging directly, there is no noticeable flattening of the spectrum. What this means is that if the central source is actually injecting relativistic particles into the shell, their spectrum must differ from that in other plerions. The magnetic field within W 50 is regular, and tangential at the spherical part of the shell.

The way in which the optical filaments are localized, which is unusual for an old remnant, enables us to draw an important conclusion about this object: the supernova shell must have existed before the relativistic jets of SS 433 turned on. In that event, the delicate optical filaments may have been produced when the shock wave engendered by the jets encountered the spherical shell formed by a previous supernova explosion; the peculiar disposition of the filaments would then be explained in a natural way.

The proposed model is supported by a number of considerations. The gas density in the filaments, as determined from the [S II] line intensity, is $n_e \approx 10^2 - 10^3$ cm^{-3} (Zealey et al., 1980; Kirshner and Chevalier, 1980). The velocity of the shock wave responsible for heating the gas in the filaments is $v_{cl} \approx 50 - 100$ km/sec, judging by the observed expansion velocity of the system of filaments and the range of gas velocity within a filament. The blue wings of filament spectral lines at the $10 - 20\% I_{max}$ level also correspond to velocities of about $100 - 150$ km/sec (Mazeh et al., 1983). Knowing the density in an [S II]-emitting region ($T_e \approx 10^4$ K) and the shock velocity in the filaments, we find from the required constancy of the gas pressure behind the front that the initial density is $n_{0 cl} = 3 - 100$ cm^{-3}. That density is too high for the interstellar medium at a height $z = 200$ pc, and most likely characterizes gas that has already been compressed by the first shock wave, which resulted from the expansion of a spherical supernova shell.

We thus observe only the brightest optical filaments at the intersection of the two shock waves. Fainter emission from the periphery of the spherical part of the shell, which is undisturbed by the dynamical pressure of the relativistic jets, is not seen because of strong interstellar absorption. Murdin and Clark (1980) have reported that the Balmer decrement of the filaments yields $A_V = 4.^m5$; in the vicinity of the central source, absorption is even greater.

Assuming a typical density $n_0 = 1$ cm^{-3} and initial explosion kinetic energy of $E_0 = 10^{51}$ erg, the adiabatic solution (9.2) yields an age of about 10^5 yr for the remnant, based on the radius $R_s = 48$ pc of the spherical part of the shell, where the jets exert negligible influence. This is consistent with our suggestion that the relativistic jets turned on only after the supernova explosion, since the supercritical accretion regime is associated with outflow from a star that has filled its Roche lobe, something which occurs on a thermal time scale, i.e., 10^5 yr at most.

The advanced age of the remnant accounts for the absence of x-ray emission from throughout the spherical shell: the temperature behind the shock front to be expected from (9.2) is $T_s = 1.3 \times 10^5$ K. The diffuse x-ray emission observed in the vicinity of the two elongated spots is entirely due to plasma heated by the relativistic jets. The brightness and temperature of this radiation are elevated in the interaction region between the jets and the spherical shell. The rate of kinetic energy loss by the jets is

$L_j = 2 \times 10^{38}$ erg/sec, sufficient to support the thermal losses of the x-ray plasma if radiation has been sustained at the present level for no more than 10^5 yr.

We stress once again that the uniqueness of the remnant W 50 stems from two factors. The first is that we are observing a stage of supercritical accretion, the duration of which is brief compared with the lifetime of a massive close binary system. The second is that the stage of Roche lobe overflow by the "normal" component of the pair began before the remnant of the first explosion had dissipated in the interstellar medium. The latter is also an improbable event, as it requires that the initial masses of the two components of the system be almost equal: the age of W 50, $t \sim 10^5$ yr, separates the final stages of evolution of the first and second components. It would therefore seem unlikely that the Milky Way harbors many more objects like W 50 and SS 433.

The nebula RCW 89 and the radio remnant MSH 15–52 (G 320.4–1.1)
This object suddenly elicited a great deal of interest when a young pulsar was discovered in it, most probably the stellar remnant of the supernova (see Section 6). We must state immediately that there is still some lingering doubt attached to the interpretation of this interesting object. The age of PSR 1509–58 indicated by the ratio of its period and period derivative is 1700 years. The age of the remnant given by its size is close to 10^4 yr, a result supported by its morphology and thermal x-ray spectrum. Furthermore, it was not immediately clear just exactly what was to be considered the remnant of the supernova explosion! Initially, the remnant was thought to be the bright filamentary nebula RCW 89 (van den Bergh et al., 1973), which coincides with a nonthermal radio source approximately 10′ across (12 – 13 pc at a distance of 4.2 kpc). This nebula displays strong lines of [N II] and [S II]; in one of its bright, dense clumps, both visible and infrared lines of Fe II have been found (Dopita et al., 1977; Danziger, 1983; Seward et al., 1983b). The $I_{1.602\,\mu m}/I_{1.644\,\mu m}$ intensity ratio implies a density $n_e = 5 \times 10^4$ cm^{-3}, which accords with the value $n_e = 10^4$ cm^{-3} given by the lines in the optical spectrum of the same clump.

The pulsar is located outside the bright nebula (see Fig. 40), and if we were to assume that it was formed in an explosion 1700 years ago, its "runaway" velocity would be close to 6000 km/sec. We can alleviate the problem by considering the entire complex of radio sources shown in Fig. 40, about 30′ in size, to be a single remnant. Arguments favoring such a hypothesis include the morphology of the region, the central location of the pulsar, the uniform spectral index throughout — $\alpha = -0.35$ (Caswell et al., 1981; Manchester and Durdin, 1983; Manchester et al., 1982), and a common distance, approximately 4.2 kpc. At visible wavelengths, there are also faint filaments connecting the bright nebula RCW 89 with the extended shell MSH 15–52 (Seward et al., 1983b).

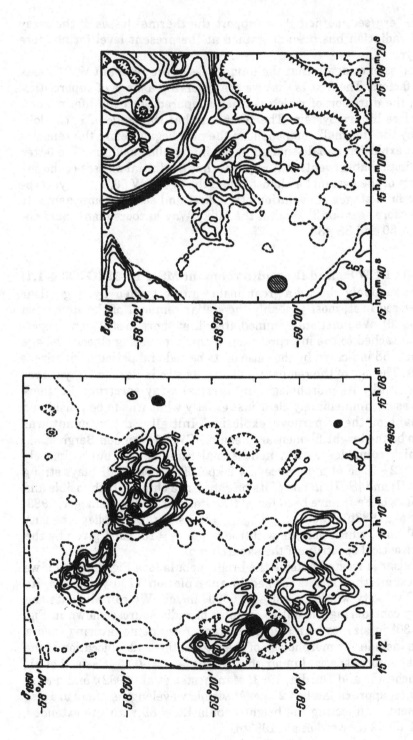

Fig. 40. The radio source MSH 15–52. Left: large-scale map. Right: central region, with the pulsar indicated by a cross (Manchester and Durdin, 1983). The nebula RCW 89 coincides with the bright region to the northwest.

But the problem of reconciling the pulsar age with that of the supernova remnant remains. According to the adiabatic solution (9.2), the diameter of the extended shell ($R \approx 36$ pc) corresponds to $t \approx 10^4$ yr, assuming only that the explosion did not take place in an anomalously tenuous medium and that it conformed to the usual energy characteristics. The age can be estimated directly using the expansion velocity; with that in mind, van den Bergh and Kamper (1984) measured the proper motion of 13 bright filaments in RCW 89. They found that the fastest-moving filaments are displaced by $\mu = 0.''07 - 0.''09/\text{yr}$, while for free expansion starting 1600 years ago, and moving away from the pulsar, one would expect $\mu = 0.''3/\text{yr}$. If the remnant was already in the adiabatic stage, it would be more than 5000 years old.

There are two compact x-ray sources surrounded by "hot spots" in the supernova remnant MSH 15–52. One is the pulsar PSR 1509–58; the extended source that surrounds it is attributable to synchrotron radiation coming from particles injected by the pulsar. The other has been identified with a bright, compact optical knot showing strong iron emission (Seward et al., 1983b, 1984). This knot clearly differs from the filaments of the old remnants, and if it were not inconsistent with the age derived from the velocity of the filaments, it would be most natural to presume that it was ejected in the explosion. The pulsar age is in better agreement with the evidence.

These inconsistencies can be dispensed with by fielding two propositions. The first is trivial: the old, extensive remnant and the young pulsar with an x-ray synchrotron-emitting nebula have nothing to do with one another, and are merely seen together in projection. The remnant MSH 15–52 and the pulsar belong to a single young stellar aggregate that includes the association Cir OB 1, some young clusters, four Wolf–Rayet stars, a great deal of dust, and H II regions (Lortet et al., 1987). The abundance of massive stars and the age of Cir OB 1 ($4 \times 10^6 - 10^7$ yr) suggest that supernova explosions are a common occurrence here, and therefore that MSH 15–52 and the pulsar might have been produced by different supernovae.

The second proposition is interesting, but so far unsupported by the evidence; it derives directly by analogy with the precessing jets of SS 433. Having noted that the boundary and certain features of RCW 89, plus the extended southeastern radio source, can be connected by straight lines passing through the pulsar, Manchester and Durdin (1983) have proposed that these two highly distinctive regions of the remnant were formed by precessing jets of relativistic plasma emerging in two directions — along the pulsar's magnetic axis, for example. So far, except for the analogy with W 50 + SS 433 and the apparent symmetry of the radio remnant, there are no observations to support this hypothesis, as there are in the case of W 50 + SS 433.

Vela XYZ

This is a system containing delicate filaments and a radio shell, within which the region known as Vela X resides — a plerion associated with the young pulsar PSR 0833–45 (see Sections 6 and 10). Relativistic particle injection is responsible for the different spectral indices in the plerion ($\alpha = -0.1$) and the shell ($\alpha = -0.6$ in the remainder of the nebula). The magnetic field structure correlates with the filamentary structure of the nebula; polarization is $p = 3 - 10\%$ in the shell, and is higher in the plerion (Milne, 1980; Lerche and Milne, 1980).

The distance of 400–500 pc has been established through the physical relationship existing between the remnant and the Gum Nebula. The radius of the shell is 15–20 pc, and the ambient gas density in the cloud component of the medium corresponds to the density of the Gum Nebula, $n_{0cl} = 5 - 10$ cm^{-3} (Jenkins et al., 1976, 1981).

Hot plasma behind the shock front radiates at x-ray wavelengths, and accordingly there is both a thermal and a synchrotron component to the emission from the remnant. According to Seward (1989), the x-ray remnant may extend farther to the southwest than previously recognized and form an eight-degree diameter shell around the pulsar.

The temperature derived from the thermal spectrum is $T_e = (1.5 - 4.3) \times 10^6$ K; a weak component at $T_e = 1.7 \times 10^7$ K is probably associated with radiation from the plerion (Kahn et al., 1983, 1985; Harnden et al., 1985). The soft x-ray component is brighter in the vicinity of optical filaments. Brightness and spectral variations occur on a typical scale of $\sim 0.1 - 1$ pc, and the corresponding density and temperature changes take place at roughly constant pressure: $P/k \sim (3 - 4) \times 10^5$ cm^{-3} K. The nature of the "plerion" nebula enveloping the pulsar is still not entirely clear. The pulsar is approximately 40' (~ 6 pc at a distance of 500 pc) from the center of the radio source Vela X; neither is it located at the center of the synchrotron x-ray nebula (see also Milne and Manchester (1986)). Pulsar proper motion measurements had failed until recently to detect any displacements at a rate that could account for the observed asymmetry (Bignami and Caraveo, 1988). However, recent optical astrometric measurements of the pulsar yield an estimated proper motion of $\mu_\alpha = -0.026 \pm 0.006$ "/yr and $\mu_\delta = -0.028 \pm 0.006$ "/yr (Ögelman et al., 1989). The magnitude of the proper motion indicates that the 11000 year old pulsar was born 7' southeast of its present location — i.e., near the center of the Vela X radio region and the bright x-ray emission region. The projected velocity of the pulsar at a distance of 500 pc is about 90 ± 11 km/sec.

The first information on the kinematics of Vela XYZ was obtained indirectly through measurements on background stars whose spectra showed absorption features in the lines of Ca, Na, O, N, S, and Si at velocities ranging from −130 to +55 km/sec (Jenkins et al. (1984) and references therein). Evidence that the absorption takes place in the supernova shell comes from the high velocity dispersion and enhanced Ca abundance, which results from its evaporation from dust grains by the passing shock wave. Direct measurements of the width of the Hα line radiated by the nebula yield $\Delta v = 310$ km/sec (Danziger et al., 1978). The [N II] and [O III] linewidth in isolated filaments is $\Delta v = 20-30$ km/sec, and this figure is also typical of random motions of individual clumps of gas relative to one another (Shull, 1983).

Puppis A

Baade and Minkowski (1954) identified with the radio source Puppis A a nebula consisting of about a dozen bright compact condensations. Present-day deep photographs display a striking variety of faint features that fill a radio shell roughly 60–80' across (Elliott et al., 1976; van den Bergh, 1978b; Goudis and Meaburn, 1978). In contrast to other old remnants, one observes compact condensations rather than delicate filaments. The brightest spectral lines are those of [N I] and [N II]; in certain knots, $I_{\text{[NII]}}/I_{\text{Hα}} = 20$, which is much higher than in other remnants (see Danziger (1983) and references therein). Most likely, the oxygen and nitrogen abundance in the optical filaments is higher than the normal cosmic value (Dopita et al., 1977).

Widely varying radial velocities of the filaments, ranging from 170 to 300 km/sec, yield a mean expansion velocity of 250 km/sec for the shell (Elliott, 1978). The bright knots typically move at 30–80 km/sec with respect to one another; the velocity dispersion within a knot is 20–30 km/sec, and the temperature is approximately 1.9×10^4 K (Shull, 1983). The compact condensations are observed against a background of weak, diffuse emission: $I_{\text{diff}}/I_{\text{comp}} = 0.5-0.8$, and the linewidth of the diffuse component is 60–80 km/sec (Shull, 1983). Broad, weak lines of [O III] and [Fe XIV] show up in the brightest eastern condensation (Clark et al., 1979).

Judging by the x-ray image (fig. 48), the explosion took place in a region with a large-scale density gradient and small-scale fluctuations, with clouds 1–2 pc across having a density $n_{0\,cl} \approx 1\,\text{cm}^{-3}$; in two bright, compact condensations to the east and north, $n_{0\,cl} \approx 10-20\,\text{cm}^{-3}$. This has also been confirmed by direct H I and CO observations (Dubner and Arnal, 1987).

The evolutionary status of Pup A has been reexamined only quite recently: it is probably a young object, and it may well belong to the class of oxygen-rich remnants whose prototype is Cas A (see Section 5). Winkler and Kirshner (1985) have discovered a new set of fast-moving filaments expanding away from the center at more than 1600 km/sec; the spectrum of the latter contains lines due to O^0, O^+, and O^{++}, while the hydrogen Balmer lines are exceptionally weak. The analogy is obvious: these fast-moving filaments can be identified with the fast-moving knots of Cas A, i.e., with knots ejected by the explosion that are enriched with the products of nuclear burning, while at the same time, the slow-moving, nitrogen-enriched filaments may be identified with the stationary flocculi, which in all likelihood were expelled during the mass outflow stage of the stellar progenitor.

A comparison of plates of the nebula taken in 1978 and 1986 has yielded the proper motion of these filaments: $\mu = 0.''1 - 0.''2/\text{yr}$ (Winkler et al., 1988). The filaments are expanding from a common center; their kinematic age is $3700 \pm 300\,\text{yr}$, apparently making this the oldest remnant in which we still observe matter from the supernova itself as yet unmixed with the interstellar gas. The velocity of these dense clumps, $1500 - 3000\,\text{km/sec}$, is less than that of the fast-moving filaments of Cas A, which probably means that deceleration of the remnant has set in. The diffuse component of the ejecta in Pup A has slowed even more: the shock front velocity given by the x-ray spectrum is $700 - 900\,\text{km/sec}$ (see Table 11), i.e., the adiabatic solution (9.2) is probably already applicable to the remnant.

In x rays and coronal lines, Pup A is one of the most carefully studied of all remnants. This enables us to analyze in detail the distribution of temperature, density, and ionization state behind a shock front in an inhomogeneous medium (see Section 8). The main results are the following. The temperature varies from $\sim 10^6$ K to $\sim 1.2 \times 10^7$ K; these variations are associated both with large-scale structure and with small-scale features of the image (Canizares et al., 1983; Aschenbach, 1988; Jansen et al., 1987. Comparison of the brightness of x-ray and optical coronal lines of [Fe X] and [Fe XIV] has divulged variations in the ionization temperature behind the front (Teske and Petre, 1987). The radiating plasma has not yet come to ionization equilibrium (Canizares et al., 1983; Winkler et al., 1983; Fischbach et al., 1988).

The x-ray emission from the remnant is consistent with both the relatively young age and the substantial oxygen and neon enrichment of the hot plasma: the ratios O/Fe and Ne/Fe are much higher than their normal cosmic values, even when it is recognized that the plasma may not be in ionization equilibrium.

Data on the radio emission from Pup A can be found in the papers by Erickson and Mahoney (1985) and Green (1988a); references to earlier work will be found in both.

Table 11, at the end of the present section, summarizes the available information on optical nebulae that are supernova remnants, including the results of x-ray observations and basic data on radio emission.

The nebulae in Table 11 account for the overwhelming majority of old Galactic supernova remnants with well-understood kinematics, and they may be used to analyze the interaction of a supernova shell with the gas of the interstellar medium. We emphasize that a consistent picture emerges among all old remnants: bright filaments and condensations move away from the center more slowly than faint ones. When the mean expansion velocity of bright filaments is of order $30-100\,\mathrm{km/sec}$, the fainter filaments or diffuse gas in these shells is found to be moving at $\sim(1-5)\times10^2\,\mathrm{km/sec}$. Within a filament, the gas velocity dispersion inferred from spectral linewidths is close to the mutual velocity of filaments in the shell. It can be seen from figs. 32 and 36 that the observed filament velocities are practically uniformly distributed over the interval between the minimum and maximum values of the "velocity ellipse." These results can only be adequately interpreted in conjunction with the x-ray data, and we therefore defer further consideration to Section 8. Here we shall deal only with the problem of the remnants' filamentary structure.

The delicate filamentary structure of old, optically visible supernova remnants is one of the most remarkable phenomena in the heavens; superb photographs of the Cygnus Loop, Vela XYZ, and Simeiz 147 stagger the imagination. The width of individual filaments is at most $2-3''$ in the plane of the sky ($0.02-0.05\,\mathrm{pc}$ in the nearest remnants), and they reach $3-5\,\mathrm{pc}$ in length; their regular morphology attests to their being a coherent entity. Oort had by 1946 posed the question of just what these openwork systems really were; we have yet to formulate a comprehensive answer.

In summarizing the observational results for optical supernova remnants, we may draw certain conclusions regarding the physical conditions inside fine filamentary structures (Lozinskaya, 1980d). To begin with, not all supernova remnants possess such structures: there exists a populous class of objects consisting solely of diffuse filaments and condensations (W 28, HB 3, HB 9, and HB 21, for example). In all nebulae that do possess this delicate filamentary morphology, one sees fainter diffuse formations within which the fine filaments are embedded. It is still not entirely clear what exactly determines whether a supernova remnant will belong to the diffuse or thin-filamentary class. Objects in both classes have the same range of linear dimensions, distance, radio brightness, and spectral index. In both groups there are both remnants that are and remnants that aren't associated with dense dust and gas clouds. No systematic differences have been detected in either the ambient interstellar gas density or the initial explosion energy. There is, however, a weak correlation of morphological type with certain interrelated parameters of the remnants, namely the x-ray luminosity, the shock velocity v_s, and the evolutionary age (see Section 9). The thin filamentary nebulae predominate among evolutionarily younger objects ($v_s \gtrsim 500-600\,\mathrm{km/sec}$, $Rn_{0i}^{1/3} \lesssim 10\,\mathrm{pc}$, $L_{0.1-5\,\mathrm{keV}} \gtrsim 10^{35}\,\mathrm{erg/sec}$), while the diffuse nebulae are more common among the older ($v_s \lesssim 100\,\mathrm{km/sec}$, $Rn_{0i}^{1/3} \approx 30\,\mathrm{pc}$, $L_{0.1-5\,\mathrm{keV}} \lesssim 10^{35}\,\mathrm{erg/sec}$).

In remnants that are completely devoid of fine filaments — specifically, W 28, HB 3, HB 9, and HB 21 — x-ray emission is concentrated near the center, rather than at the periphery. This is a result of the slowing of the shock wave: its present velocity has fallen below what would be necessary to heat the plasma to $T \gtrsim 10^6$ K, while the hot central region has not yet cooled down.

In those objects like the Cygnus Loop and Vela with a clearly delineated x-ray shell, and accordingly high temperature and shock propagation velocity, fine filamentary structure is clearly in evidence.

The geometry of the filaments — are they one-dimensional ropes or plane layers (sheets) viewed edge-on? — is still a topic of debate. The fact that thin, bright filaments coexist side by side with faint diffuse ones suggests that the former are sheets oriented edge-on to the observer. It would seem, however, that this is at variance with studies of the kinematics of old remnants. In fact, if the thin and diffuse filaments are planar structures differing solely in orientation, the linewidths due to thermal motions ought to be the same in both. But since the line of sight takes a long path through edgewise planar structures, turbulent line broadening due to relative motion of different clumps of gas in the line of sight might be stronger in the bright, thin filaments. We have demonstrated above that observationally the opposite is true.

Nevertheless, the observed kinematics can also be reconciled with the model of Hester (1987), which was first discussed by Poveda and Woltjer (1968). Here the optical emission is concentrated in thin undulatory sheets, and where the surface is tangential to the line of sight, one observes thin filaments. With this arrangement, the greater linewidth of the diffuse component of the remnant can result from the projection of several distinct portions of the undulatory sheets, whose velocity at any point is normal to the surface. Perhaps the most serious argument favoring this hypothesis is the identity between the two populations of thin filaments observed against the diffuse emission background of the Cygnus Loop. The outer "pure hydrogen" filaments are a consequence of collisional excitation of neutral hydrogen atoms at the shock front (see Section 8). It is natural to suppose that this radiation is localized within a thin layer trailing the front, where the gas has not yet been subject to radiative cooling instability, and Hester's model seems the most tenable. The system of bright "radiative" filaments in the Cygnus Loop has precisely the structure and geometry in the plane of the sky that would attest to the applicability there of the same model of thin undulatory sheets.

By measuring linewidths for elements of differing atomic weight, one can separate out the thermal and turbulent velocities of gas in the filaments. These measurements have been made for the bright condensations in the Cygnus Loop, IC 443, Vela XYZ, and Pup A, and they indicate that lines of heavier elements (N and O in particular) are broadened mainly by turbulent motion, with the filaments and condensations being characterized by small-scale cellular structure (Shull et al., 1982; Shull, 1983). Microturbulent motion of small-scale cells give rise to the observed linewidth. These cells remain spatially unresolved, but Shull derives a rough upper limit for their size by noting that the shock wave propagates at approximately 100 km/sec through the dense filaments within these remnants, and in scattering from small-scale inhomogeneities it gives rise to microturbulent velocities of about 20 km/sec. This then means that the inhomogeneities are several times smaller than the observed size of the bright filaments, i.e., $(2-3) \times 10^{-3}$ pc, judging by the optical images of the nebulae under consideration. That figure is in fact comparable to the thickness of the radiating gas layer behind the shock front in the filaments, and it is consistent both with cell formation as a result of thermal instability during the rapid radiative cooling stage, and with the shocked compression of small-cell formations that may have existed inside dense interstellar clouds prior to the explosion (Shull, 1983).

The foregoing observational effect — a multicell velocity distribution within the filaments — could also result, however, not from the "cellularity" of a set of cylindrical features, but from the "waviness" of a set of planar features oriented tangentially to the line of sight (Hester, 1987).

To sum up, we can state that the body of presently available observational data is not inconsistent with the idea of the thin filaments in old supernova remnants being a manifestation of undulatory sheets that are locally tangent to the line of sight due to warpage; there are arguments, however, that do favor cylindrically-structured filaments (Lozinskaya, 1980d; Kirshner and Arnold, 1979; Straka et al., 1986).

If the filaments are indeed planar features oriented edgewise, while the diffuse medium is made up of the same sort of planar features oriented face-on, then only very slight density variations in the surrounding medium, amounting to $20-30\%$, are needed to produce irregularities in the sheet geometry leading to the observed delicate filamentary structure.

One plausible reason for the formation of thin filaments in old remnants examined by Pikel'ner (1954) is the intersection of shock waves resulting from focusing by inhomogeneities in the interstellar medium. Bychkov (1979) has suggested that the intersection of the shock wave produced by the explosion with those engendered by the ejection of dense clumps of matter by the progenitor might be responsible for the formation of thin filaments in young remnants. In old remnants, the formation of cylindrical and planar gas structures might be associated with the focusing, reflection, and intersection of magnetohydrodynamic waves as the expanding supernova shell interacts with the cloudy interstellar medium (Sofue, 1978).

Fine-scale filamentary structure will also result from thermal instabilities encountered during the stage at which a dense, cold shell is formed. Thermal instability in the post-shock region experiencing a temperature drop from $10^6 - 10^7$ K down to 2×10^4 K will produce dense, thin layers in a homogeneous medium. Smith and Dickel (1983) have demonstrated that such a configuration is unstable, and will break up into thin parallel ropes. These features are actually observed in old supernova remnants — in the Cygnus Loop, Vela XYZ, and Simeiz 147, for example.

In the future, a more specific understanding of the nature of the filaments in old remnants will be based on observational material obtained at high angular resolution, enabling researchers to distinguish radiating gas layers typically $10^{-3} - 10^{-2}$ pc across. The most promising observations in that regard will be those of the nearest supernova remnants — the Cygnus Loop, Vela XYZ, and Simeiz 147 — made with 4- and 6-meter telescopes. Combined analysis of observations of thin filamentary structure carried out at visible, radio, UV, and x-ray wavelengths will also be necessary. Highly intriguing results have already been obtained along these lines (Fürst and Reich, 1986; Straka et al., 1986; Raymond et al., 1988).

Optical nebulae identified with supernova remnants (Table 11)

Table 11 contains the following information:

1) The usual designation of the remnant.

2) The distance (if there are no references or remarks in the Notes, the distance is obtained from the Σ–D relation plotted in Fig. 68, corrected for height above the galactic plane (Milne, 1979a)).

3) The linear radius of the remnant, as determined from the optical, x-ray, or radio image.

4) The typical shock wave velocity in the cloud component, v_{cl}, based on Hα linewidths and/or expansion velocities of bright filaments.[1]

5) The shock velocity v_s in the intercloud medium, based either on the highest-velocity features in the optical line profiles (left sub-column) or the temperature deduced from the x-ray spectrum (see Section 8) (right sub-column).

6) The characteristic initial density $n_{0\,cl}$ of the gas in the cloud component, as given by independent radio measurements at 21 cm and other wavelengths (left), or by the densities in bright filaments given by the relative intensity of [S II], and by the velocity v_{cl}, assuming constant gas pressure P_{cl} behind the radiative shock (taking $T_{[\mathrm{SII}]} = 8000\,\mathrm{K}$ and $n_e = 4n_{0\,cl}$).

7) The initial density $n_{0\,i}$ of the unperturbed gas in the intercloud medium, obtained from $n_{0\,cl}$, $v_{0\,cl}$, and v_s by requiring equality behind the shock front between the cloud and intercloud pressures P_{cl} and P_i (8.3) (left), and from the x-ray luminosity using (8.2), assuming an equilibrium plasma.

The last two columns contain information pertaining to radio emission.

8) The degree of linear polarization.

9) The spectral index.

We have given either recent or the most reliable references beneath the parameter values in each column, except for IC 443 and the Cygnus Loop, for which detailed references appear in the main text.

---

[1] It will be shown in Section 8 that bright optical filaments represent a cooling radiative shock in dense clouds, while x-ray emission originates behind the blast wave in the intercloud gas.

Table 11. Supernova Remnants.

Name	r, kpc	R, pc	v_d, km/sec	v_s, km/sec Optical	v_s, km/sec X-ray	n_{od}, cm^{-3} Radio	n_{od}, cm^{-3} [SII]+v_d	n_{oi}, cm^{-3} $P_d=P_i$	n_{oi}, cm^{-3} X-ray	Radio source p, %	Radio source $-\alpha$
1	2	3	4	5	5	6	6	7	7	8	9
HB 3	2.7–3 [1, 84]	35–50	30–50 [1]	230 [1]	340–480 [2, 67]		~1 [3]	0.04	0.06–0.02 [2, 67]		0.52–0.6 [4, 84]
HB 9	1.3–2 [3]	25–40	80 [3]	350 [3]	550 [7]		5–8 [3]	0.03–0.4	0.03 [7]	25–30 [4, 5]	0.44 /0.9 [6]
OA 184	4–5; 2 [12]; [8]	30	50 [12]	170–180 [12]			10 [3]	1–2			0.56 [11]
VRO 42.05.01	5	26/50	40–50 [9]	190–200 [9]			7–8 [3, 8]	~0.3			0.4 [10]
Simeiz 147	1.5–0.8 [8]	40	50 [19]	190 [19, 20]			4 [3]	0.3		10	0/1.2 [17, 18]
IC 443	1.5	9	100–150	600–700	700	20	5	0.5–1	0.15	6–8	0.36
Monoceros Loop	0.8 [21, 22]	29	50 [21]	100 [21]		5	3 [3]	~1		2–3 [23]	0.50 [23]
W 28	3	20	40 [25]	100 [25]	400–2000? [86]	20–25 [26]	30–40 [3]	6–7	~0.3?	15–25 [24]	0.42 [24]

Table 11 (cont'd).

1	2	3	4	5	6	7	8	9
G 65.2 +5.7	0.8 [27]	30	50-80 [27]	~500 [27] 450 [28, 29]	5 [3]	0.15 0.1-0.4 [28, 29]		~0.5 [30]
G 78.2 +2.1 (γ Cyg)	1.8	16	100 (H I) [31]	350 [3] 1000 [32]	~100 [31]	0.1-0.2 [32]		0.65 [33]
HB 21	1.2	20	35 [3]	100 [3] <200, 700 [15, 88]	5 [16]	0.6 [3]	25 [5]	0.35-0.40 [5, 13]
CTB 1	2.7 [3]	20	30-40 [3, 35]	190 [3]	10 [3]	0.2	10-17 [24, 35]	0.75 [24, 35]
MSH 15 -52 RCW 89	4.2 [36]	18		1650 [37]				0.35 [36]
Cygnus Loop	0.7	20	~80	400 390	5-10	0.2 0.15	4-25	0.84 /0.38
Vela XYZ	400-500 [39, 40]	15-20	~100 [41]	~400 [50] 400-600 [48, 49]	5-10 ~30 [51, 3]	0.15 [48, 49]	3-10 [38, 47]	0.1/0.6 [38]
Puppis A	1.8-2	12-14	250* [42]	1600 [43] ~800 [44, 45, 46]	20 [57] 15-25 [51, 3]	~0.3-1 [44, 45, 46]		0.48 [34]
Lupus Loop	0.5-0.6	20	23 (H I) [81]	390 [52, 53, 54]		0.04 [52, 53, 54]		0.3 [55]

Table 11 (conclusion).

1	2	3	4	5	6	7	8	9
W 44	3 [56]	12	10 (H I) [82, 83]	450–900 [56, 58, 85]	~20–8 [14, 83]	0.5–0.05 [56, 58]		0.3 [55]
Loop 1	0.1–0.2 (§10)	115	3 (H I) [63]	320 [59–62]	0.5 [63]	0.007 [59–62]		0.5–0.7
RCW 103	6.6 [69, 70]	9–10	100–350 [64, 65]	900 [66]		0.5		0.55 [34, 68]
RCW 86 (MSH 14–63)	3.2 [69]	24	~200 [64]	1000 [72, 73]		~0.1 [72, 73]	3 [71]	0.62 [55]
G 287.8–0.5 (η Car)	2.5 [75, 76]	7		2000 [75, 76]	6–7 [3]	0.15 [75, 76]		
G 290.1–0.8 (MSH 11–61A)	~4	7–8	50–90 [78]	~600 [74]				0.62– 0.55 [77, 55]
W 63 (G 82.7+5.4)	1.3–1.6	14–16	35–70 [79, 80]					0.7 [55]

*"Oxygen" clumps expanding at 1500–3000 km/sec [87].

Notes.

References.
1. Lozinskaya and Sitnik, 1980
2. Galas et al., 1980
3. Lozinskaya, 1980a
4. Velusamy and Kundu, 1974
5. Reich et al., 1983
6. Udal'tsov et al., 1978
7. Tuohy et al., 1979b
8. Fesen et al., 1985
9. Lozinskaya, 1978a
10. Landecker et al., 1982
11. Willis, 1973
12. Lozinskaya and Sitnik, 1979
13. Hill, 1974
14. Dickel et al., 1976
15. Davidsen et al., 1977
16. Assousa and Erkes, 1973
17. Kundu et al., 1980
18. Sofue et al., 1980
19. Lozinskaya, 1976
20. Kirshner and Arnold, 1979
21. Lozinskaya, 1971
22. Davis et al., 1978
23. Graham et al., 1982
24. Angerhofer et al., 1977
25. Lozinskaya, 1973a
26. Slysh et al., 1979
27. Lozinskaya and Sitnik, 1978
28. Snyder et al., 1978
29. Mason et al., 1979
30. Reich et al., 1979
31. Landecker et al., 1980
32. Higgs et al., 1983
33. Baars et al., 1978
34. Milne, 1979b
35. Reich and Braunsfurth, 1981
36. Manchester and Durdin, 1983
37. van den Bergh and Kamper, 1984
38. Lerche and Milne, 1980
39. Brandt et al., 1976
40. Jenkins et al., 1976
41. Jenkins et al., 1984
42. Elliot, 1978
43. Winkler and Kirshner, 1985
44. Petre et al., 1982
45. Burkert et al., 1982
46. Zarnecki et al., 1978
47. Milne, 1980
48. Hearn et al., 1980
49. Kahn et al., 1983
50. Danziger et al., 1978
51. d'Odorico and Sabbadin, 1977
52. Toor, 1980
53. Davelaar et al., 1979
54. Winkler et al., 1979
55. Milne, 1979a
56. Gronenschild et al., 1978
57. Gosachinskii and Khersonskii, 1983
58. Watson et al., 1983a
59. Iwan, 1980
60. Davelaar et al., 1980
61. Inoue et al., 1980
62. Hayakawa et al., 1978
63. Heiles et al., 1980
64. Danziger, 1983
65. van den Bergh et al., 1973
66. Tuohy et al., 1979d
67. Leahy et al., 1985a
68. Caswell et al., 1980
69. Leibowitz and Danziger, 1983
70. Ruiz, 1983
71. Milne, 1972
72. Pizarski et al., 1984
73. Nugent et al., 1984
74. Elliott, 1979
75. Becker et al., 1976
76. Bunner, 1978
77. Milne and Dickel, 1975
78. Elliott and Malin, 1979
79. Lozinskaya et al., 1975
80. Rosado and González, 1981
81. Colomb and Dubner, 1982
82. Knapp and Kerr, 1974

83. Venger et al., 1981
85. Smith et al., 1985b
87. Winkler et al., 1988

84. Landecker et al., 1987
86. Matsui and Long, 1985
88. Leahy, 1987b

Specific objects.

HB 3. Very weak radio shell at the boundary of the bright complex IC 1795/1905/1848 (Rohlfs et al., 1977). An optical nebula — thin filaments emitting in Hα, [N II], and [S II], and diffuse emission in the [O III] line (Fesen and Gull, 1983) — is visible only on the western side of the radio shell, and is absorbing radiation on the eastern side of the dense gas and dust cloud associated with IC 1805/1848. The kinematic distance to HB 3 is 3 ± 0.2 kpc, and the supernova remnant may not be physically related to the gas and dust complex, even though it is at the same distance (Lozinskaya and Sitnik, 1980). An H I cloud abuts the remnant to the north, possibly being part of an exterior shell expanding at about 30 km/sec (Read, 1981). The x-ray structure of the remnant is somewhat unusual: a bright ring approximately 30 pc across is visible against a background of diffuse emission that fills a radio pattern about 80 pc across (Leahy et al., 1985a). Leahy et al. have associated this ring-like enhancement with a spherically symmetric inhomogeneity of the interstellar medium that stems from the effects or either the wind from the progenitor or the reverse shock.

HB 9. Differs from other radio shell sources: it has a high degree of linear polarization, $p = 25$–30%, a magnetic field with regular structure (Reich et al., 1983), and a break in the radio spectrum, with $\alpha_{\nu \leqslant 1\,GHz} = -0.44$ and $\alpha_{\nu > 1\,GHz} = -0.9$ (Udal'tsov et al., 1978). The kinematic distance to HB 9 is 2 ± 0.8 kpc (Lozinskaya, 1980a), and the Σ–D relation yields a distance of 1.2–1.8 kpc. Indirect confirmation of the low density of the ambient gas listed in Table 11 comes from the lack of large-scale brightening in the optical and radio, suggesting collision with a dense cloud, and the fact that two galaxies seem to be visible right through the remnant.

X-ray emission is concentrated near a central area approximately 60′ in size; the temperature ranges from $\sim 1.4 \times 10^7$ K at the center to 4.6×10^6 K at the edge of the region, with a mean value of $9 \times 10^6 - 1.2 \times 10^7$ K (Leahy, 1987a). The spectral index between 0.4 and 1.4 GHz is $\alpha = -0.68$, and seems to be constant over the entire remnant (Leahy and Roger, 1988).

OA 184 (Sh 223). The mean radial velocity of outlying filaments yields a kinematic distance of 4–5 kpc (Lozinskaya and Sitnik, 1979). Based on absorption due to the Balmer decrement of the filaments, Fesen et al. (1985) obtained a distance of about 2 kpc. The ratio $I_{6717}/I_{6731} = 1.14$ corresponds to a density $n_{0\,cl} = 10$ cm^{-3} for the cloud component of the interstellar medium for a mean velocity $v_{cl} = 50$ km/sec (Lozinskaya, 1980a). Using the same method, Canto (1977) found $n_{0\,cl} = 115$ cm^{-3}, but he assumed an unrealistically low value for the filament velocities, $v_{cl} = 15$ km/sec, in conflict with our measurements.

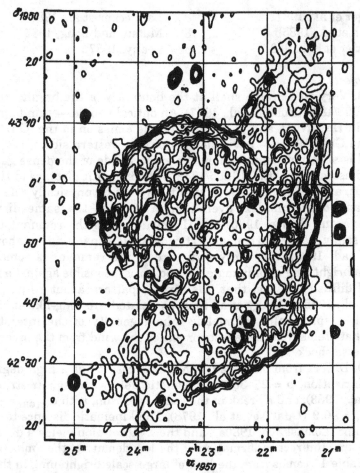

Fig.41. VRO 42.05.01, a supernova remnant at the
boundary of a dense cloud (Landecker et al., 1982).

VRO 42.05.01. A radio map appears in Fig. 41 (Landecker et al.,
1982). Judging by their similar spectra ($\alpha = -0.4 \pm 0.1$) and identical distance, as derived from the radio surface brightness of the two parts of the
shell, VRO 42.05.01 is a single entity at a distance of about 5 kpc. The
unusual shape of the shell may have resulted from an explosion at the
flattened boundary of a cloud having an abrupt drop in density.

According to the model devised by Pineault et al. (1985, 1987), the expanding remnant is "erupting" from the cloud into hot, tenuous intercloud
gas — quite possibly, into a "hot tunnel" produced by the supernova progenitor; see Chapter 3. Adopting reasonable values for the mean density of
the cloud and intercloud media ($1\,\mathrm{cm}^{-3}$ and $0.01\,\mathrm{cm}^{-3}$, respectively),
Pineault et al. obtained a good match between their model and the ob-

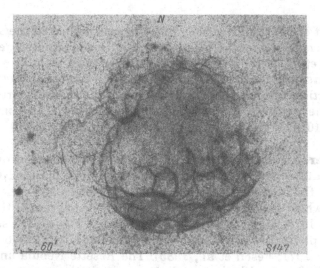

Fig. 42. The filamentary nebula Simeiz 147 (van den Bergh et al., 1973).

served structure of VRO 42.05.01, as well as with our measurements of the velocity in the eastern sector of the remnant (see Table 11).

Investigations of the kinematics of this nebula have yielded a mean expansion velocity of at most 35 km/sec for the bright filaments and a mean FWHM of the lines of about 80 km/sec, leading to the values of v_{cl} in Table 11; the velocity of the fastest-moving filaments is about 140 km/sec, giving the estimated value of v_s. In addition, the value given for $n_{0\,cl}$ refers to the filaments in the eastern sector of the shell (Lozinskaya, 1978a, 1980a). The western sector should be expanding about two or three times as fast.

Simeiz 147. Classic example of an old remnant in a relatively low-density homogeneous medium (Fig. 42). A distance of ~0.8 kpc has been obtained from measurements of the Balmer decrement of the bright filaments (Fesen et al., 1985), which determines the amount of interstellar absorption. This is not inconsistent with the value of ~1.5 kpc obtained from the Σ–D relation. Thin filaments visible in the optical are also observed in the radio, and the radio spectrum has a break, with $\alpha_{v<1\,GHz} = 0$ and $\alpha_{v\geq1\,GHz} = 1.2$ (Kundu et al., 1980; Sofue et al., 1980). The mean expansion velocity of the bright optical filaments, which reaches 30 km/sec (Lozinskaya, 1976), is in accord with the 25 km/sec expansion velocity of the outer H I shell (Assousa et al., 1975). The fastest-moving filaments are expanding away from the center at 140–150 km/sec (Lozinskaya, 1976; Kirshner and Arnold, 1979). High-velocity motion has been confirmed in the shell through observations of Doppler-shifted absorption lines superimposed on the spectrum of several background stars (Phillips and Gondhalekar, 1983). The evidence for a connection between the clouds and the supernova remnant consists of enhanced heavy-element abundances, particularly that of Ca, resulting from atoms being knocked off the surface of dust grains traversed by the blast wave. CO clouds situated near S 147 (see Scoville et al., 1977) probably have nothing to do with the supernova remnant. Fürst and Reich (1986) have detected a correlation between the radio and optical filaments, and a flattening of the radio spectrum in the filaments. The temperature of the thin filaments in the [O III]-emitting region is $(52 \pm 10) \times 10^3$ K, and in the [N II] region it is $(12 \pm 3) \times 10^3$ K (Fesen et al., 1985).

Monoceros Loop. A faint, thin-filament shell between the bright Rosette Nebula — which is excited by the stars of the cluster NGC 2244, belonging to the Mon OB2 association — and the gas and dust complex NGC 2264, which is associated with Mon OB1. A diagram of the region (Davis et al., 1978; Kirshner et al., 1978) appears in Fig. 43. The supernova remnant is physically associated with NGC 2264, lying 800 pc away (Lozinskaya, 1971; Fesen et al., 1985). The Rosette Nebula and an extended shell of neutral hydrogen that surrounds the entire complex of nebulae (approximately 130 pc in size and expanding at 20 km/sec (Gosachinskii and Khersonskii, 1982)) are located approximately 2 kpc away, and are unrelated to the remnant. The shell was probably produced by stellar wind, and possibly by supernova explosions that have taken place within Mon OB2. Wind and ionizing radiation from several of the O stars belonging to Mon OB1, which sits inside the Monoceros Loop, are apparently responsible for the diffuse [O III] shell, the filamentary supernova remnant, (which is bright in Hα and [N II]), and the synchrotron radio source in the shell. Direct velocity measurements on filaments in the remnant yield a mean shell expansion velocity of 50 km/sec; the fastest-moving filaments have a velocity of 80 km/sec (Lozinskaya, 1971). These results have subsequently been confirmed by the discovery of high-velocity features in interstellar absorption lines appearing in the spectra of two background stars (Wallerstein and Jacobsen, 1976; Cohen, 1977). Initial x-ray measurements of the remnant are also consistent with our finding of a slowly moving shock wave induced by the expansion of the shell (Leahy et al., 1985d). Decameter radio observations support the notion that O stars in the association provide an additional source of ionization for the gas in the nebula (Odegard, 1986).

HB 21. An ellipsoidal shell, first identified in the radio. The radio source is characterized by a magnetic field with regular structure and a high degree of polarization, $p = 25\%$ (Reich et al., 1983). The nebula associated with HB 21 is very faint, several of the diffuse filaments being almost undetectable against the bright Galactic background in Cygnus. Lozinskaya derived a mean expansion velocity of $v = 25$ km/sec in 1972, based on the systematic broadening of Hα as one progresses from the edge of the shell to its center; the highest-velocity features in the line profile yield $v_s = 100$ km/sec (see Lozinskaya (1980a)). The supernova remnant is surrounded by a shell of neutral hydrogen expanding at a mean velocity of 25 km/sec (Assousa and Erkes, 1973).

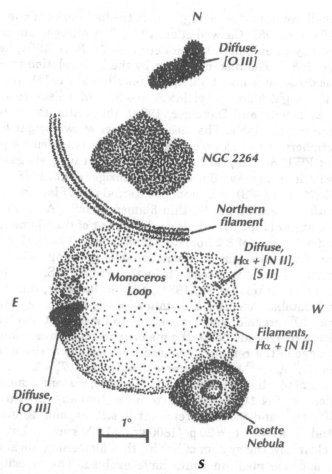

Fig. 43. Large-scale diagram of the region encompassing the Monoceros Loop, NGC 2264, and the Rosette Nebula; see text.

An enhancement of the CO emission on the eastern border of HB 21 suggests a collision of the remnant with the giant molecular cloud KH 141 (Fukui and Tatematsu, 1988). In x rays, there is a central region $\sim60'$ across, the spectrum of which corresponds to $T\sim7\times10^6$ K (Leahy, 1987b).

CTB 1. The morphology and kinematics of CTB 1 and the nearby H II regions suggest that the remnant has collided with a more or less flat-edged cloud; the kinematic distance to CTB 1 is 2.7 ± 0.5 kpc (Lozinskaya, 1980a). The synchrotron radio source and the nebula are immersed in an external H I shell of twice the size; the shell is expanding at 20–40 km/sec (Reich and Braunsfurth, 1981). The density of the bright optical knots is $n_e\sim10^2$ cm^{-3} (Fesen et al., 1985).

RCW 103. A stellar remnant — probably a neutron star — has been detected at the center of the shell (see Section 6). Nevertheless, RCW 103 is a typical shell remnant displaying none of the hallmarks of ongoing pulsar activity (Shaver, 1982; Caswell et al., 1980). The nitrogen abundance in the filaments may be enhanced (Dopita et al., 1977; Ruiz, 1983; Leibowitz and Danziger, 1983). The distance given by the Σ–D relation is 8–9 kpc; the kinematic distance is about 3.2 kpc (Caswell et al., 1975). The Hα/Hβ line ratio in the bright filaments yields $A_V = 4.5^m$ and a distance of 6.6 kpc (Ruiz, 1983; Leibowitz and Danziger, 1983); this latter estimate would seem to be the most reliable. The mean velocities of two bright filaments out at the periphery are –70 km/sec and +10 km/sec (van den Bergh et al., 1973), and the FWHM of the Hα line from the nebula suggests gas motions spanning a velocity range $\Delta v = 690$ km/sec (Danziger et al., 1978).

RCW 86 (MSH 14–63). A system of external faint filaments probably associated with the central, bright, thin-filament nebula. According to Leibowitz and Danziger (1983), the Balmer decrement of the filaments yields $A_V = 1.7^m$ and a distance of 3.2 kpc; Ruiz (1981) finds $A_V = 0.8^m$ and 1 kpc. The first pair of values seems the more reliable, agreeing as it does with the Σ–D relation. The remnant seems to be the product of SN 185 (van den Bergh et al., 1973; Pizarski et al., 1984), a point, however, that is considered largely debatable. Adopting a distance $r = 1$ kpc, the x-ray luminosity is $L_{0.2-4\,keV} = 2 \times 10^{34}$ erg/sec, which according to Pizarski et al. (1984) is at least an order of magnitude below the majority of supernova remnants. At $r = 3$ kpc, the age of the remnant deduced from its linear size and an expansion velocity $v_s \sim 10^3$ km/sec, corresponding to $T_s = 1.4 \times 10^7$ K, is ~8000 yr, assuming adiabatic expansion. Of course, we might suppose that the explosion of SN 185 took place in a tenuous medium, with $n_0 = 10^{-2} - 10^3$ cm^{-3}, and that the remnant is still expanding unimpeded with its initial velocity of $v_0 = 20$ pc/1800 yr = 11,000 km/sec, but it would then not be clear how the system of bright, thin filaments typical of an old remnant in a dense medium could have evolved. The velocity of the filaments has been measured directly only over the nebula as a whole, and seems to cover a range $\Delta v = 400$ km/sec (Danziger et al., 1978).

G 290.1–0.8 (MSH 11–61A). Thus far, we only have measurements of the FWHM of the Hα line in the bright filaments, $\Delta v = 90$ km/sec, to give us a sense of the kinematics of the shell (Elliott and Malin, 1979). The neighboring nebula MSH 11–61B is probably an H II region rather than a supernova remnant: the brightness in [S II] is low, $I_{H\alpha}/I_{[S\,II]} = 10-20$, and H 109$\alpha$ recombination radiation has been detected (Elliott and Malin, 1979; Winkler, 1979).

Lupus Loop. An extended shell about 6° across; the physically unrelated remnant of SN 1006 is seen in projection against this object. The shell is most easily identified in the radio; it is very faint optically (van den Bergh, 1976), and the kinematics of the nebula have not been investigated. Two concentric H I shells have been detected at 21 cm (Colomb and Dubner, 1982). The outer of the two is about 84 pc in radius (at a distance of 500 pc), is expanding at 12 km/sec, and may have been produced by the wind of a massive progenitor (the required wind power, $L_w = 10^{36}$ erg/sec, and duration of outflow, $t = 4 \times 10^6$ yr (see Eqs. (13.5)) are consistent with the mass loss endured by O stars). The inner shell, with a radius of 32 pc and an expansion velocity of 23 km/sec in a wind-blown bubble with a mean density $n_0 0.13 \, cm^{-3}$, can be identified as the supernova remnant.

W 44. A radio shell 27′ in size (Dickel et al., 1976). A distance of 3 kpc has been found from absorption in the lines of OH and H I (Goss et al., 1971; Myers, 1973). A broad OH absorption feature at $v_{LSR} = 42$ km/sec is probably associated with a shocked cloud. An outer H I shell expanding at 4 km/sec has also been detected (Cornett and Hardee, 1975). The nebula has not been detected optically, due to strong absorption. A dense molecular cloud abutting the remnant on the east may be genetically associated with it (Wooten, 1977; DeNoyer, 1983).

We have included W 44 in Table 11, since the x-ray observations make it possible to use it as a basis for constructing an evolutionary sequence of supernova remnants. According to Watson et al. (1983a) and Smith et al. (1985b), the interior x-ray brightness is elevated, and the spectrum corresponds to $T_e \sim (1 - 1.2) \times 10^7$ K.

8. Thermal x-ray, optical, and infrared radiation from supernova remnants: shock waves in the inhomogeneous interstellar medium

There are at least six observed components to the complicated panoply of x-ray phenomena accompanying the explosion of a supernova, and it would be difficult to exaggerate the importance of any one of them (we have already made reference in Section 6 to the x rays emitted by a collapsed stellar remnant).

1. Given a favorable orientation, one observes pulsed, nonthermal x-rays emerging from the magnetosphere of a pulsar. Pulsars have been found in the Crab Nebula, Vela X (where there are no pulsed x rays), MSH 15–52, CTB 80, and 0540–69.3.

2. A stellar remnant that is a neutron star or black hole in a close binary system will radiate x rays by accreting matter from its "normal" companion; this is observed in the W 50/SS 433 system (probably a black hole) and in the CTB 109/IE 2259+586 system (probably a neutron star).

3. The hot surface of a neutron star may be observed as a compact thermal x-ray source, regardless of orientation; see Table 9.

4. A pulsar injecting relativistic particles will be associated with extended synchrotron x-ray sources like those in 0540–69.3, the Crab Nebula, and 3C 58 (Section 4). The synchrotron nature of the radiation should be confirmed by its spectrum and linear polarization. Extended nonthermal sources are also observed around the compact x-ray sources in the older objects MSH 15–52, Vela X, and CTB 80. Extended x-ray sources with brightness increasing toward the center have been detected in the remnants G 21.5–0.9, G 291.0–0.1, G 74.9+1.2, and G 29.7–0.3, but no compact sources are visible, probably because of inadequate angular resolution in the x-ray telescope (Becker, 1983; Wilson, 1986).

It is worthwhile to examine x-ray synchrotron emission due to relativistic particles injected by a pulsar in conjunction with an analysis of plerion radio emission, which is produced by the same mechanism (see Section 10). Here we shall dwell specifically on thermal radiation, which provides the most information about the interaction of the supernova shell with the interstellar medium. Two thermal radiation components are observed:

5. Radiation from interstellar gas (or the wind of the progenitor) that is swept up and heated by the blast wave engendered by the expansion of the supernova shell.

6. Radiation from stellar matter expelled by the explosion and heated by the reverse shock wave; see Fig. 47.

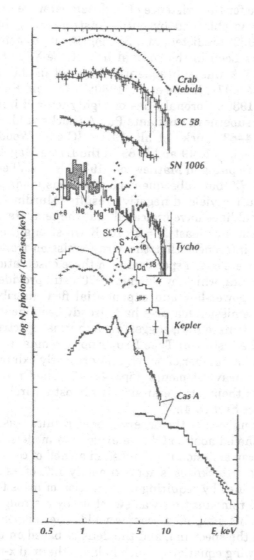

Fig. 44. X-ray spectra from the young remnants of Tycho's supernova, Kepler's supernova, SN 1006 (see Section 3), from the Crab Nebula and 3C 58 (Section 9), and from Cas A (Section 5). The spectrum of Tycho's remnant is compared with the emission from an equilibrium plasma of normal chemical composition at a temperature of $kT_e = 0.5 \, \text{keV}$. The most important heavy-element lines are indicated. The calculated spectrum for a two-temperature equilibrium plasma is shown above the spectrum of Kepler's remnant (Holt et al., 1983).

The first evidence for the existence of high-temperature plasma in supernova remnants was obtained by optical astronomers. In the early 1960's, Shklovskii (1962) predicted, and Shcheglov (1966) detected, emission from the Cygnus Loop in the coronal line of [Fe X] at 6374 Å. The [Fe XIV] line at 5303 Å was subsequently detected in the same object (Woodgate et al., 1974, 1977; Lucke et al., 1980; Teske and Kirshner,1985; Ballet et al., 1988a, 1989). Coronal lines of highly ionized iron were also found in the Galactic supernova remnants Pup A (Lucke et al., 1979; Clark et al., 1979), MSH 14–63 (Lucke et al., 1979), IC 443 (Woodgate et al., 1979), and in the remnants N 49 and N 69 in the Large Magellanic Cloud (Murdin et al., 1978; Dopita and Mathewson, 1979). [Fe XIV] emission was reported from Vela XYZ, but subsequent observations made with higher sensitivity and resolution yielded negative results (Murdin et al., 1978). Observations of coronal lines have indicated that the temperature in these fairly old objects reaches at least $T_e = 2 \times 10^6$ K (in strong [Fe XIV]-emitting regions). Initial inferences that the x-ray emission from supernova remnants is thermal were based specifically on these observations, and not at all on the x-ray spectra, which by the mid-1970's still provided no way to distinguish between power-law and exponential flux distributions. (By 1973, however, x-ray emission features had already been identified in Cas A with lines of highly-ionized iron, suggesting a thermal spectrum.)

The spectra of the historical Type I supernova remnants (Section 3), Cas A (Section 5), and a number of old objects presently exhibit numerous lines of highly ionized heavy elements (Figs. 44, 45). The Crab Nebula and 3C 58 are exceptions: their x-ray emission is almost entirely due to synchrotron radiation (see Section 4).

The first crude analysis of the observational results was carried out within the scope of the adiabatic and free expansion models, with the assumption that the x-ray emission is produced in a shell of constant density and temperature whose thickness is approximately 12% of its radius. (The thickness was determined by requiring conservation of mass for the interstellar gas inside the remnant that was swept up by a strong shock wave without radiating.) This simplified scheme yields the temperature of the hot plasma based on the spectrum, and the density based on the luminosity, through the following equations, which hold for thermal x-ray emission from the gas behind the front of a collisionless adiabatic shock wave:

$$\epsilon(T,E) = p(T,E)n_e n_p = 2.3 \times 10^{-23} T^{-1/2} e^{-E/kT} g(E,T) n_e n_p \ [\mathrm{erg \cdot cm^{-3} \cdot sec^{-1} \cdot eV^{-1}}];$$

$$T_s = 3\mu m_H v_s^2/16k; \quad n_s/n_0 \le (\gamma+1)/(\gamma-1) \approx 4; \tag{8.1}$$

$$L(T,E) = 4\pi R_s^2 \Delta R_s n_s^2 p(T,E) e^{-\sigma(E)N_H}.$$

Here ϵ, p, and L are the x-ray emissivity, cooling efficiency and luminosity in the energy range from E to $E + \Delta E$, respectively; n_e and n_p are the electron and proton number densities; $g(E,T)$ is the Gaunt factor; the term $e^{-\sigma(E)N_H}$ takes absorption into account; N_H is the number of hydrogen atoms in the line of sight; γ is the ratio of specific heats (the adiabatic index); T_s is the temperature and n_s the plasma density behind the shock front; v_s is the shock front velocity; n_0 is the unperturbed gas density; μ is the mean atomic mass; R_s and ΔR_s are the radius and thickness of the shell.

The assumption of constant gas temperature and density behind the shock front is too crude, even for relatively old shells. In the adiabatic stage (see Section 9), the temperature just behind the shock front rises with distance from the front as $T \propto R_s^{4.3}$, and the density varies as $n \propto R_s^{-9}$. Allowing for the contribution made by radiation from different layers, it is straightforward to integrate Eqs. (8.1) and obtain the mean effective temperature $T_x \approx (1.3 - 1.4)T_s$, and the luminosity

$$L_x \approx 16 n_0^2 R_s^3 \epsilon(T_s, \Delta E) \eta(T_s), \qquad (8.2)$$

where $\eta(T_s)$ is a dimensionless function that takes on values from 0.43 to 0.67 for $T_s = 8 \times 10^5 - 5 \times 10^7$ K and $\Delta E = 0.15 - 2$ keV (Rappaport et al., 1974).

The most important departures from this highly idealized scheme for young supernova remnants are as follows. First, the time scale for the ion and electron temperature of the gas behind the shock front to reach equilibrium,

$$t_{(i,e)} \approx 500 A_i T_e^{3/2} / n_e Z_i^2 \ln \Lambda \approx 3 \times 10^3 \text{ yr},$$

where A_i is the ion mass in atomic units, $\ln \Lambda \approx 30$ for $T_e = 10^7 - 10^8$ K, and $n_e \approx 10^2$ cm^{-3} (see Spitzer (1981)), may exceed the age of the remnant, resulting in plasma that is markedly nonisothermal. In fact, taking Cas A as an example, we have demonstrated that the ion temperature $T_i = 5 \times 10^8$ K corresponding to a remnant expansion velocity of ~ 6000 km/sec (the latter value has been directly measured from the Doppler shift of the lines of highly ionized elements (see Section 5), so it reveals the motion of hot plasma) is much higher than the electron temperature $T_e = (5-7) \times 10^7$ K determined from the x-ray spectrum; see also Itoh (1984).

Second, the physical conditions in young remnants are governed by the reverse shock wave propagating through the expanding ejecta, a matter touched upon repeatedly in Chapter 1. Effects of the reverse shock are substantial over the first $10^2 - 10^3$ yr following the explosion, up until the mass of swept-up interstellar gas significantly exceeds the mass of the ejecta. A thin, dense shell will form in initially uniform ejected matter under the influence of the reverse shock. The relevant dynamics and conditions for heating and subsequent radiative cooling have been analyzed in detail by Hamilton and Sarazin (1984a, b), Hamilton et al (1985), and Hamilton et al. (1986a, b). Ejecta heated by the reverse shock and circumstellar gas heated by the outgoing blast wave radiate x rays copiously, and as we have seen in Chapter 1, the observations conform quite nicely to the strictures laid down by these assumptions. The x-ray spectrum of all shell-type historical supernova remnants can be represented as radiation from an optically thin two-temperature plasma. The low-temperature plasma ($kT \sim 0.5$ keV), which is enriched in heavy elements (particularly Si and S), and which is necessary to explain the bright emission lines in the spectrum, is associated with radiation from the ejecta heated by the reverse shock; the high-temperature ($kT \sim 4 - 5$ keV) plasma, which is required to explain the continuous spectrum, is associated with radiation from the swept-up gas heated by the blast wave; see Fig. 47.

The supposition that there are both direct and reverse shocks, which heat the circumstellar gas and ejecta respectively, is not only supported by the spectrum, but by the structure of the x-ray images as well. We have already seen in Sections 3 and 5 that the emission from the ejecta and swept-up circumstellar gas in Cas A and Tycho's remnant comes from spatially distinct regions. Mass estimates for the two components of the hot plasma, $M_{sw} \sim M_{ej}$, are consistent with the assumption that these two remnants are just starting to decelerate as they interact with circumstellar gas. Small-scale knots in the x-ray images of the shells, just like the patchy structure of the optical nebulae, suggest that this scheme of things is overly simplified, inasmuch as it fails to take inhomogeneity of the ejecta and swept-up gas into consideration.

(Here, in passing, we note that if a strong shock wave propagates through a medium that is only partially ionized, as is probably the case in remnants that resemble Tycho's remnant (see Section 3), the two-component structure of the x-ray spectrum may in principle be explained differently. Collisional ionization of neutral atoms gives rise to secondary electrons of lower energy than the particles accelerated at the shock front. Itoh (1984) has calculated the temperature relaxation process occurring between ions, electrons accelerated at the front, and secondary cold electrons, and has demonstrated that when $T_e \sim 3 \times 10^8$ K at the front, which is consistent with observed radiation in the range up to 25 keV, the secondary electrons turn out to be an order of magnitude cooler. At the same time, the volume emission measure of this low-temperature component is found to be four to five times smaller than what is actually observed, and it is radiation from ejecta heated by the reverse shock that makes the dominant contribution.)

The third difference is that the hot plasma in young remnants has not yet reached ionization equilibrium. Calculations of the x-ray emissivity of plasma in ionization equilibrium have been carried out for a wide range of temperature and density encompassing the values actually encountered in supernova remnants (see Shapiro and Moore (1976), Raymond and Smith (1977), Shull (1981), Kaplan and Pikel'ner (1979), plus included references). As we concluded in Chapter 1, when these are applied to the spectra of young remnants they lead to substantial departures (frequently more than an order of magnitude) of the heavy-element abundances from solar values and to an overestimate of the plasma mass.

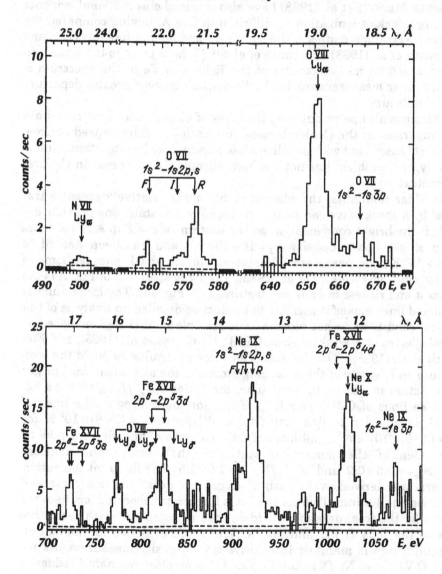

Fig. 45. X-ray spectrum of Pup A (Winkler et al., 1981), showing the brightest lines.

Anomalies in the chemical composition of young remnants are significantly ameliorated if we renounce the requirement of ionization equilibrium. That step turns out to be justified. In fact, the time needed to establish collisional ionization equilibrium of hydrogen- and helium-like ions in a tenuous plasma at $T_e \sim 10^7$ K is some hundreds to thousands of years (see the calculations of Itoh (1979), for example), which is commensurate with (or possibly greater than) the age of the historical supernova remnants. Markert et al. (1988) have also obtained observational evidence implying a lack of ionization equilibrium in Cas A, having compared the relative spectral intensities of the hydrogen- and helium-like lines of S and Ne. Jansen et al. (1988) and Smith et al. (1987) have come to the same conclusion based on measurements of the K lines of Fe in the spectrum of Cas A; similar measurements for Tycho bespeak an even greater departure from equilibrium.

We must also point out that the sizes of young supernova remnants are comparable to the Coulomb mean free path in a fully ionized plasma, i.e., effects associated with a collisionless plasma may be important. Unfortunately, this problem has not yet been adequately addressed in the present context.

In older remnants, the plasma in filaments relatively recently traversed by a shock wave can be in a nonequilibrium state; one possible example is the bright condensation on the eastern side of Pup A. That dense clump of gas has a density $n_0 = 10 - 20$ cm^{-3} and has been heated to $T_e \gtrsim 6 \times 10^6$ K; the parameter characterizing the shock passage time is $n_e t \sim 10^3$ cm$^{-3} \cdot$ yr (Winkler et al., 1983). The x-ray spectrum of Pup A is the brightest and richest in emission features (see Fig. 45). The large number of isolated lines makes it possible to conduct as detailed an analysis of the physical conditions in hot plasma as is possible in H II regions based on optical spectra (see Winkler et al. (1983), Canizares et al. (1983), and Fischbach et al. (1988)). The line ratios for some particular ion yield the temperature and density of the absorbing atoms; in the aforementioned bright condensation in Pup A, in particular, the ratios $I_{666\,eV}/I_{574\,eV}$ for the helium-like lines of O VII and $I_{817\,eV}/I_{654\,eV}$ for the hydrogen-like lines of O VIII, in addition to line ratios for Fe VII, give $T_e = (3-8) \times 10^6$ K for $N_H = (2-6) \times 10^{21}$ cm^{-2}. The lines of different ions of one element give us some idea of the ionization state; in that same condensation, $N_{O^{+6}}/N_{O^{+7}} = 0.5 \pm 0.2$ and $N_{Ne^{+8}}/N_{Ne^{+9}} = 2.0 \pm 1.5$. The different ionization temperatures derived on the basis of oxygen and neon lines imply a lack of ionization equilibrium. This conclusion is further corroborated for the x-ray plasma by the weakness of forbidden lines as compared with resonance lines of the ions O VII and Ne IX. The same evidence for ionization nonequilibrium is manifest in relative line-intensity measurements for O VII, O VIII, and Ne IX in the Cygnus Loop, an older remnant (Vedder et al., 1986).

(In much older remnants, the nonequilibrium ionization state may be due to the fact that the cooling time at $T_e \sim 10^6$ K — $t_c \sim 3nkT/L$, where L

is the cooling rate — turns out to be less than the recombination time $t_r \approx (\alpha_r n_e)^{-1}$ for the tenuous gas. Calculations by Suchkov and Shchekinov (1984) suggest that the relative abundances of C IV, N V, and O VI ions in a cooling gas in the temperature range $T_e = 10^4 - 10^5$ K are practically temperature-independent, and are far from equilibrium, but that temperature range is irrelevant to an analysis of x-ray emission.)

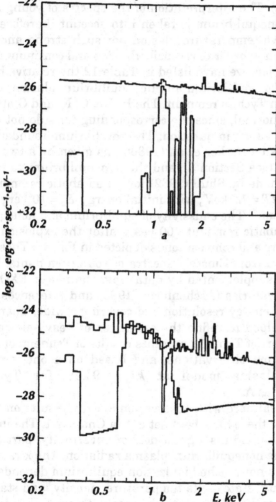

Fig. 46. The emissivity $p(T, E)$ of plasma radiation from a supernova remnant in the adiabatic stage with $kT_e = 7.2$ keV, $E_{51} = 1$, $n_0 = 1\,\mathrm{cm}^{-3}$, $t = 680\,\mathrm{yr}$, for the relative heavy-element abundances given in Table 12: **a)** total emissivity, and emissivity in iron lines for a plasma in ionization equilibrium, and **b)** for a nonequilibrium plasma (Shull, 1982).

Mass and heavy-element abundance estimates are drastically altered when one takes into account possible deviations from ionization equilibrium. Plasma that has not yet reached ionization equilibrium radiates in the lines of helium-like ions tens of times more strongly than it does in the equilibrium state (see the calculations of Gronenschild and Mewe (1982), Shull (1982), Hamilton and Sarazin (1984b), Nugent et al. (1984), and references throughout). The interpretation of the spectra of young remnants when ionization nonequilibrium is taken into account therefore often requires neither a two-temperature plasma nor such strong anomalies in chemical composition in order to reconcile the line and continuum emissivities. As an illustration, we have listed in Table 12 the relative heavy-element abundances for equilibrium and nonequilibrium plasma (compared with solar values) in Tycho's remnant. The H, He, C, N, and O abundances are assumed to be normal, since the corresponding lines do not appear in the spectral energy range in question. The equilibrium abundance comes from measurements by Becker et al. (1980b), as given by a two-temperature plasma model (see Section 3), and the nonequilibrium value derives from calculations made by Shull (1982) for an adiabatic remnant at uniform temperature $kT = 7.2\,\text{keV}$, with initial energy $E_0 = 10^{51}$ erg and ambient density $n = 1\,\text{cm}^{-3}$. The emissivity of equilibrium and nonequilibrium plasmas in an adiabatic remnant 700 years after the explosion with the indicated parameters and composition is depicted in Fig. 46. These results, which were obtained from Einstein spectra of young remnants, have now been substantially supplemented by data from the newer EXOSAT and Tenma x-ray observatories (Aschenbach (1988) and references therein). Spectra with better energy resolution and covering a wider energy range, $2-10\,\text{keV}$, have helped to refine the estimates of heavy-element abundance. The lower part of Table 12 presents results of Tsunemi et al. (1986) obtained from Tenma observations, and based on a one-temperature nonequilibrium plasma model at $kT_e = 2.9\,\text{keV}$ for Tycho and $kT_e = 3.76\,\text{keV}$ for Cas A.

We shall not deal here with observational x-ray results on the other historical remnants; these have been set out in Chapter 1. The interpretation of these data still entails a great deal of uncertainty, particularly in the calculation of the nonequilibrium plasma radiation. The x-ray luminosity of a gas that has not reached ionization equilibrium depends not only on its "instantaneous" state — its temperature, density, and stage of ionization — but also on how that state changes with time. It is important, therefore, that an adiabatic model was chosen for the foregoing calculations, one which may not yet be applicable to the historical supernovae; that the reverse shock was not taken into consideration; and that normal light-element abundances were assumed, while in Cas A and similar remnants the ejecta may consist mainly of oxygen and products of oxygen burning. Taking all this into account, it must be borne in mind that the estimates presented in Chapter 1 for the chemical composition of the shell ejected in the explosion may be in error by a factor of two or three.

Table 12. Heavy-Element Abundances in Tycho's Remnant and Cassiopeia A Derived from Einstein and Tenma Spectra

	Ne	Mg	Si	S	Ar	Ca	Fe
Einstein							
Tycho, ionization equilibrium	0.1	0.1	6.0	13.5	34.6	76.0	0.15
Tycho, ionization nonequilibrium	0.4	2.0	7.6	6.5	3.2	2.6	2.1
Tenma							
Tycho, ionization nonequilibrium			6.7^{+12}_{-2}	11^{+12}_{-4}	$6.5^{+3.3}_{-2.6}$	15^{+10}_{-6}	6.0 ± 0.2
Cas A, ionization nonequilibrium			$1.5^{+0.5}_{-0.5}$	$1.4^{+0.4}_{-0.2}$	$0.8^{+0.1}_{-0.1}$	$2.1^{+0.2}_{-0.1}$	$0.7^{+0.6}_{-0.2}$

However, the progress achieved in the last few years, both observationally and in terms of theoretical modeling of spectra, is obvious. Consideration of nonequilibrium ionization states has made it possible to reconcile mass and chemical composition estimates for the ejecta with the most up-to-date ideas concerning the supernova explosion phenomenon. The nonequilibrium spectra of most young remnants in the 0.1–5 keV range can be approximated by radiation from a one-temperature plasma. This is true both for calculations based on the Sedov solution and for numerical hydrodynamic calculations incorporating temporal variation of the ionization state; see Hughes and Helfand (1986). But wider-band EXOSAT spectra suggest that it is necessary to consider a two-temperature plasma in a nonequilibrium ionization state (Smith et al., 1987; Smith, 1988a; Jansen et al., 1988; Aschenbach, 1988). After analyzing Einstein spectra of Cas A, Markert et al. (1988) came to the same conclusion. It is natural to compare a two-temperature ionization nonequilibrium plasma with the circumstellar gas heated by the outgoing blast wave, and with the ejecta, which are heated by the reverse shock. We have already stated in Section 3 that a two-temperature nonequilibrium model that takes into account the redistribution of density and ionization state in an initially uniform, expanding ejecta (see Hamilton et al. (1986a, c)) best approximates the spectra of young type Ia supernova remnants derived from measurements made by Einstein, EXOSAT, and Tenma.

The indicated uncertainty in the interpretation of the x-ray observations pertains to the earliest stage of expansion of a remnant, when the actual parameters of the explosion are still important. In old remnants that have already "forgotten" the individual properties of the explosion and are comprised entirely of swept-up interstellar gas, analysis of the observational facts becomes simpler. The thermal x-ray emission from old remnants is governed by the interaction of the shock wave induced by the expansion of the shell with interstellar gas, and only through advances in x-ray astronomy has it become possible to correctly perceive this process. Prior to the mid-1970's, analysis of the evolution of supernova remnants was based on the kinematics of optical nebulae, and it was most natural to associate the expansion velocity of a system of optical filaments with the motion of the gas behind the shock front. By a curious coincidence, the age of the remnant IC 443 which was then determined from the linear size and the expansion velocity of the optical filaments, $v = 65 - 100$ km/sec, turned out to be exactly the same as the age of the apparently neighboring pulsar PSR 0611+22! This was taken to be a thoroughgoing confirmation of the adiabatic model for the expansion of old remnants, and there was no doubt that the pulsar and IC 443 were directly related.

But as the x-ray data piled up, our ideas changed radically. The very first observations of old remnants, including the Cygnus Loop and IC 443 (see Section 7), demonstrated that the shock velocity required to heat a plasma to the temperature inferred from the x-ray spectrum was four to five times the expansion velocity of the brightest optical filaments. This discrepancy provided the main impetus for the development of a theory of propagation of the shock wave induced by the explosion of a supernova in an interstellar medium with pronounced density fluctuations.

A number of observational factors, which are spelled out in Section 7, attest to the need to make allowance for the inhomogeneity of the interstellar gas when analyzing conditions in old remnants. The clumpy or filamentary structure of the optical nebulae and the large differences (by a factor of 2–3) in velocity of individual filaments inside a remnant can only be explained by variations in the efficiency of slowing and by differences in the radiative cooling of gas behind the shock wave. With the Cygnus Loop, IC 443, and other nebulae in the same class as examples, we have seen that the optical filaments ($T_e \sim 10^4$ K, $n_e \sim 10^2 - 5 \times 10^2$ cm^{-3}) coincide spatially with regions emitting coronal lines and a soft x-ray spectrum ($T_e = (2-10) \times 10^6$ K, $n_e \sim 0.1 - 1$ cm^{-3}). In addition, dense cold condensations radiating lines of H I, CO, H_2, and other molecular species ($T = 5 - 50$ K, $n_H \sim 10^2 - 10^3$ cm^{-3}) have been detected in those same objects. The coexistence of gaseous components with such markedly different physical properties within a single remnant can only be understood in the context of an inhomogeneous interstellar medium with sharp density contrasts.

The propagation of the shock wave induced by an explosion in a medium with small-scale clouds has been examined by Bychkov and Pikel'ner (1975), McKee and Cowie (1975), and Sgro (1975). According to the model that emerged, the x-ray emission from an old remnant is due to hot gas behind the fast shock front in the intercloud medium, while the optical emission comes from shock-heated gas in dense clouds. Figure 47 gives a schematic representation of a supernova remnant expanding into a cloudy ambient gas. When the strong shock wave labeled 1 — which is propagating at a velocity v_s through the tenuous intercloud medium — collides with a dense cloud $(n_{0cl} \gg n_{0i})$, the abrupt rise in pressure at the cloud boundary gives rise to two secondary shock waves. Wave 3 moves at velocity v_{cl} with respect to the dense gas of the cloud, and reflected wave 4 moves in the reverse direction through the hot intercloud gas. The pressure in the gas layer between them is higher than behind the front of wave 1, so the reflected wave rapidly departs from the cloud boundary. The cloud is anisotropically compressed (its interior boundary is subject to both thermal and dynamic pressure, its outer boundary to thermal pressure alone), accelerated (the cloud velocity remains below the velocity of the intercloud gas behind front 1), and engulfed in hot gas. The assumption of pressure equilibrium between the hot intercloud gas and the shocked gas in the cloud yields the simple relation

$$v_{cl}^2 n_{0cl} = \beta v_s^2 n_{0i}, \qquad (8.3)$$

where β is a dimensionless parameter of order unity. For a strong shock hitting a rigid plane, $\beta = 1, 2.5, 4.4, 6$ for density ratios $n_{0cl}/n_{0i} = 1, 10, 100, \infty$ respectively (Zel'dovich and Raizer, 1966). In the case of a spherical cloud, the reflected wave 4 is produced only in the vicinity of a head-on collision and is more rapidly dissipated. According to McKee and Cowie (1975), the limiting value for a spherical cloud is $\beta = 3.15$. As the reflected shock wave is dissipated, the cloud pressure comes to equilibrium with the pressure in the intercloud medium.

In clouds whose density is higher than some critical value n_{cr}, the gas behind front 3 cools quickly to $T \sim 10^4$ K via radiative losses, and the pressure P_{cl} of the cold gas in the cloud cannot equilibrate with the intercloud pressure P_i (Sgro, 1975). In these clouds, a thin intermediate low-pressure (P_*) layer comes into being, with higher luminosity than the cloud itself. In these dense, bright clouds, $P_{cl}/P_* \approx 3$ and $P_i/P_* \approx 10$, i.e., we typically have $P_i \approx (2-3)P_{cl}$. Thus, without separating clouds into "cold" ones, in which a layer at $T_e \sim 10^4$ K is formed as wave 3 passes, and "hot" ones, which cool radiatively only after the collapse of the shock wave, we expect pressure equality (8.3) to hold only approximately, to within a factor of 2–3. Sgro (1975) finds the typical value of the critical density by equating the time t_{cool} for radiative cooling of the cloud to the shock traversal time t:

$$\frac{t_{cool}}{t} = \frac{5 \times 10^{-6} (n_{cl}/n_i)^{-3} \beta^2 v_s^4 n_i^{-1}}{10^6 a/v_{cl}} = 1, \qquad (8.4)$$

where a is the size of the cloud. The resulting value of the critical density is then

$$n_{cr} \approx 6 \times 10^{-4} \beta^{5/7} v_s^{10/7} n_i^{5/7} a^{-2/7}. \qquad (8.5)$$

For the purposes of this estimate, we have assumed that radiative losses conform to the curve in Fig. 52, and that the remnant is in the adiabatic expansion stage. Since $t_{cool}/t \propto n_{cl}^{-3.5} v_s^5$, the changeover from $t_{cool}/t \gg 1$

to $t_{cool}/t \ll 1$ takes place quite abruptly, and the remnant ought to host two cloud populations: "cold" clouds, which cool quickly during the passage of wave 3 and radiate low-excitation optical lines, and "hot" clouds, which emit coronal lines and soft x-rays. (Naturally, there is a layer in the cold clouds with $T_e \approx 10^6$ K, but no region in the hot clouds with $T_e \approx 10^4$ K.) The hot clouds are cooler and denser than the surrounding intercloud gas, and should be observable as regions of enhanced x-ray brightness. As cooling progresses — the typical time scale being $t_{cool} \approx 7 \times 10^4 \, T_6^2/n_{cl}$ years (where $T_6 = T_e \times 10^{-6}$ K) — a hot cloud will come more and more to resemble a cold one (Sgro, 1975).

The gas in clouds is accelerated to the velocity v_{cl} during the initial ("shock") stage of development. Also important for the dynamics of clouds interacting with a supernova remnant, there may be a "post-shock" phase of smooth acceleration by the flow of hot shocked intercloud material past the clouds (McKee et al., 1978). This smooth post-shock acceleration is accompanied by the rapid evaporation of the clouds embedded in the hot gas,

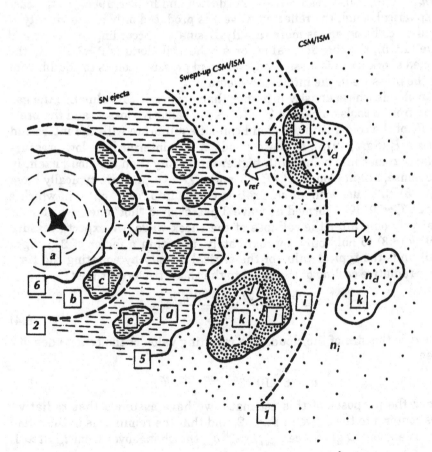

which, as we shall see in Section 9, can affect the evolution of the remnant as a whole. Low-mass clouds shocked by wave 1 at an early stage, when the radius of the remnant was 30–70% of its present value, can be accelerated by the flow of hot gas to a velocity approaching the present value v_s.

The acceleration of clouds is accompanied by the onset of dynamical boundary instabilities, primarily Rayleigh–Taylor. Perturbations grow with time, the boundary of the cloud becomes convoluted, and it eventually undergoes stratification, compression, and fragmentation (Woodward, 1976; Tenorio-Tagle and Ròzycka, 1986, 1987).

In broadest outline, we have just sketched out the interaction between the shock wave induced by the expansion of a supernova shell and a cloudy interstellar medium. The detailed hydrodynamic features of this interaction can neither be described analytically nor derived by mathematical

Fig. 47. Sketch of a young supernova remnant (with or without a pulsar) in a cloudy ambient gas (interstellar or circumstellar), based on all available observations and our best theoretical understanding. We have labeled the following: 1) blast wave propagating into the ambient gas at velocity v_s; 2) reverse shock propagating at velocity v_r into the expanding supernova ejecta; 3) secondary shock wave moving into a dense cloud with velocity v_{cl}; 4) shock reflected into tenuous intercloud gas; 5) contact discontinuity between swept-up gas and supernova ejecta. If the supernova gives rise to a pulsar, then we also have 6) a contact discontinuity between the supernova ejecta and a cloud of relativistic particles and magnetic field produced by the pulsar.

There are a number of observational manifestations of young supernova remnants in different spectral ranges. If an energetic pulsar is produced, then region (a) represents a so-called *plerion*, a source of synchrotron radio, optical, and x-ray emission which increases in brightness toward the center. Other layers also exist, with or without a pulsar. We observe thermal x-ray emission from tenuous components of the swept-up gas heated by the blast wave (i) and from the ejecta heated by the reverse shock (d); optical emission from dense clumps of ejecta (e) heated by the reverse shock, and from dense ambient clouds heated by the secondary shock wave (j). Convective amplification of the magnetic field at the contact discontinuity (5) and acceleration of relativistic electrons by the blast wave generate synchrotron radio emission with shell-like morphology; ejected and swept-up interstellar dust collisionally heated by hot x-ray plasma is a source of infrared emission. Cold portions of the ejecta as yet undisturbed by the reverse shock (b and c) and undisturbed gas in the ambient clouds (k) may be unobservable.

modeling, as either would require a large number of unknown parameters to be taken into account. Likewise, modern observations do not reveal the intricate features of this interaction, but they do enable one to analyze the overall picture of gas acceleration within supernova shells.

Let us turn now to the soft x-ray and coronal-line observational data on old remnants, which when compared with the corresponding data taken at optical and radio wavelengths provides convincing evidence of the applicability of this model to real objects. The thermal x-ray emission from a remnant is due to gas behind shock front 1 in the intercloud medium, so the outermost boundary of the x-ray shell defines the location of the front. Even when x-ray maps of the Cygnus Loop were being produced by deconvolving one-dimensional low-resolution scans, it could be seen that the hot plasma region extends 5–10' beyond the bright optical filaments. Present-day x-ray images, with angular resolution of 4", have completely confirmed this fact. The outermost boundary of the x-ray shell in the Cygnus Loop — a weak but perfectly well-defined spherical front — coincides with faint, delicate optical filaments that primarily emit the Balmer lines of hydrogen (see Fig. 31). Elsewhere, the x-ray and optical images concur only in their grossest features, rather than corresponding in detail; this is understandable, as the emission is produced by gas at different temperatures and densities. The x-ray spectrum implies a temperature $T_s \approx 2 \times 10^6$ K behind the front. The spectrum is harder near the center and in the faint, diffuse clouds, but temperature variations remain in the range $(2-4.3) \times 10^6$ K; the density of the unperturbed intercloud gas is $n_{0i} = 0.16 \, \text{cm}^{-3}$, the velocity of the shock front is $v_s \approx 400 \, \text{km/sec}$, and the total mass of gas is approximately $10^2 \, M_\odot$ (Ku et al., 1984; Charles et al., 1985). Coronal-line emission from iron can also be detected beyond the bright optical nebula (Teske and Kirshner, 1985; Ballet et al., 1988a, 1989). The ratio of the [Fe XIV] line to the continuum in the 0.4–4 keV range increases monotonically inward along the radius of the shell, in complete agreement with the temperature variation of the gas behind the front of an adiabatic shock wave (Tuohy et al., 1979c). The observed enhancement of the brightness of the x-ray continuum and the coronal lines near the optical filaments may be due to the evaporation of clouds (Charles et al., 1985). The small-scale cellular structure of the hot clouds is quite noticeable in the [Fe X] line: one sees clouds 30–60" across, with denser condensations some 10" in size inside (Teske and Kirshner, 1985) (see also Section 7).

The brightness distribution of IC 443 in x rays (Petre et al., 1983; Petre et al., 1988; Watson et al., 1983a) and coronal lines (Woodgate et al., 1979; Brown et al., 1988) is consistent with the model developed in Section 7 for a supernova explosion at the juncture of two molecular clouds. The complete picture of IC 443's x-ray emission provided by Petre et al. (1988), using four of the x-ray instruments on Einstein and HEAO 1, differs considerably from that of remnants of comparable age, for example the Cygnus Loop and Vela XYZ. The unusual x-ray morphology (the bright northeast spot instead of a shell), low surface brightness with little correlation between x-ray and optical/radio features, and pronounced spectral variations across the remnant are easily accounted for by the interaction of the remnant shell with the dense H I northeast cloud and a molecular cloud in front of the shell. According to Petre et al. (1988), the bulk of the remnant is at a temperature of about 1.2×10^7 K, with a relatively uniform inner density distribution and hardening of the spectrum due to absorption, which is attributable to increasing column density in the foreground cloud. The spectrum of the bright x-ray spot can be closely fitted either by a two-temperature plasma (2.2×10^6 K and 1.1×10^7 K) in ionization equilibrium, or by a nonequilibrium plasma with a temperature of 1.8×10^7 K. The [Fe X] emission coming from the strongest x-ray emission region shows up as discrete knots about 0.1 pc in size, with a small filling factor (Brown et al., 1988). The assumption of pressure equilibrium between x-ray and [Fe X] knots leads one to conclude that the brightest x-ray emission comes from a thin, relatively high-density shell which is not in pressure equilibrium with the rest of the remnant (Petre et al., 1988).

The 0.1–4 keV x-ray image of Pup A in Fig. 48 also suggests a large-scale density gradient in the interstellar gas, with the density typically varying by a factor of 4 over a distance of ~30 pc (Petre et al., 1983). Against a background of smooth, large-scale variations in x-ray brightness,

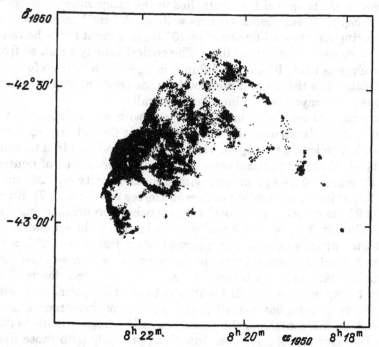

Fig. 48. X-ray image of Puppis A at 0.1–4 keV (Petre et al., 1982).

we see individual small condensations — hot clouds, in our present terminology. Among the latter is the brightest eastern condensation, whose spectrum we have referred to above. In that condensation, we also see a stratification of the radiation from hot and cold gas, confirming its identification with a dense cloud that has recently collided with a shock wave and been engulfed by the hot gas behind the front. The regions emitting lines of [Fe XIV] ($T_e \sim 2 \times 10^6$ K), [O III] ($T_e \sim 5 \times 10^4$ K), and [N I] ($T_e < 10^4$ K) are displaced successively farther and farther away from the boundary of the condensation and toward the center (Clark et al., 1979). This is the kind of structure — with a hot corona surrounding a cool, dense core — that ought to be observed in the evaporation of a dense clump engulfed in hot gas. A comparison of high-quality images of the eastern boundary of Pup A in the lines of [Fe X] and [Fe XIV] with x-ray images taken by the HRI aboard the Einstein observatory have made it possible to track the variation in ionization temperature in a thin layer of gas behind the shock front (Teske and Petre, 1987).

X-ray observations of old supernova remnants yield the shock-wave propagation velocity v_s in the intercloud medium and the density n_{oi} of the intercloud gas; these data have been summarized in Table 11, together with results of optical observations of the cloud component of the remnants. It can be seen from the table that for most of the bright nebulae, the unperturbed gas density in the clouds, as determined from the spectrum and the velocity of the optical filaments, lies in the range $n_{0cl} \sim 1 - 10 \, \text{cm}^{-3}$. According to Eq. (8.4), less dense regions with $n_{0cl} \lesssim 1 \, \text{cm}^{-3}$ cannot cool to $T_e \sim 10^4$ K during a remnant lifetime of $t \sim 10^4$ yr, and ought not to be radiating bright, low-excitation optical lines. The critical density obtained from (8.5) for the Cygnus Loop, IC 443, and Pup A is $n_{cr} \sim 50 \, \text{cm}^{-3}$. Therefore, we find in particular that the densest hot clouds in the remnant Pup A radiate simultaneously in x rays, coronal lines, and optically.

The very densest clumps of interstellar gas, with $n_{0cl} \gtrsim 100 \, \text{cm}^{-3}$, are compressed and accelerated by the shock wave, but the velocity v_{cl} turns out to be too low to ionize hydrogen. These clouds are visible in a number of remnants — IC 443 in particular — as dense condensations of neutral hydrogen; they are $\sim 0.2 - 1$ pc across, with a mean density $n_H \sim 200 \, \text{cm}^{-3}$ and 21-cm linewidth corresponding to $\Delta v \sim 50$ km/sec (see Section 7). Knots radiating at 21 cm coincide with local radio synchrotron brightening. We shall see in Section 10 that synchrotron emission from old remnants results from the enhancement of the magnetic field frozen into the compressed gas behind the shock front. In the stage at which we now find IC 443, this is most likely gas behind the shock front in the clouds. This would seem to explain the virtual identity between the optical and radio images of the remnants; but not all radio synchrotron brightening coincides with optical filaments, and the dense neutral hydrogen cloudlets discovered by DeNoyer (1978) have been identified precisely with those from which there is no bright Hα emission. A similar situation is to be expected within the framework considered here.

Theory gives us no way to predict *a priori* whether a cloud as a whole will acquire some bulk motion in a collision with a shock wave, rather than simply having internal random gas velocities increased. The end result depends on the velocity and shape of the cloud, on how nonuniform the density is, on attenuation or focusing of the shock wave in the cloud, and so on. It is difficult to take quantitative account of asymmetries in the velocity distribution within a cloud, and observations provide us with our only means of treating the collision of the supernova shell with real clouds.

Studies of the kinematics of old optical supernova remnants have shown that the mean expansion velocity of bright filaments is close to the mean gas velocity within the filaments themselves, as inferred from linewidths (see Section 7 and Lozinskaya (1980a, b)). This is what we see in the stationary condensations of the young remnant Cas A as well (Section 5). The upshot is that both the bulk and random gas velocities in a cloud increase to the same extent upon interaction with a shock wave.

The velocity v_{cl} typical of the shock wave in the clouds can be derived from the observations in two ways: using the Doppler linewidth of the filaments, or using their expansion velocity (from the center of a remnant). Since we do not know the shock wave and filament geometry, which would be necessary for a proper determination of v_{cl} in an individual filament, the observations only enable us to evaluate the mean velocity for a particular remnant. This analysis has been carried out for most of the old supernova remnants (Lozinskaya (1980a), Danziger et al. (1978), and others), and the results are given in Table 11.

Coronal-line and soft x-ray observations have made it possible to ascertain the pressure behind the shock front in the intercloud medium and in the "hot" clouds, based on the temperature and density of the radiating plasma. Likewise, based on the temperature (or the velocity) and density of the gas in the bright optical filaments, one can find the gas pressure behind the radiative shock in the clouds. There are a number of remnants in which dense molecular condensations with large velocity dispersion or anomalous chemical composition have been detected, seemingly indicative of collisions with the shock wave engendered by the expansion of the supernova shell. These observations provide us with the opportunity to estimate the pressure inside the very densest clouds interacting with the shock wave. Of course the density determined here is less reliable, as the molecular abundances relative to hydrogen behind the shock front may be anomalous, and the amount of gas compression varies over a very wide range. But estimates of the gas pressure in the best-studied galactic remnants that have been obtained in this way for the various gas components indicate, as predicted by theory, that the condition (8.3) holds to within a factor of 2–3, a result that remains valid over widely varying densities ranging from 0.1 to 10^2 cm^{-3}. Errors in the determination of density and temperature (or velocity) yield approximately the same accuracy in the pressure estimate, which suggests that given the present state of theory and experiment, the foregoing model of the interaction of a shock wave with the cloudy interstellar medium is consistent with observations of old supernova remnants.

Nevertheless, one must realize that this consistency is to be found only in the most idealized models. One look at a photograph of the Cygnus Loop is enough to show even the most inexperienced reader the large discrepancy between an actual old supernova remnant and the scheme discussed above (fast shock wave in the intercloud medium + radiative shock wave in the clouds), even when the remnant is the very one that serves as the Procrustean bed for all theoretical models. The bright optical filaments of the Cygnus Loop, IC 443, Vela, and S 147 do not even vaguely resemble the emitting regions of a shock wave encountering small dense clouds: the filaments are comparable in length to the size of the remnant, and they are preferentially convex outward, rather than inward. Furthermore, the geometry of the bright radiative filaments of the Cygnus Loop resembles that of the delicate outer filaments, which fix the location of the thin layer of gas at the nonradiative shock front. That is why the model advanced by Hester (1987) seems attractive: in that model, the optical emission from old remnants is concentrated in curved sheets which serve as tracers of the region of radiative cooling of a shock wave in a relatively dense region of the interstellar medium. The enhanced coronal-line and x-ray brightness in the same region of the shell then becomes understandable. However, that brightening can also be explained by an alternative model incorporating dense clouds and intercloud gas: x-ray and coronal-line enhancement may result from an increase in the density of hot gas due to the evaporation of dense clouds or bow-shock effects (the situation is depicted in Fig. 47).

To digress from the geometry of the optical filaments, their radiation mechanism is also comprehensible only in the broadest terms, the uncertainty here being associated with the need to take an enormous number of ill-determined parameters into account. A detailed calculation of the optical spectra of radiative shocks requires a knowledge of the velocity and geometry of the shock wave, the density distribution and ionization state of the surrounding gas (allowing also for pre-ionization by the shock wave), magnetic pressure, and the chemical composition of the interstellar medium, allowing for the destruction of dust grains by the shock wave. A sensible interpretation of the observations also requires that one take into consideration departures from ionization equilibrium, a nonisothermal state, and the presence of an incomplete recombination zone and an incomplete cooling zone — i.e., departures from steady flow conditions behind the shock front.

Many authors have calculated the emission from gas heated by a shock wave propagating at 50–200 km/sec, a typical velocity for the filaments of old supernova remnants; for example, see Pikel'ner (1954), Cox (1972a), Raymond et al. (1976), Raymond (1979), Shull and McKee (1979), Kaplan and Pikel'ner (1979), Dopita (1977), Dopita et al. (1984b), Contini et al. (1980), Balinskaya and Bychkov (1979, 1981), Binette et al. (1985), Cox and Raymond (1985), Bychkov and Fedorova (1987), and Innes et al.

(1987). We have repeatedly relied upon a number of remnants to confirm that the shock velocity deduced by comparing the observed and calculated relative intensities in the filament spectra is consistent with direct measurements of the expansion velocity of the system of filaments (see Section 7). On the whole, the "fine structure" of the nonstationary cooling and incomplete recombination zone behind the front is in accord with the observations, a matter that has been thoroughly examined in recent papers, particularly the works of Fesen et al. (1982), Raymond et al. (1988), and Bychkov and Fedorova (1987).

The latter two authors calculated the structure of the temperature-relaxation and nonstationary cooling zones for nonstationary ionization of a two-temperature atomic/ionic plasma, with due allowance for energy exchange between the electron and atomic/ionic components, photo- and collisional ionization, dielectron and radiative recombination, charge exchange, and electron impact excitation of both ionic and atomic discrete levels. The velocity range considered was $40-200$ km/sec. Their results indicated that the hydrogen lines consist of two components, collisional and recombination; the former exhibits a sharp peak at a distance of approximately 3×10^{-3} pc from the front, with a relatively broad ($\sim 3 \times 10^{-2}$ pc), weaker pedestal. The recombination component is farther removed from the front, being about 0.1 pc away for a velocity of ~ 90 km/sec and a degree of ionization $x \sim 0.5$. The structure of the forbidden-line emission reflects that of the state of ionization: [O III] lines form within comparatively narrow, hot zones near the front, where rapid cooling takes place — approximately where one observes the Hα collisional line. Lines of neutral and once-ionized atoms are emitted much farther away from the front, together with the recombination component of the Balmer lines.

For a pre-shock gas density $n_0 = 5-10$ cm^{-3}, the stationary spectrum is completely formed after $(5-8) \times 10^3$ yr, which is commensurate with the age of the Cygnus Loop and IC 443. The time for the shock wave to propagate through the filaments of these remnants is shorter still. What this means, then, is that the optical spectra of old remnants are characterized by incomplete recombination and cooling (Fesen et al., 1982). The variations in relative line intensities and the stratification of radiation detected in the Cygnus Loop, IC 443, and other old remnants (see Section 7) reflect the fact that filaments at different distances from the front are in different stages in the formation of the equilibrium emission spectrum. The "theoretical" stratification of the post-shock radiation within a single dense cloud, as we already remarked in Section 7, is of order 10^{-3} pc for a displacement of [O III] relative to [N II] and [S II], which lies below the limit of resolvability for most optical supernova remnants. Raymond et al. (1988) have demonstrated quite good agreement between profiles of the brightness, velocity, and relative optical and UV line intensity distributions and theoretical predictions for one of the "favorably" oriented isolated filaments of the Cygnus Loop, revealing convincing signs of incomplete recombination in the cooling zone. We emphasize that this work does not reconstruct the interaction of the fast shock wave with the many filaments at various distances from the front, but it is the first highly detailed treatment of the stratification of the radiation emanating from behind a radiative shock front in a single cloud.

The emission from the gas behind the shock front in the intercloud medium is observed not only in x rays and coronal lines, but in optical lines as well. We discussed in Section 7 the faint outer "nonradiative" filaments of the Cygnus Loop, whose spectrum is dominated by the hydrogen Balmer lines. The widths of Hα and Hβ in these filaments correspond to 140 and 240 km/sec. Similar delicate filaments with a Balmer spectrum have also been detected at the outer boundary of the x-ray shells of the young Tycho and SN 1006 remnants (see Section 3). The second phenomenon which is probably related to the radiation of gas behind the fast shock wave in the intercloud gas is the high-velocity wings of the Hα lines, emitted by diffuse interfilament gas in the Cygnus Loop, IC 443, and HB 9 (see Section 7). In IC 443, faint high-velocity emission has been detected coming from outside the bright filaments, but from inside the radio and x-ray images. The high-velocity wings display a more or less flat-topped pedestal; their width corresponds to a velocity approaching the value of v_s given by the x-ray spectrum.

In principle, these high-velocity wings of optical lines may be associated with the smooth post-shock acceleration of clouds by the flow of shocked hot intercloud material past the clouds. The energy source for this emission working over a typical acceleration time of $10^2 - 10^4$ yr may be ionizing photons produced in a conductive interface, which penetrate into the clouds. Because of partial ionization, the [O I] 6300 Å line should be prominent in the high-velocity gas emission; the spectrum should then also display lines of [N II], [O II], and [O III] (McKee et al., 1978). This, however, is a mechanism that has not been rigorously confirmed observationally. In fact, calculations suggest that the cloud acceleration time is comparable to the age of the remnant, and most of the cloud's mass ought to be evaporated in that time, so that the high-velocity clouds cannot be any larger than the slow, bright filaments that have recently collided with the shock wave. On the other hand, observations of the Cygnus Loop, IC 443, and HB 9 indicate that the emission region for high-velocity gas is 5–10 times as large as the region occupied by the bright filaments. Moreover, it is difficult to understand, within the framework of a smooth post-shock acceleration model, how the delicate filaments at the periphery of the Cygnus Loop and Tycho's remnant might have arisen.

A fundamentally different explanation for the faint, high-velocity Hα emission has been advanced by Bychkov and Lebedev (1979), Chevalier and Raymond (1978), and Raymond et al. (1980). They propose that it stems from the shock-induced excitation of neutral hydrogen atoms traversing the front of the collisionless shock wave 1, and that this excitation takes place prior to ionization. Now, in order to compare the ionization and collision excitation times for atoms crossing front 1, one must have an understanding of how the electron and ion temperatures vary behind the front, know the ionization state of the ambient gas (i.e., be able to take possible pre-ionization by the shock wave and ionization by ultraviolet radiation from the supernova into account), and allow for charge exchange effects between neutral atoms and fast ions — all of these being highly uncertain factors that are difficult to fold into the overall picture. Nevertheless, a rough estimate suggests that under conditions typical of supernova remnants, an average of $0.2 - 0.3$ Hα photons arrives at each neutral atom prior to ionization. The brightness per unit area of the shock wave is only a weak function of velocity, and is comparable to the observed value. Charge exchange with fast ions then gives rise to high-velocity neutral atoms, and due to collisional excitation the latter emit a broad line corresponding to the velocity v_s of the fast shock wave in the intercloud gas. There is no question that the morphology of the delicate outermost filaments of the Cygnus Loop and Tycho's remnant are more in agreement with this mechanism. The spectrum to be expected differs from that coming from post-shock accelerated clouds that are ionized by radiation from the surrounding hot gas. For the collisional excitation spectrum of partially neutral ambient gas of a normal chemical composition, the lines of He II should be approximately 100 times weaker than the Hα and Hβ lines; those of [O III] and [N II] ought to be one to two orders of magnitude fainter still. But at both UV and visible wavelengths in the aforementioned outer filaments of the Cygnus Loop, one sees the weak He II line at 1640 Å and the N V and C IV lines at 1240 and 1550 Å, respectively, as well as lines due to O II, O III, Ne V, [N II], and [S II], all of which are much brighter than predicted — an indication that the model needs to be modified, and in particular that it requires a rigorous analysis of the conditions necessary for equilibration of the electron and ion temperatures (Raymond et al., 1983; Fesen and Itoh, 1985). It is also conceivable that a significant contribution to the forbidden-line radiation may come from the very densest isolated clumps of gas, in which the shock wave — which is all but nonradiative — actually begins to cool radiatively (Fesen and Itoh, 1985). The patchy structure of [S II]- and [N II]-emitting regions in the faint, outer "nonradiative" filaments would seem to favor this view.

If the model proposed by Hester is correct — that the diffuse emission from interfilament gas represents sections of sheets that are oriented perpendicular to the line of sight, and we are seeing several different layers in projection — then the broader linewidth of Hα in the interfilament medium than within the filaments may result from the vagaries of geometry, rather than from differing physical conditions.

In any event, however, one purely empirical fact has emerged: the shock velocity in the intercloud medium may be determined not just from the x-ray spectrum of the hot plasma, but also by measuring the high-velocity wings of the hydrogen lines radiated by the optical nebulosity. The shock wave velocities inferred from the highest-velocity wings of the Hα

line in old remnants are also presented in Table 11, and these can be seen to be commensurate with the velocities derived from the x-ray spectra.

We would venture to suggest that the study of supernova remnants has now taken a third revolutionary leap forward. The first came with the dawn of radio astronomy, and the second with the launch of the Einstein x-ray observatory. The launch of IRAS, a dedicated infrared observatory (Neugebauer et al., 1984), has opened up yet another promising and information-packed new field of study. Firstly, we can now widely observe the emission coming from a heretofore unstudied component of a remnant — the shock-heated dust. Secondly, the cooling of plasma at $T_e \gtrsim 10^6 - 10^7$ K via radiation from dust dominates that due to atomic processes, or in other words, energy losses due to infrared radiation from dust constitute the main energy sink for the shock wave.

IRAS observations have now encompassed 70 galactic supernova remnants: IR emission has been detected from 16, no relation has been established between the remnant and the observed IR radiation in 24, and no emission has been detected above the background in the remaining 30 (Arendt, 1988). Of the nine supernova remnants observed in the Large Magellanic Cloud, four have been detected in the infrared (Graham et al., 1987b).

We have already made reference to IRAS data in discussing specific supernova remnants. Here we shall examine the overall properties of the infrared emission, based on the data of Dwek (1987, 1988), Dwek et al. (1987), Braun (1987), and Graham et al. (1987a, b).

The four bands covered by IRAS (12, 25, 60, and 100 μm) capture radiation from the following constituents of the interstellar and interplanetary medium: cold dust (dominant at 100 μm), zodiacal light (dominant at 12 – 25 μm), dust heated directly by UV stellar radiation within H II regions and by the background UV — the so-called "IR cirrus" — (25 – 60 μm), and dust heated by inelastic collisions with hot-plasma electrons and ions (25 – 60 μm). A certain contribution also comes in the form of line radiation from S I, S II, S III, S IV, Fe II, Fe III, O I, O II, Si I, and Ne II, among others; galactic and extragalactic point sources contribute as well, as does infrared synchrotron emission.

The principal IR emission mechanism in virtually all supernova remnants is thermal emission from shock-heated dust. Many supernovae erupt in OB associations, however, so that remnants tend to lie near OB stars (examples include IC 443 and the Monoceros Loop). Furthermore, the fast shock wave is itself a source of ultraviolet radiation. Radiative heating might therefore be substantial in certain remnants; in particular, it dominates all other sources in the remnant N 186D in the Large Magellanic Cloud (Graham et al., 1987b). The IR emission from remnants can come from shock-heated interstellar dust, from dust that has condensed in the metal-rich supernova ejecta, or from dust that has condensed in the wind of the progenitor — more specifically, in the atmosphere of the supergiant.

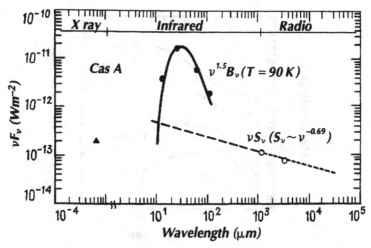

Fig. 49. Power spectrum radiated by Cas A at x rays and in the infrared and radio (Dwek, 1987a). Points denote IRAS results; the solid curve has been plotted for a temperature of 90K.

Methods for extracting the IR emission of the remnant from the galactic background and discriminating between the radiatively and collisionally heated components have been considered widely — e.g., see Braun (1987), Braun and Strom (1986a, b), and Dwek (1987). The IR synchrotron contribution can be easily identified by extrapolating the radio synchrotron emission.

The Crab Nebula is the only remnant in which IR emission does not dominate the emission at other wavelengths, and in which the IR tends to be synchrotron radiation. The latter property was inferred by extrapolating the synchrotron flux to millimeter wavelengths (Mezger et al., 1986). Infrared emission dominates in all other remnants, both young and old. At wavelengths detectable by IRAS, the old object IC 443 emits approximately 10^{37} erg/sec, while it gives off about 10^{33} erg/sec in the radio (10 MHz – 10 GHz), and 10^{35} erg/sec at x rays (0.1–4 keV) (Mufson et al., 1987). In Cas A, a young remnant, we see a similar picture (Dwek, 1988); see Fig. 49. Arendt (1988) has found that young and old remnants have different IR spectra; see Fig. 50.

The fact that dust heated by collisions with hot plasma is the source of IR emission in all remnants except the Crab is most convincingly affirmed by the correlation between the x-ray and IR maps of individual remnants (see Figs. 31a, b, c). Further evidence comes from energy estimates. A rigorous quantitative analysis of the IRAS data is a highly complicated matter. Besides the uncertainties associated with nonequilibrium of the plasma behind the shock front and the inhomogeneity of the interstellar medium, which make the interpretation of the x-ray and optical observations quite difficult, the IR bands introduce their own peculiar problems.

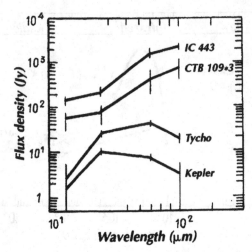

Fig. 50. IRAS infrared spectrum of young (Tycho, Kepler) and old (IC 443, CTB 109) remnants (Arendt, 1987).

The IR emissivity of dusty plasma depends on the temperature, density, chemical composition, and size of dust grains, and though the temperature can easily be determined from the IR spectrum (using $F(\nu) \propto \lambda^{-1}$), as can the density of the dust, by assuming that the average dust/gas ratio is constant throughout the Galaxy, the size distribution and chemical composition of the dust grains are altered by the shock wave. The same inelastic collision process between fast electrons and ions that heats the dust also tends to induce sputtering and sublimation of the grains, changing both their size distribution and composition; they may also become electrically charged, enabling them, among other things, to interact with the interstellar magnetic field. Dwek (1987) has thoroughly analyzed the shock-induced heating of dust and subsequent radiative cooling, as applied to supernova remnants. He has obtained a cooling function for dusty plasma containing silicate/graphitic grains spanning a range of sizes, as a function of the temperature of the hot plasma that is collisionally heating the dust. It turns out that in the temperature range $T_e \gg 10^6 - 5 \times 10^6$ K, plasma cooling mediated by collisions with dust and IR emission from dust proceeds more efficiently than cooling by means of radiation from gas undergoing free–free, free–bound, and bound–bound transitions.

This is illustrated by Fig. 51, where the dashed curve shows the ratio of the dust-emission cooling function to the gas-emission cooling function over a range of temperature. If the source of infrared emission from supernova remnants is the collisional heating of dust by hot plasma, then the observed ratio of infrared flux detected by IRAS (F_{IR}) to the x-ray flux ($F_{0.2-4\,keV}$) ought to correspond to the theoretical ratio of the cooling functions for dust and gas radiation over the corresponding energy range. In Fig. 51 (Dwek et al., 1987), we see the theoretical temperature dependence of the ratios $IR/X \equiv \Lambda_d/\Lambda_{0.2-4\,keV}$ and $IR/X \equiv \Lambda_d/\Lambda$, where Λ_d is the cooling

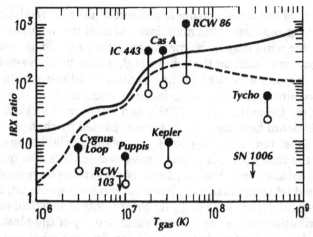

Fig. 51. Theoretical ratio of the radiative cooling functions of dust and gas, Λ_d/Λ, for a dusty nonequilibrium plasma (dashed curve), and the same ratio, $\Lambda_d/\Lambda_{0.2-4\,keV}$, for gas emission in the 0.2–4 keV band (solid curve). Open circles show the observed flux ratio between the IRAS and Einstein bands, $F(IR)/F(0.2-4\,keV)$, for a number of supernova remnants; filled circles represent the same ratio, but with ionization nonequilibrium of the x-ray plasma taken into account (Dwek et al., 1987).

function associated with gas–dust collisions and radiation from dust, Λ is the cooling function for radiation from gas, and $\Lambda_{0.2-4\,keV}$ is the cooling function for gas emitting in the range 0.2–4 keV. Open circles denote observed flux ratios $F_{IR}/F_{0.2-4\,keV}$ (filled circles show values with ionization nonequilibrium of the x-ray plasma taken into account).

The observational data plotted in Fig. 51 attest to the fact that the IR luminosity of all galactic remnants (with the possible exception of SN 1006 and RCW 103) greatly surpasses their x-ray luminosity. In the LMC, Graham et al. (1987b) observe $IR/X = 12$ for N 63A and N 49, and $IR/X = 4$ for N 49B; in N 186D, $IR/X = 2100$, suggesting that radiative heating of dust predominates there. For the Magellanic Clouds, where the metallicity is half what it is in our own Galaxy and the relative dust abundance is one-quarter the Galactic value, the theoretical ratio is $IR/X = \Lambda_d/\Lambda_{0.2-4\,keV} \approx 15$ at $T_e = 6 \times 10^6$ K. These crude estimates serve to indicate that collisional heating of dust by the hot post-shock plasma may be the principal energy source for the IR emission from supernova remnants at wavelengths observable by IRAS. Deviations of the observed IR/X flux ratio from its theoretical value are due to the oversimplification inherent in the numerical model. For computational purposes, it has been assumed that the dust and x-ray plasma share the same region of space, and that the dust is not depleted by the shock wave. Moreover, we decided above that ionization nonequilibrium of the x-ray plasma is difficult to take correctly into consideration. In young remnants, the abundance of dust in the supernova ejecta or the progenitor wind may lie below the galactic mean; in old remnants located in molecular clouds, on the other hand, it may be above-average. It is quite difficult to properly take IR line radiation falling within the IRAS bands into account, although there is some evidence that it may be substantial; e.g., see Dinerstein et al. (1987) and Graham et al. (1987a).

Since collisional heating of dust is accompanied by changes in the size, chemical composition, and charge of the grains, calculations of the IR emissivity of the dusty plasma in supernova remnants is also quite complicated and uncertain. Nevertheless, IRAS observations have enabled us to determine the dust temperature T_d from the infrared spectrum, and knowing the theoretical dependence of T_d for grains with a specified composition and size distribution on the temperature and density of the plasma responsible for collisionally heating the dust, we can find out what the properties of the hot plasma are. These estimates are most reliable for young remnants, whose IR spectrum may be satisfactorily approximated by radiation from dust at uniform temperature throughout (Dwek, 1987, 1988). This approach has been used for Cas A, Tycho's remnant, and Kepler's remnant to find the mass of gas behind the blast wave front ($1.2 - 1.5 M_\odot$) and the mass of the ejecta heated by the reverse shock ($0.3 - 0.4 M_\odot$ at $T_d = 85 - 100$ K) (Braun, 1987).

We see, then, that plasma diagnostics for young remnants based on their thermal x-ray and IR emission yield consistent results, notwithstanding the difficulty entailed in interpreting specific observations.

9. The evolution of supernova remnants

The large-scale dynamics of an old remnant is governed by the propagation of a shock wave in the intercloud medium (shock front I in Fig. 47). In other words, if the gas has been concentrated into moderately small, dense clouds separated by large intercloud distances, so that the initial collisional deformation of the shock front is small compared with the shell radius and energy losses in the cloud component may be neglected, then the clouds will exert a negligible influence on the dynamics of the remnant.

We therefore begin by considering the evolution of a remnant in a homogeneous intercloud medium; then, following the treatment by McKee and Ostriker (1977), we allow for the evaporation of the clouds into hot gas within the remnant, increasing the density and thereby altering the course of evolution. To conclude, we examine the results of a numerical analysis of remnant evolution in a medium with small-scale cloud structure, as well as the first attempts to take dust-mediated cooling into account.

Numerical methods for analyzing the propagation of shock waves induced by a supernova explosion, which make due allowance for radiative cooling, magnetic fields, thermal conductivity, and evaporation, have demonstrated on the one hand that the process is exceedingly complex, and that we are far from a comprehensive understanding. On the other, it has become quite clear that the overall development of a remnant can be broken down into several idealized stages, which to a first approximation can be satisfactorily described by simple self-similar solutions.

The initial phase of evolution may be characterized by the free expansion of the ejected shell at a velocity $v_0 = (5-10) \times 10^3$ km/sec, with essentially no deceleration. In this phase, the initial explosion energy E_0 is almost all manifested by kinetic energy of the ejected matter; thermal energy comes to 2 or 3% of E_0. Slowdown of the ejecta effectively begins when the circumstellar gas that is swept up by the expanding shell attains a mass M_0 equal to that of the ejecta. The radius and age of the remnant at that point are $R_s = (3M_0/4\pi\mu m_H n_0)^{1/3} \approx 2$ pc and $t \approx R_s v_0^{-1} \approx 180$ yr in a medium with mean density $n_0 = 1\,\mathrm{cm}^{-3}$, and with $M_0 = 1\,M_\odot$, $v_0 = 10^4$ km/sec.

Deceleration of a supernova shell in the interstellar medium was first considered by Oort (1946). Starting with the expression for momentum conservation,

$$\left(M_0 + \frac{4}{3}\pi R_s^3 \rho_0\right)v_s = M_0 v_0,$$

Oort derived the relations that describe, in a completely satisfactory manner, the evolution of a significantly decelerated shell ($4\pi R^3 \rho_0/3 \gg M_0$):

$$R_s = \left(\frac{3M_0 v_0}{\pi \rho_0}\right)^{1/4} t^{1/4} ; \quad v_s = \frac{3M_0 v_0}{4\pi \rho_0 R_s^3}; \quad R_s = 4v_s t. \tag{9.1}$$

From here on, we use the following notation: $\rho_0 = n_0 \mu m_H$, R_s is the radius and v_s the velocity of the shock front, T_s is the ion temperature of the shocked gas behind the front, and $\gamma = c_P/c_V$ is the ratio of specific heats.

The next important step was taken by Shklovskii (1962), who proved that the supernova explosion phenomenon in the interstellar medium could be likened to a powerful point explosion in a gas with constant heat capacity, and that the self-similar Sedov (1957, 1981) solution (see also Taylor (1950)) was applicable to this problem; this solution had been validated in nuclear explosions in the earth's atmosphere. The self-similar solution describes the evolution of a remnant at the adiabatic stage, when, on the one hand, the mass of swept-up interstellar gas is several times that of the ejecta, and on the other, radiative energy losses are still negligible by comparison with the initial energy E_0. At that stage, the motion is described by the equations (Shklovskii, 1962, 1976a)

$$R_s = (2.02 E_0/\rho_0)^{0.2} t_{(sec)}^{0.4} (cm) = 0.34 (E_{51}/\mu n_0)^{0.2} t_{(yr)}^{0.4} (pc),$$

$$T_s = 2.27 \times 10^{-9} \mu v_s^2 = 1.5 \times 10^{10} (E_{51}/n_0) R_{(pc)}^{-3} (K), \tag{9.2}$$

$$v_s = 0.4 R_s/t, \quad E_{51} \equiv E_0 \times 10^{-51} (erg).$$

Numerical calculations by Chevalier (1974) have shown that in the adiabatic stage, about 70% of the initial energy of the ejecta has been transformed into thermal energy of the swept-up interstellar gas: $E_T = \epsilon E_0$, $\epsilon = 0.72$.

The remnant continues to expand adiabatically up to the moment at which radiative cooling begins in earnest, when the gas temperature behind the front reaches the value that corresponds to the maximum on the radiative loss curve, $T_* \sim (5-6) \times 10^5$ K (see Fig. 52). The age t_{cool}, radius R_{cool}, and expansion velocity v_{cool} of a supernova remnant entering into the intense radiative cooling stage are, according to Falle (1981) (see also Falle (1987))

$$t_{cool} = 2.7 \times 10^4 E_{51}^{0.24} n_0^{-0.52} (yr),$$

$$R_{cool} = 20 E_{51}^{0.295} n_0^{-0.409} (pc), \tag{9.3}$$

$$v_{cool} = 280 E_{51}^{0.055} n_0^{0.111} \text{ (km/sec).}$$

Estimates made by different workers differ somewhat, which is basically a result of the differing radiative loss curves that are assumed: Chevalier (1974) gives $R_{cool} = 19 E_{51}^{0.29} n_0^{-0.41}$, while according to Cox (1972b), $R_{cool} = 25 E_{51}^{0.29} n_0^{-0.41}$; see also Blinnikov et al. (1982). By the time a remnant has arrived at this stage, approximately $0.5 E_T$ has been radiated away, and a cold, dense shell has been formed, containing about half the mass of the swept-up gas. The development of a shell takes place on the unstable part of the radiative loss curve, and is therefore essentially a very rapid avalanche process. The cavity bounded by the thin, cold shell contains hot, low-density gas that continues to expand adiabatically. The evolution of the remnant following the formation of the cold shell is described quite well by the "snowplow" model (Cox, 1972b). The law of motion governing the shell at that stage can be derived from the adiabatic expansion condition imposed on the interior hot gas,

$$\frac{dE_T}{dt} = -4\pi R_s^2 P \frac{dR}{dt}; \quad \frac{4}{3}\pi R_s^3 P = (\gamma - 1)E_T, \tag{9.4}$$

and the equations for the mass and momentum of the shell:

$$M = \frac{4}{3}\pi R_s^3 \rho_0, \quad \frac{d(Mv_s)}{dt} = 4\pi R_s^2 P, \quad \frac{dR}{dt} = v_s. \tag{9.5}$$

Here P is the pressure of the hot gas, the thickness of the shell is assumed to be small compared with its radius, and the mass of the hot gas is assumed to be much less than that of the shell. The system of equations (9.4), (9.5) has the solution (McKee and Ostriker, 1977; Blinnikov et al., 1982)

$$R_s = 38(\epsilon E_{51})^{5/21} n_0^{-5/21} \left(\frac{t}{10^5 \text{ yr}}\right)^{2/7} \text{ (pc);} \quad v_s = \frac{2}{7}\left(\frac{R_s}{t}\right), \tag{9.6}$$

where $\epsilon = E_T/E_0 = 0.2 - 0.35$.

A comparison of this result with (9.1) reveals that as simple as they are to derive, Oort's laws of motion for old remnants are quite accurate. Chevalier (1974) obtained a similar expansion law for the shell numerically, $R_s \propto t^{0.31}$. According to McKee and Ostriker (1977), expansion continues until the gas pressure inside the remnant is balanced by the pressure

P_0 of the ambient interstellar gas. This pressure-balance requirement then yields the maximum radius of the remnant,

$$R_{max} = 55E_{51}^{0.32}n_0^{-0.16}\tilde{P}_{04}^{-0.20} \text{ (pc)}, \tag{9.7}$$

which is attained at time

$$t(R_{max}) = 8.3 \times 10^5 \, E_{51}^{0.31}n_0^{0.27}\tilde{P}_{04}^{-0.64} \text{ (yr)},$$

where $\tilde{P}_{04} = 10^{-4}\,P_0/k$. The maximum lifetime of a supernova remnant before it completely dissipates in the interstellar medium is (McKee and Ostriker, 1977)

$$t_{max} = R_{max}\left(\frac{P_0}{\rho_0}\right)^{-1/2} = 7 \times 10^6 \, E_{51}^{0.32}n_0^{0.34}\tilde{P}_{04}^{-0.70} \text{ (yr)}. \tag{9.8}$$

There have been a number of numerical analyses of the interaction of a supernova with the interstellar medium, and these span a wide range of initial conditions (see Chevalier (1974), Mansfield and Salpeter (1974), Falle (1975a, 1981), Chieze and Lazareff (1981), and Cioffi et al. (1988), among others). The numerical methods employed have revealed a number of intriguing features inaccessible to self-similar solutions, not the least of which is the process whereby a thin, dense shell is formed via radiative cooling of the gas behind the front. Figure 53 shows the behavior of gas density, temperature, and velocity obtained by Mansfield and Salpeter

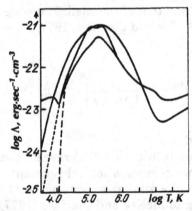

Fig. 52. Radiative losses calculated by various authors (see Falle (1975a)).

(1974), plotted as a function of distance from the center of the explosion for various epochs. The calculations were carried out for the standard model ($M_0 = 1 M_\odot$, $E_0 = 3 \times 10^{50}$ erg, $n_0 = 1$ cm^{-3}) with no magnetic field; below a temperature of 10^3 K, cooling was neglected. Taking dust-mediated radiative cooling into account for $T \leq 10^3$ K makes the shell denser and colder.

The curves corresponding to the earliest stage ($t = 10^3$ yr, $M(t) = 10 M_\odot$) show the reverse shock quite well: a sharp interior maximum in the density, coinciding with a temperature minimum and velocity directed toward the center of the explosion. The onset of the reverse shock is associated with radiative cooling and the "collapse" of the matter expelled by the explosion. At time $t = 3 \times 10^4$ yr, as can be seen here, there is already a cold, dense shell that contains roughly 30% of the overall mass of

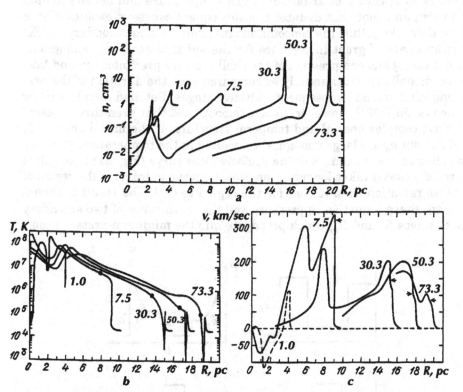

Fig. 53. Density (a), temperature (b), and velocity (c) as a function of distance from the center in supernova remnants of different ages (Mansfield and Salpeter, 1974). The curves are labeled by the remnant age in thousands of years; points on the curves in panel **b** show the mean temperature, and arrows in panel **c** give the mean velocity of the shell. The dashed curve in panel **c** is the velocity at 1/10 scale.

the remnant. When the age of the remnant approaches $t = 7 \times 10^4$ yr, the mass of the shell reaches $10^3 M_\odot$, its thickness is approximately 10^{-4} of the radius, and 60 to 70% of the initial energy E_0 has been radiated away. The shell has been squeezed between two layers of copiously emitting gas: the inner one is a layer of hot gas that is at the interface with the cold shell, from whence comes more than 70% of the total energy radiated during the formation of the shell at $t \approx 3 \times 10^4$ yr, and the outer one is a layer composed of swept-up interstellar gas heated by the shock wave, which accounts for approximately 80% of the energy radiated at the later stage $t \approx (7-8) \times 10^4$ yr (Mansfield and Salpeter, 1974).

The moment at which the transition from adiabatic expansion to the formation of a cold shell takes place is of special interest. It is at that point when the greatest part of the energy of the remnant is radiated away; there is an abrupt redistribution of gas temperature and density behind the front, an event that dictates the subsequent course of evolution. It is more than likely that the concomitant instabilities and secondary shocks that arise are of great importance for the enhancement of the magnetic field. Cooling and compression of the shell increase precipitously, and take place virtually instantaneously as compared with the duration of the preceding adiabatic and subsequent radiative stages. Detailed calculations by Falle (1975a, 1981), however, have demonstrated that even this "instant" displays complex spatial and temporal structure. If the initial energy E_0 and density n_0 are large enough, then cooling at the temperature T_* corresponding to the maximum of the radiative loss curve (Fig. 52) takes place so rapidly that it takes longer for an acoustic wave to traverse the region of intense radiation than it does for the region to cool. The result is then a sudden, sharp reduction in pressure and the appearance of two secondary shock waves S_2 and S_3, which propagate into the minimum-pressure zone

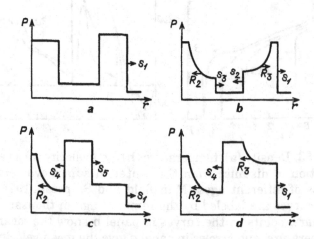

Fig. 54 Secondary shocks resulting from thermal instability at the instant of collapse of a cold shell (Falle, 1981).

behind the main shock S_1 (illustrated schematically in Fig. 54a), two rarefaction waves R_2 and R_3 (Fig. 54b), and two stronger shocks S_4 and S_5 engendered by the collision of S_2 and S_3 (Fig. 54c). Shock front S_5 overtakes and amplifies the main shock S_1, and the rarefaction R_3 (Fig. 54d) weakens S_4. The condition for the formation of the multiple secondary shocks can be derived by equating the hydrodynamic scale time t_{hyd} ($t_{hyd} \propto R_s/c \propto R_s/v_s$, where c is the speed of sound in the gas behind the front and v_s is the shock velocity) to the radiative scale time t_{cool} ($t_{cool} \approx P/L$, $L = n^2 \Lambda(T) \propto \rho^2 T^{-\alpha}$, where $\Lambda(T)$ is the radiative loss function (cooling rate coefficient), L is the cooling rate per unit volume, and α determines the functional dependence of Λ on T). If $T \propto v_s^2$ and v_s conforms to (9.2), we then have $t_{hyd}/t_{cool} \propto (E_0 n_0^2)^{-2\alpha/11}$. The secondary waves come into being when $E_{51} n_0^2 \gtrsim 10^{-5}$ (Falle, 1981), a condition satisfied by virtually all supernova remnants.

Figure 55 (Falle, 1981) shows the structure of a shell, including the secondary waves induced by thermal instability (the figure displays the dependence of expansion velocity v/v_s, density ρ/ρ_s, gas pressure P/P_s, and magnetic pressure P_{mag}/P_s on distance from the center R/R_s; all parameters have been normalized to their values at the shock front). The standard adiabatic solution (9.2) is shown in Fig. 55a; Fig. 55b clearly depicts the region of minimum pressure; Figs. 55c and 55d display the shell structure for the stage at which the two strong shock waves are formed (corresponding to Fig. 54c), and Fig.. 55e corresponds to Fig. 54d.

Falle (1981) has demonstrated that the process described above can occur repeatedly: as soon as the wave S_5 overtakes the main shock wave S_1, the temperature behind the front will again rise above T_*, and the previous scenario will be repeated. In Fig. 55h, which illustrates the age-dependence of expansion velocity, it is clear that this process can recur three or four times. It will terminate when the velocity v_s of the main shock wave falls below $v_* \approx 100$ km/sec, which corresponds to the temperature at the maximum of the radiative loss curve. Subsequent cooling is "quiescent," taking place at constant pressure, and no further multiple secondary shocks are formed. This stage corresponds to Figs. 55f and 55g, when 98% of the mass is enclosed within a cold shell whose thickness is 15% of its radius. The shell is much thicker than that calculated by Mansfield and Salpeter (1974), a consequence of the fact that Falle has taken magnetic pressure into account, which prevents further compression of the gas. Magnetic pressure, which is negligible in the initial free expansion and adiabatic stages, turns out to be comparable to gas pressure in the dense shell, due to enhancement of the field resulting from the freezing-in of field lines during collapse of the shell, and as a result of tangling brought about by the secondary waves. During the quiescent radiative cooling stage, the magnetic and gas pressures balance one another. The inner part of the shell is contiguous at $R/R_s \approx 0.85$ to a transition zone in which the gas rapidly cools from $\sim 10^5$ K to 10^3 K, while its velocity and pressure remain unchanged. Within the transition zone is hot gas that is bounded at

$R/R_s \sim 0.8$ by the reverse shock S_4. Inwards of S_4 we find very hot $(T \gtrsim 10^6 \text{ K})$ tenuous gas, which by virtue of its low density expands with essentially no cooling. The outer part of the dense shell is bounded by a thin layer with $\Delta R/R_s \sim 0.01$, within which gas heated by the shock wave cools from $\sim 10^4$ to $\sim 10^3$ K. This layer may be responsible for the optical emission from old remnants. The calculations upon which Fig. 55 is based rested on the assumptions that the initial expansion is governed by the solution (9.2), that heat conduction is negligible, that the unperturbed interstellar gas is immobile, homogeneous, and fully ionized, that it has a temperature of 10^4 K and a density $\rho_0 = 1.7 \times 10^{-24} \text{ g/cm}^3$, that the magnetic field is $H_0 = 10^6$ G, and that $E_0 = 3 \times 10^{50}$ erg. The presumption of full ionization is based on the suggestion that in the early stages, the gas ionized by the supernova outburst has not yet managed to undergo recombination, while in the later stages, the surrounding gas is ionized by the hard photons emanating from the shock wave.

Within the framework of the adiabatic solution (9.2), there exists a strong temperature and density gradient behind the shock front (Fig. 55a). The temperature gradient brings about heat conduction, which is difficult to treat quantitatively since it is inhibited by the magnetic field. Heat conduction can only be an important factor in young supernova remnants or the interior of the hot region; within the shell, where the field is strong and badly tangled, it can be neglected (Falle, 1981). If heat conduction is in fact important, then the temperature inside the remnant will equalize at

$$T_h = \frac{2\mu E_T}{3kM} = 1.2 \times 10^{10} R^{-3} E_{51} n_h^{-1} \text{ (K)}, \qquad (9.9)$$

where E_T is the thermal energy, M is the total mass of the remnant, and n_h is the density of the hot gas.

An effect likely to be important in the early stages is associated with heat conduction — the evaporation of the cold clouds embedded in the hot gas of the remnant (Cowie and McKee, 1977; McKee and Cowie, 1977; McKee and Ostriker, 1977). The effects of evaporation on the dynamics of a remnant can only be treated crudely. Moreover, it is sometimes unclear whether the evaporation of clouds outstrips condensation, and just what the rate of evaporation might be (Doroshkevich and Zel'dovich, 1981). Nevertheless, we present here the outcome of the analysis by McKee and Ostriker (1977), which rests upon the following simplifying assumptions. The temperature and density of the hot gas are taken to be uniform, with $\gamma = 5/3$; the shock velocity is proportional to the isothermal speed of sound in the hot gas, $v_s \propto \alpha c_h = \alpha(0.7E_0/2\pi R_s^3 \rho_h)^{1/2}$, and only two effects dictate the influence that the clouds have on the dynamics: the increase in density of the hot gas, and the changing value of α, from $\alpha = 1.68$ for the adiabatic solution to $\alpha = 2.5$ during intense evaporation. Given these assumptions, the law of motion for the shell can be derived from mass and energy conservation:

$$\frac{dM}{dt} = 4\pi R^2 \rho_0 v + N_{cl} \dot{M}_{evap} V; \quad E_T = 1.5 \rho_h c_h^2 V, \qquad (9.10)$$

where $\dot{M}_{evap}$ is the mass of gas evaporated from one cloud, V is the volume of the remnant, and N_{cl} is the number of clouds per unit volume. If at an early stage the amount of gas evaporated exceeds the swept-up gas (i.e., the second term in (9.10) is greater than the first), then putting $R_s \propto t^{\eta}$, one finds that $\eta = 3/5$ rather than $\eta = 2/5$, as in adiabatic expansion. Allowing for evaporation, the motion of the shell is given by

$$R_s = 0.18 \left(\frac{\alpha}{\eta}\right)^{2/5} \left(\frac{E_{51}}{n_h}\right)^{1/5} t_{(yr)}^{2/5} \text{ (pc)}, \qquad (9.11)$$

where we use the notation (McKee and Ostriker, 1977)

$$\eta = \frac{2}{5}\left(\frac{1 + x^{5/3}}{2/3 + x^{5/3}}\right),$$

$$x \equiv 0.065 R_s \Sigma^{1/5} n_0^{3/5} E_{51}^{-2/5}, \quad \frac{n_h}{n_0} = 1 + x^{-5/3}$$

The early $t^{3/5}$ dependence is the result of variation in n_h due to evaporation (see McKee and Ostriker (1977)). Here the constant $\Sigma \equiv \alpha a^2 / 3 f_{cl} \varphi$; φ represents the evaporation efficiency; a is the cloud size, and f_{cl} is its filling factor. The quantity n_h/n_0 is large when x and Σ are small; when $x = 1$, the density of the evaporated gas equals that of the unperturbed medium.

The late stage in the development of a shell, when evaporation has ceased, is characterized by a time-dependency exponent $\eta = 0.28 - 0.31$ (see Eq. (9.6)). If the late stage is dominated by evaporated gas, rather than swept-up gas, the shell will expand at constant velocity, and $R_s \propto v_s$. The actual dynamics of the late-stage shell will obviously lie somewhere between the limiting cases $\eta = 1$ and $\eta = 0.3$, but evaporation plays only a minor role (McKee and Ostriker, 1977).

Everything that has been said thus far refers to the evolution of a remnant in a homogeneous medium; cloud evaporation has been considered from a strictly formal standpoint as the source of the hot gas in the

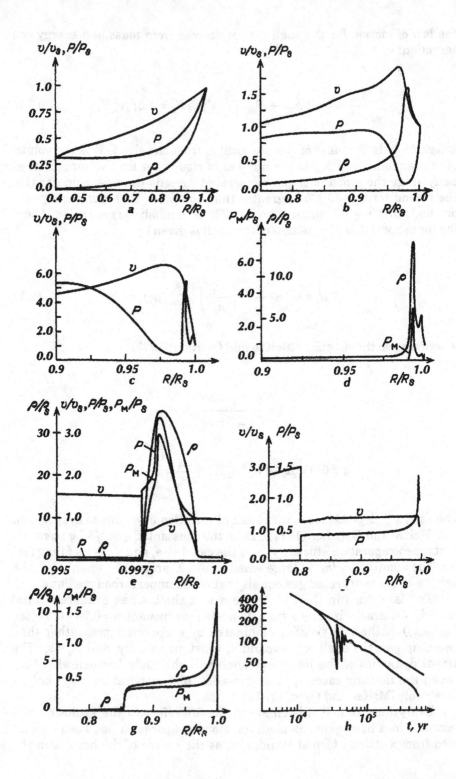

remnant. The effect of inhomogeneities in the interstellar medium on shell development can be formalized in two ways, through the large-scale density gradient, and small-scale fluctuations — i.e., dense cloudlets.

The large-scale gradient affects evolution of the remnant only if the scale length of the gradient is comparable to the size of the shell. This is a situation that is encountered fairly often in the interstellar medium:

1) when an explosion takes place near the boundary of a dense cloud;

2) when a supernova remnant reaches ~200 pc in size and starts to "feel" the gas-density gradient in the galactic disk;

3) when a massive star rapidly loses mass prior to exploding, and a spherically symmetric density distribution is established around it.

In the first two instances, the density distribution can be represented by $\rho(z) = \rho_0 e^{-z/H}$. For this distribution, there exists the self-similar solution found by Kompaneets (1960) for a powerful point explosion in the earth's atmosphere. In that case, the radius and velocity of the shock front are given by

$$R_s(\theta) = R_0\left(1 + \frac{cR_0}{H}\cos\theta\right), \quad v_s(\theta) = v_0\left(1 + \frac{2cR_0}{H}\cos\theta\right), \quad (9.12)$$

where v_0 and R_0 are the velocity and radius for the standard adiabatic so-

Fig. 55. Structure of a supernova shell during the unstable stage of cooling (Falle, 1981). Shown here are the velocity v/v_s, density ρ/ρ_s, gas (P/P_s) and magnetic (P_{mag}/P_s) pressure, all as functions of the distance from the shock front R/R_s at different times.

Panel **a** corresponds to the solution (9.2). Parameter values at the front are as follows for the other panels in the figure:

b) $(t = 3.2 \times 10^4$ yr$)$ $R_s = 16.4$ pc, $v_s = 93$ km/sec, $P_s = 2 \times 10^{-10}$ dyn·cm^{-2}, $\rho_s = 6.3 \times 10^{-24}$ g·cm^{-3};

c, d) $(t = 3.56 \times 10^4$ yr$)$ $R_s = 16.7$ pc, $v_s = 30$ km/sec, $P_s = 2.5 \times 10^{-11}$ dyn·cm^{-2}, $\rho_s = 4.8 \times 10^{-24}$ g·cm^{-3};

e) $(t = 3.7 \times 10^4$ yr$)$ $R_s = 16.8$ pc, $v_s = 130$ km/sec, $P_s = 3.68 \times 10^{-10}$ dyn·cm^{-2}, $\rho_s = 6.5 \times 10^{-24}$ g·cm^{-3};

f, g) $(t = 9 \times 10^4$ yr$)$ $R_s = 22$ pc, $v_s = 52$ km/sec, $P_s = 6.7 \times 10^{-11}$ dyn·cm^{-2}, $\rho_s = 5.6 \times 10^{-24}$ g·cm^{-3}.

Panel **h** show the velocity of the shock front with thermal instability taken into account; the straight line corresponds to the adiabatic solution (9.2).

lution (9.2); θ is the polar angle reckoned from the direction of the gradient, and $c = 0.186$ is a constant (we have already taken advantage of this solution in Section 7 to interpret the observations of the nebula IC 443). Two-dimensional numerical calculations of the evolution of a remnant in a medium with a plane density gradient (Yorke et al., 1983; Falle et al., 1984) also predict shell structure consistent with that observed in IC 443 and other objects located near the periphery of dense clouds.

For a spherically symmetric density distribution, $\rho = \rho_0 R^{-u}$, there is another self-similar solution (Sedov, 1981; Aizenberg, 1977):

$$R_s = \left(\frac{E_0}{\rho_0}\right)^{1/(5-u)} t^{2/(5-u)},$$

$$(9.13)$$

$$v_s = \frac{2}{5-u}\left(\frac{E_0}{\rho_0}\right)^{1/(5-u)} t^{-(3-u)/(5-u)}.$$

The first attempt to numerically model a remnant in an inhomogeneous medium with small-scale density fluctuations — compact dense clouds immersed in the intercloud gas — was made by Cowie et al. (1981a), in a paper that represents a qualitative step forward in the study of the evolution of a remnant in the interstellar medium. The authors did not simply restrict their considerations to evaporation-induced density variations in the hot gas, as had been done previously — they included energy losses due to evaporation, compression, and acceleration of clouds, and also allowed for heat conduction and radiative cooling of clouds. The calculations took account of mass, energy, and momentum transfer between clouds and the intercloud gas, as well as the sweeping up and evaporation of clouds in the hot intercloud gas. The assumed initial conditions incorporated the parameters of a three-phase interstellar medium, which itself is the product of the collective action of supernovae on gas in the Galaxy, a subject about which we shall have more to say in Section 17. A corresponding assumption was that most of the volume of the galactic disk ($f_h \approx 75\%$) is occupied by hot, tenuous gas ($T_h = 4.5 \times 10^5$ K, $n_h = 2 \times 10^{-3}$ cm^{-3}) enveloping numerous cold clouds ($T_c = 80$ K, $n_c = 40$ cm^{-3}), which themselves occupy but a small fraction ($f_c \approx 2\%$) of the volume and are surrounded by warm, partially ionized coronae ($T_w = 8000$ K, $n_w = 0.2 - 0.3$ cm^{-3} $f_w \approx 20\%$). The calculations were carried out under certain simplifying assumptions, namely that the intercloud medium is homogeneous; rather than a broad spectrum of cloud sizes, a standard cold cloud and a standard warm cloud were used, modeled as spatially separated short cylinders; collisions between clouds were not allowed; magnetic-field and cosmic ray pressure were not taken into consideration. Six models were calculated, with a reasonable set of values assumed for the initial explosion energy E_0, the temperature and density of the intercloud gas, and the size, density, and filling factor of the clouds. As might be expected, even such an idealized approach revealed differences from the evolution of a remnant in a homogeneous medium; we take one model as an example.

Model number 1 of Cowie et al. (1981a) has $E_0 = 3 \times 10^{50}$ erg, $n_h = 2.4 \times 10^{-3}$ cm^{-3}, $T_h = 1.4 \times 10^5$ K, and evaporation efficiency $\varphi = 1$; cold and warm clouds have radii $a_c = 1.6$ pc and $a_w = 2$ pc, surface densities $N_c = 2.7 \times 10^{20}$ cm^{-2} and $N_w = 2.2 \times 10^{18}$ cm^{-2}, and volume filling factors $f_c = 0.02$ and $f_w = 0.27$; thermal conductivity is low, and has been ignored. In Fig. 56, the density, temperature, and intercloud gas velocity distributions in the remnant have been plotted as functions of the radius at different instants in time. The large size of the shells is not surprising; we have already noted that in a model with small clouds embedded in tenuous gas, shell dynamics are dictated by shock propagation in the intercloud medium, and the density of the latter is more than two orders of magnitude lower than the standard value $n_0 = 1$ cm^{-3} assumed in the work of Falle (1981) and Mansfield and Salpeter (1974). A reverse shock is apparent in curve a, which corresponds to an early stage in which the mass of swept-up and evaporated gas has reached approximately 40 times the mass of the ejecta, as in the case of a homogeneous medium. The high temperature behind the front right at the very outset, $T_s \gtrsim 10^8$ K, leads to the sweeping out and rapid destruction of warm clouds through vigorous evaporation. This then increases the density of hot gas on the periphery of the young remnant to ~ 0.1 cm^{-3}, which is close to the density of the warm component of the interstellar medium.

Evaporation of cold clouds also contributes to a certain extent, but they are only weakly entrained by the shock wave and distributed over the bulk of the remnant more uniformly. Curve b corresponds to a time at which the ejecta are finally indistinguishable from the rest of the swept-up and evaporated gas, but radiative energy losses are still negligible (this is the analog of the adiabatic stage (9.2)). Curves c and d represent the stage following shell formation. Here, in contrast to a supernova in a homogeneous medium, it is quite clear that rather than beginning out on the periphery, cooling starts inside the remnant, at the inner boundary of the region receiving the preponderance of the warm clouds that have been swept out. Here there is a precipitous collapse of a cold, dense shell, which accumulates swept-up gas from without, and cooling, evaporated gas from within. Curve d corresponds to a shell that has formed at a distance of 55 pc from the center, while the radius of the shock front is 157 pc.

Figure 57 shows the evolution of the energy reserves of this same remnant (Cowie et al., 1981a). Before rapid losses set in, the energy of the remnant is concentrated in the form of thermal and kinetic energy of the swept-up intercloud gas. The kinetic component is virtually constant, and

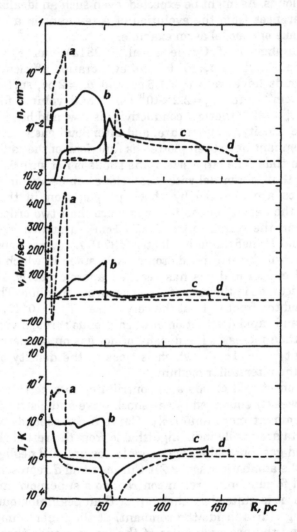

Fig. 56. Structure of a supernova remnant in a medium with dense, small-scale clouds: variation of density, velocity, and temperature of the intercloud gas with distance from the center of the shell. Computed behavior for four choices of the age and the radius of the shock front (model no. 1 of Cowie et al. (1981a)):

a) $t = 10^4$ yr, $R_s = 17$ pc; **b)** $t = 10^5$ yr, $R_s = 50$ pc; **c)** $t = 7.7 \times 10^5$ yr, $R_s = 140$ pc; **d)** $t = 10^6$ yr, $R_s = 160$ pc.

is approximately the same (~30%) as in the case of adiabatic expansion into a homogeneous medium. The principal energy loss mechanisms are associated with radiation from the intercloud gas and compression of clouds. The clouds subsequently radiate away the energy of compression at optical wavelengths, and later on, in the infrared. Compressive and radiative energy losses from clouds grow quickly, and in the final stages of evolution they dominate all other sources. About 3–4% of the explosion energy is transformed into kinetic energy of clouds accelerated by the shock wave. Note also that the total energy of the remnant can increase slightly over time, due to the influx of energy carried by the swept-up gas.

To what extent does the shock-wave energy loss engendered by collisional heating of dust and reradiation in the infrared modify the picture of supernova remnant evolution laid out thus far? At first glance, since we concluded in Section 8 that plasma cooling via dust radiation at $T_e \gg 5 \times 10^6$ K may exceed previous estimates by a factor of 10–20, it would not be too surprising if the preceding considerations were entirely demolished. But upon closer examination (e.g., see Graham et al. (1987b)), it turns out that the modifications in the evolutionary model are not nearly

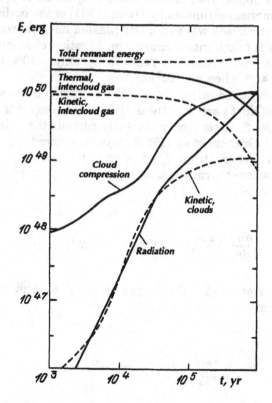

Fig. 57. Distribution of energy for a supernova at different times (Cowie et al., 1981a; see text).

so severe. During the free-expansion stage, the shell does not cool, and the effects of dust are negligible overall. Dust radiation augments radiative energy losses, and speeds up the transition of the remnant from the adiabatic stage to the radiative snowplow stage. According to Graham et al. (1987b) (see also Cox (1972b)), remnant dynamics start to be dominated by cooling (the temperature of an element of mass drops faster due to radiation losses than due to expansion) at time

$$t_{dyn} = \frac{4}{5}\left(\frac{n_i + n_e}{n_i n_e}\right)\frac{3kT_s}{8\Lambda}.$$

(9.14)

If the cooling rate coefficient Λ^{dust} encompasses dust-mediated cooling, and the temperature T_s behind the front is governed by (9.2), then $t_{dyn}^{dust} \sim 10^4 E_{51}^{0.18} n_0^{-0.64}$ yr. When the density of the interstellar gas is high, the presence of dust can significantly curtail the adiabatic expansion stage. But at high densities, the erosion of dust grains via sputtering in the hot plasma behind the shock front is also enhanced, which reduces dust-induced cooling. Numerical estimates by Dwek (1981) of the cooling and evolution of an adiabatic shock wave in dusty plasma have shown that when dust-grain erosion is taken into account, the duration of adiabatic expansion in a homogeneous medium is reduced by only $\sim 10\%$ in all when $n_0 = 1\,\mathrm{cm}^{-3}$ and by 40% when $n_0 = 10^2\,\mathrm{cm}^{-3}$.

More recent, improved calculations of the cooling of dust-laden plasma via shock heating of dust particles (Dwek, 1987) also suggest that the presence of dust with $n_0 \lesssim 10\,\mathrm{cm}^{-3}$ only modestly curtails the adiabatic stage (Dwek et al., 1987). Graham et al. (1987b) have compared t_{dyn}^{dust} as obtained from (9.14) with the lifetime t_{grain} of a dust particle of size a, $t_{grain} = a(da/dt)^{-1}$, where thermal sputtering can be represented by (Draine and Salpeter, 1979a, b)

$$\frac{da}{dt} \sim 10^{-6} n(1 + 0.002 T_7^{-3})^{-1}\,\mu\mathrm{m/yr},$$

(9.15)

and have found from (9.2), for a gas density n_0 and temperature $T_7 \equiv T_s \times 10^{-7}\,\mathrm{K}$, that

$$\frac{t_{grain}}{t_{dyn}^{dust}} \sim 3a_{-5}(E_{51} n_0^2)^{-2/11}.$$

(9.16)

Clearly, then, the lifetime of 0.1-μm dust grains at $n_0 < 20\,\mathrm{cm}^{-3}$ is much longer than the dynamic time, and the adiabatic expansion stage of a remnant in a homogeneous medium with a typical density of $n_0 \sim 1\,\mathrm{cm}^{-3}$ can turn out to be brief.

In a two-phase medium ($n_{cl} \sim 20\,\mathrm{cm}^{-3}$, $n_i \sim 10^{-2}\,\mathrm{cm}^{-3}$), the lifetime of a dust grain undergoing thermal sputtering in the plasma behind the shock front in the intercloud medium is $t_{grain} \sim 10^5$ yr, while $t_{dyn}^{dust} \sim 5 \times 10^4$ yr. As the temperature drops to $T_s \lesssim 10^6$ K, energy losses due to dust and gas cooling become approximately equal — although dust-mediated cooling appears to hasten the transition of a remnant to the snowplow stage (Graham et al, 1987b).

The propagation of a secondary shock wave in clouds has little to do with dust-mediated cooling, as dust grains in a dense medium have a short lifetime. Therefore, even the grossest estimates indicate that cooling via radiation from dust speeds up the evolution of a supernova remnant — but also that it does not determine it completely. Indeed, the propagation of a shock wave in a dust-laden plasma is an extremely complicated, poorly understood process. The passage of a shock wave leads to dust-grain erosion, a process that modifies the chemical composition of the post-shock gas,

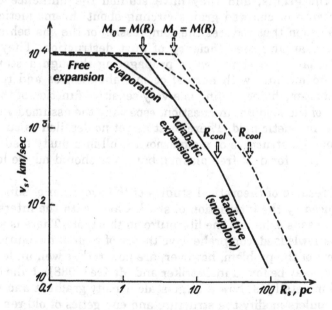

Fig. 58. "Theoretical" dependence of shock velocity v_s on radius R_s. The free-expansion, snowplow with strong evaporation, snowplow, and radiative cooling stages are indicated. Arrows point to the beginning of deceleration and the instant of cold shell formation; see text. The solid curve corresponds to $n_0 = 1\,\mathrm{cm}^{-3}$ and the dashed curve to $n_0 = 0.1\,\mathrm{cm}^{-3}$.

thereby altering its radiative cooling; that in turn changes the structure of the shock wave itself. Dust-grain erosion is due not only to thermal sputtering produced by collisions with fast ions in the high-temperature plasma, but to nonthermal sputtering resulting from betatron acceleration of charged grains and grain–grain impacts in the magnetic field. The charge on the grains, which determines the intensity of erosion, derives from a great many process, including photoelectric emission, secondary electron emission, ion and electron sticking, and grain–plasma drift.

McKee et al. (1987) have considered in great detail the structure of a shock wave in dusty plasma when these shock-induced dust destruction mechanisms are taken into account. In contrast to previous work analyzing the destruction of dust by a shock wave, they devoted particular effort to the role of electric charge, which notably alters the efficacy of erosion. Grains carrying less charge experience less plasma drag, and are more easily accelerated by the betatron process; fast-moving dust grains are efficiently destroyed. In their 1987 paper, McKee et al. have included dust dynamics in the hydromagnetic equations for the flow of gas and dust behind the shock front; they have examined the rate of dust destruction as a temporal function of the change in temperature, plasma density, and gyrovelocity of the grains; and they have studied the influence of the "thermal" pressure of charged grains circling about the magnetic field lines, a phenomenon that reduces the compression of the gas behind the front, thereby changing the efficiency of dust destruction. They have demonstrated that for a shock wave propagating through a standard cloud–intercloud medium with normal cosmic abundances and relative dust content, dust-mediated cooling is a very sensitive function of the specific properties of the medium in question, especially the assumed value of the interstellar magnetic field. There exist as yet no detailed calculations of the evolution and structure of a supernova shell in a dusty medium to match existing ones for dust-free plasma, but these should not be long in coming.

Our brief résumé of theoretical studies of the evolution of supernova remnants induced by the interaction of shock waves with the interstellar medium by no means exhausts the literature on this topic. There is still no universally acknowledged, comprehensive theory of remnant evolution. Individual aspects of the problem, however, are now rather well understood (a recent survey may be found in Ostriker and McKee (1988)). Falle (1988) has examined in detail just how a large-scale density gradient and small-scale dense cloudlets modify the structure and energetics of old remnants in transition from adiabatic expansion to the radiative (snowplow) phase. Likewise, McKee (1988) and Band and Liang (1988) have looked carefully into the evolution of supernova remnants in a medium with a spherically symmetric, "multilayer" density distribution, which might result from UV radiation effects, a stellar wind of varying intensity, or the possible ejection of a slow-moving shell by the progenitor (see Lozinskaya (1988a) and Section 16). These particular solutions are important in their own right, and make it possible to understand some aspects of the phenomenon. The abundance of "theoretical" possibilities attests to the importance of a strictly observational approach to the problem.

Two of the highest priority observational problems remain: 1) to clarify whether the observed supernova remnants form an evolutionary sequence selected from a homogeneous class of objects, and 2) to identify an empirical law for the evolution of remnants and compare it with the standard theoretical models.

The "theoretical" evolution of a supernova shell, incorporating the most important of the stages discussed above — free expansion, expansion with cloud evaporation, adiabatic expansion, and expansion with strong radiation — has been drawn in Fig. 58 as a plot of shock front velocity v_s vs. radius R_s. Theoretical evolutionary tracks have been plotted for the remnant of a standard explosion ($M_0 = 1 M_\odot$, $v_0 = 10^4$ km/sec, $E_{51} = 1$) in a homogeneous medium for two values of the density, $n_0 = 1\,\text{cm}^{-3}$ and $n_0 = 0.1\,\text{cm}^{-3}$. The small clouds have been taken into account only insofar as they constitute an evaporative source of hot gas. The arrows in Fig. 58 mark characteristic points, namely the end of free expansion, which corresponds to $M_0 \sim M_{swept}$, and the formation of a dense shell, which is dictated by (9.3).

The first attempt to construct an empirical evolutionary sequence for

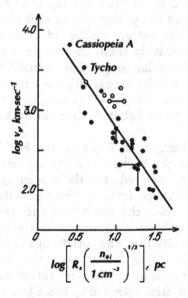

Fig. 59. Empirical evolutionary sequence for Galactic supernova remnants (filled circles) and those in the LMC (open circles) (Lozinskaya, 1980a, b). The "reduced" radius $R_s(n_{0i}/1\,\text{cm}^{-3})^{1/3}$ takes account of the different density in the ambient gas in the vicinity of the remnant.

supernova remnants and compare it with theory was made by the present author in 1975. Later on, when the kinematics of most of the supernova remnants in the northern sky had been investigated, Lozinskaya (1980b) constructed the empirical evolutionary sequence shown in Fig. 59, based mainly on the optical and x-ray observations of supernova remnants summarized in Table 11.

The evolutionary sequence that we obtained displayed v_s as a function of the "reduced radius" of the shell, $R_s(n_{0i}/1 \text{ cm}^{-3})^{1/3}$. We introduced this new parameter because of the necessity of taking the wide range of interstellar gas densities in the vicinity of remnants into consideration. Attempting to understand the evolutionary sequence of supernova remnants without allowing for the density of the medium would lay one open to errors as serious as mistaking a Chihuahua for a young Saint Bernard. The analogy is not a new one, but the point is well taken. What we have really demonstrated above is that the development of a shell at any stage except during the initial free expansion depends on the density of the surrounding gas, and the supernova remnants that are actually observed are located in regions whose density differs by orders of magnitude. As a case in point, we mention the 30–40 pc shell of W 28, which is located in a dense region (mean intercloud density $n_{0i} = 5-6 \text{ cm}^{-3}$; mean cloud density $n_{0i} = 40-50 \text{ cm}^{-3}$), and Loop I (the North Polar Spur), 250 pc across, which we are thoroughly convinced is the remnant of a supernova that exploded in tenuous gas with $n_{0i} \approx 0.01 \text{ cm}^{-3}$ (see Section 11). Of these two remnants, there can be no doubt that the younger in an evolutionary sense is the extensive Loop I, which has not yet embarked upon an intense radiation phase, while the comparatively small shell W 28 is a highly evolved one now in its post-shell formation stage. The reduced radius of a remnant makes direct allowance for the actual effectiveness of deceleration of the expanding shell by the ambient gas — it amounts to the radius divided by the normalized mean distance between atoms in the surrounding interstellar gas.

The shock front velocity v_s in Table 11 and Fig. 59 is determined by the x-ray spectrum and the highest-velocity motions of the optical filaments. Both mechanisms referred to in the preceding section — radiation by gas behind the fast shock front in the partially ionized intercloud medium, and emission from the surface of the post-shock accelerated clouds — can give rise to faint Hα emission at a velocity close to v_s. Formally, however, the velocity that one gets from the high-velocity wings of Hα yield only a lower limit on v_s. Estimates of the shock velocity based on interpretation of the x-ray data are likewise rather uncertain (see Section 8). Nevertheless, the velocities obtained by these two methods differ by less than the mean errors in the plotted points.

Figure 59 suggests that all of the well studied galactic remnants may be looked upon as forming an evolutionary sequence of similar objects, whose slowing in the interstellar medium is governed by how much gas they sweep up. The open circles in the figure indicate remnants in the

Table 13. Comparison of Theoretical Models with Observations

Models	η	$(\eta - 1)/\eta$
Adiabatic stage	0.4	−1.5
Early stage with evaporation	0.6	−0.67
Radiative cooling	0.3	−2.3
Late stage dominated by evaporated gas	1.0	0
Observations		
Mean value	0.41	−1.45
95% confidence interval	0.37 to 0.45	−1.24 to −1.68

Magellanic Clouds with measured expansion velocities. Bearing in mind that we have observed only those remnants in the LMC that are the very brightest x-ray emitters, and therefore that they probably have the highest kinetic energy, we may infer that they too belong to the same evolutionary sequence. The straight line is a least-squares fit to the old galactic objects (we have not included the historical remnants, as for the most part they may still be in the free expansion stage); its slope yields $v_s \propto (R_s n_{0i}^{1/3})^{-1.45}$. The observations have all been assigned the same accuracy. The typical measurement errors in v_s are perhaps 30%; in measurements of $R_s n_{0i}^{1/3}$, they reach 30–50%. The least reliable estimates of velocity in HB 21 and of the shell radius for CTB 1 are plotted as short line segments in Fig. 59. The purely empirical evolutionary sequence of supernova remnants that we obtained in 1980 can be compared with standard theoretical models. The results of that comparison are given in Table 13. Using the general notation $R \propto t^{\eta}$ and $v \propto R^{(\eta-1)/\eta}$, we present the values of η and $(\eta-1)/\eta$ for the various idealized stages of evolution discussed above, as well as results obtained by statistical processing of the observations.

We see from Table 13 that the standard adiabatic model best fits the empirical evolutionary sequence, to within the observational errors. If we assume *a priori* that there exists a break in the vicinity of $R_s n_{0i}^{1/3} \sim 20\,\mathrm{pc}$, corresponding to the onset of strong radiative cooling, then the observational data may, with equal certitude, represent an early evaporative stage before the break, and a radiative cooling stage afterwards. The strictly observational sequence, however, is not statistically reliable enough to corroborate the existence of the break.

Unfortunately, it is difficult to count on a substantial improvement in the accuracy of observations in the near future. The radius estimate — i.e., the distance to a given object — suffers from the greatest uncertainty. Distances have thus far been determined from the radio surface brightness or mean radial velocity of a remnant (or of a related cloud, group of bright stars, etc.) and the galactic rotation law; improved accuracy would come as a surprise. The only qualitatively more accurate method for assessing distance is to measure both the radial and tangential velocity of some filaments within a remnant. Such measurements have as yet only been carried out for Cas A, the Crab Nebula, and 3C 58. The distance to the Cygnus Loop has also been determined by comparing radial and tangential velocities, but using mean values rather than the velocities of individual filaments, thereby increasing possible systematic errors.

At current levels of accuracy, we thus see that it is entirely feasible to use the standard adiabatic model to estimate shell parameters. In any event, the error committed in applying that model to a later stage of evolution is no greater than the error induced by the low accuracy of the radius estimate. Having concluded in the preceding section that pressures in the intercloud and cloud media behind the shock front are equalized to within a factor of order 2 (see Eq. (8.3)), we may evaluate E_0 in two ways: using (9.2) and the shock velocity v_s, along with the density of the intercloud medium n_{0i}, or using the parameters v_{cl} and n_{0cl} of the cloud medium in Eq. (8.3). For the remnants plotted in Fig. 59, the mean initial kinetic energy of the explosion is $E_0 = (4 \pm 1.5) \times 10^{50}$ erg (Lozinskaya, 1981b). Individual values deviate from the mean by a few times the latter, and some objects may deviate by an order of magnitude.

Viewed in the context of the adiabatic model, the age of the oldest remnants in Fig. 59 is $(1-2) \times 10^5$ yr; for the most highly evolved objects, some $(5-8) \times 10^3\, M_\odot$ of interstellar gas has been swept up. One may then ask whether we can observe shells at yet later stages of evolution. For interstellar gas that exerts a pressure $P_0/k \approx 3500$ K·cm^{-3}, with $E_{51} = 0.4$, the largest radius and age that one obtains from (9.7) and (9.8) are $R_{max} \approx 50$ pc and $t_{max} \approx 10^7$ yr in a medium with density of order $0.2-1$ cm^{-3}. The gas comprising such a very old shell would have a mass of approximately $(1-5) \times 10^4\, M_\odot$. If it lacked any exciting stars, it would be invisible optically, since its expansion velocity $v_s \lesssim 50$ km/sec would be too low to ionize gas, and the recombination time for the initially heated and ionized gas would already have expired. If there were indeed exciting stars — for example, if the explosion took place in an OB association — then one might expect an extensive shell complex of ionized gas and dust to form in the later stages of evolution. We note here that such objects have actually been observed, and we shall consider them in Chapter 4. Even with no sources of ionizing radiation, the oldest supernova remnants may still be observable in the radio as extended H I shells (see Heiles (1979, 1984), Gosachinskii and Khersonskii (1983), and Bychkov (1986)).

10. Synchrotron emission from supernova remnants

All supernova remnants are radio synchrotron sources, a fact that had been reliably established by 1960. The origin of the relativistic electrons and magnetic field responsible for the radiation, however, is still not entirely clear, even now. More to the point, what *is* clear is that in different types of remnants at different stages of evolution, a small number of mechanisms can account for the generation of magnetic fields and relativistic particles. Supernova remnants harboring a compact stellar remnant as well — a pulsar — will also emit x-ray and optical synchrotron radiation.

The radio emission from supernova remnants has defined an extremely active area of research. The vigor of the attendant information flow stems from an important contrast between radio signals, on the one hand, and optical and x-ray signals on the other: for the former, absorption impairs virtually none of our observations of galactic objects.

Present-day catalogs of nonthermal radio sources that have been found to be galactic supernova remnants contain close to 155 objects (van den Bergh, 1983; Green, 1984b, 1988a). In addition, 45 radio synchrotron sources (supernova remnants) have been identified in the Large and Small Magellanic Clouds (Mills et al., 1984; Mathewson et al., 1984; Mathewson et al., 1985), nine in M33 (D'Odorico et al., 1982; Goss and Viallefond, 1985), and ten in M31 (Dickel et al., 1982; Dickel and D'Odorico, 1984).

Research on radio emission from supernova remnants has been pursued along a number of different lines. Historically, the first approach was to measure the spectral flux density, and a remarkable fact had already been gleaned by 1961, in the early days of radio astronomy — that the radio flux from Cas A was decreasing by approximately 1% per year. Measurements of the secular fading of Cas A have continued: data obtained by Ivanov et al. (1982b) showed a reduction in flux at 31 cm of $\Delta S = 0.92 \pm 1\%$ per year over the period 1964–72, and $\Delta S = 0.41 \pm 0.08\%$ per year in the period 1972–81. Similar results have been obtained by Baars et al. (1980):

$$\Delta S = [(0.97 \pm 0.04) - (0.30 \pm 0.04) \log \nu (\text{GHz})] \ (\%/\text{yr}).$$

(See also Vinyaikin et al. (1980), which includes further references to the literature.)

Tycho's remnant, another shell remnant, has also exhibited a secular decrease in radio flux: $\Delta S = 0.4 \pm 0.5\%$ per year according to Dickel and Spangler (1979) and Ivanov et al. (1982b), and $\Delta S = 0.23 \pm 0.19\%$ per year according to measurements by Strom et al. (1982). Radio emission from the Crab Nebula, a young plerion, is also on the wane. Aller and Reynolds

(1985) report a falloff in brightness of $\Delta S = 0.167 \pm 0.015\%$ per year between 1968 and 1984. Ivanov et al. (1982a) propose a mean rate of decrease of $\Delta S = 0.2 \pm 0.06\%$ per year from 1967 to 1981, although their value may possibly be ascribable to a more rapid short-term fade. The falloff in radio luminosity of a young plerion predicted by Reynolds and Chevalier (1984) is 0.26% per year for the Crab Nebula, and about 0.13% per year for 3C 58 (see below).

In contrast, observations of 3C 58 suggest that in the period from 1965 to 1985, it brightened by $0.32 \pm 0.13\%$ per year at 408 MHz (Green, 1987), by $0.24 \pm 0.15\%$ per year at 1.6 GHz (Aslanyan et al., 1987), and by $0.284 \pm 0.046\%$ per year at 8 GHz (Aller and Reynolds, 1985b). The remnant may possibly be passing through a phase involving an irregular jump in radio luminosity associated with the passage of the reverse shock — see Figs. 65, 67 — but the point is debatable.

Radio spectra have been constructed for more than 100 galactic remnants, spanning the frequency range from 10 to 10^4 MHz, and these are fit quite well by a power law of the form $S_\nu \propto \nu^{-\alpha}$. The spectral index for different remnants lies in the range $0 \le \alpha \le 0.8$ (Fig. 60). It has been reported that α depends on height above the galactic plane, but follow-up studies have failed to provide confirmation (Lerche, 1980; Clark, 1976).

In some objects, the spectral index varies with frequency. The low-frequency cutoff observed in Cas A is due to free–free absorption in dense, partially ionized clouds, or in the tenuous intercloud medium. In certain remnants, including the Cygnus Loop, Simeiz 147, HB 9, 3C 391, G 41.1–0.3, and W 49B, there is a break in the spectrum in the vicinity of $\nu \approx 10^3$ GHz, and the change in the spectral index is $\Delta\alpha = 0.5$ (Udal'tsov et al., 1978; Sofue et al., 1980; Fürst and Reich, 1986).

The radio spectrum also changes as a remnant evolves. We can view these changes directly in the young remnant Cas A: the secular decrease in flux is frequency-dependent, and the spectrum is flattening out in the course of time (Dickel and Greisen, 1979; Vinyaikin and Razin, 1979; Vinyaikin et al., 1980). In the frequency range $0.3-30$ GHz, the spectral index of Cas A is $\alpha = 0.770$ (epoch 1980), and is decreasing by $\Delta\alpha = 1.3 \times 10^{-3}$ per year (Baars et al., 1977).

It is conceivable that the spectral index depends rather weakly on radius, i.e. age, in the oldest remnants (Sakhibov and Smirnov, 1982; Berkhuijsen, 1986), but the observed dependence may also be partly due to selection effects. Fading of the radio brightness with increasing size depends on α (see Eq. (10.12) below), so sources with small α are visible longer against the galactic radio background.

In most of the radio remnants, linear polarization amounts to $3-5\%$ ($20-25\%$ in selected objects), and the magnetic field is quasiregular. An analysis of all polarization measurements presently available (Milne, 1987, 1988) suggests that a radially directed magnetic field predominates in young shell remnants. In older shells (that are not too close to the galactic plane), the polarization vector is typically tangential, although large-scale radial structures are observed as well. The magnetic field of the Crab Nebula displays small-scale cellular structure (Swinbank and Pooley, 1979; McLean et al., 1983; Velusamy, 1985), which tends to corroborate the picture of magnetic field lines being "held back" by the system of optical filaments. Fluctuations in the magnetic field have a typical size of more than 0.01 pc (Wilson et al., 1985a). Henbest (1980) has observed this fine cellular structure in the distribution of brightness and linear polarization in Tycho's remnant, with the cells being at most 0.1 pc across. Correlation between the magnetic field structure and that of the optical filaments is also observed in old remnants, such as Vela XYZ (Milne, 1979b, 1980; Lerche and Milne, 1980).

We have now briefly recounted the basic properties of radio emission from supernova remnants that are satisfactorily explained by synchrotron radiation, namely the observed power-law spectrum (with a weakly time-varying spectral index), the decrease in flux density with time, and the linear polarization. The theory of synchrotron radiation is well understood, and many expositions can be found in the literature (e.g., see Ginzburg and Syrovatskii (1965), Kardashev (1962), Kaplan and Pikel'ner (1963), and Shklovskii (1976a)). Here we shall present only the most fundamental relations required for our subsequent treatment.

If the electron energy distribution is described by a power law

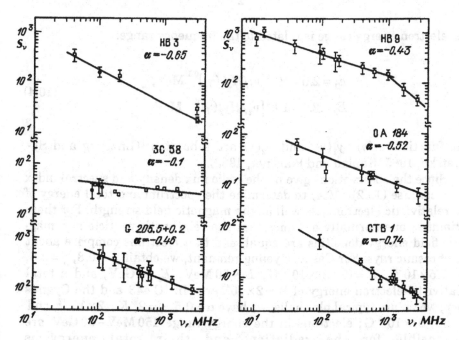

Fig. 60. Radio spectra of several plerions and shell remnants (V. A. Udal'tsov, private communication). Spectra of the Crab Nebula and G 0540–69.3 appear in Fig. 24.

$$N(E)dE = KE^{-\gamma}dE \tag{10.1}$$

over a wide enough energy range $E_1 - E_2$ and the field is homogeneous, then the emissivity is

$$\epsilon(\nu) = a(\gamma)\frac{e^3}{mc^2}\left(\frac{3e}{4\pi m^3 c^5}\right)^{\frac{\gamma-1}{2}} H_\perp^{\frac{\gamma+1}{2}} K\nu^{-\frac{\gamma-1}{2}}, \tag{10.2}$$

where m and e are the mass and charge of the electron, $H_\perp$ is the magnetic field component perpendicular to the electron velocity, and $a(\gamma)$ is a dimensionless function. In a completely randomized field, the emissivity takes the same form, but $a(\gamma)$ becomes some other dimensionless function, and $H_\perp$ is replaced by H. The radio spectral index is governed by the relativistic-particle power spectrum:

$$\alpha = \frac{\gamma-1}{2}. \tag{10.3}$$

The electron energy range is related to the frequency range:

$$E_1 = 2.5 \times 10^{-4}\left[\nu_1/Hy_1(\gamma)\right]^{0.5} \text{ MeV},$$
$$E_2 = 2.5 \times 10^{-4}\left[\nu_2/Hy_2(\gamma)\right]^{0.5} \text{ MeV}. \tag{10.4}$$

The functions $a(\gamma)$, $y_1(\gamma)$, and $y_2(\gamma)$ are tabulated (Ginzburg and Syrovatskii, 1965; Kaplan and Pikel'ner, 1963).

Since the observations give us the radio flux density and spectral index α, we can use (10.2)–(10.4) to determine the spectrum and total energy of the relativistic electrons, as well as the magnetic field strength. For these estimates, one normally assumes that the relativistic-particle and magnetic-field energy densities are equal, and that electrons comprise about 1% of cosmic rays. For Cas A, a young remnant, we obtain $\alpha = 0.8$, $\gamma = 2.6$, $K = 7.5 \times 10^{-12}$, $H = (2-3) \times 10^{-4}$ G, $E_1 \approx 50$ MeV, $E_2 \approx 8$ GeV, and a total relativistic-electron energy of $W \sim 2 \times 10^{48}$ erg. In IC 443 and the Cygnus Loop, which are typical old shells, we have $\alpha = 0.5$, $\gamma = 2$, $K = 2 \times 10^{-12}$, and $H = (1-3) \times 10^{-5}$ G; electrons in the energy range 150 MeV – 30 GeV are responsible for the radiation, and their total energy is $W = 5 \times 10^{48} - 10^{49}$ erg.

In a homogeneous magnetic field, the degree of linear polarization is

$$p = (\gamma + 1)/(\gamma + 7/3), \tag{10.5}$$

which yields 70% for $\gamma = 2$. The radiation emerging from a totally random-ized field is unpolarized. Somewhat of an intermediate situation prevails in real supernova remnants: the field is quasihomogeneous, with typical uni-form-field cell sizes of $0.1-1\,\mathrm{pc}$, and there is 3–30% linear polarization.

The most important processes producing secular changes in the radio synchrotron emission from supernova remnants are synchrotron losses and adiabatic expansion (Kardashev, 1962). Synchrotron radiation losses de-pend on the energy and magnetic field as $dE/dt \propto H_\perp^2 E^2$. This halves the energy of an electron in a time

$$t = 5 \times 10^8\, H_\perp^{-2} \left(\frac{E}{mc^2} \right)^{-1} \text{sec.} \tag{10.6}$$

If no new relativistic particles are injected, the electron spectrum in a ran-domized field will have a cutoff at

$$E_b(t)\,[\mathrm{eV}] \approx 8.3 \times 10^6\, H^{-2}\,[\mathrm{G}] t^{-1}\,[\mathrm{yr}], \tag{10.7}$$

and the radiation spectrum will be cut off at

$$\nu_b\,[\mathrm{Hz}] \approx 3.4 \times 10^8\, H^{-3}\,[\mathrm{G}] t^{-2}\,[\mathrm{yr}]. \tag{10.8}$$

The synchrotron radiation intensity below and above the break in the spec-trum will be

$$I_{\nu < \nu_b} \propto H^{\frac{\gamma+1}{2}} \nu^{-\frac{\gamma-1}{2}}; \quad I_{\nu > \nu_b} \propto H^{-2} \nu^{-\frac{2\gamma+1}{3}} t^{-\frac{\gamma+5}{3}}. \tag{10.9}$$

If the injection of particles with the spectrum (10.1) continues for a time t, synchrotron losses will lead to a pileup of electrons near energy E_b. The radiation spectrum will have a break at ν_b with a change $\Delta\alpha = 0.5$ in the spectral index:

$$I_{\nu < \nu_b} \propto \nu^{-\frac{\gamma-1}{2}} t; \quad I_{\nu > \nu_b} \propto \nu^{-\frac{\gamma}{2}}. \tag{10.10}$$

The adiabatic expansion of a cloud of relativistic particles is accompanied by a decrease in flux, but the shape of the spectrum remains unchanged. In the absence of particle injection, we have

$$E = E_0 \left(\frac{R_0}{R} \right), \quad N(E, R) = K_0 \left(\frac{R_0}{R} \right)^{\gamma+2} E^{-\gamma}. \tag{10.11}$$

If the magnetic flux remains constant due to freezing-in of the field lines, the field will vary as a function of radius as $H = H_0 (R_0/R)^2$, and the radio spectral flux density will vary as

$$S_\nu \propto R^3 K_0 \left(\frac{R_0}{R} \right)^{\gamma+2} H_0^{\frac{\gamma+1}{2}} R_0^{-(\gamma+1)} \nu^{-\frac{\gamma+1}{2}} \propto R^{-2\gamma}. \tag{10.12}$$

When particle injection does take place during adiabatic expansion, the spectrum at time t takes the form

$$N(E, t) = KE^{-\gamma} \frac{t}{\gamma} \left[1 - \left(\frac{t_0}{t} \right)^\gamma \right] \left(\frac{t_0}{t} \right)^3 \propto t^{-2}, \tag{10.13}$$

and the radio flux varies as

$$S_\nu \propto R^{-\gamma} \propto t^{-\gamma}. \tag{10.14}$$

We have seen that over a characteristic time $t \approx 10^4$ yr, the size of a supernova remnant will increase from 1 pc to 20–40 pc. The resultant adiabatic cooling of the cloud of relativistic particles then reduces the radio brightness by two to three orders of magnitude. For comparison, some estimates of typical synchrotron loss times at a frequency of 1 GHz are $t \approx 10^6$ yr for $H = 10^{-4}$ G and $t \approx 5 \times 10^7$ yr for $H = 10^{-5}$ G. If synchrotron losses are the only operative effect, they will only be significant at very high frequencies, usually when the remnant is $10^4 - 10^5$ years old. But because of the expansion of the shell, the break in the spectrum due to synchrotron losses may be shifted down to radio frequencies.

The decline in the radio flux of Cas A, which was predicted by Shklovskii (1960b), goes as $S_\nu \propto R^{-5.08}$ for $\alpha = 0.77$ if the remnant is a freely expanding cloud of relativistic plasma (see Eq. (10.12)). We have seen in Section 5 that the actual observable state of affairs in Cas A is much more complicated; nevertheless, the measured decrease in flux is in moderately good agreement with the expected value. Assuming that the remnant is in the adiabatic stage, we expect a rate $\Delta S/S = 0.67\%$ per year. This is a gross estimate; in actuality, the radio knots are most likely expanding within an amorphous shell, which reduces their brightness. Since the knots contribute approximately 30% to the total flux from Cas A, the effect can be substantial. Some knots will have brightened in this same time frame, so more accurate estimates are not possible. Taking account of the reverse shock, which is important in the early stages of shell deceleration, further reduces the discrepancy between the predicted and measured decrease in flux density.

Evolutionary changes such as those described above can be detected in certain remnants within a human lifetime, but only for the youngest — and therefore the most rapidly evolving — objects. The late stage of evolution can only be studied statistically, using observations of a large number of supernova remnants. Before proceeding with that analysis, we wish to dwell upon an important new fact that has been established within the last decade. We showed in Chapter 1 that the physical basis for differences among young remnants lies in the presence or absence of a pulsar — a source of relativistic particles and magnetic fields. The most clear-cut of these distinctions should show up precisely in the synchrotron emission from those remnants.

Perhaps the most interesting recent achievement in the realm of radio observation of supernovae is the firmly established fact that radio remnants fall into two distinct classes — classical shells and the so-called Crab-like remnants or plerions. The latter term is derived from the Greek πληρης or πληθωρη, meaning *filled*, and it refers to remnants whose brightness increases toward the center. Composite remnants — a plerion within a shell — have also been observed; see below.

The most complete surveys of the data on plerions are those of Caswell (1979), Weiler and Panagia (1980), Weiler (1983), Wilson (1986), and Helfand and Becker (1987).

The modern classification scheme for radio remnants is basically as follows. Objects are classed as shells if:

1. The radio brightness increases from the center toward the periphery, and the radio map forms a shell, either complete or incomplete;

2. The spectral index is greater than $\alpha = 0.2$; the mean is $\bar{\alpha} = 0.5$.

3. There is weak linear polarization, $p = 3 - 15\%$.

4. The magnetic field displays quasiregular structure.

The plerions are those objects for which:

1. Brightness increases toward the center, and shell structure is lacking;

2. The radio spectrum is fairly flat, with $\alpha = 0 - 0.3$. Weiler and Panagia (1980) give a mean value $\alpha = 0.13 \pm 0.1$ based on those objects with

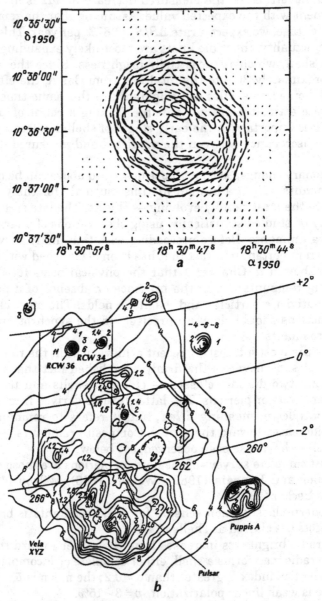

Fig. 61. Radio images of the plerion G 21.5–0.9 (**a**) and the composite remnants Vela XYZ (**b**), G 326.3–1.8 (Weiler, 1983) (**c**), and CTB 80 (Wang and Seward, 1984) (**d**). The lighter curves in panel **d** are radio isophotes, while the heavier ones are x-ray isophotes (see text). The shell remnant Puppis A is visible in panel **b**.

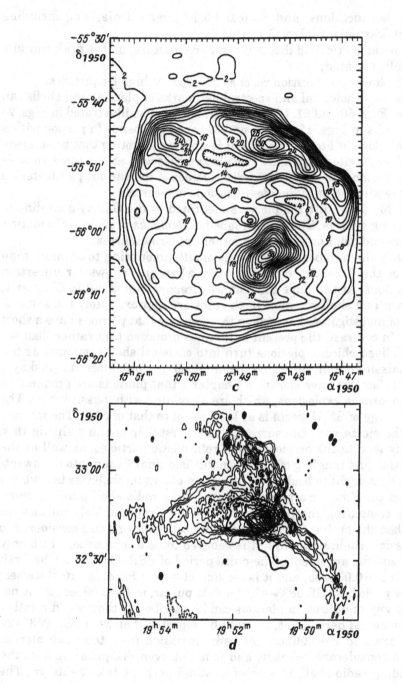

reliable identifications, and $\alpha = 0.26 \pm 0.18$ over all plerions, including some with less reliable identifications;

3. The magnetic field displays regular structure, and is preferentially tangentially oriented;

4. The linear polarization reaches 20–30% at high frequencies.

These morphological and spectral hallmarks of plerions and shells can be seen in Figs. 60 and 61. Classical plerions are also illustrated in Figs. 23 and 28, shells in Figs. 17, 21, 29, and 31, and composites (a plerion with a shell; see below) in Figs. 38 and 40. It is well worth noting that the properties of plerions cited above are not always observed at the same time — there exist remnants with intermediate values of the various parameters, a fact that renders some of the identifications less certain.

In Table 14 we list the plerions identified in the Galaxy according to the foregoing criteria. The column labelled *Remarks* provides information on pulsars and point or extended nonthermal x-ray sources.

Clearly the number of plerions is small, amounting to no more than 5–10% of the Galactic remnants, even when one allows for uncertain identifications. At the same time, as we already pointed out in Chapter 1, the plerion and shell remnant birth rates are approximately the same. A number of investigators have taken this to imply that plerions have a short lifespan. In contrast, the present author has proposed that rather than being short-lived objects, plerions turn into classical shell remnants as the radio emission from the central enhanced region fades (Lozinskaya, 1980b). In fact, we have shown in Chapter 1 that plerions are produced in type II supernova explosions, which are associated with massive stars. The kinetic energy of SN II ejecta is at least equal to that of SN I. The interaction of the ejecta with the surrounding interstellar medium should then inevitably lead to the acceleration of relativistic particles, as well as the compression and tangling of the magnetic field lines frozen into the swept-up gas, i.e., it ought to lead to the formation of a radio shell (see below). We estimated the time for a plerion to turn into a radio shell purely empirically, by comparing the pulse periods of the Crab and Vela pulsars (in 1979, when the author carried out the project, Vela was still considered to be a classical shell; nowadays, it is believed to be a composite, which only strengthens the argument). The pulse period of PSR 0531+21 in the Crab Nebula is $P = 0.033$ sec, and it is slowing at a rate $\dot{P} = 4.23 \times 10^{-13}$ sec/sec; the pulse period of PSR 0833–45, the Vela pulsar, is $P = 0.09$ sec. If the notable activity of pulsars in plerions can be ascribed to their rapid rotation and large period derivatives, it might be expected that were PSR 0531+21 to slow down to $P \approx 0.09$ sec, the radio emission from the Crab plerion would be considerably weaker, and it might even disappear against the surrounding radio shell; this process would perhaps take $t \sim 10^4$ yr. The Vela XYZ remnant is 1.2×10^4 years old, and it possesses a fully formed radio shell. Naturally, this is a rough estimate, since the formation of a

Table 14. Galactic Plerions

Name	Distance, kpc	z, pc	Ang. size, arcmin	Linear size, pc	α	$L_{10^7-10^{11}\,Hz}$ 10^{34} erg/sec	Notes	Refs.
G 5.27−0.90	6	−90	~1.2	~2	0.2		Associated with the shell G 5.4−1.2 and PSR 1758−24?	1
G 20.0−0.2	≈4		10	~12		1.8		14, 15
G 21.5−0.9	5.5−9.3	−85	1.3	2	0	1.8		2, 3, 6, 11
G 24.7+0.6	4.4	50	30×15 (15×10)	40×20	0.17		Size of brighter core in parentheses	3, 4
G 27.8+0.6	1.9−2.3	20	50×30 (10×8)	≈30×20	0.3/1		Size of brighter core in parentheses	3, 4
G 54.09+0.24			2×1.2		0.1−0.2			10, 12
G 57.5−0.3 4C 21.53	2−10	−10 to −50	~1	0.5−3			Compact radio source	5
G 74.9+1.2 CTB 87	11−12	230	9.4×6	30×20	0.24	5−7	X-ray brightness increases toward center. Compact source; extragalactic	7, 8, 9, 13
G 130.7+3.1 (3C 58)	3−4.5	140	10×6	11×7	0.09	2.5−6	Compact radio, x-ray source	see Section 4
G 184.6−58 (Crab)	2	200	5×7	3×4	0.26	18	Pulsar	see Section 4
G 328.4+0.2 MSH 15−57	9[3] 19[8]	30 70	6×5	16×14 35×30	0.24	5.7 26	Possible plerion	3

1. Caswell et al. (1987).
2. Davelaar et al. (1986).
3. Weiler and Panagia (1980).
4. Reich et al. (1984).
5. Purvis (1983).
6. Weiler (1983).
7. Wilson (1983).
8. Sakhibov and Smirnov (1982).
9. Geldzahler et al. (1984).
10. Reich et al. (1985).
11. Becker and Szimkowiak (1981).
12. Velusamy and Becker (1988).
13. Green and Gull (1989).
14. Becker and Helfand (1985).
15. Helfand et al. (1989).

shell depends on the density of the interstellar medium, and the fading of a plerion depends on pulsar energetics. Nevertheless, it is consistent with the more rigorous calculation to which we shall come below.

Assuming, then, that plerions can turn into shell remnants, there ought to exist objects in a transitional stage between the two. Such objects have in fact been detected, and their properties are summarized in Table 15. In each column of that table, the first row refers to the plerion, and the second to the shell; three components have been identified in CTB 80 (see below).

In Fig. 61, there is one representative of the composite-remnant class, Vela XYZ. That object has long been studied in great detail, inasmuch as it is large and bright in visible light, radio, and x rays (see Section 7). As estimated in two independent ways — from the pulsar spin-down rate and from the expansion velocity and shell radius — it is approximately 12000 years old. Weiler and Panagia (1980) have concluded that the Vela X region is a plerion, while regions Y and Z comprise the associated shell. Indeed, the radio spectrum of Vela X is flatter than that of Vela Y and Z (see Table 15); the linear polarization of Vela X is about 20%, while in the remainder of the shell it is several percent at most; taken in isolation, the source Vela X exhibits characteristic plerion morphology; and the pulsar, with its associated nonthermal compact x-ray source, is located within Vela X. The pulsar is rather far off-center, however, with respect to both the shell and plerion, although until recently, no proper motion had been detected, raising a number of questions; see Bignami and Caraveo (1988). The very recent proper-motion measurements by Ögelman et al. (1989) (see Section 7) indicate that the pulsar was born near the center of Vela X.

The extended synchrotron x-ray source in the Vela X region is smaller than the radio-emitting region because of synchrotron energy losses. A compact x-ray nebula 0.1 pc in diameter (designated Vela A in Section 6) has been detected by the Einstein observatory, and Ögelman et al. (1989) recently discovered an optical counterpart of the x-ray nebula. Optical nebulosity centered around the pulsar can be seen in B and V, and is characterized by wisp-like structure. The size of the compact optical nebula, $(2-6) \times 10^{17}$ cm, is comparable to that of the compact x-ray spot, and the optical and x-ray brightness can be related by a power law.

An even more complicated composite remnant is CTB 80. As can be seen in Figs. 61d and 61e, in the radio there are at least three components of different sizes (Angerhofer et al., 1980, 1981; Wong and Seward, 1984; Strom et al., 1984; Strom, 1987): a bright central core about 50″ in size with a flat radio spectrum ($\alpha \sim 0$) and strong polarization, imbedded in a flat plateau (measuring approximately $10' \times 6'$, $\alpha \sim 0.4$); beyond the plateau, there are two extensive jets about 40′ long. At a distance of approximately 2 kpc (see Kulkarni et al. (1987) and references therein), the linear dimensions of the three components are about 0.6 pc, 6×4 pc, and $\sim 20-30$ pc, respectively. The polarization falls off and the radio spectrum becomes steeper at the periphery. The bright core at the center exhibits a

Name	Distance, kpc	z, pc	Ang. size, arcmin	Linear size, pc	α*	$L_{10^7-10^{11}\,Hz}$* 10^{34} erg/sec	$L_{0.3-4\,keV}$* 10^{35} erg/sec	Notes: compact sources (x-ray, radio, and γ)	Refs.
G 0.9+0.1	10	20	2	6	0	7.0	<0.4		1
			8	20	0.45	2.8	<1		
G 6.4–0.1 W 28 (?)	3	–5	1	1	0.2	0.1		Compact radio, x-ray (and γ?) source	2; see §7
			~45	40	0.4	7	1–6		
G 16.37+0.08			4		0.65–0.70				7
			1		0.15				
G 27.4+0.0 (?)	3–26	0	—	—	—			Compact and extended x-ray sources	3
			4	4–38	0.45				
G 29.7–0.3 Kes 75	20	–100	0.5	3	0.25	1.2	40	None; x-ray brightness increases toward center	1, 4, 6
			3	18	0.6	5	<5		
G 34.6–0.5 W 44 (?)	3	–25	—	—				None; x-ray brightness increases toward center	see §7
			27	25	0.3	7.4			
G 39.7–2.0 W 50 (?)	5.2	–200	120×65	~180×96	0.3–0.4	—	1.2 (two "spots")	SS 433	see §7
			—	—		6			
G 68.9+2.8 CTB 80	3	150	0.8	0.7	0	~0.1	—	PSR 1951+32	see §§5, 10
			10×6	9×5	0.33	0.2	0.3		
			~40	~38	0.84	1.5	—		
G 263.9–3.3 Vela XYZ	0.5	–30	210×110	31×17	0.1	2.5	0.1	PSR 0833–45	see §§7, 10
			260	37	0.6	0.3	3		
G 320.4–1.2 MSH 15–52	4	–85	~8	~10	—	<0.3	2	PSR 1509–58	see §7
			30	3.5	0.3	3	4		
G 326.3–1.8 MSH 15–56	5	–160	15×8	22×12	0.1	7			
			~3.6	~50	0.4	6.6			
G 327.4+0.4 Kes 27 (?)	6	40	—	—	—	—		Compact and extended x-ray sources	
			20	35	0.6	2			
G 332.4–0.4 RCW 103 (?)	6	–40	—	—	—	—		Compact x-ray source	
			10	18	0.5	2			

Remarks: The first line in each row refers to the plerion, and the second to the shell. A question mark (?) denotes an object whose identification as a composite is either uncertain, or has been made only on the basis of its having a compact central source; see also Helfand and Becker (1987).

1. Helfand and Becker (1987).
2. Weiler (1983).
3. Kriss et al. (1985).
4. Becker and Helfand (1984).
5. Blair et al. (1984b).
6. Becker et al. (1983).
7. Helfand et al. (1989).

shell-like distribution of radio emission; the magnetic field in the shell is regular and oriented tangentially.

The shell structure of the central core is also easily visible optically (Blair et al., 1984b; Fesen and Gull, 1985; Strom and Blair, 1985; van den Bergh and Pritchet, 1986). filaments bright in [O III] form a central ring that coincides with the radio shell. That same bright shell is visible in Hα + [N II], as well as a faint filamentary nebula (elongated in the same direction as the radio plateau) 40″×75″ across.

The filaments have low radial velocities, about 30–40 km/sec. But in the bright western filament, a displacement of 2.″5 has been measured in

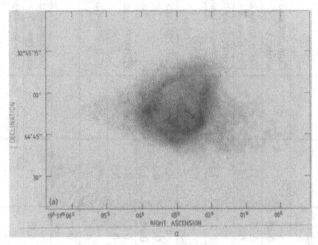

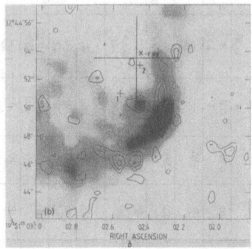

Fig. 62. Central region of CTB 80. Radio image at λ = 20 cm (**a**) and an enlargement of the vicinity of the pulsar — the "hot spot" and compact x-ray source (**b**); see Strom (1987).

28 years, corresponding to a velocity of ~500 km/sec at 2 kpc (Strom and Blair, 1985). New optical observations — spectrophotometry, imagery, and kinematics of the compact core of the peculiar supernova remnant — have been obtained by Kulkarni (1989) and Whitehead et al. (1989).

Both compact and extended x-ray sources have been detected in CTB 80, as shown in Fig. 61e (Becker et al., 1982; Wong and Seward, 1984). At ~2 kpc, the luminosity of the extended x-ray nebula is $L_{0.2-4\,keV} = 1.4 \times 10^{34}$ erg/sec, and that of the compact source is $L_{0.2-4\,keV} \sim 10^{34}$ erg/sec. Recent EXOSAT measurements in the 0.05–6 keV range (Angelini et al., 1988) show that the x rays conform well to a power law with an exponent of 0.9, which is typical of Crab-like supernova remnants (early Einstein data yielded a steeper spectrum because of overestimated column density of the absorbing gas; see Angelini et al., 1988). With the power-law spectrum and a column density of $N_H = 2 \times 10^{21}$ cm^{-3}, the EXOSAT measurements yield a total remnant luminosity $L_{0.05-6\,keV} = 10^{34}$ erg/sec.

The identification of the compact x-ray source with a pulsar has been admirably confirmed. In 1987, Strom discovered a compact radio source within the x-ray error box, having a very steep spectrum — $\alpha \sim -2$ or steeper — and a high degree of polarization, $p \approx 30\%$ at 20 cm. These properties were most suggestive of an association between the compact source and a pulsar; Strom's paper concluded with the remark that "... an essential element — pulsed emission — has yet to be detected." The pulsar was found in that same year, when Clifton et al. (1987) discovered PSR 1951+32. According to Kulkarni et al. (1988) and Fruchter et al. (1988), $P = 39.53 \pm 0.001$ msec, $\dot{P} = (5.92 \pm 0.06) \times 10^{-15}$ sec/sec, and the dispersion measure is DM $= 50 \pm 8$ pc·cm^{-3}. (The dispersion measure yields a pulsar distance of about 1.4 kpc; previous distance estimates to CTB 80 were in the neighborhood of 3 kpc. It is therefore reasonable to adopt a distance of ~2 kpc.) The pulsar's characteristic age is $t \sim P/2\dot{P} \sim 10^5$ yr.

The brightest feature of the shell at the center of the core of CTB 80 is the so-called hot spot (Strom, 1987), which may possibly be an analog of the central wisp in the Crab, and which identifies the "frontal" thermalization zone of the cloud of pulsar wind as it impacts the ejected gas. In both remnants, the pulsars are moving directly toward their respective impact zones, at ~125 km/sec in the Crab and approximately 300 km/sec in CTB 80 (the latter estimate was obtained indirectly via scintillation measurements, which revealed abnormally rapid fluctuations (Fruchter et al. (1988)). The elongated shape of the plateau is associated with east-to-west motion of the pulsar.

More recent work has completely elucidated the nature of this unusual supernova remnant. During the IRAS survey of SNR's, Fesen et al. (1988) discovered a 64′-diameter shell-like supernova remnant encompassing CTB 80 and centered 30′ east of CTB 80's core. A relationship between the extended radio jets of CTB 80, the infrared shell, and the central core containing the pulsar is suggested by the remarkable positional coincidence, together with similar distance and age estimates. Fesen et al. proposed the following model. About 10^5 years ago, a supernova explosion generated an expanding shell and a rapidly rotating pulsar with a westward space velocity of about 300 km/sec. This westward motion is consistent with scintillation measurements (Fruchter et al., 1988), and explains the radio morphology of the core, the plateau, and the eastern ridge, and possibly the rapid optical changes in the core. The pulsar's location right near the shell's western edge suggests that the extended northern and southwestern ridges result from the injection of pulsar-generated relativistic particles into the compressed magnetic field of the very old, slowly expanding shell.

W 28, which we discussed in Section 7, may possibly also be a composite remnant. Within the nonthermal shell lies the compact radio source G 6.6–0.1, which has a flat spectrum ($\alpha = 0.2$). High-resolution measurements have demonstrated that this compact source consists of a $3'' \times 7''$ core and a fainter halo $45''$ across. The spectrum of the core has $\alpha = 0.1 \pm 0.1$, and the degree of polarization is 10%. A compact x-ray source has been detected $1.'7$ (1.2 pc) from the core (Andrews et al.,, 1983). In addition, Shull et al. (1989) have suggested a possible interaction between the supernova remnant W 28 and PSR 1758–23, which has $P = 0.415$ sec (Manchester et al., 1985b), but there is no positional coincidence of the pulsar with the compact radio and x-ray sources.

No more than half the composite remnants display a clear-cut shell + plerion structure; the rest are assigned to the class by virtue of the observation within the shell of a pulsar, compact x-ray source, or extended x-ray source with a hard spectrum and enhanced central brightness. Examples of this sort of identification are the pair MSH 15–52 and W 50, as discussed in Section 7. It is also conceivable, as we pointed out in Section 6, that in specific instances one has a random coincidence, as projected onto the plane of the sky, of diverse compact and extended objects.

Recently, nonthermal radio sources have been found which, while they are possible supernova remnants, fit into none of the three categories discussed above — shell, plerion, or composite. The objects in question are G 357.7–0.1 and G 5.3–1.0 (now known as G 5.4–1.2) (Shaver et al., 1985; Becker and Helfand, 1985; Helfand and Becker, 1985). Both radio sources display highly unusual axisymmetric structure with perpendicular ridges of emission (radio filaments). Polarization measurements suggest regular field structure, with the field lines preferentially oriented along the bright filaments, and $p \sim 10\%$; the radio spectrum has $\alpha \approx 0.25 - 0.45$. Both objects are far away, but the distances are poorly determined; characteristic dimensions along their major axes are in the $20 - 50$ pc range. In both, the compact radio source is right on the symmetry axis, but beyond the extended source, which is brighter in the direction of the compact source. The integrated luminosity in the range $10^7 - 10^{11}$ Hz is $10^{35} r_{10}^2$ erg/sec for G 357.7–0.1 and $2 \times 10^{35} r_{10}^2$ erg/sec for G 5.3–1.0, where r_{10} is the distance in units of 10 kpc.

No sooner were these objects discussed than they were immediately likened to the narrow jet of relativistic particles sometimes seen to emerge from binary systems. It is certainly conceivable that future detailed studies will blunt the novelty of these remnants.

Recent observations of G 5.3–1.0 have been carried out by Caswell et al. (1987); see also Shull et al. (1989). They have shown that the compact source G 5.27–0.90 (the "Goose's Head") is connected eastward by a thin bridge of emission to a fan-shaped emission region of G 5.3–1.0 approximately ~30′ long, with a sharp western edge. Together with a fainter arc of radio emission of about the same size located ~30′ to the east, this region may comprise a single shell. The 125-msec pulsar PSR 1758–24 (Manchester et al, 1985b) is situated near the compact source, and they may possibly be related, although the pulsar is quite faint and its position is poorly known.

The suggestion by Fesen et al. (1988d) that a rapidly moving pulsar may rejuvenate a related old supernova remnant by injecting relativistic particles into the old shell (as in CTB 80) raises an interesting possibility. According to Fesen et al. (1988d) and Shull et al. (1989), the G 5.3–1.0/G 5.27–0.90/PSR 1758–24 complex looks like a more evolved version of the CTB 80 interaction between a fast-moving pulsar and an old supernova remnant. In fact, in the case of G 5.3–1.0, a pulsar and its cloud of relativistic plasma — the plerion G 5.27–0.90 — has already penetrated the old, slow-moving shell and left it behind.

There are still a number of other supernova remnants that seem odd (see, e.g., Caswell (1988)); one example is G 179.0+2.7 (Fürst and Reich, 1986). The nature of G 18.95–1.1 has also long been unclear; new observations by Patnaik et al. (1988) indicate that it is a normal shell-like supernova remnant with a central source perhaps resembling the accreting binary SS 433.

How does modern theory account for the origin of the magnetic field and relativistic particles in plerions and shell remnants? As paradoxical as it may sound, although the latter have been studied since the dawn of radio astronomy, while plerions were identified as a class only recently, it is the plerions whose radio emission now seems to be better understood. The point is that in plerions, the source of particles and fields is obvious, while in shells, it is not entirely clear where they originate, although a number of particle-acceleration and field-enhancement mechanisms have been considered.

The main energy source for the radio and x-ray synchrotron emission from plerions is a pulsar. The evidence includes brightening toward the center, rather than toward the rim, rapid brightness fluctuations associated with the ejection of clouds of relativistic plasma in the central wisps of both the Crab Nebula and the core of CTB 80, and the concentric pattern of magnetic field lines about PSR 0531+21.

In plerions, the extended sources of hard x rays are smaller than the radio synchrotron-emitting region, and the pulsar is centered on the x-ray source, not the radio source. This makes sense if the moving pulsar continuously furnishes relativistic electrons: electrons radiating x rays have lifetimes of years, while radio-emitting electrons have lifetimes of centuries (Eq. (10.6)).

Without dwelling on the still controversial relativistic-particle injection process in a pulsar, let us examine (following Pacini and Salvati (1973)) the synchrotron emission from a source driven by the continuous injection of relativistic electrons and magnetic fields, bearing in mind that the required energy comes from spin-down of a young pulsar due to magnetic dipole radiation. The rate of rotational energy loss from a neutron star is given by

$$L(t) = \frac{2L_0}{(1+t/\tau)^\beta},$$ (10.15)

where $\beta = (n+1)/(n-1)$, the braking index $n \equiv \ddot{\Omega}\Omega/\dot{\Omega}^2$, Ω is the pulsar angular velocity, t is the pulsar age, and τ is its characteristic time for period decay. The spin-down parameter β ranges between 1.8 and 2.7; $\beta = 2$ for a dipole, and $\beta = 2.3$ for the Crab pulsar.

Three stages can be identified in the evolution of a plerion. The first ($t < R_0/v$, where R_0 is the initial stellar radius and v is the plerion expansion velocity) lasts a few days, and is of no interest to us here. The second stage spans the period $R_0/v < t < \tau$; the plerion radius increases, but the pulsar has not yet slowed significantly, and it is supplying energy at some fixed efficiency. The third stage ($t > \tau$) is characterized by a reduction in the amount of energy injected.

The total magnetic field energy of an expanding plerion, $W = H^2R^3/6$, is governed by the equation

$$\frac{dW}{dt} = \frac{L_0}{(1+t/\tau)^\beta} - W\frac{v}{R},$$ (10.16)

where R is the plerion radius, and the velocity v is taken to be constant; the second term takes into account the decrease in energy upon expansion. Integration of (10.16) with $\beta \geq 2$ yields the magnetic field of the nebula in the three stages:

$$H_I^2 = H_0^2 + \frac{6L_0 t}{R_0^3}, \quad H_{II}^2 = \frac{3L_0}{v^3 t^2}, \quad H_{III}^2 = \frac{6L_0\tau^2}{(\beta-1)(\beta-2)v^3 t^4}.$$ (10.17)

At a given stage, the radio emission from a plerion is determined by the influx of field and particle energy from the pulsar, the dissipation of the field due to expansion, and the falloff in relativistic electron energy due to synchrotron and adiabatic losses. The injected particle energy spectrum takes the usual form (10.1) over the energy range $0 \le E \le E_{max}$. The quantity K_0 is determined by the requirement that

$$L_0 = \int_0^{E_{max}} EN(E)dE.$$

It can be assumed here that $N(E) = 0$ for $E > E_{max}$ and at $t = R_0/v$, since particles injected during the initial stage I are subject to very rapid synchrotron losses. By making use of Eqs. (10.6)–(10.13), Pacini and Salvati obtained the energy spectrum of particles in phase II, assuming a constant plerion expansion velocity:

$$N(E,t) = \frac{K_0}{\gamma} tE^{-\gamma}, \quad E < E_b = \frac{2v^3 t}{3c_1 L_0},$$

(10.18)

$$N(E,t) = \frac{K_0 v^3 t^2}{3c_1 L_0 (\gamma - 1)} E^{-\gamma - 1}, \quad E > E_b.$$

Here c_1 is a constant. This is clearly the canonical form of the spectrum when low-energy particles lose energy principally through adiabatic expansion, and high-energy particles, through synchrotron emission. The difference is that E_B, the break in the energy spectrum, moves to higher and higher energies with time. This is related to the fact that the magnetic field varies in a different way here: we have (10.17) instead of $H = H_0 (R_0/R)^2$ as when magnetic flux is conserved. The plerion spectrum at stage II takes the form

$$S_\nu(t) = \frac{K_0 c_2}{2\gamma} \left(\frac{3L_0}{v^3} \right)^{\frac{\gamma + 1}{4}} c_2^{\frac{\gamma - 2}{2}} t^{\frac{1 - \gamma}{2}} \nu^{\frac{1 - \gamma}{2}}, \quad \nu < \nu_b,$$

(10.19)

$$S_\nu(t) = \frac{K_0}{2(\gamma - 1)} \left(\frac{3L_0}{v^3} \right)^{\frac{\gamma - 2}{4}} c_2^{\frac{\gamma - 2}{2}} t^{\frac{2 - \gamma}{2}} \nu^{\frac{\gamma}{2}}, \quad \nu > \nu_b.$$

The value of the constant c_2 is dictated by the requirement that the radiation from relativistic electrons have its maximum at a frequency $\nu = c_2 HE^2$.
In stage III, energy injection is moderated:

$$N(E,t) = \frac{K_0}{(1+t/\tau)^\beta} E^{-\gamma} \approx K_0 \left(\frac{\tau}{t}\right)^\beta E^{-\gamma}. \tag{10.20}$$

The break in the energy spectrum

$$E_B = \frac{t^3}{M\tau^4}, \quad M = \frac{3L_0 c_1}{2(\beta-1)(\beta-2)v^3\tau^2} \tag{10.21}$$

shifts rapidly with time. Pacini and Salvati have distinguished two cases in the evolution of the energy spectrum.

Case A: the maximum energy of the injected particles is no greater than the energy at the synchrotron break at time $t = \tau$: $E_B(\tau) > E_{max}$, and only adiabatic losses are important. In the energy range $0 < E < E_{max}(\tau/t)$, particles injected before or after time $t = \tau$ display an energy spectrum at the time of observation equal to

$$N(E,t) = \frac{K_0\tau^\gamma}{\gamma}t^{1-\gamma}E^{-\gamma} \quad \text{or} \quad N(E,t) = \frac{K_0\tau^\gamma}{\beta-\gamma}t^{1-\gamma}E^{-\gamma} \tag{10.22}$$

respectively. The spectrum of high-energy particles with $E_{max}(\tau/t) < E < E_{max}$ is governed solely by "fresh," newly-arrived particles, and takes the form

$$N(E,t) = \frac{K_0\tau^\beta E_{max}^{\beta-\gamma}}{\beta-\gamma}t^{1-\beta}E^{-\beta}, \tag{10.23}$$

i.e., it depends not on the injected spectrum, but on the rate at which the pulsar is slowing down.

Case B: here $E_B(\tau) < E_{max}$, and the energy spectrum is determined by adiabatic and synchrotron losses. There are three relevant energy ranges, for which the power-law exponent k in the spectrum $N(E) \propto E^{-k}$ is equal to γ, $(\beta+3\gamma)/4$, and $(\gamma+1)$. After reaching time $t_* = (M\tau^4 E_{max})^{1/3}$, corresponding to $E_{max} = E_B$, evolution is again governed solely by adiabatic losses. The power-law exponents pertinent to the frequency and time dependence of the synchrotron emission flux density are given in Table 16.

Naturally, any specific plerion may differ substantially from the idealized picture presented here. The most serious simplifying assumptions are as follows:

1. The power law for the energy spectrum of injected relativistic particles has been assumed constant, although in actual pulsars it most certainly varies with time.

Table 16. Synchrotron Emission from a Plerion in the Adiabatic Stage of Development (Pacini and Salvati, 1973). Frequency breakpoints: $v_{cr} \propto t^{-4}$, $v_b \propto t^4$. **Flux density:** $S_v \propto t^q v^\alpha$.

	q	α	v
Case A	-2γ	$(1-\gamma)/2$	$v < v_{cr}$
	-2β	$(1-\beta)/2$	$v > v_{cr}$
Case B $t < t_*$	-2γ	$(1-\gamma)/2$	$v < v_{cr}$
	$-(\beta+3\gamma)/2$	$(4-\beta-3\gamma)/8$	$v_{cr} < v < v_b$
	$2-\beta-\gamma$	$-\gamma/2$	$v > v_b$
Case B $t > t_*$	-2γ	$(1-\gamma)/2$	$v < v_{cr}$
	$-(\beta+3\gamma)/2$	$(4-\beta-3\gamma)/8$	$v_{cr} < v < v_b(t_*)t_*^4/t^4$
	-2β	$(1-\beta)/2$	$v > v_b(t_*)t_*^4/t^4$

2. The assumptions made about the way in which the magnetic field varies are highly arbitrary. In general, we still know little about the mechanism responsible for generating the magnetic field and relativistic particles from a neutron star.

3. The pulsar has been assumed to be the sole source of energy, and the plerion has been treated as unrelated to the ejected shell and swept-up circumstellar gas, although their interaction has important effects on the dynamics of the cavity filled with the "pulsar wind," a relativistic plasma plus magnetic field.

The problem has been solved by Reynolds and Chanan (1984) in a somewhat more general form. They take the variation of plerion radius with time to be $R \propto t^\eta$ (Pacini and Salvati take $\eta = 1$) and the variation of the magnetic field to be $H \propto t^{-b}$; they do not require equality of the energy provided to the field and particles by the pulsar, and they assume a constant ratio between the two.

How do the observed plerions fit into this idealized picture? Estimating the characteristic spin-down time τ from the braking given by magnetic dipole radiation,

$$P^2 = P_0^2 \left(1 + \frac{2\dot{P}_0 t}{P_0}\right), \tag{10.24}$$

we can relate τ to the observed parameters P, $\dot{P}$, and t:

$$\tau \sim \frac{P_0}{2\dot{P}_0} = \frac{P}{2\dot{P}} - t. \tag{10.25}$$

For PSR 0531+21 in the Crab Nebula, $t = 930$ yr, $P = 0.03$ sec, and $P = 4.2 \times 10^{-13}$ sec/sec, which yields $\tau \approx 300$ yr. Integrating (10.15) with $\beta = 2$, it can be shown that the pulsar will transfer approximately 50% of its rotational energy to the plerion in a time τ. This then means that the transition from stage II to stage III characterizes the transition from the plerion formation stage to the adiabatic expansion stage (Weiler and Panagia, 1980). Comparing this with the results in Section 9, it is clear that the characteristic time τ is close to the time t that it takes for the gaseous shell ejected in the explosion to evolve from the free expansion stage into the adiabatic expansion stage, $t \leq 10^3$ yr.

The rotational energy losses of five pulsars associated with supernova remnants are greater than the total synchrotron luminosity of the corresponding plerions at radio, optical, and x-ray wavelengths.

The rotational kinetic energy of the Crab pulsar is approximately 10^{49} erg; comparing this with the mean kinetic energy of matter ejected in SN II explosions, $E_0 \sim 5 \times 10^{50}$ erg, it is clear that the presence of a pulsar should not seriously affect shell kinematics, even initially; the same is true — all the more so — in the late stages of evolution. This would explain why the evolutionary sequence of supernova remnants obtained in Section 9 comprises a unique set of objects, despite the existence of two types of outburst. In the Crab Nebula itself, because of the low expansion velocity of the filaments, rotational energy losses of the pulsar are comparable with the kinetic energy of the ejected shell, which is why the pulsar wind, as we have shown in Section 4, is ultimately the decisive factor in the kinematics and radiation of the nebula.

As soon as the critical frequency ν_{cr} in the synchrotron spectrum shifts down to the radio, the plerion brightness falls off rapidly with frequency and time. The plerion ceases to be observable at radio wavelengths, turning first into a composite and then into a shell source. If the ambient gas density is high, this transformation precedes the complete dissipation of the plerion by a wide margin, and these are then the objects that are observed as composite radio remnants.

The x-ray emission from plerions is also synchrotron radiation. In composite remnants, in addition to synchrotron x-ray emission from relativistic electrons injected by the pulsar, one also sees thermal x-ray emission from the plasma behind the blast wave front (see Section 8).

The synchrotron mechanism behind the x-ray emission of plerions and its connection with the central pulsar is confirmed by a filled-center brightness distribution, a power-law spectrum (in those objects in which it has been studied), and, in the Crab Nebula, strong linear polarization (Becker, 1983).

As yet, optical synchrotron radiation due to relativistic particles injected by a pulsar has only been detected in one object besides the Crab — the remnant 0540–69.3 in the LMC (see Section 4). Although there is still no rigorous quantitative interpretation of observational results for this nebula due to a paucity of data, one can still make some crude estimates.

The overall slope of the synchrotron spectrum of 0540–69.3 (Fig. 24) yields $\alpha = 0.8$ all the way from the radio to x rays, which corresponds to a relativistic particle energy spectrum $N(E) \propto E^{-2.6}$. The plerion radius at the $0.1 I_{max}$ level is 4″, corresponding to a linear radius of 1 pc and a radiating volume $V \sim 10^{56}$ cm^{-3}. Equations (10.2)–(10.4) yield a magnetic field strength $H \sim 2 \times 10^{-4}$ G and a total particle+field energy $W \sim 10^{47}$ erg, which is at most 2% of the pulsar rotational energy lost over a time span $t \sim P/2\dot{P} = 1.7 \times 10^3$ yr at a loss rate $L \sim 1.5 \times 10^{38}$ erg/sec (see Chanan et al. (1984)). (An analysis carried out by Reynolds (1985) that takes the interaction of the pulsar wind with the ejected shell into consideration (see below) yields a higher magnetic field strength in the pulsar cavity, but the estimates remain self-consistent.) At $H = 2 \times 10^4$ G, the synchrotron loss time for particles radiating x rays, as given by (10.6), is about 20 years. Since there is no break in the spectrum between the x-ray and optical ranges, it would not be surprising if that whole range were dominated by synchrotron losses and the corresponding energy spectrum of injected particles took the form $N(E) \propto E^{-1.6}$, which is a close approximation to the radio spectrum of the Crab Nebula. The predicted break in the spectrum of 0540–69.3 for $H = 2 \times 10^{-4}$ G and $t \sim 10^3$ yr comes at $\nu_b \sim 10^{14}$ Hz. Allowing for synchrotron losses and extrapolating the derived optical and x-ray spectrum to radio frequencies, one can see that the predicted flux density $S_{1GHz} \sim 0.1$ Jy is only 10% of the observed value (Chanan et al., 1984). Such is the contribution to the radio emission from particles injected by the pulsar; the remaining $\sim 90\%$ of the flux is probably emitted from the shell. According to Reynolds (1985), the initial rotational energy of the pulsar was $(1.5 - 4.2) \times 10^{49}$ erg, and the initial rotation period was $P_0 = 30 \pm 8$ msec, which comes fairly close to the properties of the Crab pulsar.

Since adiabatic cooling of relativistic particles has a significant influence on the luminosity of a plerion, making allowance for the actual dynamics of the cloud of relativistic plasma that interacts with the ejecta and swept-up interstellar gas is the next essential step in an investigation of the evolution of remnants of this type; this step was taken by Reynolds and Chevalier (1984), who examined two possibilities for the relationship between the particles injected by the pulsar and the magnetic field. In the first, as in the calculations of Pacini and Salvati, their interaction was neglected, so that field and particle energies vary independently in the course of evolution (the non-equipartition model, "NE"). In the second (the equipartition model, "E"), particle interactions with the field permanently enforce approximate energy equality between the two. The ejecta are also modeled in two ways: as a homogeneous shell, or a rapidly expanding shell (of mass M_s) with a slowly expanding interior layer of denser core matter (the mass of the core $M_c < M_s$); see Fig. 63. The calculations for the two-component ejecta were carried out for a velocity of the outer edge of the core matter moving at a velocity $v_1 = 300$ km/sec, and the inner and outer edges of the shell moving at $v_2 = 3000$ km/sec and $v_3 = 8000$ km/sec.

At the beginning of evolution, the cloud of injected relativistic particles plus field expands in a homogeneous medium of core matter, sweeping the latter up into a shell. As long as the pulsar provides the required energy, the expansion accelerates and the mass of the swept-up material increases. When the influx of energy abates, the cloud of relativistic particles will no longer overtake the ejecta, but it will continue to expand unimpeded until it encounters the reverse shock wave. If the swept-up matter of the core, which is expanding and accelerating, overtakes the high-velocity shell prior to the appearance of the reverse shock, their interaction will bring about a complicated redistribution of density and velocity (the latter quickly becomes constant). After passage of the reverse shock, the pressure in the ejecta swept up by the pulsar wind equalizes with the pressure behind the blast wave front in the adiabatic stage, and the radius of the cloud of relativistic plasma varies with time as $R \propto t^{0.3}$, i.e., expansion rapidly slows down.

These purely dynamical variations are superimposed on the time-variant rate of injection, and are accompanied by the aforementioned "pure" variations in the energy spectrum of the relativistic electrons and magnetic field resulting from adiabatic and synchrotron losses.

Reynolds and Chevalier have identified four stages in the evolution of a cloud of relativistic particles that interacts with ejected matter. In the first stage, the cloud expands in matter from the core (or in homogeneous ejecta) until such time as it either sweeps it up entirely or the rate of energy injection declines at time $t = \tau$. If the former case transpires first, then the second stage begins, wherein a shell of fixed mass continues to be accelerated by the pulsar wind. The third stage begins at time $t = \tau$; if the ejecta are not swept up completely prior to the beginning of meaningful pulsar slowdown, the plerion will pass directly from stage I to stage III. The beginning of stage four is signified by passage of the reverse shock.

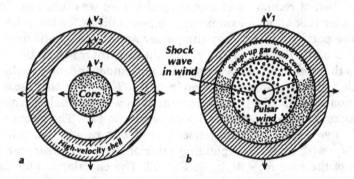

Fig. 63. Two-component ejecta: core plus high-velocity shell (**a**); interaction between pulsar and ejecta (Reynolds and Chevalier (1984).

The time at which the reverse shock comes into being is given by (Reynolds and Chevalier, 1984)

$$t_{rev} \approx 10^4 \left(\frac{M_0}{15 \, M_\odot} \right)^{5/6} E_{51}^{-1/2} n_0^{-1/3} \text{ yr.} \tag{10.26}$$

For $M_0 = 1 M_\odot$ and $n_0 = 0.1 - 1 \text{cm}^{-3}$, we have $t_{rev} \approx 10^3$ yr. Compression of the cloud of relativistic particles and magnetic field by the reverse shock enhances the synchrotron losses and alters the plerion luminosity at time t_{rev}. The luminosity jumps after a monotonic decline, and then starts to fall

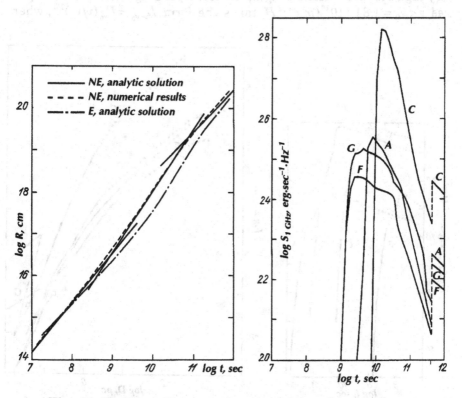

Fig. 64. Plerion radius as a function of time for two numerical models (Reynolds and Chevalier, 1984). In Figs. 64–67, E denotes energy equipartition between magnetic field and relativistic particles, and NE denotes nonequilibrium. Solid curves denote the analytic solution.

Fig. 65. Plerion radio luminosity: F) homogeneous ejecta (NE); A) core and shell ejecta (NE); G) homogeneous ejecta (E); C) core and shell ejecta (E) (see caption to Fig. 64). The dashed line corresponds to the sharp jump associated with shock-wave passage.

off again more gradually. The final results obtained by Reynolds and Chevalier are summarized in Figs. 64–67. Figure 64 displays the time variation of plerion radius; Fig. 65 shows the variation of synchrotron radio luminosity for two cases, equipartition of magnetic field and relativistic particle energy (E) or the lack of it (NE), and for two models of the ejecta, homogeneous shell (curves F and G), and shell plus core (curves A and C).

The extremely rapid intensity rise in the early stage results from two effects. At the very beginning, in the course of the first year or two of the pulsar's existence, we have for the energy of electrons radiating at frequency v the expression $E_v \equiv 7 \times 10^{-10} (v/H)^{1/2} < m_e c^2$ by virtue of the very high magnetic field strength; the corresponding spectrum at frequencies $v < v_0 = 1.81 \times 10^{18} (m_e c^2)^2 H$ takes the form $L_{v<v_0} = L_{v_0}(v/v_0)^{1/3}$, where

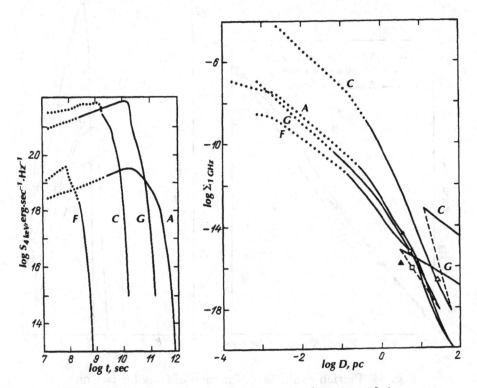

Fig. 66. Plerion x-ray luminosity as a function of time (same notation as Figs. 64, 65). The strong absorption region is indicated by dotted curves.

Fig. 67. Reynolds and Chevalier's Σ–D relationship for plerions (same notation as Figs. 64, 65). Dotted curves indicate strong absorption. Symbols show observational results for known plerions: the Crab Nebula (filled circle), G 29.7–0.3 (filled triangle), G 21.5–0.9 (open circle), 3C 58 (open square), and G 74.9+1.2 (open triangle).

L_{v_0} is the canonical synchrotron luminosity (Reynolds and Chevalier, 1984).

The second effect takes place over a longer period, and consists of free–free absorption within the shell, which quickly falls off as the ejecta expand over a characteristic time

$$t_{f-f} \approx 500 T_4^{-3/10} \left(\frac{M_c}{2\,M_\odot} \right)^{-1/5} \left(\frac{v_1}{300\,\mathrm{km/sec}} \right)^{-1} v_9^{-2/5} \ \mathrm{yr}. \qquad (10.27)$$

Here T_4 is the temperature of the freely expanding ejecta in units of 10^4 K, and v_9 is the frequency in units of 10^9 Hz. (There is much less absorption in the rapidly expanding shell than in the core.) Consequently, as long as the matter making up the ejecta is neither too hot nor too clumpy, the plerion will be invisible at radio wavelengths for the first 100–200 years. It is possible, however, for the ejecta to become clumpy as a result of Rayleigh–Taylor instability at its interface with the pulsar wind right at the start of plerion expansion (see Bandiera et al. (1984)), and for the shell to become transparent much sooner.

We emphasize here that the present discussion centers on the radio visibility of a plerion. The pulsar itself ought to make its appearance considerably sooner. The explosion of SN 1987A launched an enormous number of studies that predicted both how and when the pulsar and its cloud of relativistic plasma would appear over the full range of the electromagnetic spectrum; see Section 2. Time will tell just how this actually comes to pass.

The jump in luminosity at the late stage $t \approx t_{rev}$ is associated with the reverse shock wave. Since the reverse shock abruptly changes the luminosity of the cloud of relativistic particles in a rather complicated way, the time span from t_{rev} to $2t_{rev}$ is indicated only schematically in Fig. 65, by a dashed line.

Figure 66 shows the behavior of the synchrotron x-ray luminosity of a plerion, the calculations having been carried out for the same models. The dotted portions of the curves represent the strong absorption regime prior to the time

$$t_{(\tau_x=1)} = 60 \left(\frac{h\nu}{4\,\mathrm{keV}} \right)^{-4/3} \left(\frac{M_c}{2\,M_\odot} \right)^{1/2} \left(\frac{v_1}{300\,\mathrm{km/sec}} \right)^{-1} \ \mathrm{yr}, \qquad (10.28)$$

and an earlier time for homogeneous ejecta.

Once the wide-ranging and nonmonotonic change in plerion radio and x-ray luminosity is taken into consideration, it becomes quite clear that the observed ratio L_x/L_{radio} cannot serve as an age indicator, as has been assumed by many authors (Reynolds and Chevalier, 1984).

Despite the significant differences among the various models, the overall trend of the temporal variation of radio luminosity remains the same: for $t < \tau$ and $\nu > \nu_B$, $L_\nu \propto t^{b(2-\gamma)/2}$, i.e., the luminosity grows with time; however, free–free absorption can mask this growth. (Here b is the exponent that defines the variation of magnetic field strength, $H \propto t^{-b}$.) Above the cutoff in the relativistic particle spectrum, the luminosity varies as $L_\nu \propto t^{1-b(\gamma+1)/2} \propto t^{-0.6}$ up to time $t = \tau$; afterwards, the luminosity drops abruptly: $L_\nu \propto t^{-2}$ to $L_\nu \propto t^{-4}$. In stage IV, after the reverse shock has passed, the plerion luminosity changes slowly: $L_\nu \propto t^{-1}$.

Figure 67 shows the predicted dependence of radio surface brightness $\Sigma_\nu \equiv L_\nu / 4\pi r^2$ on linear diameter D, the so-called Σ–D relationship. This relationship has played an important role in the study of supernova remnants, since on the one hand, it constitutes a strictly observational test of the theory of radio synchrotron emission from remnants, and on the other, it provides an empirical method for estimating the distance to individual remnants. The different numerical models are indicated in Fig. 67 by the same symbols as in the previous figures; the dotted curves signify those regions in which absorption precludes observations of the plerions. The curves also carry time markers. The abrupt jump late in the development $(t \approx t_{rev})$ is associated with the reverse shock wave. Plerions having more or less reliable distances are noted on the theoretical Σ–D relationship. The theoretical predictions are clearly quite satisfactorily consistent with the observations, notwithstanding the arbitrariness involved in choosing the many parameters governing the dynamics of the cloud of relativistic particles, the behavior of the magnetic field, and the spectrum of the injected particles.

The weakest link among the simplifying assumptions that have been made shows up immediately when one compares theory with experiment: the energy spectrum of the injected particles is poorly represented by the simple power law (10.1); see Reynolds and Chevalier (1984) and Reynolds and Chanan (1984). The observed radio luminosity of plerions is consistent with the values that have been calculated assuming a simple energy spectrum, but the x-ray luminosity is then too high. In other words, the spectral index of synchrotron radiation over the range from radio to x rays is inconsistent with theoretical predictions. This problem can be side-stepped by postulating the existence of a two-component spectrum of injected relativistic particles: relativistic electrons produced at the pulsar surface are most likely to be responsible for the radio emission, while high-energy electrons possibly produced near the light cylinder or accelerated by shock waves in the pulsar wind zone would radiate x rays (Reynolds and Chevalier, 1984; Reynolds and Chanan, 1984).

It must be noted immediately that there is still a genuine dearth of observational data with which one might construct an empirical evolutionary Σ–D sequence for plerions. There are few Galactic plerions whose distance has been measured, and not all of them have been adequately studied in x rays. Apart from those in the Large Magellanic Cloud, we know essentially

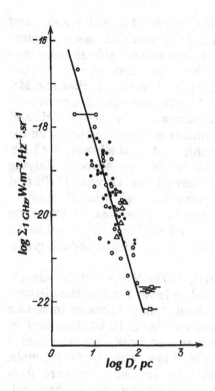

Fig. 68. Empirical relationship between radio brightness and supernova remnant diameter, as developed by the author. Empty circles represent Galactic remnants (these have been used to find the straight line corresponding to Eq. (10.30)); filled circles are remnants in the LMC, M33, and M31; squares are large-scale Galactic radio loops (see Section 11); triangles represent the shell components of the Galactic composite remnants W 28, G 326.3+1.8, CTB 80, and Vela XYZ.

nothing at all about plerions in nearby galaxies whose distances are ostensibly known.

The opposite is true of investigations of shell supernova remnants. Observations of a great many Galactic and extragalactic objects of this type have enabled researchers to put together a completely satisfactory empirical Σ–D sequence that reflects the falloff in radio brightness with increasing diameter. At the same time, the generation of relativistic particles and magnetic fields in shell remnants is still a matter of debate; in all likelihood, they come about in different ways in young and old objects.

In 1960, Shklovskii derived the relation

$$\Sigma \propto D^{-4\alpha-4} \tag{10.29}$$

to describe the variation of radio brightness during adiabatic expansion of a spherical cloud of relativistic plasma. The relation makes it possible to estimate the distance to a supernova on the basis of its observed angular size and radio brightness if the constant of proportionality has been determined using calibrators at known distances, and if one makes the important assumption that all of the remnants comprise an evolutionary sequence of homogeneous objects. The latter assumption is most certainly untrue; nevertheless, radio observations provide a reasonably reliable — and in most instances the only — means of estimating distance. The objects utilized to calibrate the Σ–D relationship are those whose distance has been determined by comparing radial and tangential velocities of filaments; by observing 21-cm and other radio lines in absorption; from the photometric distance to the exciting stars of nearby H II regions; and from the brightness of historical supernovae at maximum light.

A number of authors, including Poveda, Woltjer, Milne, Ilovaisky, and Lequeux, have constructed an empirical Σ–D relationship based on carefully selected calibrators. The number of sources has never been more than 10 to 15, and a function of the form $\Sigma \propto D^{-\beta}$ has been obtained, with β in the range from 2.7 to 4.5, depending on the author. Using the Parkes and Molonglo radio telescopes, Clark and Caswell (1976) compiled the most complete catalog of radio remnants. The resolution was $3-4'$ at 408 MHz and 5 GHz, and a number of new distance estimates were derived based on the 21-cm absorption line. Their Σ–D relationship contained a break: $\beta = 3$ for $\Sigma_{408\,MHz} \geq 3 \times 10^{-20}$ W/m^2/Hz/sr, and $\beta = 10$ for fainter sources. Relying mainly on Clark and Caswell's catalog, Caswell and Lerche (1979) and Milne (1979a) introduced distance corrections that took account of height above the galactic plane; the resulting function was of the form $\Sigma \propto D^{-3} \exp(-|z|/175\,pc)$ in the first paper, and $\Sigma \propto D^{-4} \exp(-|z|/66\,pc)$ in the second. The corrections allowed for gas-density and magnetic-field gradients perpendicular to the galactic plane.

Modern catalogs (Clark and Caswell, 1976; Milne, 1979a; Green, 1984b) contain about 40 calibrators, but only a lower limit on the distance is known for half of them. The errors in the distance estimates to certain remnants may reach $50-100\%$, a good example being IC 443 (see Section 7). The distances to the historical supernovae that come from estimates of their magnitude at maximum light are also of low accuracy; for example, the distance to Kepler's supernova has been revised downward from $10-12$ kpc to 3 kpc (Section 3). Sakhibov and Smirnov (1983) have collected all of the recent distance estimates obtained by different means, but even though they have significantly augmented the number of calibrators, their conclusions are not entirely incontrovertible. In particular, their substantial revision of the distance to the Cygnus Loop seems somewhat dubious.

In this situation, every new calibrator with a reliably measured distance becomes an important element in the ultimate refinement of the empirical relationship. Investigations of the kinematics of optical supernova remnants have enabled the present author to ascertain the kinematic distance (determined from the mean velocity, with expansion of the shell taken into account) to six new calibrators, and to improve the distance to three more. Figure 68 shows the empirical Σ–D relationship with these new calibrated remnants included (Lozinskaya, 1981).

These new measurements can be seen to have extended the relationship into the least well-determined area of the plot, containing faint, extended objects. The figure also includes large-scale Galactic radio loops, the properties of which will be discussed in the next section. The least-squares straight line passing through the Galactic supernova remnants is given by

$$\Sigma_{1\,GHz}[W/m^2/Hz/sr] = 1.4 \times 10^{-15} D^{-3.4} \text{ pc}. \qquad (10.30)$$

Here the 95% confidence limits are $\beta_1 = -2.9$, $\beta_2 = -5.7$ (Lozinskaya, 1981). Our data have not corroborated the break found by Clark and Caswell in the vicinity of $D \approx 30$ pc, which was probably attributable to the paucity of large calibrators.

In Fig. 68, we have also used the data of Milne et al. (1980), Sabbadin (1979), and Berkhuijsen (1983) to plot supernova remnants in nearby galaxies: the Large Magellanic Cloud, M31, and M33. These are substantially the same as the remnants in the Milky Way, attesting to the similarity of supernovae in these four galaxies.

The implication of Fig. 68 is that the empirical Σ–D relationship, which is based primarily on shell remnants, may be looked upon as a homogeneous evolutionary sequence. In light of our suggestion that plerions turn into shell remnants, it is not surprising that the shells of composite remnants conform to the very same dependence of brightness on distance. As a check, we have plotted the brightness of the shell components of composite supernova remnants whose distance is known: Vela XYZ, W 28, CTB 80, and G 326.3–1.8 (these are shown by triangles). These are fit very nicely by the evolutionary sequence of "classical" shell remnants, thereby substantiating the idea that plerions turn into shell remnants as their centrally-peaked source fades and they sweep up interstellar gas. It is therefore conceivable that among the more than one hundred old Galactic shell remnants, 30–40% might once have been plerions, inasmuch as that is the fraction of plerions among the historical remnants (see Chapter 1).

The purely observational evolutionary sequence for radio remnants that is obtained without invoking any theoretical models is not consistent with the predicted relationship (10.29). Using $\alpha \approx 0.5$, the mean value for shell remnants, the predicted slope of the Σ–D relationship is $\beta \approx 6$ instead of the observed $\beta \approx 3.5$. This implies that the assumptions made in deriving (10.29) provide a poor approximation to the actual situation in supernova remnants. Those assumptions are:

1) the injection of relativistic electrons ceases at the beginning of evolution;

2) the remnant consists of an adiabatically expanding cloud of relativistic particles frozen into a random magnetic field;

3) the magnetic field goes as $H \propto R^{-2}$.

To bring theory into conformity with experiment, it is necessary to introduce some sort of mechanism that will amplify the magnetic field and accelerate relativistic particles. Specifically, the observations are better fit by a shell model in which the total energy of the field plus particles is conserved, i.e., energy is pumped in to replenish losses due to adiabatic expansion (Shklovskii, 1976b). The observed "flattening" of the synchrotron spectrum of Cas A is also suggestive of particle acceleration, since all of the changes in the energy spectrum described at the beginning of this section lead to a spectral index that is either constant or increases with time.

A magnetic field amplification mechanism for young shell remnants has been proposed by Gull (1973). His numerical solution of the equations of gas dynamics that describe the evolution of young shells shows that when the ratio of swept-up matter to ejected matter is 0.1–1, a region prone to Rayleigh–Taylor instability arises behind the shock front, at the contact surface between the ejecta and swept-up gas. The instability gives rise to strong convective motions, primarily in the radial direction. Convection enhances the magnetic field by tangling the field lines frozen into the gas of the shell. In this scenario, the original field could be the interstellar field, which is amplified severalfold when the gas at the blast wave front is compressed. Field amplification continues in the convection zone until the magnetic field energy density equals the kinetic energy of convective motion. When a supernova shell expands into a homogeneous medium — the case considered by Gull — instability occurs only at the contact surface between the ejecta and the swept-up gas. In a medium with an inhomogeneous density distribution, there is a broad turbulent-motion zone engendered by Rayleigh–Taylor instability (Shirkey, 1978). The tangling-induced field-amplification zone then reaches out to about 1/3 the radius of the young remnant. The enhancement of synchrotron emission within that region may possibly account for the shell structure seen in radio maps of Cas A, Tycho, SN 1006, and Kepler (see Chapter 1).

It is interesting that according to Gull's calculations, a shell will reach its maximum brightness,

$$\Sigma_{max} \propto n_0^{(7+3\alpha)/6} M_0^{-(7+3\alpha)/6} E_0^{(1+5\alpha)/2},\tag{10.31}$$

at time

$$t_0 \sim 150 n_0^{-1/3} M_0^{5/6} E_{51}^{-1/2}.\tag{10.32}$$

For the typical values $n_0 = 1\,\mathrm{cm}^{-3}$, $E_0 = 5 \times 10^{50}$ erg, and $M_0 = 1 - 2\,M_\odot$, the brightness-enhancement time for a shell remnant, $t_0 \sim 200 - 400\,\mathrm{yr}$, turns out to be equal to the plerion formation time, $\tau \approx 300\,\mathrm{yr}$. Both kinds of objects subsequently fade as a result of adiabatic expansion.

The acceleration of relativistic particles in young remnants probably takes place at the blast wave front, and may also be associated with the turbulent-motion zone. At least three possible cosmic-ray acceleration mechanisms have been discussed in the recent literature: statistical type-I Fermi acceleration with multiple crossing of the shock front, statistical type-II Fermi acceleration, and acceleration via plasma waves. The most complete surveys of the relativistic-particle acceleration problem may be found in the papers by Drury (1983), Ginzburg (ed.) (1984) (the book *Cosmic-Ray Astrophysics*), Blandford and Eichler (1987), Blandford (1988), and Berezhko and Krymskii (1988). Current opinion holds that the predominant acceleration mechanism in young supernova remnants is Fermi type I at the shock front.

Fermi acceleration ensures that the relativistic particles will have the observed energy spectrum. When particles with high enough energy interact with a shock front, they are accelerated to still higher energies. Initial particle energies must be at least equal to the thermal energy at the front, so that they can traverse the front essentially undeflected and without significant energy loss; an initial energy of about 10 keV is sufficient, corresponding to a shock velocity of ~ 1000 km/sec. If particles are scattered efficiently, they will cross the front numerous times. For this to occur, it is necessary that there be a source of turbulence, which gives rise to randomly moving magnetic scattering centers. The turbulence may be associated with the aforementioned convective layer or with instability resulting from the motion of dense clumps of ejected matter through a tenuous region of the shell (Chevalier et al., 1976).

One can thereby account for the formation of radio shell sources at the early deceleration stage intermediate between free and adiabatic expansion. At the onset of the adiabatic stage, the same processes probably operate to enhance relativistic particles at the shock front and to amplify the magnetic field by entwining it in a turbulent layer; later on, compression of the interstellar magnetic field with particles frozen into the swept-up gas also becomes important. Reynolds and Chevalier (1981) have carried out a numerical calculation of the synchrotron emission to be expected of a shell remnant subject to these factors.

They assumed that the accelerated-particle energy density is proportional to the pressure behind the front, and that the accelerated-particle spectrum has the canonical form (10.1); the dynamical behavior of the remnant is described by the standard adiabatic solution (9.2). Then, knowing the radial dependence of the relativistic electron density and the tangential and radial components of the magnetic field (for the swept-up and tangled constituents separately), Reynolds and Chevalier constructed integrated radio brightness contours for the remnant. Comparison with radio maps of Tycho and SN 1006 indicates that turbulent entanglement of the magnetic field better fits the observations for these young objects. When the magnetic field is amplified via entanglement, the predicted Σ–D relationship takes the form $\Sigma \propto D^{-(7+3\alpha)/2} \propto D^{-4.4}$, and the decrease in radio flux density for Tycho becomes $\Delta S/S = 0.25\%$ per year, which is in accord with the observations. For compression of the swept-up field, the predicted relationship is $\Sigma \propto D^{-2}$, which provides a poorer fit to the observed sequence.

In the late stages of evolution, significant particle acceleration no longer takes place at the shock front (Fedorenko, 1981). The synchrotron emission from old shells can be attributed to the swept-up interstellar magnetic field containing particles frozen into the gas behind the shock front. The theory of this process was developed by van der Laan (1962); see also Shklovskii (1976a).

The density at an adiabatic shock front is increased by a factor of four; subsequent compression of the gas and field is induced by radiative cooling, so the van der Laan mechanism is effective only in old supernova remnants in the radiative stage.

Bychkov (1978b) has proposed a modification of the van der Laan mechanism that is applicable to the adiabatic stage as well, whereby the synchrotron radiation from an adiabatic remnant is linked to the field and particles frozen into compressed gas behind the shock front in dense cloudlets, where the gas is radiating strongly. It is difficult to estimate the amount of synchrotron radiation expected from dense filaments because of the need to take into account the geometry of the filament and magnetic field, as well as the rate at which high-energy particles escape the compressed-field region. According to Bychkov's calculations, the particles can be isotropized by plasma oscillations excited by the compression of the field in the time it takes for the shock wave to compress the filament. As a result, one can establish between the relativistic particle pressure P_{cl} and the gas density the relation $P_{cl} \propto \rho^{4/3}$. Under these conditions, for $\alpha = 0.5$, the volume emissivity is $\varepsilon_\nu \propto (\rho/\rho_0)^{4/3}(H/H_0)^{3/2}$, and the two-dimensional compression of a flux tube (as applied to actual model filaments) is

$$H \propto P_{mag}^{1/2} \propto (\rho/\rho_0)^{2/3}, \quad \varepsilon_\nu \propto P_{mag}^{7/4} \propto \rho^{7/3},$$

where P_{mag} is the magnetic energy density. The predicted dependence, $\Sigma \propto \varepsilon_\nu R \propto D^{-17/4}$, fits the observations quite well.

Here we underscore the fact that we have been dealing with dense interstellar clouds. Having discovered the complete correlation between the radio and Hα maps of IC 443's optical filaments, Duin and van der Laan (1975) went on to elaborate the van der Laan mechanism; in particular, they studied the compression of the swept-up field and relativistic particles due to thermal instability induced by cooling of the gas behind the blast wave front. Their estimates were used to show that the nonthermal radio emission from IC 443 and similar remnants could be due either partially or entirely to the aggregate emission from dense clouds formed as a result of thermal instability.

Everything that we have said thus far is summarized in Table 17, where we outline the manner in which the magnetic field and relativistic particles responsible for the synchrotron emission from the different types of supernova remnants (with and without pulsars) are generated at different stages in their evolution. The duration of each of the successive stages is of course only indicative, as it depends on the density of the ambient medium (here we use $n_0 = 1\,\text{cm}^{-3}$), pulsar energetics, and the initial explosion energy.

The Σ–D relationship in Fig. 68 has no breaks at all that would correspond to a transition from one developmental phase to another. In all

Table 17. Generation of Radio Synchrotron Emission by Supernova Remnants

Type	Stage of evolution			
	$\lesssim 10^2$ yr	$10^2 - 10^3$ yr	$10^3 - 10^4$ yr	$10^4 - 10^5$ yr
Plerion (SNR with pulsar)	Field and particles injected by pulsar; strong free–free absorption	Field and particles injected at constant rate; plerion observable	Injection falls off due to pulsar spin-down; enhanced emission from swept-up field plus particles; plerion plus shell	Injection negligible; radiation from swept-up field plus particles; observable as radio shell source
Shell (SNR without pulsar)	Field enhancement and particle acceleration by shock wave in wind of pre-SN?	Field enhancement and particle acceleration in convective layer at boundary of ejecta	Radiation from compressed interstellar field and particles behind shock front in dense clouds	Radiation from field and particles in swept-up shell during radiative stage

probability, they are concealed by the considerable spread in the points, which is due, among other things, to disparate densities in the ambient interstellar medium (which determine the dynamics, and therefore also determine the observed radio brightness of the remnant), and errors in the distance estimates, i.e., the linear sizes.

The rapid improvement in observational methods of radio astronomy in recent years has exerted an exceedingly productive influence on supernova science: many new, faint radio remnants have been discovered; well-known, more familiar remnants have been studied with arcsecond angular resolution; the ranks of extragalactic remnants have quickly become populated. Unfortunately, it has still proven impossible to say that this growth in the sheer volume of information has resulted in a substantial improvement in our "naïve" assumptions, as summarized in Table 17. More to the point, we have become more acutely aware of their naïveté.

Actually, despite our attempts to pigeonhole the multitude of known remnants according to some system, however complex, we keep encountering objects that are unclassifiable within any single framework. Furthermore, our present classification scheme is beginning to come apart at the seams — in the last few years, this has happened with regard to both the structure and the evolution of supernova remnants, based on radio observations.

Instead of the naïve representation of remnants as quasispherical shells, two geometrical models have been proposed; each claims to be applicable to the majority of all remnants.

Kesteven and Caswell (1987, 1988) have suggested a "barrel-shaped" structure for remnants, with an axisymmetric brightness distribution. Out of 70 radio maps examined in detail, they have identified 21 clear-cut barrel-shaped remnants, 23 further possible candidates, and a total of only four spherically symmetric shells. Barrel-shaped remnants are encountered frequently both among young and old objects. The two most striking examples are the remnant of SN 1006 and G 296.5+10.6 (PKS 1209–52), which have been investigated in detail by Roger et al. (1988) and are shown in Figs. 69a and 69b. In G 296.5+10.6, we see not only large-scale axial symmetry of both the structure and the brightness distribution, but symmetry of the delicate radio filaments on opposite sides of the "barrel." At the center of the remnant is a compact x-ray source — possibly a neutron star (see Table 9) — but no sign of x-ray synchrotron radiation or "plerionic" radio emission. In x rays there is thermal emission with $T_e \sim 1.7 \times 10^6$ K and $n_0 \sim 0.3\,\mathrm{cm}^{-3}$; in no way does this object differ from typical shell supernova remnants in the adiabatic stage (see Kellett et al. (1987) and references therein). Optical filaments are visible on the northern side of the remnant, probably demarcating the collision zone between the shell and a dense cloud.

Possible reasons for the barrel-shaped asymmetry may be, firstly, asymmetric supernova ejecta (for example, toroidal or cylindrical); secondly, an asymmetric density distribution for the circumstellar gas; and thirdly, regular structure of the interstellar magnetic field. It is conceivable that all three factors are important at different stages of development. The latter two offer little in the way of explaining the asymmetry of the young remnant of SN 1006, the deceleration of which has not advanced very far; see Section 3. The first is scarcely relevant to old objects that have long ago hidden any traces of the original explosion. Nevertheless, the observations actually suggest that supernova explosions can be asymmetric (see Chapter 1), just as the density distribution about the progenitor often shows axial asymmetry (see Chapter 3).

We have recently carried out numerical simulations of two striking barrel-like remnants, SN 1006 and G 296.5+10.0 (Bisnovatyi-Kogan et al., 1990). The young remnant of SN 1006 was shown most probably to have been formed by an anisotropic explosion, with enhancement of ejecta in the equatorial plane in an ambient medium with a plane-stratified gas distribution. The barrel-like morphology of G 296.5+10.0 can be accounted for by a spherical explosion occurring in a tunnel with a density that falls off with distance from the plane of symmetry. The explosion point is situated on the symmetry axis, but is displaced from the plane of symmetry.

Manchester (1987) has proposed that most radio remnants are characterized by bipolar or biannular asymmetry, and that "barrels" are special cases in terms of spatial orientation (see Fig. 69c). Biannular or bipolar asymmetry of remnants might result from bipolar asymmetry of the ambient medium associated with effects due to the progenitor (see Chapter 3), bipolar ejection of matter from the supernova, or activity of a stellar remnant that emits relativistic particles in the form of two precessing beams (as in SS 433). The presumption of precessing outflows of relativistic plasma in fact ameliorates some of the difficulty involved in interpreting Vela XYZ and G 320.4–1.2 (MSH 15–52), two remnants containing pulsars (see also Figs. 40, 61).

Highly interesting results have been obtained in investigations of small-scale features appearing in radio maps of remnants. The most impressive of these are the polarized radio observations of Cas A, mapped by Braun et al. (1987) at the VLA with 1."3 resolution. For the first time, the fragmentation and change in velocity of several radio knots were detected over a period of 2 – 5 years; likewise, maps showing the direction of the polarization vector clearly indicated for the first time that the field could indeed be amplified by turbulent motion in the shells of young remnants. The magnetic field lines are entrained by the motion of a dense radio knot, piercing the diffuse shell, which has already begun to decelerate.

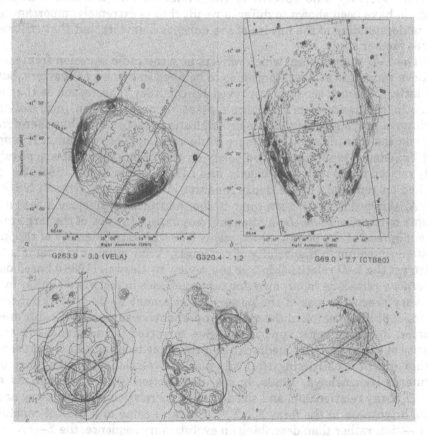

Fig. 69. a) The young barrel-shaped remnant of SN 1006; **b)** the old barrel-shaped remnant G 296.5+10.6 (Roger et al., 1987); **c)** radio remnants discussed above with double-ring structure (Manchester, 1987). See also Figs. 40 and 61.

Comparison of the fine structure of arcsecond-resolution optical and radio maps of remnants clears the way to an understanding of both the field-amplification and relativistic particle acceleration mechanism, on the one hand, and the delicate filaments of supernova remnants, on the other. A detailed analysis of the mechanism governing the acceleration and diffusion of relativistic particles has begun, based on VLA observations of the structure of radio filaments in the plerions 3C 58 and the Crab Nebula and the shell remnant SN 1006 (Reynolds, 1988a). The first steps have also been taken in studying the thin radio filaments of the Cygnus Loop (Straka et al., 1986) and Simeiz 147 (Fürst and Reich, 1986). The radio filaments have been shown to coincide with the optical, and they are at most 2–3″ wide over a length of 5–10′; as in the optical, the bright radio filaments are observed against a background of fainter diffuse radio emission. In Simeiz 147, the radio spectra of the filaments and the interfilament medium have been shown to differ, a result that is extremely important to our understanding of the nature of the delicate filaments, but that still requires further study.

The Σ–D relationship, which reflects how the radio emission from supernova remnants evolves, continues to be filled in by new calibrators, but the latter have not sensibly improved its accuracy. The reason is that it takes in remnants from different classes immersed in an interstellar medium of varying density. A major emphasis in recent work has therefore been on the identification of a more uniform sample of remnants. Huang and Thaddeus (1985) have relied solely on shell remnants that are physically related to CO molecular clouds, and Allakhverdiev et al. (1984b) have worked on remnants in the dense medium in spiral arms. Green (1984b) and Allakhverdiev et al. have come to the inauspicious conclusion that attempts to improve the Σ–D relationship are useless in view of the large inherent spread of the points, which stems from the varied density of the interstellar medium in the vicinity of remnants. A fair number of attempts to construct the Σ–D relationship (see the review by Caswell (1988)) based on supernova remnants in our own and other galaxies in fact yield similar but not very precise values of the power-law exponent $\beta \sim 3$–4. Berkhuijsen (1986, 1988) has compiled the most comprehensive summary of calibrated sources, including radio supernovae. She corrected for differences in the density of the interstellar medium using the same technique as the present author did in constructing the "kinematic" evolutionary dependence of remnants (Lozinskaya, 1980a, b; see also Section 9); she constructed a Σ_x–D_x x-ray relationship; and she derived a correlation between x-ray and radio brightness in the form $\Sigma_r \propto \Sigma_x^{0.69}$. She also came to a radical conclusion — that rather than describing an evolutionary sequence, the Σ–D relationship reveals differences in the parameter $E_0 n_0^2$, which actually determines the position of a remnant in the diagram. There is still no question that the statistical study of remnants is far from being exhausted.

It would seem, in particular, that conformity to a single slope over 10 orders of magnitude in Σ and three orders of magnitude in D is difficult to explain simply by variations in circumstellar density (differences in E_0 amount to less than an order of magnitude, a result consistent both with theory and with observations of supernovae and their remnants). Even if one eliminates the radio supernovae, where the progenitor wind is suspected to exert an influence, one is still left with Σ varying over five orders of magnitude, which according to Berkhuijsen can be accounted for by locating the supernovae in a medium with a wide-ranging interstellar density. Note that there are two well-studied remnants in the Galaxy that are expanding in regions of the interstellar medium with widely disparate densities: Loop I ($n_0 = 0.007\,\mathrm{cm^{-3}}$) (see Section 11), and W 28 ($n_0 \approx 10 - 40\,\mathrm{cm^{-3}}$). The ratio of their diameters (see Table 11) is consistent with this density difference: $D_{\mathrm{(Loop\,I)}}/D_{\mathrm{(W\,28)}} \propto [n_{\mathrm{(Loop\,I)}}/n_{\mathrm{(W\,28)}}]^{-0.2} \approx 5 - 6$ for adiabatic remnants; see Eq. (9.2).

The radio surface brightness of these remnants differs by a total of only two orders of magnitude, and it would be far-fetched to think that within the galaxies of the Local Group there are many remnants immersed in a medium of substantially higher or lower density. It seems likely that the observed falloff in brightness with increasing diameter is a reflection not just of differences in $E_0 n_0^2$, but of evolutionary changes as well. The adiabatic expansion of shells, accompanied by acceleration of relativistic particles at the shock front and magnetic field enhancement via convective entanglement and/or radiative compression of the gas behind the front provides a satisfactory explanation of the empirical Σ–D relationship; see, for example, Duric and Seaquist (1986) and Bychkov (1984). Furthermore, an inhomogeneous circumstellar medium, just like the existence of different types of supernovae, increases the spread of points in the Σ–D diagram by virtue of variations in the evolutionary tracks of individual remnants.

The evolution of plerions is of considerable interest. Since the theory advanced by Reynolds and Chevalier (1984) demonstrated the strong dependence of plerion luminosity on a large number of poorly known parameters, the construction of an empirical evolutionary sequence for supernova remnants with pulsars has taken on a certain urgency.

The following questions require a strictly observational answer.

1. Do plerions, or at least the known supernova remnants with pulsars, represent a unique evolutionary sequence? In other words, are the pulsars formed by supernovae all the same?

2. Can the evolutionary age of a plerion be determined from its observed properties?

3. Do composite remnants represent a stage intermediate between plerions and shells, as suggested by Lozinskaya (1980b), or are there two types of supernovae, one forming plerions, and the other, composite remnants?

4. To put question 3 differently, can we assume that 30 - 40% of the old shell supernova remnants in the Milky Way and other spiral galaxies were once plerions?

Many authors have attempted — unsuccessfully, however — to construct an empirical sequence for the x-ray or radio luminosity of plerions (or the ratio of the two) as a function of linear diameter (or age); see, for example, Becker et al. (1982), Becker (1983), Wilson (1986), and Helfand and Becker (1987). This problem has also been analyzed by Lozinskaya (1987), who has shown that out of three parameters — plerion size, age t, or the rate of pulsar rotational energy loss $\dot{E}$ — no one in isolation can serve as a good indicator of evolutionary age. Indeed, plerion luminosity depends heavily on adiabatic losses, i.e., on the expansion rate, and it also depends in a complicated way on the interaction between the pulsar wind, the ejecta, and the interstellar gas. The luminosity depends not only on the instantaneous injection rate, but also on the age, and so on. We have therefore proposed the combination $\dot{E}t/V$ as an indicator of the evolutionary age of a remnant containing a pulsar, where V is the overall volume of the radio-emitting region of the plerion.

If in fact the mean volume density of pulsar-injected energy is an indicator of the evolutionary phase, then the empirical relationship between L_x and $\dot{E}t/V$ found by Lozinskaya (1986) on the basis of the four remnants with pulsars known at the time may be considered to be evolutionary as well. (The total synchrotron luminosity is a more representative characteristic, but it has not been reliably determined for all plerions with pulsars.)

There is no question that the relationship identified in 1986 on the basis of the four objects then known does not inspire a great deal of confidence. We note, however, that a fifth object — CTB 80 — belongs there too, if we accept the age to be $t \sim 10^5$ yr, as given by Fesen et al. (1988d). But even this gross speculation suggests that the explosion of SN 1181 was accompanied by the formation of a less energetic pulsar than the one in the Crab Nebula, Vela, MSH 15–52, or 0540–69.3. We can then frame an answer to at least one of the questions above, with observational backing: supernova explosions can give rise to dissimilar pulsars.

In Section 11, we shall be concerned with a number of questions that bear directly on the propositions that we have outlined involving the synchrotron radiation from remnants; they will require, however, that variations in the density of the interstellar medium in the vicinity of the remnant be taken into account (so far we have implicitly considered the evolution of a radio remnant in a homogeneous medium with a mean density $n_0 \sim 1\,\mathrm{cm}^{-3}$).

11. The Galactic supernova rate; large-scale radio loops

The estimation of the Galactic supernova rate — the determination of the mean interval between outbursts, based on counts of the number of radio supernova remnants — is a fundamental problem that can be addressed after the laws governing the evolution of shells have been established.

Systematic searches for extragalactic supernovae have detected approximately 600 outbursts in galaxies of different morphological types (see Section 1). An analysis of these supernovae enables one to ascribe to a galaxy of given type, with known mass and luminosity, a predicted supernova rate, with an expected error of at most 50%. Extrapolation of these estimates to our own stellar system ($L = 2 \times 10^{10} \, L_{B_\odot}$) leads to predicted rates for SN Ia, SN Ib, and SN II of about $0.6h^2$, $0.8h^2$, and $2.2h^2$, respectively, where $h \equiv H/100$ km·sec^{-1}/Mpc (van den Bergh et al., 1987). For $h = 0.75$, the predicted total number of supernovae in the Galaxy is about two per century — i.e., the interval between outbursts is $\tau \approx 50$ yr. Capellaro and Turatto (1988) have arrived at the same figure.

Seven historical supernovae have been observed in the Galaxy. In Fig. 70, we have plotted the location of those that have occurred in the last thousand years, projected on the galactic plane. The distances are as given in Chapter 1; Cas A is indicated by an asterisk. All of the historical supernovae are clearly confined to a galactocentric sector with an opening angle of less than 50°, and are at most 5 kpc from the sun. This distance is consistent with the naked-eye detectability of a supernova with $m_V \approx 1$ over the course of approximately two weeks, for $M_{max} = -18$ and absorption $A_V = 1^m$/kpc (see Tammann (1982)).

We can also estimate how many unobserved supernovae there may have been within the same 50° sector, but more than 5 kpc from the sun (either toward the center or the anticenter). Comparison of the galactocentric distances of the historical supernovae with the radial distribution of supernovae in S galaxies (Fig. 71) suggests that the Galactic outbursts have been observed right in the range in which the supernova rate is elevated. The distribution in Fig. 71 suggests that perhaps 20–30% of the supernovae that have occurred at $r \gtrsim 5$ kpc

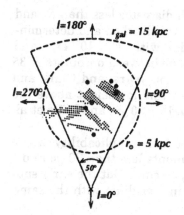

Fig. 70. Historical Galactic supernovae of the last thousand years. Cas A is designated by an asterisk.

Fig. 71. The radial distribution of supernovae in S galaxies (Tammann, 1982).
The filled areas represent regions corresponding to historical Galactic supernovae.

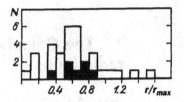

have gone undetected. Tammann (1982) has
used these results to find that there have been approximately 60 Galactic
supernovae in the last thousand years, or in other words that the mean interval between outbursts is $\tau \approx 16$ yr.

A number of different authors have come to the same or similar conclusions regarding the supernova rate, based on the historical supernovae.
Bearing in mind as well that observers have probably missed supernovae
that were in the daytime sky at the time of explosion, and still fairly close
to the sun by the time they had faded significantly, a somewhat arbitrary
correction has been introduced to take into account the incompleteness of
the historical records and of observations in the densely populated regions
near the galactic center. Tammann (1977, 1978) and Clark and Stephenson
(1977) have thereby found that the mean time between Galactic supernova
outbursts is $\tau = 10-15$ yr. Correcting for observational selection effects in
a different way, Pskovskii (1978b) has used the same observational data to
arrive at $\tau = 50-100$ yr; Shklovskii (1960a) obtained $\tau = 60$ yr. The large
differences among the various authors are understandable, inasmuch as
their conclusions are based on only a small fraction of all events, and the
rates that have been deduced are determined not so much by what has
been observed as by what has been unobservable.

Milne (1970) has developed a method to estimate the Galactic supernova rate more reliably through counts of nonthermal radio sources
(remnants). At the basis of his method is the self-evident expression

$$N(<D) = t(D)/\tau, \qquad (11.1)$$

where $N(<D)$ is the number of remnants with diameter less than D, and
$t(D)$ is the age of a shell of diameter D. Using this method and determining $t(D)$ via the standard adiabatic model with $E_0 = 10^{51}$ erg and
$n_0 = 1 \text{cm}^{-3}$, Downes (1971) obtained a mean time between outbursts of 35
years, Ilovaisky and Lequeux (1972) obtained 50 years, and Clark and
Caswell (1976) obtained 150 years; after correcting for height above the
galactic plane, Caswell and Lerche (1979) found $\tau = 80$ yr, and Mills et al.
(1984) obtained $\tau = 25$ yr.

The results described in Section 9 support the applicability of the
standard adiabatic solution to supernova remnants less than 30 pc in diameter in a medium with a typical density $n_0 = 1 \text{cm}^{-3}$. But the early supposition that all radio remnants are evolving in a medium with the same

to the galactic plane, as assumed by Milne (1979a) and Caswell and Lerche (1979), is open to serious objections. The second requisite assumption in estimating τ through radio counts — that the population of observed supernova remnants is homogeneous — also requires some proof, since there exist two types of supernova.

The author examined these two factors — the inhomogeneous density distribution of Galactic gas and the heterogeneity of supernova remnants — in an estimate of the supernova rate (Lozinskaya, 1979a), and concluded that the first must be taken into account. Let us estimate here the Galactic supernova rate, including a rough correction for selection effects related to large-scale fluctuations in the Galactic gas density. Observations of the x-ray background and absorption in the O VI line, on the one hand, and an analysis of the effects of supernovae on the gaseous component of the Galaxy (see Section 17), on the other, suggest that a sizable volume of the Galactic disk is occupied by hot, low-density gas; we shall have more to say about the origin of that gas in Chapter 4. In the meantime, we use the data of McKee and Ostriker (1977), Myers (1978), and Cox (1988) on the temperature T, density n, and filling factor f of the "hot," "warm,", and "cold" gas in the galactic disk:

$$\{T_h, n_h, f_h\} = \{5 \times 10^5 \text{ K}, \ (3-15) \times 10^{-3} \text{ cm}^{-3}, \ \sim 0.5\},$$

$$\{T_w, n_w, f_w\} = \{(5-8) \times 10^3 \text{ K}, \ 0.2-0.5 \text{ cm}^{-3}, \ \sim 0.5\},$$

$$\{T_c, n_c, f_c\} = \{100 \text{ K}, \ 10-50 \text{ cm}^{-3}, \ 0.02\}.$$

The boundaries between these regimes are sharp. All of the bright optical supernova remnants discussed above are located in the warm or cold components of the Galactic gas. The fact that a large fraction of the galactic disk is filled with hot, tenuous gas affects the estimate of τ obtained from counts of nonthermal radio sources in two ways. The first stems from the fact that supernova remnants last longer in a dense medium, since according to the adiabatic model (9.2), $t(D) \propto n_0^{1/2}$. Given the sharp division of Galactic gas into dense and tenuous regions, we have instead of (11.1) a different relation between the number of radio-emitting remnants and the supernova rate:

$$N(<D) = \frac{t_w(D) f_w}{\tau} + \frac{t_h(D) f_h}{\tau}, \tag{11.2}$$

where $t_{w,h}$ is the age of a remnant of diameter D immersed in the warm or hot components of the disk. To a first approximation, remnants immersed in the cold component can be ignored, due to the negligible volume occupied by that component compared with the other two. Equation (11.2) holds for a uniform distribution in the Galactic disk and constant supernova rate, and assuming that the warm and hot regions are typically larger than the remnants.

We make use of the catalog of Milne (1979a), which contains 125 Galactic supernova remnants. At distances $r \leq 5 \, \text{kpc}$ from the sun, for which the catalog is complete, and which comprises 10% of the volume of the galactic disk, there are $N(< 30) = 25$ sources less than 30 pc in diameter; accordingly, in the entire Galaxy, there are $N(< 30) \approx 250$. Taking $E_0 = 5 \times 10^{50}$ erg, and obtaining $t_{w,h}$ from (9.2), the density and filling factor given above for warm and hot gas yield $\tau = 25 - 30 \, \text{yr}$ if the supernovae are uniformly distributed about the Galactic disk. SN II and SN Ib are inherently associated with dense gas and dust complexes, since their massive progenitors cannot migrate very far from their place of birth over a lifetime of $10^6 - 10^7 \, \text{yr}$. The selection effect referred to above therefore applies mainly to SN Ia. Note, however, that extensive low-density cavities may exist in the vicinity of OB associations, having been formed by the joint activity of stellar wind and supernova explosions; see Section 18.

We first estimated the supernova rate based on the number of radio remnants in 1979, when hot gas was thought to occupy up to 80% of the Galactic disk. Consequently, we obtained $\tau \sim 15 - 20 \, \text{yr}$. A more conservative figure is currently accepted for the volume of the hot component (see Cox (1988) and Section 17), $f_h \sim 0.5$ or even less. Our estimate has changed accordingly: $\tau \sim 25 - 30 \, \text{yr}$ (or more, if $f_h \sim 0.2$), which is consistent with the rate extrapolated on the basis of extragalactic supernovae.

The second factor affecting the estimate of τ based on radio counts is that in a low-density medium, remnants of a given size are less radio-bright than in a higher-density environment. In fact, the synchrotron emission from young remnants, which is associated with pulsars in plerions or field enhancement and relativistic-particle acceleration at a shock front in a shell remnant, increases over the first $t \approx 300 \, \text{yr}$, and then rapidly fades upon expansion of the remnant after $t \approx 10^3 \, \text{yr}$ (see Section 10). The concomitant falloff in brightness due to adiabatic cooling then depends solely on the extent of the relativistic plasma cloud. The subsequent contribution to synchrotron emission from relativistic particles in the interstellar magnetic field is governed by the density of the swept-up interstellar gas. It can easily be shown that radio synchrotron emission from the compressed interstellar magnetic field is negligible in remnants less than 30 pc in diameter that are immersed in the hot component of the Galaxy's gaseous disk. Inasmuch as these remnants contain no more than $2 - 3 \, M_\odot$ of swept-up interstellar gas, the shell will not yet have been decelerated, and the shock velocity in the intercloud medium will be close to the initial velocity of the ejecta, $v_0 = (5 - 10) \times 10^3 \, \text{km/sec}$. The cooling time of such a shock wave, even in dense clumps of gas with $n_0 \approx 10 \, \text{cm}^{-3}$, is about $10^5 \, \text{yr}$, which is much greater than the age of a remnant with $D \leq 30 \, \text{pc}$ in the hot medium: $t_h = R/v \approx 2 \times 10^3 \, \text{yr}$. On the other hand, in shells of that size located in the warm or cold component of the interstellar medium, $t_{cool} < t_{w,c}$ for densities $n_0 \gtrsim 1-3\,\text{cm}^{-3}$, and radiation from the compressed interstellar magnetic field behind the radiative shock wave in the dense clouds makes a significant contribution to the radio emission (see Section 10).

This second effect, which relates to the lower radio brightness of remnants in a hot, tenuous medium, substantially alters the supernova rate derived from radio source counts, but it is difficult to make proper allowance for it; one tends to overestimate the linear dimensions of remnants in a hot medium, which are derived from their radio surface brightness, and thereby to underestimate $N(<D)$. In the limiting case, wherein radio remnants with $D \le 30\,\text{pc}$ in a hot medium generally fall below the threshold of detectability, the second term in Eq. (11.2) can be dropped.

Old remnants immersed in the hot component of the gas in the Galactic disk may possibly become "radio invisible" for a certain length of time. They are probably observable in the earliest stages, when a radio shell can be produced via the interaction of the blast wave with the wind of the progenitor. This phase is essentially a continuation of the radio supernova phenomenon (see Sections 1 and 2), and may last, depending on the parameters of the progenitor outflow, for $t \lesssim 10^2 - 5 \times 10^2\,\text{yr}$, with $R \lesssim 1-2\,\text{pc}$. They reappear in the late stages of expansion, when the mass of the swept-up interstellar gas reaches $10^2 - 10^3\,M_\odot$, and synchrotron radiation from relativistic particles in the swept-up interstellar magnetic field becomes significant, first behind the radiative shock front within the cloud medium, and eventually in the intercloud medium. Straightforward estimates suggest that in a medium with density $n_0 \approx 0.01\,\text{cm}^{-3}$, the corresponding remnants have a diameter of $200-300\,\text{pc}$. Do we actually see such objects in the Milky Way? Conceivably, we may identify them with large Galactic radio loops such as Loops I, II, III, and IV (Lozinskaya, 1979a).

In the earliest days of radio astronomy, when the first low-resolution radio maps of the Milky Way were produced, the North Polar Spur was discovered — an extended arc among the radio isophotes near $l = 30°$, comparable in size and brightness with the Galactic map itself. Further investigations revealed that this was not a unique phenomenon. It was found that the NPS is the brightest part of a small circle on the celestial sphere having a diameter of about 116° and a width of 10°, which was thereupon designated Loop I. Three more similar large-scale ring structures could also be discerned among the Galactic isophotes: Loop II (the Cetus Arc), Loop III, and the less distinct Loop IV, which was located inside Loop I (see Fig. 72). The angular size of the latter three loops is $91 \pm 4°$, $65 \pm 3°$, and $40 \pm 2°$ (Berkhuijsen, 1973). The radio emission from the loops is synchrotron radiation, as indicated by their spectrum ($\alpha = 0.5 - 0.7$) and strong linear polarization.

The idea of identifying the loops with old supernova remnants that were nearby and therefore of large angular size was proposed as early as

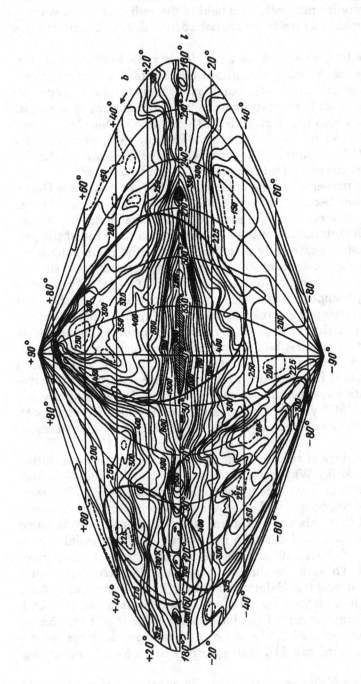

Fig. 72. Large-scale Loops (Berkhuijsen, 1971) superimposed on 150-MHz radio isophotes (Landecker and Wielebinski, 1970). The centers of the loops are marked by crosses.

1960 (Hanbury-Brown et al., 1960). Alternative hypotheses link them to large-scale structure in the Galactic magnetic field induced, for example, by the passage of spiral density waves (see Sofue et al. (1974) and Sofue (1976), which contains further references).

One fundamental criterion bearing on the nature of these loops is their distance and true linear size. The first estimate of the distance to Loop I was made by Lozinskaya (1964). A comparison of the radio continuum brightness of the North Polar Spur with the Galactic brightness in the 21-cm line at different radial velocities made it clear that there was a deficit of neutral hydrogen at low radial velocities along the line of maximum brightness of Loop I. Neutral gas is concentrated toward the rim of Loop I, and outside it. This then suggested a physical connection between Loop I and nearby hydrogen in the solar neighborhood high above the Galactic plane, yielding a distance comparable to the thickness of the gaseous disk, i.e., about 100–200 pc. The link between the NPS and solar-neighborhood hydrogen, and our estimate of its distance, have been subsequently confirmed. It was found that Loop I is surrounded at a distance of approximately 5° by a shell of neutral hydrogen with a density of about $2 \, cm^{-3}$, which is expanding at about 3 km/sec (Heiles et al., 1980). A comparison of the polarization of light from background stars behind Loop I with the direction of polarization of its radio synchrotron emission yields a distance of approximately 100 pc (Spoelstra, 1972).

X-ray analysis provides yet another distance estimate for Loop I. The mere detection of soft x-ray emission, in and of itself, is an important argument favoring the identification of the Loops with supernova remnants (see Shklovskii and Sheffer (1971)). A soft x-ray source is centrally located in Loop I; its spectrum yields a temperature $T_e \approx (3-4) \times 10^6$ K, and the density of the radiating plasma is approximately $0.01-0.02 \, cm^{-3}$ (Hayakawa et al., 1978; Iwan, 1980; Schnopper et al., 1982). The fact that the x rays from Loop I are not absorbed by the dense cloud of ρ Oph means that the Loop is the nearer of the two objects — i.e., it is situated at most 160–200 pc away (Davelaar et al., 1980). We also point out here that soft x-ray enhancement is seen in the vicinity of Loop II.

Spoelstra (1972) has found that Loops II, III, and IV are 175 ± 65, 200 ± 65, and 210 ± 75 pc away by comparing the observed radial distribution of radio brightness with the distribution calculated for the interstellar magnetic field plus swept-up relativistic particles in the shell. These estimates presuppose a particular "theoretical" model of the Loops.

The distance to Loop II based on 21-cm observations is 100 pc (Johnson, 1978); using interstellar absorption lines in the spectrum of background stars, Bates et al. (1983) obtained a distance between 30 pc and more than 700 pc, and an expansion velocity of about 100 km/sec.

The most likely linear sizes of Loops I, II, III, and IV are 230, 170, 200, and 210 pc, respectively, with perhaps ±50% error.

What are the main arguments against the identification of loops with supernova remnants? first of all, there is their anomalously large size.

"Normal" supernova remnants in both our own and other galaxies dissipate in the interstellar medium, reaching sizes $\lesssim$ 100 pc (see Section 9). Sofue et al. (1974) have also remarked upon the gap in the angular size distribution of radio remnants between the "normal" remnants, which have a diameter of less than 5°, and structures larger than 40°, which include the Loops.

A second argument involves the lack of optical nebulosity, an invariable hallmark of old supernova remnants. Very faint wisps of nebulosity have only been identified with Loop II. Johnson (1978) has obtained extremely deep plates of the region and detected many faint optical filaments, both narrow and diffuse (see also the references in his paper). The system of optical filaments and H I clouds associated with Loop II have an expansion velocity close to 13 km/sec. Very faint diffuse Hα emission may also have been detected from the vicinity of Loop III (Elliot, 1970).

Thirdly, the low expansion velocity of the H I shell in Loop I, which is typical of the late stages of remnant evolution, is inconsistent with the high temperature of the x-ray emitting plasma, which is more typical of the adiabatic stage. The radial brightness distribution of the x-ray emission from Loop I is also characteristic of the adiabatic, rather than the radiative, stage of remnant development (Davelaar et al., 1980). Attempts to explain this contradiction have included repeated heating of an old ($t \approx 2 \times 10^6$ yr), slowly expanding remnant by a new outburst (Borken and Iwan, 1977; Heiles et al., 1980), and a single, anomalously powerful explosion with initial energy $E_0 = 10^{52} - 10^{54}$ erg (Hayakawa et al., 1977).

These problems can all be disposed of by assuming that Loops I–IV are the remnants of supernovae that exploded in the hot component of the gaseous disk of the Galaxy. Indeed, as we have shown above, an old, radio-bright remnant in a medium of density $0.01\,\mathrm{cm}^{-3}$ ought to be just the same size as these loops. Furthermore, since the boundary between the hot and warm components of the interstellar medium is a sharp one, the observed gap in the size distribution of remnants between "normal" objects in the warm medium and extensive shells in the hot medium is just what is to be expected!

The fact that the Loops are located in a tenuous medium accounts for their optical faintness. According to the model discussed in Section 8, the lack of optically bright filaments results from the tenuous hot gas having no dense condensations, wherein a shock wave would be radiating strongly even in the adiabatic expansion stage.

Finally, the inconsistency between the low expansion velocity of the H I shell and the high temperature of the plasma in Loop I may simply be explained by the neutral shell being unrelated to a supernova explosion, instead having been swept up by the wind from the Scorpius–Centaurus OB association (Davelaar et al., 1980). In Fig. 73 we have an overall picture of the region comprising Loop I, the cavity produced by stellar wind, the shell of neutral gas, and the low-density neighborhood of the sun. Here the low density of the ambient gas in the vicinity of Loop I can be ac-

Fig. 73. Loop I, the remnant of a supernova explosion inside the cavity produced by the wind from the Sco–Cen association; from Davelaar et al. (1980) (see also Bochkarev (1987)).

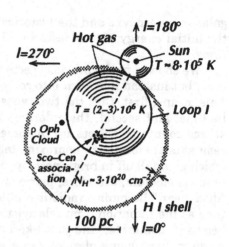

counted for in a natural way by assuming that the explosion took place inside the cavity swept out by the stellar wind of the association (these structures are examined in Chapter 4).

We have already shown in Fig. 68 that Loops I–IV lie quite close to the continuation of the Σ–D relationship based solely on Galactic supernova remnants (with no Loops included!). In Section 9, we also demonstrated that Loop I is a "normal" remnant that sits in the middle of the evolutionary sequence depicted in Fig. 59, if we make allowance for the anomalously low density of the ambient interstellar gas indicated by the x-ray luminosity of the remnant. Here we have two independent arguments suggesting that Loops I–IV are old supernova remnants in tenuous, hot gas. We should emphasize that according to this interpretation, despite the Loops' large size, they are by no means the oldest known remnants. On the contrary, in an evolutionary sense, they are comparatively young objects entering upon the adiabatic stage, since the requirement that $M_{swept\,up} \sim M_{ejected}$ can only be satisfied in a tenuous medium by a large shell, with $R \gtrsim 100\,pc$.

We should also point out that modern radio maps of the Milky Way are littered with what may be interpreted as spurs and ejecta among the lower-level isophotes. We are not suggesting that all of these are associated with supernova explosions. Instead, spurs oriented perpendicular to the galactic plane, for example, may be due to the expulsion of magnetic field and relativistic particles in regions traversed by spiral density waves (Sofue, 1973).

Estimates of τ derived from counts of radio remnants using (11.2) are only valid insofar as the supernova remnants upon which they are based comprise a homogeneous group of objects. This is most certainly not the case, given the existence of plerions, shells, and composite remnants; nevertheless, the present estimates will not be substantially altered if the birthrate, lifetime, variation of radius with time, and initial energy of plerions, shells, and composites are not too grossly different.

According to the data in Chapter 1, plerions are produced by SN II explosions, and shells by SN Ia and by objects like Cas A. They have approximately the same frequency of occurrence, judging by information on extragalactic supernovae and the historical supernovae in our own Galaxy, and the initial energy of the ejecta in all of these events also tends to be the same.

We showed in the previous section that plerions are not short-lived objects, but instead first turn into composites, and then into shell remnants. This means, first, that the two types of remnant have the same expected lifetime, and second, that plerions are automatically included in radio source counts of objects up to some given diameter. Plerions and shell remnants are formed in approximately the same amount of time, following which both fall off in brightness by virtue of adiabatic expansion. In a hot, tenuous medium, therefore, plerions — just like shell remnants — tend to "disappear," becoming radio-invisible for an extended period, until such time as the radiation from relativistic particles in the interstellar medium begins to dominate. And just like remnants with no pulsar in a tenuous medium, very young plerions can also pass through a brief initial stage of being a composite object, in which a radio shell is formed at the boundary between the pulsar wind and the ejecta, or between the ejecta and the wind from the progenitor (we have already noted this possibility in Section 4 with regard to the Crab Nebula).

If the plerion-to-composite scenario is correct, there ought to exist old (or more properly, large) plerions with no evidence of an outer shell that are located within the hot, tenuous component of the Galactic disk. Such plerions do indeed exist (see Table 14), and an estimate of the local density (for example, based on 21-cm observations at a velocity corresponding to the most distant absorption feature in the line of sight to the remnant) would afford an extremely important test of the proposed scenario.

There is yet another condition required for the validity of our estimate of the mean time between supernova explosions, namely that the remnants in question be expanding adiabatically. This has been assumed in relying on the empirical evolutionary sequence — the dependence of expansion velocity on the radius of the shock wave (see Section 9). A count of the number of remnants up to some maximum diameter provides an independent check on the applicability of adiabatic expansion to most supernova remnants. The most interesting such counts have been taken in nearby galaxies, especially in the Large Magellanic Cloud (which is face-on), since the counts are then uncorrupted by distance errors to the individual remnants. Using the most recent data for the Magellanic Clouds, Mills et al. (1984) found $N(<D) \propto D^{12}$ instead of the behavior $N(<D) \propto D^{2.5}$ expected of radio remnants in the adiabatic stage of development. The flatter $N(<D)$ curve suggests that the remnants are expanding faster than predicted by the adiabatic law (9.2) over a substantial fraction of their lifetime. This may possibly be associated with an extended period of expansion with evaporation (see Section 9). It is also conceivable that the free expansion stage is fairly long if most of the supernovae in the Small Magellanic Cloud exploded in a tenuous medium.

Finally, there is another more trivial possibility — that observational selection effects that alter the apparent distribution $N(<D)$ have been dealt with improperly. Hughes et al. (1984a) and Green (1984b) maintain that proper correction for observational selection enables one to reconcile the $N(<D)$ count results in the LMC with the assumption of adiabatic expansion. According to Fusco-Femiano and Preite-Martines (1984) and Berkhuijsen (1987, 1988), allowance for density inhomogeneities in the interstellar medium makes the observed slope of the $N(D)$ relation steeper, or in other words closer to that predicted by adiabatic expansion.

In the near term, the statistics of Galactic supernova remnants will improve markedly, thanks to the work of Reich et al. (1988), who have compared a new, deeper radio survey with the IRAS data.

Several extremely young remnants have also been found recently, possibly having been produced in the last thousand years. These include G 11.2–1.1 (Trushkin, 1986b), the plerion G 54.09+0.24, which is at most 1000 years old, and the shell G 70.68+1.20, which may be just 100 years old (Reich et al., 1985). The identification of the latter object with such a young remnant, however, is still somewhat debatable (Green, 1986). According to Becker and Resen (1988), it is a pre-main sequence star with a bright optical and radio emission shell.

Bearing in mind all of the foregoing remarks, it must still be recognized that the corrected estimate of the Galactic supernova rate based on the number of radio sources is a crude one, at best. Nevertheless, taking account of the large fraction of the galactic disk occupied by hot, tenuous gas reduces the time between outbursts determined by radio source counts.

The value $\tau \approx 25-30$ yr obtained above is consistent with estimates based on the historical supernovae, and with the currently accepted pulsar birthrate, assuming that pulsars are produced only in SN II explosions, which comprise no more than 50% of all supernovae in the Galaxy.

Why then, if the Galactic supernova rate is so high, have no supernovae been seen in the Milky Way in the past 380 years? As long as we have in mind the detection of naked-eye historical supernovae, i.e., those no farther away than $5-6$ pc, their absence may just be a statistical fluctuation. If the distribution of supernova events in space and time is governed by a Poisson distribution, the probability of observing k supernovae in a time t, given a mean rate $\nu = 1/\tau$, is $P(k,t) = e^{-\omega t} (\nu t)^k / k!$. With $\tau_{0,1} = 250-300$ yr for a region $r \leq 5$ kpc, which comes to 10% of the volume of the Galactic disk, the probability of observing 0, 1, or 2 outbursts in 380 years is $20-27\%$, $33-35\%$, or $23-25\%$, respectively, without counting weak outbursts like Cas A. This means that the lack of an observed Galactic supernova over the past four centuries is not badly at odds with the time between explosions estimated from radio source counts, but the next supernova to appear in our skies is long overdue! (A numerical experiment carried out by Clark et al. (1981), modeling the naked-eye observation of Galactic supernovae, has accurately "reproduced" the historical situation. Two series of random numbers were used to assign polar coordinates to "supernovae" in the Galactic disk out to a radius of 16 kpc, uniformly distributed in time with a mean rate of 1 SN per 20 years, over an "observation" period of 1.8×10^5 yr. The number of detectable supernovae, i.e., those at $r \leq 6$ kpc, ranged from 2 to 15 per thousand years, with an average of $6-8$ per thousand years.)

The instrumentation accessible to modern observers promises a sharp increase in the number of detectable supernovae in the near future. For example, whereas supernovae buried in dense clouds might otherwise have been undetectable and gone unnoted in the historical chronicles, these would now show up in the infrared. A supernova in a cloud with a typical density $n_H \sim 10^5$ cm^{-3} should give a brief burst in the far infrared, corresponding to a luminosity of $(2-3) \times 10^8 L_\odot$, in the first few years after the explosion, with a more sustained secondary outburst (about $10^7 L_\odot$) at the time of shell formation, i.e., 20 years after the explosion (Shull, 1980). That radiation could be detected by current infrared telescopes out to 1 Mpc. Thus, the current picture may well be clarified by extragalactic monitoring programs and, perhaps even more interesting, by the far-infrared monitoring of galactic molecular clouds. It has also become clear in recent years (see Section 1) that radio observations of all types of supernovae, unimpeded by interstellar absorption, are quite promising. For the time being, however, there is essentially no systematic monitoring effort under way, and it is difficult to say just how complete present-day observations of galactic supernovae are in the radio and infrared. It seems likely that in the not-too-distant future it will also be possible to monitor gravitational radiation, gamma rays at $E \gtrsim 100$ MeV, and neutrino radiation in the range $E \sim 10-20$ GeV and $E \gtrsim 100$ GeV, all of which might signal the explosion of a supernova (see, e.g., Berezinskii et al., 1984). The explosion of SN 1987A has spectacularly confirmed the utility of such observations.

Chapter 3
The Interaction of Stellar
Wind with Interstellar Gas

While the study of supernovae has enjoyed a long and illustrious history, the study of stellar wind and its effects on the interstellar medium is of more recent vintage. The wind from early-type stars was discovered by Morton in 1967, and subsequent observations confirmed that all high-luminosity stars earlier than type B2 are characterized by strong outflows of matter (see Snow and Morton (1976) and references therein). Stellar wind was immediately added to the arsenal of techniques to be used to account for a number of kinematic and morphological features of H II regions and planetary nebulae, and was a frequent topic of discussion at the Thirty-fourth Meeting of the International Astronomical Union in 1968. Vorontsov-Vel'yaminov and Matthews related the filamentary structure of certain planetary nebulae to wind effects. In discussions at that meeting, Menon ascribed to the influence of stellar wind the supersonic gas velocities then recently discovered by Shcheglov and Gershberg in "normal" H II regions, and that idea was exhaustively developed by Pikel'ner in a subsequent series of papers.

The ensuing two decades have witnessed the discovery of a multitude of observational manifestations of the influence exerted by stellar wind on the ambient interstellar gas, as well as the development of an orderly theory of the phenomenon. It is now clear that the outflow of matter in large part dictates the evolution of massive stars (including their ultimate explosion as a supernova). This in itself is an important problem that we shall not touch upon here.

We have had a great deal to say in the preceding chapters about the interaction of the expanding shell of a supernova with the wind of the progenitor. In the present chapter, we consider the interaction of stellar wind with interstellar gas. In Section 12, we briefly summarize the mass-loss parameters and outflow mechanisms of early-type stars, and sketch out a list of phenomena whose physics is largely dictated by stellar wind. Our present theoretical understanding of the process whereby the wind interacts with the interstellar medium is described in Section 13. In Sections 14, 15, and 16, we analyze a new class of emission nebulae — the ring nebulae that surround the strongest sources of stellar wind, Wolf–Rayet (WR) and Of stars. We shall rely on these nebulae to demonstrate, in the first place, that stellar wind is not the sole reason for the formation of shell-like H II regions around O stars; ring nebulae may also form as a result of the ejection of a shell, or as an H II region evolves, if one takes into account the photoionization evaporation of small cloudlets. Most likely, the observed ring nebulae result from a combination of processes, and studies of the kinematic and physical conditions in shells shed light on the main ones. Secondly, we shall see that a strong stellar wind determines the physical state of the close-in medium to the same degree and in a manner similar to the sudden ejection of matter in a supernova explosion.

12. Stellar Wind

Before proceeding to an analysis of wind effects on interstellar gas, let us dwell for a moment on methods for estimating the mass outflow rate $\dot{M}$ and the wind velocity v_w. The fact that certain stars are losing matter was well known long ago. As early as 1926, Milne predicted mass loss due to acceleration by radiation pressure in lines, and in 1929, Beales explained the characteristic spectral line shapes of P Cyg and a number of WR stars (emission lines with an absorption feature in the blue wings) in terms of resonant scattering in an outflowing atmosphere; in 1953, at the Second Workshop on Cosmogony in Moscow, Mustel' suggested looking for shell-like nebulae swept up by outflowing matter from O stars. Observations in the visible revealed mass outflow in WR, Of, and ζ Ori- and ζ Pup-type supergiants.

Ultraviolet observations of stellar spectra from above the earth's atmosphere ushered in a new era in stellar wind research. The first measurements showed that the Si IV (1394–1403 Å) and C IV (1548–1551 Å) resonance lines in the spectra of a number of O stars in the Orion complex have P Cyg profiles, and that the mass outflow velocity determined from the blue edge of the absorption feature is three times the escape velocity. An abundance of observational data — spectra of many dozens of early-type stars — was obtained with the orbiting Copernicus (OAO-3) observatory and the International Ultraviolet Explorer (IUE) (see the reviews by Conti (1978) and Conti and Underhill (1988), and references therein). Bright emission lines have painted an accurate picture of the characteristic P Cyg line profile: the resonance lines of Si IV, C IV, and N V 1240 Å, and subordinate lines of O IV 1342 Å, O V 1371 Å, and N IV 1579 Å. The wind velocity obtained from the absorption boundary in the profiles of these lines, the so-called terminal velocity, is shown in Table 18. The discrepancies among individual estimates taken from different lines in the spectrum of a single star are at most 30–50%.

It is much more difficult to estimate the mass-loss rate $\dot{M}$ observationally. In general, the outflow intensity can be determined using the same resonance lines via the relation

$$\dot{M} = 4\pi R^2 \rho(R) v(R), \tag{12.1}$$

where $\rho(R)$ is the density and $v(R)$ the wind velocity at a distance R from the center of the star. Observations of lines from different elements yield the distance dependence of the velocity, but the distance dependence of the density is not directly observable, requiring instead the specification of a model stellar atmosphere, chemical composition, and degree of ionization. In other words, the density depends on a large number of poorly known parameters. Mass-loss rates have been obtained for about 70 O stars from Copernicus and IUE observations of P Cyg resonance profiles by Gathier et al. (1981), Olson and Castor (1981), Garmany et al. (1981), and Garmany and Conti (1984). Because of the large model uncertainty, it is very important to be able to derive the mass-loss rate for a given star by different methods.

A more reliable method of estimating $\dot{M}$ is based on infrared and radio observations of free–free emission from the outflowing atmosphere. Free–free absorption is significant in these spectral ranges, and a layer at optical depth $\tau = 1$ lies in the wind region. In the infrared, the emitting surface at $\tau = 1$ lies deeper; its velocity is not well-determined, and a certain amount of the flux, perhaps $10-15\%$, comes directly from the star itself. The wind is responsible for radio emission far from the stellar surface, where the wind velocity is constant and can reach v_w, as determined from the blue edge of ultraviolet absorption features. The mass-loss rate obtained from the radio flux density is

$$\dot{M} = 2.5 \times 10^{-6} \left(\frac{S_{10\,\mathrm{GHz}}}{1\,\mathrm{mJy}} \right)^{0.75} \left(\frac{T_e}{10^4\,\mathrm{K}} \right)^{-0.075} \left(\frac{v_w}{10^3\,\mathrm{km/sec}} \right)$$

$$\times \left(\frac{r}{1\,\mathrm{kpc}} \right)^{1.5} \left(\frac{v}{10\,\mathrm{GHz}} \right)^{-0.45} \bar{\mu}_e \bar{Z}^{-0.5}\ M_\odot/\mathrm{yr}, \tag{12.2}$$

where r is the distance to the star, $\bar{\mu}_e$ is the mean atomic mass per electron,

$$\bar{\mu}_e = \sum_i X_i m_i \Big/ \sum_i X_i Z_i,$$

and $\bar{Z}$ is the mean charge,

$$\bar{Z} = \sum_i X_i Z_i^2 \Big/ \sum_i X_i Z_i$$

(see Felli and Panagia (1982), and references therein). Here the greatest uncertainty comes from $\bar{\mu}_e$ and $\bar{Z}$, i.e., the chemical composition and state of ionization of the plasma of the wind, as well as the distance to the star. Especially promising are coordinated observations at a number of radio and infrared frequencies. The first to undertake such observations were Felli and Panagia (1982), who obtained comprehensive information on outflows from WR stars, including $\dot{M}$, the radius of the stellar atmosphere, and the dependence of wind velocity on distance from the surface; see also Blomme and van den Bergh (1988).

Bieging et al. (1989) recently carried out a survey of radio emission from Galactic OB stars. According to their data, all highly luminous O4f and O7–8 If stars are losing mass at a rate of the order of 10^{-5} $M_\odot/\mathrm{yr}$.

The implied by evaluating $\dot{M}$ by the radio method is that the radio emission is entirely thermal. It has recently been recognized that the radio emission from some of these luminous stars (including WR stars) has a nonthermal component (this problem has been discussed by Abbott et al. (1986)).

Since all methods are to some extent model-dependent, comparison of mass-loss rates obtained for the same stars by different means provides an important test of simplifying assumptions. A comparison of mass-loss rates obtained from visual, UV, IR, and radio measurements clearly divulges systematic differences as large as a factor of 3–5; see Figs. 4-1 and 4-2 in Conti and Underhill (1988).

Results from a great many recent observations are summarized in Table 18 and Figs. 74 and 75. The table lists the mass-loss rate and wind velocity of the stars, about which we shall have more to say in Sections 14–16; for comparison, we have also provided data for a few stars of other spectral types and luminosity classes. Obviously, the WR and Of stars are the ones with the most powerful stellar wind: $\dot{M} = 10^{-5} - 10^{-4}$ $M_\odot/\mathrm{yr}$ at a velocity of $v_w = 10^3 - 4 \times 10^3$ km/sec. Without exception, all of the observed high-luminosity O stars show mass outflow. Stars in which stellar wind has been detected are plotted on the Hertzsprung–Russell diagram in Fig. 75; Wolf–Rayet stars have been left out, as their position on the diagram is uncertain (see Section 15).

Several hundred stars have now been observed in the luminosity range $500 - 5 \times 10^7$ $L_\odot$, with mass-loss rates ranging from 10^{-9} to 10^{-4} $M_\odot/\mathrm{yr}$ and wind velocities of 10 to 10^4 km/sec. The most complete collection of data on stellar wind has been compiled by de Jager et al. (1987); see also Conti and Underhill (1988) and references therein. Abbott et al. (1986) have estimated the wind velocity and mass-loss rate of 40 galactic WR stars based on radio measurements using the VLA. All measured $\dot{M}$ lie in the range $8 \times 10^{-6} \leq \dot{M} \leq 8 \times 10^{-5}$ $M_\odot/\mathrm{yr}$. For certain WR carbon-sequence sub-

Table 18. Outflow Rates for Stars of Different Spectral Types

Star	Spectral type	$\log \dot{M}$ $M_\odot/yr$	v_w 10^3 km/sec	References
HD 50896	WN5	−4.39	2.6	1
HD 192163	WN6	−4.70	2	1,5
HD 191765	WN6	−4.42	2.3	1
HD 92740	WN7	−4.40	2.1	1
BAC 209	WN8	−4.66	(1.5)	1
HD 165763	WC5	−4.22	3.9	1
HD 192103	WC8	−4.49	1.8	1
9 Sgr	O4f	−4.60	3.4	2
ζ Pup	O4f	−5.45	2.7	2
λ Cep	O6 ef	≤ −5.12; −5.4	2.6	2,3,11
HD 151804	O8 If	−5.03	1.7	2
α Cam	O9.5 I	−5.46	1.9	2
ε Ori	B0 Ia	−5.50	2.0	2
HD 56925	WN5	−5.7 to −7.7	2.5	6
HD 89358	WN5	−6.47	3.5	7
HD 148937	O6 f	−7.0	2.6	8
BD +60°2522	O6.5 IIIf	−6.0 to −7.0	1.8	9
68 Cyg	O7.5 IIIf	−5.5	3	10
ρ Leo	B1 Iab	−5.72	1.6	2
χ² Ori	B2 Ia	−6.36	1.1	2
η CMa	B5 Ia	−6.12	0.5	2
μ Sgr	B8 Ia	−5.62	0.5	2
HD 152408	O8 If	−5.0	1.8	3
HD 163758	O6.5 Iaf	−5.2	2.6	3
HD 101413	O8 V	−6.9	2.9	3
HD 54662	O7 III	−6.7	2.5	3
HD 42088	O6.5 V	−6.9	2.6	3

cont'd

classes, the value of $\dot{M}$ was approximately twice as high fairly recently (Cassinelli and van der Hucht, 1987). According to van der Hucht et al. (1986), the outflow rates are $\dot{M}(WN) = (1-12) \times 10^{-5}\ M_\odot/yr$, $\dot{M}(WC) = (2.5-15) \times 10^{-5}\ M_\odot/yr$. Garmany and Conti (1985) have shown that early-type stars in the Magellanic Clouds have mass-loss rates similar to or lower than those in the Galaxy, and an average wind velocity that is 600–1000 km/sec lower. This is in accord with the theory of radiation-driven wind, which predicts a significant decrease with metal abundance for both $\dot{M}$ and v_w (see Kudritzki et al., 1987b).

Table 18. Outflow Rates (cont'd)

Star	Spectral type	$\log \dot{M}$ $M_\odot/yr$	v_w 10^3 km/sec	References
9 Sge	O8 If	−5.2	2.2	3
HDE 303308	O3 Vf	−5.6	3.4	3
	M5 II–III	−8.7 to −9.2	0.01	4
α Boo	K2 IIIp	−10.1	0.04	4
β Gem	K0 III	−9.9	0.04	4
μ UMa	M0 III	< −9.8	0.015	4
β Cru	G5 III	< −9.3	0.05	4
μ Cep	M2 Ia	< −7.5	0.02	4

References: 1) Barlow et al. (1981); 2) Casinelli et al. (1981);
3) Garmany et al. (1981); 4) Drake and Linsky (1983);
5) Dickel et al. (1980); 6) Schneps et al. (1981); 7) Johnson (1982a);
8) Bruchweiler et al. (1981); 9) Johnson (1982b); 10) Kumar et al.
(1983);
11) Abbott et al. (1984).

Recent observations of radio and IR free–free emission for well-established thermal sources, coupled with recent analysis of the wind ionization state and chemistry yield a WR mass-loss rate in the range $10^{-5} - 10^{-4}$ $M_\odot/yr$, with no significant differences between the mean mass-loss rates of a) single and binary WR stars, b) WN and WC stars, and c) WN and WC subclasses (see Willis (1990), and references therein).

Other analyses of mass outflow include one of a large sample of OB stars (Wilson and Dopita, 1985), of M-type giants and supergiants (Wannier and Sanai, 1986), and of stars in the later stages of evolution (Knapp and Morris, 1985).

Infrared observations have shown that in virtually all WC9–10 stars, about 50% of WC8 stars, and several WC7 stars, outflow is accompanied by the formation of dust, with $\dot{M}_{dust} \sim 10^{-8} - 10^{-6}$ $M_\odot/yr$ (Williams et al., 1987; van der Hucht et al., 1987).

Several attempts have been made to determine the empirical dependence of the mass-loss rate on stellar parameters. Garmany et al. (1981) obtained

$$\log \dot{M} = -7.15 + 1.73 \log(L/10^5 L_\odot) \qquad (12.3)$$

for O stars, and Abbott et al. (1980) obtained a similar dependence, $\dot{M} \propto (L/L_\odot)^{1.8}$. Lamers (1981) and Chiosi (1981) found empirical relationships between mass-loss rate, luminosity, radius R, and stellar mass M_* of the form $\dot{M} \propto L^{1.42} R^{0.61} M_*^{-0.99}$ in the first paper, and $\dot{M} \propto L^{0.72} R^{2.5} M_*^{-2.5}$ in the second. Snow (1982) extrapolated these relationships to later-type stars, comparing them with the observed mass loss from 22 B stars, most of which have emission lines in their spectra. It was found that the best agreement with the observations is obtained with Eq. (12.3), extrapolated to lower luminosities. A larger sample of stars in the solar neighborhood at $r \leq 2\,\text{kpc}$ was employed by van Buren (1985). For stars with a mass-loss rate exceeding $\dot{M} = 10^{-8}\ M_\odot/\text{yr}$, he obtained

$$\dot{M} = 2 \times 10^{-13} (L/L_\odot)^{1.25}\ M_\odot/\text{yr},$$
$$\log v_w = -25.2 + 16.23 \log T_{\text{eff}} - 1.70 (\log T_{\text{eff}})^2. \tag{12.4}$$

Wilson and Dopita (1985) have shown that the farther an OB star has traveled off the main sequence, the higher its outflow rate, and they have

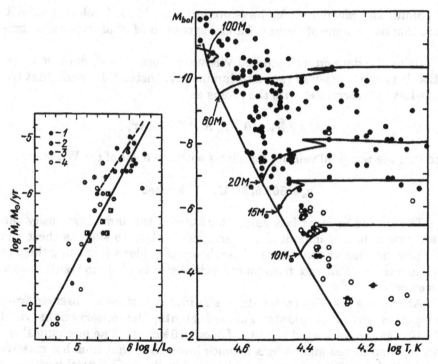

Fig. 74. Mass loss from high-luminosity stars (Garmany et al., 1981). 1) Of stars; 2) O(f) stars and OB stars of luminosity classes I and II; 3) O((f)) stars and stars of luminosity class III; 4) main-sequence OB stars.

Fig. 75. Stars with mass outflow plotted on the Hertzsprung–Russell diagram (filled circles). Empty circles represent stars with no mass loss detected in ultraviolet spectra (Abbott, 1982).

found an empirical relationship between $\dot{M}$ and the parameters of the star:

$$\dot{M} = (0.7 \pm 0.3) L/v_w c \qquad (12.5)$$

(c is the speed of light).

According to Abbott et al. (1986), the mass-loss rate of WR stars correlates with stellar mass and luminosity as

$$\dot{M}(\text{WR}) \sim 7 \times 10^{-8} (M/M_\odot)^{2.3} \ M_\odot/\text{yr}$$

and

$$\dot{M}(\text{WR}) \sim 7 \times 10^{-14} (L/L_\odot)^{1.6} \ M_\odot/\text{yr}.$$

Smith and Maeder (1989) have produced an $\dot{M}$ vs. L relation for WR stars that has a slope of about 0.7 and a spread in $\dot{M}$ of an order of magnitude.

Using the data on well-studied WR stars, Doom (1988) finds no correlation between mass-loss rate and luminosity. Instead, he finds that the mass-loss rate correlates with the radius as

$$\dot{M} = 3.34 \times 10^{-6} (R/R_\odot)^{1.18} \ M_\odot/\text{yr},$$

and that the terminal velocity correlates with the mass of the WR star as

$$v_w = 602.6 (M/M_\odot)^{0.616} \ \text{km/sec}.$$

The spread in mass-loss values for stars of the same luminosity can reach two orders of magnitude, so empirical relations such as these can only provide observational tests of the theory of stellar wind; using them to estimate the outflow rate from an individual star is likely to result in serious error.

Among the massive cooler stars, significant outflow is observed from red giants and supergiants. For red giants, the observations yield $\dot{M} \sim 10^{-6} \ M_\odot/\text{yr}$ at a wind velocity of about 10 km/sec. The mechanical energy injected by red-giant winds is much lower than that from hot massive stars, but the density of matter in the wind in the stellar neighborhood is comparatively high, due to its low velocity.

There is observational evidence for mass outflow from the central stars of planetary nebulae: the central stars that lose mass most rapidly are population II Wolf–Rayet stars[1] and O VI stars (WC2–WC11 in the classi-

---

[1]Population I Wolf–Rayet stars are massive, high-luminosity stars; population II Wolf–Rayet stars are low-mass central stars of planetary nebulae.

fication scheme of Heap(1982)). Central stars of these spectral types have an effective temperature of about $(1-2) \times 10^5$ K and a mass of order $0.5-0.6 M_\odot$; outflow is suggested by a P Cyg line profile or broad optical lines, as in the central stars of NGC 6826 and NGC 6891. It has been found that stellar wind is a widespread phenomenon among the central stars of planetary nebulae. IUE observations have revealed a great many ultraviolet lines with P Cyg profiles, particularly lines of C IV, O IV, O V, Si IV, N IV, and N V, in central stars of types WR, Of, O VI, with pure continuous spectra, and absorption-spectrum types O6, O7, and others as well (in all, in 21 out of 38 nebulae studied) (Heap, 1983; Perinotto, 1983). The outflow velocity, as derived from absorption edges, is $\sim (2-3) \times 10^3$ km/sec, i.e., the same as in population I O stars; this is three to four times the escape velocity. In the planetary nebulae studied in most detail, NGC 6543, NGC 2371, and IC 3568, the mass-loss rate of the central star ranges from 4×10^{-9} to 7×10^{-7} $M_\odot$/yr. Complete data, including further references, have recently been published by Pauldrach et al. (1988).

A wind that powerful, as in the case of population I stars, has a profound effect on the evolution of the central star, on the one hand, and on the morphology, kinematics, and spectrum of the nebula, on the other. Furthermore, only 10–15 years after stellar wind burst upon the astrophysical scene, questions were raised as to whether the "instantaneous" ejection of a shell (typically over a period of 10^3 years) was really necessary for the creation of a planetary nebula around a red giant. The formation of a planetary nebula, in general, may result from the interaction of the "slow" wind from its red giant ($v_w \sim 10$ km/sec, $\dot{M} \approx 5 \times 10^{-5} M_\odot$/yr) with the follow-on "fast" wind of its fully evolved central star. In this model, the fast wind sweeps up matter from the slow, producing the shell structure of the nebula (see Kwok (1983), Kahn (1983), Volk and Kwok (1985), Balick and Preston (1987), and included references).

The momenta carried by the slow and fast winds are comparable, and the interaction between the two may create multi-shell and other complicated structures seen in planetary nebula. The interaction of the fast wind with asymmetrically distributed material in the slow wind or with a surviving remnant of the pre-stellar disk may lead to the formation of prolate and bipolar planetary nebulae (for example, see Balick and Preston (1987) and references therein).

The typical time of formation of a planetary nebula is at least 10^4 years. In fact, fairly recent observations have shown that the immediate precursors of low-mass planetary nebulae — Miras — lose mass at a rate between 7×10^{-7} and $6 \times 10^{-6} M_\odot$/yr (Knapp et al., 1982). Mass outflow from the precursors of planetary nebulae with massive central stars — high-luminosity infrared objects — is an order of magnitude greater. Kwok (1983) found an empirical relationship for the outflow from stars on the asymptotic giant branch: $\dot{M} = 10^{-13} (L/L_\odot)^2 (M/M_\odot)^{-2}$ $M_\odot$/yr for Miras to infrared objects. That rate of mass loss suffices for the slow outflow of the shell of a star with $M_{init} < 6 M_\odot$ (we have seen in Section 6 that this repre-

sents the upper limit on the boundary between the initial mass of a pre-
supernova and the progenitor of a planetary nebula). But apart from the
issue of whether or not the shell was ejected rapidly, an analysis of the
physical conditions in planetary nebulae must take into account both the
interaction of the shell with matter previously expelled as stellar wind by
the red giant, and outflow from the central star.

The outflow mechanism varies among stars of different spectral types.
For hot, massive stars like OB supergiants, Wolf–Rayet stars, and Of
stars, it is mainly controlled by radiation pressure in resonant lines. Gas is
accelerated via momentum transferred through resonant scattering in
strong ultraviolet lines. These stars emit UV radiation in sufficient quantities to accelerate matter in an extended atmosphere; radiation pressure
will accelerate the outflowing gas efficiently if the latter is initially already
transonic. Observations obtained with the Copernicus spacecraft resulted
in a surprise from the standpoint of the "cold" wind model, with a wind
temperature close to the effective temperature of the stellar surface. It
turned out that in addition to spectral lines of C III, C IV, N III, and the
like, the spectrum also exhibited lines due to highly ionized species such as
N V and O VI, which requires a special explanation, inasmuch as the
abundance of those ions as given by the cold wind model ought to be negligible. Several attempts have therefore been made to modify the model (see
the review by Casinelli et al., 1978a, b). The common thread running
through all of them is that below the cold gas region, there is assumed to
be a spatially distinct layer of hot coronal gas above the photosphere, with
a thickness of $\Delta R \approx 0.1R$, that can be heated to a temperature
$T \approx 5 \times 10^6 \, K \gg T_{eff}$ by various acoustic and magnetohydrodynamic instabilities at the surface. In that event, there ought to be some hope of detecting
x rays from the hot layer.

In fact, observations of 20 supergiants ranging from type O4f to A2 I
have revealed that they are all soft x-ray sources: the O supergiants have a
luminosity of more than $10^{32} \, erg/sec$, the B supergiants, more than
$10^{31} \, erg/sec$, and the WR stars, between 10^{32} and $10^{33} \, erg/sec$. But the x-
ray results are in poor agreement with the model of a thin, hot corona between the photosphere and a cold wind, as there is no low-energy cutoff in
the spectrum due to absorption in overlying layers. Observations of WR
stars in binary systems — V 444 Cygni, in particular — make it possible to
localize the hot wind region more precisely. Periodic variations in the x-ray
flux having the same period as the optical eclipses suggest that the radiating plasma is located no farther from the star than $40 \, R_\odot$, but no closer
than $\sim 10 \, R_\odot$, since no low-frequency cutoff has been observed. If the coronal gas occupies a significant fraction of the outflowing atmosphere, it can
be cooler, $T \approx 2 \times 10^5 \, K$. The wind may possibly be heated by turbulence-in-
duced shock waves or by rapid rotation at the surface of the star: due to the
falloff in density with distance, radial flows originating at the surface are
accelerated. The outermost layers of the wind may be heated to
$T \approx 2 \times 10^5 \, K$ by radiation pressure operating on C III ions, with the latter

being accelerated to approximately 10^3 km/sec relative to the plasma of the wind (Vil'koviskii, 1981).

Most of the basic observational properties of hot-star stellar winds, at least for massive, luminous OB stars, can be accounted for by the theory of radiation-driven winds (Kudritzki et al., 1987b). Kudritzki et al. (1988) present a complete summary of much of the theoretical work accomplished thus far (see also the references included).

The gas in the outer atmosphere of late-type stars is heated to $T_e \sim 10^5$ K due to the development of magnetohydrodynamic instabilities and the damping of wave perturbations. Because of the comparatively low gravity in the atmosphere of a giant, and especially in the atmosphere of a supergiant, outflow can begin at temperatures even that low. Another important acceleration mechanism is radiation pressure on dust grains in a cool stellar atmosphere; macroturbulent motions in the chromosphere also contribute, to a certain extent. One source of stellar wind in low-mass main sequence stars — the sun, in particular — is hot coronal gas. The coronal temperature of close to 10^6 K is high enough that the pressure of the hot gas can result in hydrodynamic outflow. Observations of main sequence stars have shown that their x-ray emission is correlated both with the level of turbulence and the efficiency of magnetic field generation at the surface; there is probably significant heating in the internal layers of the outflowing corona of F and G stars. In earlier-type stars, heating most likely occurs in the lowest layer of the corona, which is isolated from the rapidly expanding wind.

These, then, are the most general current ideas on the reasons for mass loss by stars of various spectral types. The detailed theory of stellar wind is quite complex, and has yet to be fully worked out. Nevertheless, a great deal of purely empirical data has been accumulated, and it provides a fairly complete picture of mass outflow from the surface of stars in the form of stellar wind.

So far, we have dealt with observations of the wind phenomenon in the outflowing atmospheres themselves. A much broader range of phenomena relates to the impact of stellar wind upon interstellar gas. The scale of these phenomena can be assessed from the rough figures given in Table 19, where for comparison we have also listed data for a typical supernova explosion. Clearly, the influence of the wind from early-type stars on interstellar gas is comparable, in terms of energy, to the effect of a supernova, if one takes into account the frequency of outbursts and the birthrate of Galactic O stars. On the scale of a single OB association, the mechanical energy injected by stellar wind can exceed the energy of supernova outbursts, and distinguishing between the shells formed by the wind from an association and by supernovae is no easy matter (see Section 18).

Historically, the first observations of wind effects on the ambient interstellar medium were of supersonic gas motion in diffuse emission nebulae (by no means the simplest phenomenon upon which to base a theoretical model of wind interaction with the interstellar gas, but a starting point,

Table 19. Stellar Wind Energetics

Type of star (or association)	$\dot{M}$ $M_\odot/yr$	v_w km/sec	L_w erg/sec	Duration of wind, yr	Total energy input, erg	Size of "disturbed" region, pc
O5 I	10^{-4}	3000	3×10^{38}	10^5	10^{51}	10–50
O9 V	10^{-8}	1000	3×10^{33}	10^7	10^{48}	10–50
Of, WR	5×10^{-5}	2000–3000	$10^{37} - 10^{38}$	10^5	$3 \times 10^{49} - 3 \times 10^{50}$	10–50
Central star, planetary nebula	10^{-7}	3000	3×10^{35}	10^4	9×10^{46}	0.1–1
Red giant	10^{-6}	10	10^{32}	10^6	10^{46}	10
OB association	10^{-4}	2000–3000	2×10^{38}	10^7	$\sim 6 \times 10^{52}$	100–1000
Region of star formation	$10^{-2} - 10^{-7}$	3–100	$10^{33} - 2 \times 10^{36}$	10^4		0.1–1
Supernova explosion	$M = 0.1 - 1 M_\odot$	$\sim 10^4$	$t_{SNR} \sim 10^4 - 10^5$		$\sim 10^{51}$	10–100

nevertheless). Making use of a Fabry–Perot etalon to observe in Hα, Shcheglov (1963) and Gershberg and Shcheglov (1964) detected gas motions in the Orion Nebula, NGC 6618, and NGC 6523 at speeds of some tens of km/sec. This was completely incomprehensible in light of the classical theory of photoionization and emission from H II regions surrounding O stars. The pioneering observations of Shcheglov and Gershberg were completely confirmed by subsequent spectroscopic and interferometric measurements. Velocities of up to 200–240 km/sec were found in the Orion Nebula (see Taylor and Münch (1978) and Goudis et al. (1984), which include further references); splitting of the Hα, [N II], and [O III] lines corresponding to velocities of several tens of km/sec was observed in the H II regions IC 1318, NGC 7000, IC 5070, M8, M16, M17, in the nebulae surrounding η Car and 30 Dor, and in many others (e.g., see Elliott and Meaburn (1975a, b), Walborn and Hesser (1982), Goudis and Meaburn (1976), Meaburn (1977, 1981), Elliott et al. (1978), and Canto et al. (1979)). Ring-like H II regions with an outer shell of neutral gas expanding at $\lesssim 50$ km/sec have been found at the boundaries of dense gas–dust complexes (Meaburn, 1979).

New planetary nebulae have been discovered whose morphology and kinematics attest to the influence of the stellar wind of the central star (see Arkhipova and Lozinskaya (1978b), as well as a number of papers in IAU Symposia Nos. 103 and 131 on planetary nebulae). In 1965, nebulae of a new type were identified — ring-like shells surrounding WR stars (Johnson and Hogg, 1965; Smith, 1968; see also Sections 14 and 15). More recently, a wide variety of ring-like shells have been identified around Of stars (Lozinskaya, 1982; see Section 16). As a result of stellar wind and multiple supernova explosions, shells and supershells are formed around OB associations, a subject we shall discuss in Chapter 4.

It has become quite obvious that rapid, nonstationary mass loss plays a decisive role in star-forming regions and in young, pre-main sequence stars. The interaction of stellar wind with the ambient interstellar gas in regions of star formation in dense molecular clouds is the most efficient mechanism, and is the best-studied (e.g., see the collection edited by Roger and Dewdney (1982), *Regions of Recent Star Formation*). Observations in the lines of CO, H_2, OH, H_2O, SiO, and other species have revealed supersonic gas flows moving at 10–300 km/sec. The wind exerts an influence for $10^3 - 10^4$ yr, and the mass-loss rate varies from $10^{-7} - 10^{-6}$ $M_\odot$/yr to $10^{-2} - 10^{-3}$ $M_\odot$/yr. The outflow mechanism in protostellar sources is fundamentally different from that in hot stars, losses via mechanical energy are hundreds of times the stellar luminosity, and the wind often displays bipolar asymmetry.

This, then, is the array of astronomical objects whose behavior is governed by the interaction between stellar wind and the ambient gas.

Note also that as a star evolves, its mass-loss rate varies. As a hot star passes into the giant or supergiant stage, there is a sudden reduction in wind velocity, although the decrease in the rate of outflow is negligible; the net result is an increase in the density of the wind. This means that in the later stages of evolution, a massive star will be surrounded by a multilayer shell of matter, both outflowing and swept up by the wind. Furthermore, any "instantaneous" ejection of matter, be it the expulsion of the outermost hydrogen shell or a supernova explosion, will interact with that multilayer structure in a most complicated manner. We presently have no way to analyze the interaction in full detail; to make matters worse, stars move with respect to the interstellar medium, and the interstellar gas is inhomogeneous. As a first approximation, therefore, we consider below the properties and physical characteristics of those objects for which one can elucidate the interaction between the wind and the interstellar gas in an environment more or less unencumbered by complicating factors.

13. Interaction of stellar wind with interstellar gas

The interaction of a strong stellar wind with interstellar gas was first examined by Mustel' (1959a, b), Pikel'ner (1968), and by Pikel'ner and Shcheglov (1968). The basic proposal by Pikel'ner laid the foundation for our current understanding: the wind sweeps up gas, forming a cavity in which the wind expands freely, and which is surrounded by a layer of shock-heated, decelerating wind and a layer of swept-up gas. Avedisova (1971) obtained a solution of the equations of motion for two stages — adiabatic expansion and strong radiation — in the development of a shell. A self-similar solution for the interaction of the wind with the interstellar gas was also found by Dyson and de Vries (1972) and by Dyson (1973), who took into account the inhomogeneous density of the medium and temporal variations in the power of the wind. Numerical solutions have been obtained by Falle (1975b), who devoted particular attention to the avalanche collapse of the cold shell. Falle demonstrated that even a slight inhomogeneity in the interstellar medium will ultimately destroy the regular structure of a shell and lead to leakage of hot gas through the holes in the shell; he also considered the influence of the interstellar magnetic field. Ionization conditions in shells swept out by stellar wind have been analyzed by Steigman et al. (1975). Dyson (1977) and Canto et al. (1979) considered a shell formed in a medium with a strong density gradient — for example, at the boundary of a molecular cloud. Pikel'ner (1973) analyzed wind flow past dense condensations in planetary nebulae and H II regions. Dyson (1978) and Taylor and Münch (1978) have used the wind to explain the formation of dense neutral condensations in H II regions.

The structure and evolution of shells produced by stellar wind have been investigated in the most detail by Castor et al. (1975) and Weaver et al. (1977), who obtained analytic and numerical solutions for the initial adiabatic stage, and the intermediate and late stage of strong radiative cooling. Their work incorporated thermal conductivity and the evaporation of cool gas into the hot-wind region, and they analyzed shell stability.

In a series of papers, Rózyczka and Tenorio-Tagle (1985) carried out two-dimensional hydrodynamic calculations that took account of fluctuations in wind power and gas density. They considered a medium with small-scale fluctuations and a large-scale density gradient, and paid special attention to instabilities during shell formation and the stage at which hot gas leaks through the shell.

Let us consider an idealized picture, based on these studies, of the interaction of wind from an early-type star with the ambient interstellar gas. The "ideal" conditions assumed are that outflow begins at time $t = 0$ and is steady and spherically symmetric, with constant power $L_w = 0.5 M v_w^2$, and that it occurs in a homogeneous medium with number density (of atoms) n_0. The multilayer shell formed about a source with a strong wind is diagrammed in Fig. 76. The innermost cavity a is filled with wind freely expanding at a velocity v_w; the wind density varies with distance from the star as $\rho(R) = \dot{M}(4\pi R^2 v_w)^{-1}$. The region of freely expanding wind is separated from the layer of decelerated wind by the shock front I; the layer b consists of gas from the wind that has been heated at the shock front I, and gas evaporated from shell c. Layer c consists of interstellar gas swept up and heated by the shock wave; that gas is separated from the unperturbed interstellar gas by shock front II, and from the hot wind by the contact discontinuity R_c (from here on we denote the radius and shock velocity at fronts I and II by subscripts 1 and 2, respectively).

Three idealized stages can be identified in the evolution of this multilayer wind-blown bubble; as for supernovae, there exist self-similar solutions of the equations of motion for all three. In the first, radiative energy losses are negligible everywhere within the shell, and the system expands adiabatically. In the second, radiative losses from layer c become comparable to the energy injected by the wind. The swept-up gas collapses abruptly due to radiative cooling, and a cold dense shell forms. The latter is observed at optical and radio wavelengths as a ring emission nebula, but the layer b of hot gas is still expanding adiabatically. In the third and final stage, emission from layer b induces its collapse into a thin layer, and the freely expanding wind a interacts directly with the swept-up interstellar gas c.

Let us look at the basic parameters of the bubble swept out by the wind during these three developmental phases, following Castor et al. (1975) and Weaver et al. (1977). The early stage of adiabatic expansion of the system is relatively brief. During that stage, the equation of motion of the shell possesses a self-similar solution; the independent parameters are the power of the wind L_w and the density $\rho_0 = \mu m_H n_0$ of the unperturbed gas in the region d. The solution takes the form

$$R_2(t) = \alpha (L_w/\rho_0)^{1/5} t^{3/5}, \quad \alpha = 0.88. \tag{13.1}$$

The contact discontinuity between the hot wind b and the shell c occurs at a distance $R_c \approx 0.86 R_2$, and moves at velocity $v(R_c) \approx 0.86 v_2$. The

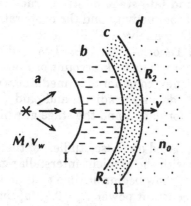

Fig. 76. A bubble swept out by stellar wind. a) Freely expanding wind; b) hot decelerated wind; c) shell of swept-up interstellar gas; d) ambient interstellar gas. I and II are the shock fronts in the wind and ambient gas, respectively, and R_c is the contact discontinuity.

conditions prevailing within layer b are virtually isobaric. The pressure of the hot gas — with a high speed of sound that is comparable to the wind velocity — is uniform, and is equal to the dynamic pressure of the wind at the front I. The kinetic energy of the gas in layer b is negligible, and the thermal energy is $E_b \approx 5/11 L_w t$. The energy of layer c is $E_c \approx 6/11 L_w t$, with about 40% going into kinetic energy and about 60% into thermal. Conditions in layer b, i.e., at $R_1 \lesssim R \lesssim R_c$, are described by the equations

$$v = 0.2 \frac{v_w R_1^2}{R^2}, \quad \rho \approx 0.35 \frac{\dot{M}}{R_1^2 v_w},$$

$$P \approx 0.07 \frac{\dot{M} v_w}{R_1^2}, \quad R_1 \approx 0.74 \left(\frac{\dot{M}}{\rho_0} \right)^{3/10} v_w^{1/10} t^{2/5}. \tag{13.2}$$

By analogy with supernova remnants, the time t_{cool}, representing the transition from the adiabatic to the shell-formation stage, can be ascertained by equating the duration of radiative cooling — at a rate comparable to the input of wind energy — to the kinematic age of the shell:

$$t_{cool} \approx 1.7 \times 10^3 (\dot{M}_6 v_{2000}^2 / n_0)^{1/2} \text{ yr}. \tag{13.3}$$

The radiative losses here are taken in accordance with Fig. 52, and $\dot{M}_6 = (\dot{M}/10^{-6}) \, M_\odot/\text{yr}$, $v_{2000} = (v_w/2000) \, \text{km/sec}$.

For the "standard" values $\dot{M}_6 = 1$, $v_{2000} = 1$, and $n_0 = 1 \, \text{cm}^{-3}$, we obtain $t_{cool} \approx 2 \times 10^3$ yr. As soon as the gas begins to cool, it will do so all at once. The sudden drop in pressure behind front II causes it to slow down, and it accelerates the contact surface; the shell c collapses, and makes its appearance optically. Just as in supernova remnants, the transition from the adiabatic to the shell stage is accompanied by multiple secondary shock waves which are gradually damped, endowing the shell with a rather complex temperature and density distribution (see Section 9). The system typically settles down to a steady state after about 10^4 yr.

The mass of gas contained within shell c is completely determined by the amount of interstellar gas swept up: $M_c = 4\pi R_2^3 \rho_0 / 3$. Thermal conductivity at the contact surface R_c leads to evaporation of the cold shell, furnishing an influx of matter to the hot region b. The loss of matter is insignificant for the shell, but in layer b, the evaporated gas can actually dominate the gas of the wind: $M_{evap} \gtrsim \dot{M} t$. The mass influx due to evaporation is

$$\dot{M}_{evap} \approx \frac{16}{25} \frac{\pi \mu}{k} c T_b^{5/2} R_2 \approx 3 \times 10^{19} T_{b6}^{5/2} R_2 \frac{30}{\ln \Lambda} \text{ g/sec}. \tag{13.4}$$

Here the term $cT_b^{5/2}$ corresponds to the "classical" thermal conductivity (see Spitzer (1981)), $c = 1.2 \times 10^{-6}$ erg/(cm·sec·K$^{7/2}$), and $\ln \Lambda$ is the Coulomb logarithm:

$$\ln \Lambda = 30 + \ln n_e^{-1/2} T_6^{-1/2} \text{ for } T > 4 \times 10^5 \text{ K}; \ T_6 = (T_b/10^6) \text{ K}.$$

In the shell stage, $R_1 \ll R_c \approx R_2$, and most of the cavity is occupied by hot wind at constant pressure; this forms a hot cushion that transmits the kinetic energy of the outflowing matter to the swept-up interstellar gas.

The equation of motion in the shell stage also has a self-similar solution; rather than taking the wind power as the independent variable, one takes the pressure of the hot wind at the shell (this is usually referred to as the *energy-driven bubble* or *energy conservation stage*). The solution takes the form

$$R_2(t) = 0.76 \left(\frac{L_w}{\rho_0} \right)^{1/5} t^{3/5} = 28 n_0^{-1/5} L_{36}^{1/5} t_6^{3/5} \text{ pc,}$$

$$v_2(t) = 16 n_0^{-1/5} L_{36}^{1/5} t_6^{-2/5} \text{ km/sec,}$$

(13.5)

where $L_{36} = (L_w/10^{36})$ erg/sec and $t_6 = (t/10^6)$ yr. During this second stage, the temperature at the center of layer b is given by

$$T_b = 1.6 \times 10^6 n_0^{2/5} (\dot{M}_6 v_{2000}^2)^{8/35} t_6^{-6/35} \text{ K,}$$

(13.6)

and the density at the center of layer b is

$$n_b = 0.01 n_0^{19/35} (\dot{M}_6 v_{2000}^2)^{6/35} t_6^{-22/35} \text{ cm}^{-3}.$$

(13.7)

Bremsstrahlung from the hot gas in layer b is generated mainly at x rays and UV; the luminosity is

$$L_{rad} = 3.8 \times 10^{33} n_0^{18/35} (\dot{M}_6 v_{2000}^2)^{37/35} t_6^{16/35} \text{ erg/sec.}$$

(13.8)

In the early stages, radiative losses from hot gas are clearly negligible compared with the mechanical energy L_w injected by the wind, but they increase with time as $t^{16/35}$.

If the outflow goes on for long enough, there may come a time at which $L_{rad} \gtrsim L_w$, whereupon layer b will collapse, due to a drop in both energy and pressure. At that point $R_1 \approx R_2$, and the wind will interact directly with the cold, dense shell c. The motion of the shell in this third stage (the so-called *momentum-driven bubble* or *momentum conservation stage*) (Steigman et al., 1975; Avedisova, 1971) is given by

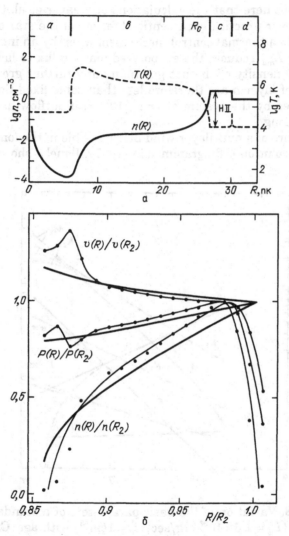

Fig. 77. a) Temperature $T(R)$ and density $n(R)$ in a "standard" bubble (Weaver et al., 1977), with $L_w = 1.3 \times 10^{36}$ erg/sec and $n_0 = 1\,\text{cm}^{-3}$ at $t = 10^6$ yr; **b)** velocity $v(R)/v(R_2)$, pressure $P(R)/P(R_2)$, and density in $n(R)/n(R_2)$ the outer shell, normalized to values at front II. The solid curves show the analytic solutions; points are computed values (Rózyczka, 1985).

$$R_2(t) = \left(\frac{3\dot{M}v_w}{2\pi\rho_0}\right)^{1/4} t^{1/2}. \tag{13.9}$$

We point out here that the calculations of Weaver et al. (1977) imply that $L_{rad} \leq L_w$ over practically the entire lifetime of the star on the main sequence, due to a thermal control mechanism whereby an increase in the radiated power L_{rad} reduces the evaporated mass influx to layer b, and a decrease in the density of the hot gas limits any further growth of L_{rad}. Energy losses of layer b are therefore less than some fixed limit $\sim 0.9 L_w$, which is reached by the system after $t \gtrsim 10^7$ yr, if outflow has been sustained for that long.

The structure of a multilayer wind-blown bubble in the longest-lasting shell phase of evolution is diagrammed in Fig. 77. Panel a shows numerical

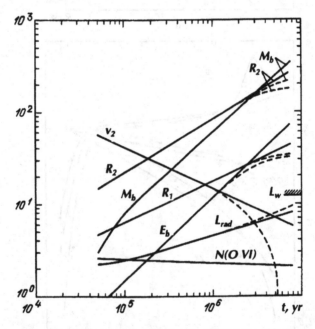

Fig. 78. Variation of the basic parameters of a standard bubble ($L_w = 1.3 \times 10^{36}$ erg/sec, $n_0 = 1\,\mathrm{cm}^{-3}$) with age. Calculations by Weaver et al. (1977) for two cases: negligible external pressure (solid curves), and wind in an H II region, the pressure of which can be substantial at the later stage (dashed curves). Units as follows: R_1, R_2 (10^{18} cm); v_2 (km/sec); mass of gas in layer b, $M_b = 10^{33}$ g; energy of hot wind, $E_b = 10^{49}$ erg; $L_{rad} = 10^{35}$ erg/sec (same notation as in text and Fig. 65).

results obtained by Weaver et al. (1977) for the "standard" model $\dot{M}_6 = 1$, $v_{2000} = 1$, $n_0 = 1\,\mathrm{cm}^{-3}$, corresponding to an age of 10^6 yr. Weaver's calculations allowed for the curvature of the shock-wave and contact discontinuity, radiative losses throughout the system, thermal conductivity, and evaporation at the contact surface. It is clear from the figure that 10^6 yr after the beginning of outflow, front I has progressed out to 6 pc from the star, and front II has reached 27 pc; most of the region is filled with hot, shocked wind. Taking thermal conductivity and evaporation at the interface between hot and cold gas into account has revealed an intermediate layer not shown by Fig. 76. The structure of that transitional layer is governed by the electron thermal conductivity, and the temperature varies from 10^4 to 10^6 K at a distance of about 2 pc. About 40% of the thermal energy is radiated from the layer by virtue of the collisional excitation of resonant ultraviolet lines of highly ionized atoms — in particular, the lines of O IV. The other 60% goes into evaporating the cold shell into region b, where the newly evaporated gas mixes with the gas of the wind; the mass of the evaporated gas far exceeds that of the wind material.

In Fig. 77b we see the structure of the outer region from $0.85R_2$ to R_2 (Rózyczka, 1985) — the behavior of the velocity, density, and pressure as functions of the distance from the center.

The way in which the parameters of this standard bubble change with time is shown in Fig. 78. Solid curves give the variation of radius and shock front velocities for shock waves I and II, as well as the mass, energy, and luminosity of the hot gas for a shell expanding into a neutral medium, where the external pressure can be neglected. The dashed curves display the same parameters for a bubble in ionized gas, the pressure of which becomes substantial in the later stages. The numerical results plotted in Fig. 78, which take into account thermal conductivity, evaporation, and radia-

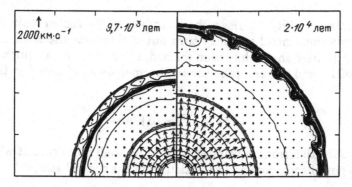

Fig. 79. Structure of a wind-blown bubble. Two-dimensional numerical solution by Rózyczka and Tenorio-Tagle (1985) for wind velocity (arrows) and contours of equal density. Deformation of the outer dense shell is due to small-scale density fluctuations.

tive losses, yield for the radius of front I the function $R_1(t) \propto t^{0.44}$, and for front II, $R_2(t) \propto t^{0.58}$; these expressions are quite close to the simple self-similar approximations. An "accurate" numerical solution yields an expression for R intermediate between (13.5) and (13.9), due to allowance being made for radiative losses of the hot gas in the early stages; it also yields a velocity v_2, energy, and mass of hot gas somewhat lower than the "self-similar" values.

The results of two-dimensional modeling of the structure and evolution of a bubble that takes account of small-scale density perturbations of the swept-up gas are shown in Fig. 79. Distortion of the regular structure of the outer shell is due to density fluctuations; the numbers refer to the age of the bubble in the corresponding panel. Rózyczka and Tenorio-Tagle (1985) carried out a similar series of two-dimensional calculations for a range of wind models having variable power in a medium with density variations. Clearly, small-scale inhomogeneities have only a minor effect on shell dynamics, and do not lead to shell fragmentation.

The gas density in a cold shell, assuming that it contains all the homogeneous gas swept up by the wind, is

$$
\begin{aligned}
n_c &= (m_{\mathrm{H}} \mu v_2^2 / k T_c) n_0 \\
&= 3.2 \times 10^4 \mu (\dot{M}_6 v_{2000}^2)^{2/5} t_6^{-4/5} n_0^{3/5} T_c^{-1} \ \mathrm{cm}^{-3},
\end{aligned}
\tag{13.10}
$$

where T_c is the gas temperature in the shell, and $\mu = 0.65, 1.30, 2.36$ for the H II, H I, and H_2 regions, respectively. The total surface density of the gas in the shell behind front II is

$$
N_2 = n_0 R_2 / 3 \approx 3 \times 10^{19} (\dot{M}_6 v_{2000}^2)^{1/5} n_0^{4/5} t_6^{3/5} \ \mathrm{cm}^{-2}.
\tag{13.11}
$$

As the surface density N_2 rises, the boundary of the region ionized by radiation from the central star may turn out to lie inside the shell swept up by the wind. After shell c has been formed, the swept-up gas quickly cools to $T \approx 8000\,\mathrm{K}$, and one can take the isothermal speed of sound in the shell to be

$$
c_{\mathrm{II}} = (kT / \mu_{\mathrm{H}})^{1/2} \approx 10 \ \mathrm{km/sec}.
$$

The ionization front is trapped within the shell if the rate at which the swept-up gas absorbs ionizing radiation from the central star,

$$
S_{abs} = 4 \pi R_2^2 \Delta R \alpha_{(2)} n_c^2,
$$

is equal to the rate S at which ionizing photons are produced, where

$$\alpha_{(2)} = 3.1 \times 10^{-13} \text{ cm}^{-3} \cdot \text{sec}^{-1}$$

is the recombination coefficient, and

$$\Delta R = (R_2/3)(n_0/n_c)$$

is the thickness of the shell.

In that event, the criterion for trapping the ionization front is

$$S = (4\pi/3)\alpha_{(2)}n_0^2(v_2^2/c_{II}^2)R_2^3. \tag{13.12}$$

Using (13.12) and (13.5), we may write for the ionization-front capture time

$$t = 0.44 \times 10^6 n_0^{-1} L_{36}^{-1} S_{48} \mu \text{ [yr]}, \tag{13.13}$$

where $S_{48} = S/10^{48} \text{ sec}^{-1}$; see also Wisotzki and Wendker (1989).

For a star of type O7 III ($S_{48} = 11$) in a medium with $n_0 = 1 \text{ cm}^{-3}$, and taking $L_{36} = 1$, we have $t = 5 \times 10^{36}$ yr for the ionization front to be captured by the cold shell. If the ionization front is trapped, shell c will have an outer layer of H I (or H_2) with a temperature $T \approx 80$ K and 200–400 times the density of the interior ionized layer.

When the expansion velocity of the dense shell falls to the speed of sound c_0 in the ambient unperturbed gas, the shell will dissipate in the interstellar medium. For our now-familiar standard bubble, the dissipation time is approximately equal to the lifetime of the star on the main sequence ($t_{MS} \approx 8 \times 10^6 S_{48}^{-1/4}$ yr).

The approach discussed thus far is applicable to an idealized steady, spherically symmetric wind with constant terminal velocity and mass-loss rate in a homogeneous interstellar medium. In analyzing the interaction of a strong wind — from a Wolf–Rayet star, for example — with the cloudy interstellar medium, one must pay attention not so much to the small-scale density fluctuations (just as for supernova remnants, very small dense clouds do little to distort the dynamics of the shell) as to the overall recoil effect associated with the evaporation of clouds by ionizing radiation. That effect, which has to do with the asymmetric expulsion of gas from the evaporating side of a cloud, leads to the acceleration of the latter in the opposite direction (Oort and Spitzer, 1955). As a result, even before the strong wind from the WR stage has turned on, ionizing radiation from a main-sequence O star will have produced a spherical cavity about the star that is filled with the gas that has evaporated; it will be essentially free of clouds, which turn out to be concentrated toward the periphery (Elmegreen, 1976). McKee et al. (1984) find a cavity radius (having investigated the contribution of clouds with differing masses conforming to a standard mass spectrum $N(M_{cl}) \propto M_{cl}^{-2}$) of $R_c^* = 1.05(t^*)^{4/7}$, where the dimensionless radius and age are $R_c^* \equiv R/R_{St}$ and $t^* \equiv t/t_{St}$, respectively, t_{St} is the characteristic dynamic time for the Strömgren sphere

$$t_{St} = R_{St}/c_{II} \approx 3 \times 10^6 \left(S_{48}/\langle n \rangle^2 \right)^{1/3} \text{ yr},$$

and $\langle n \rangle$ is the mean density obtained by filling the empty cavity with the matter in the cloud. The rate of change of the radius is identical to the expansion of a fully evolved H II region in a homogeneous medium of density $\langle n \rangle$. If intercloud gas is taken into account, the numerical coefficient is reduced somewhat. Over the time t_{MS} (i.e., prior to the onset of the WR stage, when the wind energy output increases tenfold), the radius of the cavity filled with homogeneous evaporated gas surrounding a type O4–O9 star reaches $R_c(t_{MS}) \approx 56 \langle n \rangle^{-0.3}$ pc. The velocity of the clouds at a distance R_c is $v_c \approx 6.8 \times \left(S_{48}/\langle n \rangle \right)^{1/4}$ km/sec (McKee et al., 1984).

The wind from a Wolf–Rayet star acts upon this symmetric, two-layer H II region, and allowance for the initial state of the ambient interstellar gas alters the dynamics of the shell produced by the stellar wind. If the outflow is weak, with $L_w^* \equiv L_w / 4\pi R_{St}^2 \langle n \rangle m_H c_{II}^3 \ll 1$, the radius of the shell swept up by the wind will be R_c at most, and Eqs. (13.5) and (13.9) will remain valid, since then the wind will be interacting with homogeneous evaporated gas. If the wind is strong, $L_w^* \gg 1$, the wind-blown shell will retain a cloud layer on the outskirts of the cavity, and will expand rapidly while conserving momentum (see Eq. (13.9)). That expansion will continue out to a radius of approximately $R^* \sim 0.14(L_w^*)^{0.5}$, after which the shell dynamics will be governed entirely by the dynamics of the evaporating and accelerating clouds: $R_2 \sim R_c$.

When $L_w^* \sim 1$, the wind-blown shell expands out to the cloud boundary — i.e., the hot wind fills the cavity. The evaporated gas penetrates the hot wind region due to Rayleigh–Taylor instability at the shell boundary, which speeds up radiative cooling. Shell expansion slows down, the shell no longer encompasses a cloud layer, and evolution proceeds at a rate intermediate between (13.5) and (13.9) (McKee et al., 1984).

The situation in WR and Of stars is actually more complicated, since their precursors — massive O stars — also undergo mass loss via stellar wind. In the first stage, it is necessary to consider the combined effect of reactive evaporation of clouds and stellar wind, while in the second stage, the wind is an order of magnitude more powerful. Furthermore, the preceding discussion has ignored stellar motion, which distorts the symmetry of both the two-layer H II region and the shell swept out by the wind. The present picture is therefore necessarily a rather crude one; as we shall see in Sections 14 and 15, however, ring-like H II regions that fit the picture completely have been observed about strong sources of stellar wind.

The theory of the interaction of stellar wind with the interstellar medium has developed in a number of different directions. Firstly, attempts have been made to assess the way in which the power in the wind and in ionizing radiation varies as a star evolves. This has been particularly important for the shells around young clusters, such as the Rosette and η Carinae, where allowance for the "history" of the outflow yields better agreement with observations of ring-like H II regions (Dorland et al., 1986).

Secondly, the inhomogeneity of the ambient gas has been taken into consideration. The structure of a wind-blown bubble in a cloudy medium and the analytic solution of the equations of motion have both been examined (Hanami and Sakashita, 1987). The density gradient has been introduced into the treatment of small bubbles; as a bubble grows to the point where it is larger than a typical cloud, these effects show up principally through an increase in the density of the hot wind, due to evaporation. This accelerates radiative cooling, and the shell enters the snowplow stage that much sooner.

Wolff and Durisen (1987) have embarked upon a new series of numerical calculations of the structure and evolution of a wind-blown bubble in a two-phase medium consisting of clouds and intercloud gas. They have included mass exchange between cold clouds and hot gas in the bubble resulting from both thermal and ionization evaporation, energy and momentum exchange between the two phases, cloud motion due to momentum transfer, evaporation and condensation, and entrainment by hot gas, and secondary shock waves in dense clouds. It would not be surprising to find that these numerical models modify the structure and evolution of a bubble swept out by stellar wind to the same extent and in the same way as was found by Cowie et al. (1981) for supernova remnants (see Section 9).

Thirdly, an analysis of thermal conductivity in the strong temperature gradient behind a shock front in the wind modifies the structure and evolu-

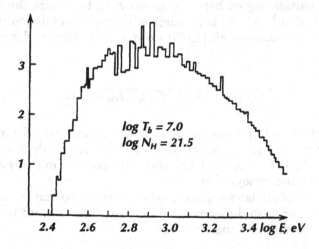

Fig. 80. X-ray spectrum expected from a wind-blown bubble. Calculations for NGC 6888 by Bochkarev and Lozinskaya (1985). The spectrum is labeled by the mean temperature in hot gas layer b and the column density of absorbing atoms.

tion of a wind-blown bubble significantly, even in a homogeneous medium (see Dorland et al. (1986) and Dorland and Montmerle (1987)).

Since the thermal flux is determined by the temperature profile at a thermal conductivity front, the dissipation of wind energy in plasma with a strong temperature gradient is enhanced by a mechanism known as delocalization of the conductive heat flux. These authors have shown that the wind loses a significant fraction (if not most!) of its energy within a relatively narrow dissipative layer behind the shock front, a process that profoundly affects the efficiency of stellar-wind energy transfer to the interstellar gas, and that is most clearly manifested in soft x-ray emission.

What, then, are the most salient observational features of the theory of the interaction of stellar wind with the gas of the interstellar medium, as we have sketched it here?

The hot region b in Fig. 76 ought to radiate in soft x rays. The x-ray spectrum of a hot wind has been calculated by Weaver et al. (1977) and Bochkarev (1985). Bochkarev and Lozinskaya (1985) have calculated the spectrum and x-ray flux expected of about a dozen of the most observationally promising ring nebulae produced by the wind from WR and Of stars. Figure 80 shows their results, namely the predicted spectrum of the nebula NGC 6888, assuming the hot plasma to be in ionization equilibrium . The integrated 44–70 Å x-ray luminosity is $L_{44-70\,\text{Å}} \approx 2 \times 10^{33}$ erg/sec; i.e., it is three orders of magnitude less than the wind power L_w (recall that the x-ray luminosity of supernova remnants ranges from 10^{35} to 10^{36} erg/sec). The x-ray emission from the hot gas in the bubble produced by the wind from isolated WR stars, in particular NGC 6888, has in fact been detected (see Section 15).

The transition region between layers b and c, where the gas is at a temperature of $10^5 - 10^6$ K, is a source of ultraviolet emission in the O VI resonance line. Castor et al. (1975) estimate the surface density in that layer to be

$$N(\text{O VI}) \approx 3.4 \times 10^{16} X_O n_0^{9/35} L_{36}^{1/35} t_6^{8/35} \text{ cm}^{-2}, \qquad (13.14)$$

where X_O is the relative oxygen abundance (by number of atoms). For a standard shell, this yields $N(\text{O VI}) \approx (1-2) \times 10^{13}$ cm^{-2}, which is fully consistent with the O VI resonance line strength observed in the spectra of hot stars (Jenkins and Meloy, 1974).

The effects of stellar wind are most clear-cut at optical wavelengths. It turns out that prior to the immersion of the ionization front in the swept-up shell, the thickness of the latter is

$$\Delta R_2 \approx \frac{R_2}{3} \left(\frac{v_2^2}{c_{\text{II}}^2} \right). \qquad (13.15)$$

The radius of the ionization front is given by

$$R_i^3 = R_{St}^3 - R_2^3\left(\frac{v_2^2}{c_{\mathrm{II}}^2} - 1\right),$$ (13.16)

where $R_{St} = 31(S_{48}/n_0^2)^{1/3}$ pc is the Strömgren radius of the unperturbed H II region. The emission measure in the central part of the shell is

$$E_{m0} = 2n_0^2\left[\frac{v_2^2}{c_{\mathrm{II}}^2}\frac{R_2}{3} + (R_i - R_2)\right]\mathrm{cm}^{-6}\cdot\mathrm{pc},$$ (13.17)

while the maximum emission measure in the "tangential" direction is

$$E_{m\,max} = 2n_0^2\left[\frac{v_2^3}{c_{\mathrm{II}}^3}(2/3)^{1/2}R_2 + (R_i^2 - R_2^2)^{1/2}\right]\mathrm{cm}^{-6}\cdot\mathrm{pc},$$ (13.18)

and the relative limb brightening of a bubble for a just-trapped ionization front is

$$\frac{E_{m\,max}}{E_{m0}} \approx 2.45\frac{v_2}{c_{\mathrm{II}}};$$ (13.19)

see Wisotzki and Wendker (1989).

Objects with emission measure that high appear as bright optical nebulae with ring-like morphology and a thermal radio spectrum, and observations of such nebulae — shells surrounding isolated stars with strong outflow — enable one to investigate experimentally the interaction of stellar wind with the interstellar medium. It is now also clear that the IRAS data present a new opportunity to study wind-blown bubbles. Van Buren and McCray (1988) have shown that even the IRAS "quick-look" surveys divulge a great many ring-like and arc-like IR features associated with massive hot stars. Several well-known optical ring nebulae around WR and Of stars are also clearly seen in the infrared; their IR emission comes from interstellar dust swept up by stellar wind and radiatively heated by stellar UV radiation. In the next three sections, we proceed to analyze nebulae associated with WR and Of stars.

14. Ring nebulae around Wolf–Rayet stars

A new class of emission nebulae — extended shells around Galactic population I Wolf–Rayet stars — was identified in 1965 (we note here, as in Section 12, that population I Wolf–Rayet stars are massive, high-luminosity stars, while population II Wolf–Rayet stars are low-mass central stars of planetary nebulae). These ring nebulae, formed by stellar wind, thus joined a number of other well-established classes of objects: diffuse H II regions, planetary nebulae, and supernova remnants. The first investigators to draw attention to this new class were Johnson and Hogg (1965), who estimated nebular masses on the basis of radio flux measurements, and suggested that such shells are formed when interstellar gas is swept up by a strong stellar wind. Smith (1968) found seven ring nebulae surrounding Galactic WR stars in the Palomar atlas; subsequent searches tripled the number of stars with ring nebulae. The most complete survey of Galactic WR stars with ring nebulae is the one carried out by Chu (1981), Heckathorn et al. (1982) and Chu et al. (1983) (see also Table 20). This new type of nebula came in for thorough study throughout that period; more recently, a new wave has rolled in, since it has been realized that shell ejection may be a natural process in the evolution of massive close binaries that takes place at the stage when such a system consists of a helium star with a compact relativistic companion.

  The nebula NGC 6888, surrounding the WN6 star HD 192163, is the best-known and best-studied object of this type. This nearby, comparatively bright nebula, with clearly delineated shell structure and delicate filamentary morphology, is illustrated in Fig. 81. The optical spectrum of NGC 6888 shows strong lines of [N II], [S II], and [O III], and the relative intensity $I_{H\alpha}/I_{[N\ II]}$ ranges from 0.5 to 2.2 among the various filaments, averaging 1.5 (Parker, 1964, 1978; Esipov and Lozinskaya, 1968). The typical gas density in the bright filaments is $400-500\ cm^{-3}$. The intensity variations, particularly in the [O III] line, are due to variation of the ionizing flux with distance from the central star. The nitrogen abundance in the filaments of the shell is two to three times the normal cosmic value, and the helium abundance is also above the norm (Parker, 1978; Kwitter, 1981), which in all likelihood is related to the enrichment of the interstellar gas by matter shed by the Wolf–Rayet star (see below).

  The nebula has a thermal radio spectrum (Johnson and Hogg, 1965; Lozinskaya, 1970). Observations at high angular resolution show that the fine details of the knots and filaments in the radio and optical are virtually identical. The radio spectral index of both the nebula as a whole and the individual condensations is $\alpha = 0.1$. Various authors have estimated the mass of the nebula to be $4-20\ M_\odot$, based on the radio flux density, and assuming an optically thin shell (Johnson and Hogg, 1965; Wendker et al.,

1975; Smith and Batchelor, 1970). The large range is probably due to the uncertainty in the filling factor; the most reliable estimate is $4.6\,M_\odot$, which is obtained at the highest angular resolution (Wendker et al., 1975).

Courtes (1960) was able to obtain an interferometric image of a large region in Cygnus that also included the nebula NGC 6888, and he remarked upon the high internal velocities of the gas, $\Delta v = 80\,$km/sec. In 1967, 1975, and 1985, the present author carried out extensive series of Fabry–Perot interferometric observations, obtaining more than 100 monochromatic and interference photographs in Hα, [N II], and [O III], uniformly covering the whole field of the nebula. Several of these images are shown in Fig. 81, and they illustrate the shell morphology of the nebula, as well as individual bright filaments and knots seen against the fainter amorphous emission from the interfilament medium. Our measurements show the integrated brightness of the nebula in Hα to be $(4-7) \times 10^{-10}$ erg$\cdot$cm$^{-2}\cdot$sec; Wendker et al. (1975) have obtained $(5-6) \times 10^{-10}$ erg$\cdot$cm$^{-2}\cdot$sec. The interference observations yielded the radial velocity distribution over the entire nebular image. If we assume a prolate ellipsoid as a spatial model, then we find that the major axis of the ellipsoid is inclined $47 \pm 10°$ to the plane of the sky, and that the expansion velocity of the system of filaments ranges from 55 km/sec along the minor axis to 110 km/sec along the major axis (Lozinskaya, 1970; also see below). These results have been confirmed independently: Whittet et al. (1979) and Huber et al. (1979) obtain 50–90 km/sec for the expansion velocity of the shell, based on the Doppler shift of interstellar absorption lines in the spectrum of HD 192163. The mean linewidths in the outlying filaments of the nebula are $\Delta v([\text{N II}]) = 28 \pm 3\,$km/sec and $\Delta v([\text{H}\alpha]) = 43 \pm 2\,$km/sec (Lozinskaya, 1980c). Knowing the linewidths for two elements of different atomic mass, and assuming that due to their similar ionization potentials, the Hα- and [N II]-emitting regions overlap, we obtain a mean temperature $T_e = 19000 \pm 4000\,$K over the nebula. Wendker et al. (1975) give $T_e = 16000 \pm 6000\,$K, having compared the thermal radio brightness with the brightness in Hα. It is important to note that both methods determine the temperature independent of the assumed chemical composition of the filaments, inasmuch as mass loss from a WR star can substantially modify

Fig. 81. Images of NGC 6888 obtained by the author with a short-focus image-intensifier system; 50-cm, 60-cm, and 125-cm telescopes were used. Panels **a**, **b**, and **c** were obtained with filters centered at Hα, [N II] λ 6584, and [O III] λ 5006, respectively. Panels **d**, **e**, and **f** are Fabry–Perot images of the central ring in Hα and [N II], and of part of the central ring in [O III]. The three interference fringes are due to radiation from the near and far sides of the thin expanding shell, and from intervening H II regions.

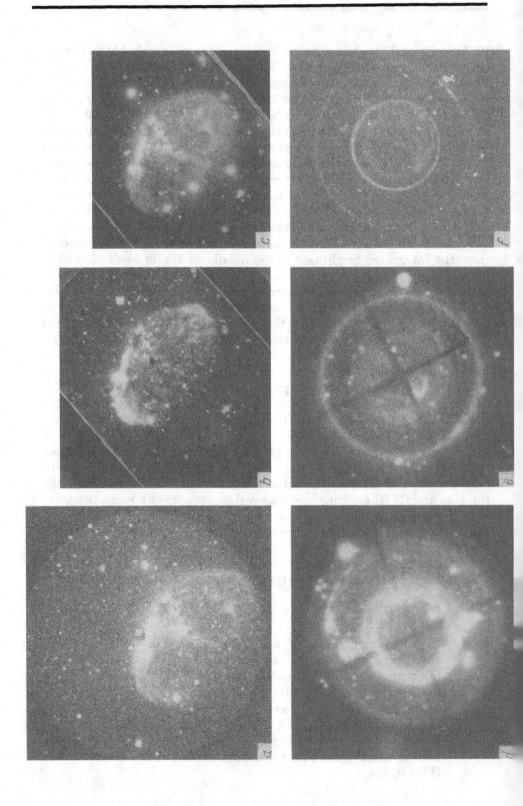

the relative heavy-element abundance in its immediate neighborhood.

All of the foregoing results, and particularly the spectrum, temperature, and expansion velocity, serve to characterize the bright filaments and knots. Long-exposure images taken by the present author in 1970 also revealed faint Hα and [N II] emission from the diffuse medium between the filaments. Wendker et al. (1975) have shown that a similar situation prevails in the radio as well: there, bright condensations are embedded in diffuse gas. An analysis of the Hα and [N II] line profiles in the bright filaments and the diffuse medium (Lozinskaya, 1980c) suggests that both diffuse gas and the knots are concentrated in the shell, and do not fill the nebula uniformly: the diffuse component displays a brightening toward the periphery of NGC 6888, but not toward the center. The contrast of filaments against the diffuse background in Hα is twice as high as in [N II] 6584 Å. The velocity spread of the emitting atoms in the diffuse medium is much greater than in the bright filaments and knots. The gas between the filaments emits a faint, broad, flat-topped line whose full width is $\Delta v_{\text{Hα}} = 250 \pm 50$ km/sec and $\Delta v_{\text{N II}} = 210 \pm 60$ km/sec. The diffuse background is quite homogeneous, and could scarcely result from the integrated effect of faint, unresolved filaments and knots.

Calculations by Bochkarev and Lozinskaya (1985) have demonstrated that the most promising place to search for x-ray emission from bubbles swept out by stellar wind is NGC 6888 (see Table 24). Archival data from the Einstein observatory show that x rays have actually been detected from the nebula (Bochkarev et al., 1988; Bochkarev, 1988), but the luminosity, $L_{0.2-4\,\text{keV}} = (4-5) \times 10^{32}$ erg/sec, is an order of magnitude too low. The 0.2–4 keV isophotes, overlaid on an Hα image of the nebula, are shown in Fig. 82. The x-ray brightness distribution — two bright regions in opposing zones along the major axis — may in principle be associated with bipolar mass loss from the central star. We note in this regard that the optical image, bearing in mind the bright central "transverse" filaments, may result from the projection of a dumbbell-shaped surface (due to a bipolar wind or bipolar ejection) onto the plane of the sky, something that would make the reconstruction of the spatial geometry of the shell from its projection on the plane of the sky and its radial velocity distribution rather interesting.

That reconstruction was recently revisited by the present author. We carried out an extensive new series of interferometric observations of NGC 6888 in the [O III] line (Lozinskaya et al., 1988), confirming our previous results in Hα and [N II] which were obtained with the same angular (~10″) and spectral (~15 km/sec) resolution. Three components of the line were identified — radiation from nearby and background H II regions, and from the two sides of the expanding shell. Despite differing morphology, the large-scale kinematics of the gas emitting in the three lines turned out to be identical. Our velocity measurements were supplemented by data acquired by Treffers and Chu (1982) (in Hα at 2′ resolution) and Johnson and Songsathaporn (1981) (in [N II] at 1′ resolution). The projected shape

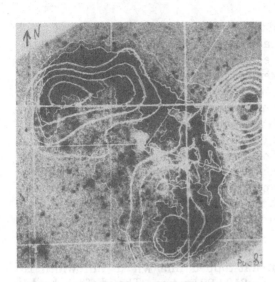

Fig. 82. X-ray isophotes (0.2–4 keV) of NGC 6888 (heavy curves) superimposed on an [N II] optical image (lighter curves). The bright source to the northwest is a foreground star (Bochkarev et al., 1987).

of NGC 6888 in the plane of the sky and the radial velocity field were approximated by prolate or oblate ellipsoids and the surface of revolution of a lemniscate; the expansion velocity was assumed to be constant or proportional to the radius of the shell, and radial or normal to the surface. The best match to the observations was obtained for a prolate ellipsoid inclined 20–40° to the plane of the sky, and expanding at 88±12 km/sec. In Fig. 83, by way of illustration, we have plotted the radial velocity of only those filaments situated along the major axis of the nebula in a band 2' wide, based on measurements by Lozinskaya (1970, 1983) and Lozinskaya et al. (1988), as well as the theoretical velocity ellipse for such an ellipsoid. Neither the dumbbell model of the shell nor even the "triple cavity" proposed by Johnson and Songsathaporn have been confirmed.

Radial velocity measurements along the major axis of NGC 6888 have recently been carried out with arcsecond resolution in the Hα and [N II] lines using an échelle spectrograph (Chu, 1988; Marston and Meaburn, 1988); on the whole, they have confirmed our previous conclusions, and have very reliably revealed velocity variations that reflect both systematic expansion and local fluctuations. In addition, Marston and Meaburn detected the effect referred to above (see Lozinskaya, 1980c): the velocities of the brightest dense clumps are lower than those of the faint diffuse emission. This picture suggests that the shell is being accelerated by stellar wind: the dense clumps accelerate less readily than the tenuous interfilament gas. The small-scale fluctuations on the velocity ellipse constructed by Marston and Meaburn (1988) are probably a projection effect attributable to the "squeezing down" of dense condensations by a shock wave.

Are the observations consistent with the proposition that the nebula NGC 6888 was produced by the interaction between the stellar wind of HD 192163 and the ambient gas? At a distance of 1.3 kpc (van der Hucht et al., 1981), the linear dimensions of the shell are 6.8×4.5 pc, and when its inclination to the plane of the sky is taken into consideration, the maximum radius is $R_2 \approx 4$ pc, while the expansion velocity of the system of bright filaments is $v_2 = 88$ km/sec. One can find the density of the ambient gas, $n_0 = 1-2$ cm^{-3}, by assuming that all of the gas swept out of the body of the nebula, the mass of which is $M_{swept} = 4.6\, M_\odot$, has been concentrated into the thin shell. The wind power required to produce such a shell, according to (13.5), is $L_w \sim 10^{37}$ erg/sec, and it must act for a time $t \sim 2 \times 10^4$ yr. We see from Table 18 that the rate of mass loss from the central star HD 192163 is $2.3 \times 10^{-5}\, M_\odot$/yr, and the wind velocity is $v_w = 2000$ km/sec. Accordingly, $L_w = (3-4) \times 10^{37}$ erg/sec, so the wind does indeed have enough power to form the nebula by sweeping up ambient gas. Over the lifetime of the nebula, the star has lost half a solar mass of matter in the form of stellar wind, and the total kinetic energy imparted by the wind is 2×10^{49} erg. The shell appears optically at $t \approx 10^3$ yr, as given by (13.3); the adiabatic expansion stage thus comprised a negligibly brief episode in the evolution of the nebula. Making use of Eqs. (13.17) and (13.18), we expect the emission measure to be $E_{m\,center} = 800$ cm$^{-6} \cdot$pc and $E_{m\,periph} = 10^4$ cm$^{-6} \cdot$pc, which is consistent with the optical and radio brightness of the nebula. The shell is thicker than indicated by the "theoretical" value $\Delta R = 0.01$ pc, most likely because of the inhomogeneity of the swept-up gas, which deforms the shock front.

It must be emphasized here that although there is no doubt that the nebula is dominated by a strong stellar wind, a recent analysis of elemental abundances, morphology, and nebular excitation shows that the shell swept up by the wind may actually contain stellar ejecta instead of interstellar gas. The fact that NGC 6888 is situated inside the extensive cavity around Cyg OB1 (see Chapter 4) also suggests that the shell contains mostly stellar material.

All of the bright knots and filaments discussed above most likely represent stellar material. The boundary of the shock wave induced by the strong stellar wind in the interstellar gas is actually represented by the faint, delicate outer filamentary shell visible in the [O III] line. As early as 1978, Parker discovered very faint radial and tangential filaments beyond the bright, dense nebula. Recent excellent images of the nebula (see Dufour (1989) and references therein) show that it appears much more circular in [O III], with a radius of about 4 pc. The outer wind-shock boundary of NGC 6888 in [O III] is smooth, and does not correlate with bright knots and filaments. The λ 5007 [O III]/Hβ ratio in the outer filaments is in excess of 20 in some regions. According to Dufour (1989), the high ratio is produced by the shock wave at the wind/interstellar gas boundary, which has a very significant incomplete recombination zone due to UV ionizing radiation from the central WR star. (The "pure" oxygen spectrum of the faint outer filaments of NGC 6888 resembles that of the "non-steady flow" shocks observed in the outer parts of the Cygnus Loop (Hester et al., 1983; Raymond et al., 1988) — another supernova remnant — although the reasons for incomplete recombination are different in the two objects.

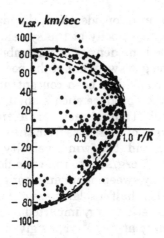

Fig. 83. "Velocity ellipse" of the nebula NGC 6888 (Lozinskaya et al., 1988). The various curves have been computed for models of an extended ellipsoid of rotation, expanding at 88 ± 12 km/sec and inclined 20–40° to the plane of the sky.

It seems reasonable to identify with the shock wave in the interstellar gas the faint, diffuse nebular emission in the broad wings of the Hα and [N II] lines. By analogy with the pattern observed in old supernova remnants (see Section 8), it would seem that the faint, broad component of the lines corresponds to gas behind shock front II, which propagates through the low-density intercloud medium, while the bright filaments define a region in which the density of the ejected gas is enhanced.

The detection of x rays from NGC 6888 has essentially provided the first direct proof of the existence of hot plasma within the cavity swept out by the wind. Since reconstruction of the true geometry of the shell based on the radial velocity field does not confirm a dumbbell-shaped structure, one might expect x-ray attenuation at the center of the nebula to be related to absorption. Indeed, confirmation comes from the detection of a dense H I cloud there (Johnson and Songsathaporn, 1981), and the bright filaments at the center, which probably mark the collision zone between a shock wave and that dense cloud. The x-ray luminosity of NGC 6888 is an order of magnitude lower than predicted by the wind-blown bubble model of Weaver et al. (1977); this may be evidence in favor of the rapid dissipation of wind energy via delocalization of the conductive heat flux in a layer with a strong temperature gradient (Dorland et al., 1986; Dorland and Montmerle, 1987). Another possibility might be that hot wind is passing through breaches in the cold shell (Falle, 1975b; Lozinskaya, 1983). Infrared emission from NGC 6888 has been detected by IRAS (Marston and Meaburn, 1988; van Buren and McCray, 1988). Supernova remnants have taught us that properly estimating hot-plasma energy losses via infrared radiation from dust is an extremely complicated process. The situation is even more complex in NGC 6888, where it is difficult to distinguish between the radiation from shock-heated dust, that due to dust heated by ultraviolet radiation from the Wolf–Rayet star, and IR line emission in the IRAS bands. According to van Buren and McCray (1988), the infrared emission from NGC 6888 is dominated by lines.

The nebula NGC 2359 (associated with the star HD 56925) is another object of the same class that has been studied in detail. The nebula is a complete filamentary elliptical shell embedded in a diffuse, irregular H II region (see Fig. 84). The shell has a mean diameter of 5′, and the filaments are less than 2–3″ across; the diffuse H II region is two to three times the size of the shell.

Detailed investigations of NGC 2359 have been particularly interesting, as it is a ring nebula that is physically associated with a dense molecular cloud. The first optical spectroscopic and interferometric observations of the filamentary shell were carried out by Esipov and Lozinskaya (1971) and Lozinskaya (1973b). Those observations revealed that the Hα line in the nebula has a three-component structure, due to emission from the nearest Galactic spiral arm, the diffuse H II region, which is ionized by a WR star, and the filamentary shell (Lozinskaya, 1973b). Neither the linewidth ($\Delta v = 33 \pm 3$ km/sec) nor the variation of the radial velocity from point to point (at most 5–7 km/sec in the diffuse nebula) are inconsistent with normal motion within an H II region characterized by subsonic velocities. The mean radial velocity, $v_{LSR} = 52 \pm 4$ km/sec, implies a kinematic

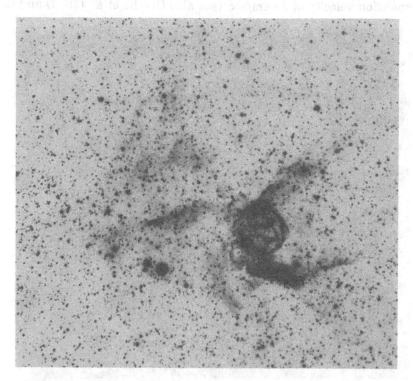

Fig. 84a. NGC 2359, a ring and diffuse nebula surrounding the Wolf–Rayet star HD 56925 (Palomar Sky Survey red plate).

distance of 4 ± 0.5 kpc. According to van der Hucht et al. (1981), the photometric distance to the central star is 5.3 kpc; the corresponding radius of the filamentary shell is approximately 4 pc. In the filamentary shell, it has been possible, as in NGC 6888, to identify two components: a bright, relatively narrow line ($\Delta v = 30-40$ km/sec FWHM) that sits on top of a faint, flat-topped pedestal (with $\Delta v = 150-200$ km/sec at intensity $\leq 0.1 I_{max}$). The fact that the shell has a small angular extent has thus far made it impossible, as in NGC 6888, to associate the broad and narrow components with filaments and the interfilament medium. But since NGC 2359 also shows faint diffuse emission from the gas between the filaments (Schneps and Wright, 1980), it would seem reasonable by analogy with NGC 6888 that that is precisely where the broad, faint component comes from.

Our measurements indicate that the expansion velocity of the filamentary shell is 35 ± 15 km/sec, as given by the Doppler shift of the narrow component (and 50 ± 25 km/sec as given by the shift of the center of the faint, broad component). Subsequent observations have confirmed the first value: although Pismis et al. (1977) and Treffers and Chu (1982) have found an expansion velocity of only $15-18$ km/sec, Schneps and Wright (1980) have reliably detected splitting of the [O III] line corresponding to an expansion velocity of 30 km/sec (see also Goudis et al. (1983) and Chu

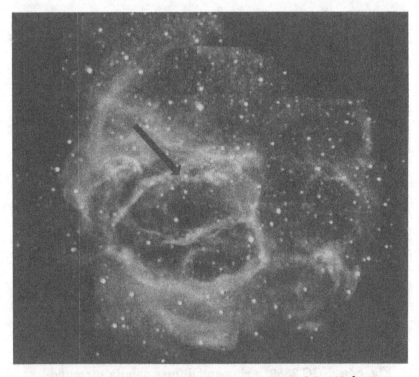

Fig. 84b. Hα image of G 2.4+1.4 (Dopita and Lozinskaya, 1990).

(1988)). The gas density in the bright filaments given by the [O II] and [S II] lines lies in the range $10^2 - 2.5 \times 10^3$ cm^{-3}, and the lines of [O III] yield a temperature of 12,000 K (Esipov and Lozinskaya, 1971; Peimbert et al., 1978; Schneps et al., 1981). The nitrogen abundance in the filamentary shell may possibly be three times the normal value (Talent and Dufour, 1979; see below). Comprehensive investigations of NGC 2359 in the CO lines (Schneps et al., 1981) have demonstrated that the nebula is interacting with a dense cloud of neutral gas with $v_{LSR} = 54$ km/sec. The evidence for a physical association between the cloud and the shell NGC 2359 consists of the anomalously large CO linewidth for a cold molecular cloud (close to 4 km/sec), which may be a stellar wind effect; an ionization front is advancing along the boundary of the cloud. The density of the molecular hydrogen in the cloud is 10^3 cm^{-3}, and it has a mass of about 10^2 $M_\odot$. The radio structure of the region is fully consistent with the optical map, displaying thin radio filaments that augment the optical shell to the east. Besides the central WR star, the nebula may also house another source of ionizing radiation with a somewhat weaker wind. The latter may be responsible for a secondary shell structure and a radio brightening in the southern part of NGC 2359. The mass of ionized gas in the shell nebula, as determined from the radio flux density, is 16 $M_\odot$, and the mean gas density within the shell is $n_e = 70$ cm^{-3} (Schneps et al., 1981).

It is quite plausible that the mass-loss rate from the star HD 56952 is sufficient to produce a thin filamentary shell of radius ~ 4 pc, expanding at approximately 35 km/sec in a medium with a mean density of $3-4$ cm^{-3} (the density has been estimated under the reasonable assumption that all of the gas swept up by the wind has been compacted into a thin shell). The velocity of the stellar wind is $v_w = 2500$ km/sec, based on the absorption feature of the ultraviolet spectral lines (Johnson, 1980). Making use of (13.5), we find that the required outflow rate is $\dot{M} = (2-3) \times 10^{-7}$ $M_\odot$/yr, two orders of magnitude below the upper limit on mass loss, $\dot{M} \leq 7 \times 10^{-5}$ $M_\odot$/yr, which comes from the upper limit on the radio flux density from the star, which is 5 kpc away (see Schneps et al. (1981)). This estimate may well be close to the actual rate of outflow. If the nebula were in the later momentum-driven stage of development, as described by (13.9), the required mass loss would be $\dot{M} = 3 \times 10^{-6}$ $M_\odot$/yr (and at most 2×10^{-5} $M_\odot$/yr, if we take into consideration the uncertainty in the distance and expansion velocity).

Recent imaging in [O III] and Hβ, along with high spatial resolution spectroscopy (see Dufour (1989) and references therein) show that the thin filamentary shell is visible only in [O III] and H I lines — with the [O III]/Hβ ratio as high as 15 — but not in [S II] or [N II]. The explanation advanced by Dufour is that the wind-induced shock wave is propagating through a preionized H II region and is "backlit" by WR ionizing radiation, which prevents complete recombination of the gas swept up by the shock. The modern interpretation of the two-component system (H II plus wind-blown bubble) is that of a blister-type H II region previously formed by the progenitor O star which is now being swept up by the strong stellar wind of the WR star.

The nebula G 2.4+1.4, surrounding the WO star Sand 4 (WR 102; LSS 4368). In attempting to find an optical counterpart to the x-ray source Sgr X-1, Blanco et al. (1968) discovered an intense ultraviolet source associated with a faint nebula 5′ across. Any relationship to Sgr X-1 was subsequently ruled out, but the object itself turned out to be extremely interesting. Freeman et al. (1968) have spectroscopically identified the object as a WR star — the most extreme WC class (very broad lines of He II, O III–VI, C III–IV, with FWHM = 70–207 Å). The classification adopted for the star is WC4pec–WC5pec (van der Hucht et al., 1981). Our present understanding is that the star Sand 4 belongs to a separate WO sequence, and is the hottest and most energetic WR star known (Barlow and Hummer, 1982). With a stellar wind velocity $v_w = 5500-5700$ km/sec (Torres et al., 1986; Barlow and Hummer, 1982; Dopita et al., 1990), the mechanical luminosity of the star is ten times that of a "normal" WR star.

The ring nebula G 2.4+1.4 was regarded as a supernova remnant because of its nonthermal radio spectrum, even though the WR star was recognized as a source of excitation and stellar wind (Johnson, 1975, 1986; Treffers and Chu, 1982; Chu et al., 1983). An alternative interpretation — a wind-blown bubble around the WO star — was also considered (Chu et al., 1983; Green and Downes, 1987). Neither model fitted the observational data completely. In particular, it remained unexplained why the WO star — the most powerful source of UV radiation and stellar wind known — was situated on the periphery of the bright, symmetrical ring nebula, rather than at its center.

In 1988, we undertook a detailed study of the WO star and ring nebula (Dopita et al., 1990; Dopita and Lozinskaya, 1990), using three instruments: the coudé échelle spectrograph of the 1.8-m Mt. Stromlo telescope to map the velocity field in the 5007-Å line of [O III]; the double-beam spectrograph of the 2.3-m telescope at Siding Spring in imaging mode to obtain deep Hα and [O III] images, and to obtain long-slit spectra (300 line/mm gratings) of the star and ring nebula in the 3400–8200 Å range; and the Fabry–Perot infrared grating spectrometer of the Anglo-Australian Telescope to obtain 1.0–2.5 μm spectra of the star. We established that the nebula G 2.4+1.4 is at least twice as large as had been previously suspected, with a radius of 220±20″ toward the southeast, and 390±30″ toward the northwest (3.1±1.2 pc and 5.5±2.0 pc at a distance of 3±1 kpc); see Fig. 84b. The derived expansion velocity of the near side of the shell is 48±3 km/sec (faster than previous thought), whereas the far side of the shell shows no systematic expansion velocity. We suggest that the WO star is located at the edge of a dense cloud that is stalling the expansion of the far side. The bright, previously recognized 5′ nebula represents the southeastern part of the bubble swept up in the cloud, while the newly-discovered faint filamentary northwestern enlargement represents a wind-blown bubble that has burst into the nearby low-density gas. The nebula is there-

fore a "blister" bubble expanding out of this dense cloud. The ellipticity of
the ring of bright filaments enables us to estimate the inclination of the
near surface of the cloud to be $35 \pm 10°$. In this model, the WO star is actually situated near the center of symmetry of the bubble, and is surrounded
by the most highly ionized gas, which emits in the He II lines. Bubble expansion in the direction of the strong density gradient has stimulated the
development of a Rayleigh–Taylor instability, giving rise to the characteristic scalloping of the northwestern part of the nebula; see Fig. 84b.

The nebula M 1–67, surrounding the star 209 BAC (= WR 124;
Merrill's star) was discovered by Minkowski, and initially classified as a
planetary nebula (PK 50+3° in the well-known catalog of Perek and Kohoutek (1967)). Citing a careful analysis of spectroscopic and photometric
observations, Cohen and Barlow (1975) argued persuasively that this nebula is in fact a ring-like shell around a population I WR star, and the object
was subsequently dropped from catalogs of planetary nebulae. Merrill
(1938) observed that the star and nebula had an anomalously high radial
velocity, $v_☉ \sim 200$ km/sec. Subsequent spectroscopic, interferometric, and
photometric investigations substantiated the velocity measurements, and
demonstrated that the excitation level was too low for a planetary nebula
(see Cohen and Barlow (1975) and the references therein). Detailed studies
of the spectrum and kinematics of M 1–67 have also been made more recently (Pismis and Recillas-Cruz, 1979; Glushkov et al., 1979; Barker,
1978; Chu and Treffers, 1981c; Solf and Carsenty, 1982). The nebula is a
low-excitation H II region with $T_e = 8000$ K; the only ionization state of
oxygen lines observed in the spectrum is O^+, and the intensity ratio
$I_{[O III]}/I_{Hβ} \leq 0.03$; furthermore, there are no [O I] lines. The gas density in
the filaments and knots is 1000 ± 300 cm^{-3}, and the absorption is
$A_V = 4.7 \pm 0.''6$. The radio and optical images of the nebula are identical,
and the radio spectrum is thermal; the mass of ionized gas, as obtained
from the radio flux density, is $0.5\,M_☉$ (Felli and Perinotto, 1979). Based on
the optical emission and the filling factor of the shell, Solf and Carsenty
(1982) obtained a similar value, $0.8\,M_☉$. Rather than showing delicate filaments, the nebula is patchy (Fig. 85). Bright clumps and knots, typically
$10''$ across, are concentrated into a thin shell with $\Delta R/R \approx 0.05$. If the star
is assumed to be a population I type WN8, its photometric distance is 4.3
kpc (Cohen and Barlow, 1975), which corresponds to a shell radius of 0.9 pc
and a thickness of 0.05 pc. The shell is expanding at 42 km/sec; the mean
expansion velocity of the clumps on the outskirts is somewhat lower (Solf
and Carsenty, 1982). Pismis and Recillas-Cruz (1979) found individual
condensations moving at −80 to +113 km/sec.

There is no unambiguous interpretation of the kinematic results for
M 1–67. Pismis and Recillas-Cruz (1979) account for the observed velocities by means of an anisotropic ejection of matter approximately 6000
years ago; Chu and Treffers (1981a) argue that ejection took place twice,
approximately 60,000 and 200,000 years ago. According to Johnson (1980)
and Solf and Carsenty (1982), the nebula is in fact a cavity swept out by

stellar wind. This is suggested by the morphology of M 1–67: it is essentially a spherical shell. One immediate reservation, however, is that the wind in this model sweeps up matter ejected by the star, not interstellar gas (see Section 15), the evidence being that the high velocity of the shell as a whole is equal to the peculiar velocity of the star. If interstellar gas were being swept up, the rapidly moving source of wind would form a highly elongated cavity with the star located off-center, rather than a spherical one. Furthermore, the energy required for the wind to accelerate initially stationary interstellar gas to the speed of the star turns out to be 20–30 times the energy needed to sweep out the cavity. If the wind from 209 BAC has swept out ejected gas comoving with the star, then (13.5) yields a power for the outflow of $L_w = 3 \times 10^{35}$ erg/sec, and an age of approximately 10,000 years for the shell, taking $n_0 = 10 \, \text{cm}^{-3}$. For a wind velocity $v_w = 1500 \, \text{km/sec}$, which comes from measurements of the ultraviolet spectrum of 209 BAC (Johnson, 1980, 1982a), the required mass-loss rate is $\dot{M} = 4 \times 10^{-7} \, M_\odot/\text{yr}$, which is much lower than the observed rate (see Table 18).

An attempt has been made by van der Hucht et al. (1985) to reconsider the identification of M 1–67 with 209 BAC, again classifying the object as a planetary nebula. Their justification comes from IRAS data that suggest that the nebula's infrared emission comes from dust heated to about 100 K. This is precisely the sort of radiation, coming from dust at temperatures $T = 60 - 200 \, \text{K}$, that is observed in planetary nebulae (Pottasch et al., 1984; Khromov, 1985). For population I WR stars, either no emission is observed to come from circumstellar dust, or it corresponds to $T = 1000 \, \text{K}$ at 100 AU from the nucleus; in the associated ring nebulae, and in RCW 58 and NGC 6888 in particular, the radiating dust has been heated to $30 - 40 \, \text{K}$ (van der Hucht et al., 1985; Marston and Meaburn, 1988; van Buren and McCray, 1988).

Nevertheless, basing the reclassification of the star 209 BAC and its

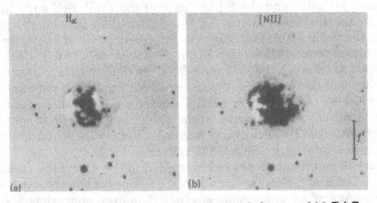

Fig. 85. The ring nebula M 1–67 around the star 209 BAC as seen in the light of Hα (a) and [N II] (b) (Chu and Treffers, 1981a).

associated nebula solely on the available infrared data, with no additional independent confirmation, would seem to be premature, considering the fact that the dust-formation mechanism is poorly understood. Specifically, the dust in shells ejected by population I WR stars is probably essentially the same as that in planetary nebulae whose central stars are population II WR stars. More than likely, these two classes of shells ejected by stars differ only in the scale of the phenomenon, and it would not be surprising to find that the temperature of the dust, which is governed by the radiation from the central star, depends on the radius of the shell. The absorption quoted above, $A_V = 4.^m7$, is in poor agreement with the photometric distance to 209 BAC, $r = 500 - 800$ pc, the latter having been obtained under the assumption that the star is a population II WR.

We shall not dwell upon observational results for other ring nebulae associated with WR stars, especially since not all of them have been so well studied. Table 20 lists the most important parameters for Galactic WR stars with ring nebulae; only reliably identified nebulae have been included. Apart from these, Heckathorn et al. (1982) have published a list containing ten more probable identifications. The columns in the present table are as follows:

1) designation of nebula;

2) central WR star (an asterisk denotes an object containing additional exciting stars);

3) spectral type of star;

4) best distance estimate (with due regard to the photometric distance to the star determined by van der Hucht et al. (1981) and the kinematic distance to the nebula obtained by different workers);

5) angular size of the nebula (Heckathorn et al. (1982));

6) linear size;

7) expansion velocity;

8) peculiar velocity of the star, or systematic velocity of the nebula (if no velocity is given, it is less than 30 km/sec);

9) height above the galactic plane;

10) type of nebula, in the classification scheme of Chu (1981) (W = wind-blown bubble; E = stellar ejecta; R_a and R_s = amorphous and shell H II regions, respectively);

11) mass of the nebula derived from thermal radio flux density, brightness in Hα, or gas density (from observations of the S II lines), combined with the size and filling factor;

12) period and mass of any compact companion of the WR star (Cherepashchuk, 1982).

Table 20 is followed by extensive notes and remarks. It lists no extended shell complexes around OB associations containing Wolf–Rayet stars, such as NGC 3372, RCW 113, Sh 11, Sh 54, or Sh 157. Although three of the starred objects in Table 20 do contain additional sources of ionizing radiation, the WR stars are most certainly the dominant source of stellar wind.

The information in Table 20 comes mainly from data obtained by Chu et al. (1983), Heckathorn et al. (1982), and sources cited in those papers. Subsequent studies are reflected in the remarks following the table.

It is clear, just from the objects considered above, that stellar wind may not be the only process responsible for the formation of ring nebulae. The effects of WR stars upon the ambient interstellar gas show up in the form of ionizing radiation, stellar wind, and "instantaneous" ejection of a shell. If one of these processes is dominant, it will determine the morphology and kinematics of the surrounding H II region. Chu (1981) has proposed a natural classification scheme for H II regions associated with WR stars that is based mainly on morphology and kinematics, and we shall conform to that scheme for the sake of consistency.

The most populous group of nebulae excited by WR stars contains the amorphous H II regions — i.e., the classical diffuse emission nebulae (here denoted by R_a). Most of the H II regions identified with WR stars in the catalog of van der Hucht et al. (1981) fall into that category. Stellar wind effects may show up as broad, faint wings in the lines of the optical spectrum, corresponding to supersonic velocities, which have been detected in many "normal" H II regions. A lack of ring structure in these nebulae may be due to the inhomogeneity of the medium, the presence of several exciting stars, or observational selection effects, which prevent one from detecting a faint shell against a bright emission background.

The shell structure of nebulae around WR stars may result from the motion of ionization and shock fronts during expansion of a fully evolved H II region, photoionization evaporation of small cloudlets, or the sweeping up of circumstellar gas by stellar wind (see Section 13). If the first two processes dominate, geometrically thick ring-type H II regions display a diffuse morphology (denoted by R_s in Table 20), while if wind is the main determinant, one sees thin filamentary shells surrounding wind-blown cavities (denoted by W). Classification of these two types of nebulae must basically rely on their kinematics, as R_s nebulae tend to expand at velocities $\lesssim 10\,\mathrm{km/sec}$, while W nebulae are supersonic. A fourth type, denoted by E, is made up of matter thrown off by the star. Ejected shells can be characterized by their patchiness and their recognizable distribution of gas-clump velocity.

We note here that this classification scheme is not unique, inasmuch as all three processes compete simultaneously in any real nebula. That much is clear in the nebulae discussed above: in NGC 2359, we see an extended R_a nebula with a filamentary W-type shell; in M 1–67, we are most probably looking at ejected matter that has been swept up by stellar wind (see also Section 15).

Table 20. Ring Nebulae Associated with Wolf–Rayet Stars

Nebula	WR Star	Spectral type	r, kpc	Angular diameter, arcmin	Linear diameter, pc
1	2	3	4	5	6
Sh 308	HD 50896	WN5	1.5	40	17
NGC 2359	HD 56925	WN4	5.3	5(30)	8(45)
NGC 3199	HD 89358*	WN5	3.8	16×18	16×18
NGC 3372	HD 92740	WN7+abs	2.3	30	20
Anonymous	HD 92809	WC6	2.5	30	20
RCW 58	HD 96548	WN8	2–3	9	6
Anonymous (around θ Mus)	HD 113904	WC+O9I	1.8	80×45	30×45
RCW 78	HD 117688*	WN8	7.5	39	86
RCW 104	HD 147419	WN4	2.6–3	~18	~16
Anonymous	HD 187282	WN4	5.5	50×60	97×78
Anonymous (near Sh 109)	HD 191765	WN6	1.6	17	8
NGC 6888	HD 192163	WN6	1.3	12×17	4.5×6.8
Anonymous (near Sh 132)	HD 211564	WN3	3–4	17 and 36	36 and 17
Anonymous (near Sh 132)	HD 211853*	WN6 + O	3–4	~20	~20
M 1–67	209 BAC	WN8	4.3	1.5	1.9
Anonymous	HD 115473	WC5	2–4	~40	~44
NGC 6357	HD 157504	WC6	1.7	36	18
Anonymous (near Sh 54)	HD 168206 (CV Ser)	WC8+O8–9	2	4 and 9	2.4 and 11
G 2.4+1.4	LSS 4368	WO1	~3–5	~20	~20–30

v_{exp} km/sec	v_{pec} km/sec (LSR)	z, pc	Type of nebula	Mass of nebula $M_\odot$	Compact companion P^d; $M/M_\odot$	Remarks
7	8	9	10	11	12	13
60–80	+33	−300	$W + R_s$	30–40	3.75; 1.3	R_s shell (Sh 103 + Sh 104)
35		−12	$W + R_a$	16		
18(60)	−8	−61	W	160		
		−34	$W?$			Part of Carina complex
20–30		−2	W			
~87		−300	$E + W$	3–6	4.8; 0.3	Ejecta in swept-up shell
≤7 (9 for H I)		−83	R_s			Inside an H I bubble
	−90	+20	R_a			
25		−77	W	650		
		−360	R_s		3.85; 1	
50	+50	+43	W	40–80	7.4; ~1.5	
55–110	+100	+50	W	5	4.5; ~0.5	
		−60	$W + R_a$	~20		Double shell
		−60	R_s	~20 ·		
42	+180	+260	E	~1	2.36; 1–1.6	
12		+240	R_s			
		+24	R_s			
		+60	$W + R_a$			Double shell
~60–80		≲ 100	W			Dopita and Lozinskaya, 1990

Notes and remarks.

Sh 308. By examining high-resolution UV spectra of HD 50896 and of 15 early-type field stars located nearby in the plane of the sky, Howard and Phillips (1986) and Nichols-Bohlin and Fesen (1986) have discovered high-velocity components of several interstellar lines. These authors detected high-velocity motion far beyond the boundaries of the bright shell of Sh 308, but inside the outer shell of Sh 103/104. The origin of the high velocities is not entirely clear. A –30 km/sec feature has been interpreted as arising in the wind-blown bubble Sh 308, while –75 to –80 km/sec and –125 to –30 km/sec systems are related to absorption in a previously unknown old SNR. No genetic relationship between this hypothetical remnant and the compact companion of HD 50896 has been established. It is also possible that the high-velocity gas is associated with the outflow of hot gas through a rift in the bright central shell. Since similar high-velocity absorption systems have been reported in the direction of several OB associations (see Chapter 4), they may also be related to stellar wind and UV radiation from other OB stars in those groups. In any event, the enigmatic nature of multilayer high-velocity large-scale structures is still to be explained.

NGC 3199. Deep photographs show a complete elliptical shell with bright filaments on the western side; the WR star is located near the bright side of the shell (Whitehead et al., 1988). The shell is expanding in an inhomogeneous medium. Internal velocities are high (ranging from +45 to – 82 km/sec relative to the stationary component of the line), but they are irregular (Chu, 1988).

Anonymous nebula around HD 92809. This object has been investigated in detail by Georgelin et al. (1986). The optical shell-like nebula is located within a void swept out inside a CO cloud.

RCW 58. Smith et al. (1984) have detected high-velocity absorption lines in the spectrum of HD 96548, due to absorption in the ring nebula. To account for the observed correlation of velocity with the stage of ionization of the absorbing element, Smith et al. proposed a model involving a wind-blown bubble, in which hot gas from the WR wind has been mixed with cold clumps of matter ejected by the stellar progenitor at an earlier stage of evolution. In deep photographs obtained with the 3.9-m Anglo–Australian Telescope, RCW 58 clearly consists of clumps and filaments elongated outwards from the WR star (representing stellar ejecta) and of two thin arc-shaped filaments (representing the wind-blown shell) (Smith et al., 1988b). Detailed studies of the kinematics of the nebula by Chu (1988) and Smith et al. (1988b) have revealed a systematic expansion of the shell involving clumps moving at 90–100 km/sec (velocities inside the shell were previously though to be random).

Anonymous nebula around θ Mus. de Nicolau and Niemela (1984) have shown that this nebula is located inside an H I shell approximately 200 pc across that is expanding at 9 km/sec. Up to $\sim 10^4 \, M_\odot$ of neutral hydrogen have been evacuated from the bubble. In all likelihood, the H I bubble has been swept out by the wind of the WR progenitor while the latter is still on the main sequence; see below.

Anonymous nebula near Sh 54. This ring nebula is probably associated with the H II complex Sh 54, which is excited by stars belonging to the association Ser OB2. Deep photographs in the light of Hα, [S II], and [O III] (Gonzales and Rosado, 1984) clearly delineate a double shell. The inner shell features thin filamentary structure, and radiates only in the lines of [S ii] and Hα, with $I([S\,II])/I(H\alpha) \approx 0.8$, which is typical of shocked excitation. The outer shell is incomplete, and radiates in Hα, [S II], and [O III]. Apart from being excited by the WR star, the latter probably also receives a significant amount of ionizing radiation from the Of star HD 168112.

RCW 104. Goudis et al. (1988) have demonstrated that the kinematics of this nebula are irregular — in addition to line splitting corresponding to expansion velocities of 20–25 km/sec, they see clumps retreating from the center at 50 to 120 km/sec.

The nebula surrounding θ Mus (HD 113904) is observed to lie within a cavity in the neutral hydrogen distribution (de Nicolau and Niemela, 1984). It has recently been made clear that this is probably a fairly common occurrence. Three more H I cavities have been identified around the WR stars HD 88500 (Cappa de Nicolau et al., 1986), HD 156385 (Cappa de Nicolau et al., 1987), and HD 197406 (Dubner et al., 1990). The neutral hydrogen hole around HD 156385 also contains the optical shell RCW 114. The star HD 197406 (spectral type WN7) is surrounded by a nearly complete ring-shaped H I shell (36 pc in size at a distance of 4.5 kpc; $v(\text{expansion}) = 6$ km/sec and $M(H\,I) = 1200 \, M_\odot$). Embedded in this shell there is a small (7–8 pc) hole centered on the star with a similar expansion velocity — 6 km/sec — and evacuated H I mass of about $8 - 10 \, M_\odot$. The large H I shell is also visible in 60- and 100-μm IRAS images.

Three other stars are of type WC; their cavities are typically about 100 pc in size, with expansion velocities of about $8 - 12$ km/sec, and the neutral hydrogen deficiency corresponds to a mass $M(H\,I) = 6 \times 10^3 - 10^4 \, M_\odot$. More H I shells around WR stars have been reported by Niemela and Cappa de Nicolau(1990); Lozinskaya and Sitnik (1988) have pointed out several WR and Of stars in Cygnus that are located in CO holes.

Extensive H I cavities have a kinematic age of $t_{kin} \sim 3 \times 10^6$ yr, much greater than the duration of the WR stage. They are therefore most likely to be associated with the wind from the WR progenitor, a massive O star.

A recent review of the observational data (Lozinskaya, 1988c) leads to the conclusion that all *W*- and *E*-type WR rings are located within large,

Table 21. Chemical Abundances in Ring Nebulae around WR Stars (12 + log(X/H)) (see also Chu (1990)).

	He	O	N	S	Ne	Ar	log(N/O)	Ref.
Sh 308	11.17	8.54	8.28	>6.24	8.08	>6.39		1
	11.09	8.20	8.00				−.20	5
NGC 3199	11.04	8.52	7.83	>6.77	7.91	>6.36		1
	10.98	8.51	7.55				−.96	5
RCW 58	11.40	8.49	8.07	>6.27				1
	11.36	8.72	8.42				−.03	5
NGC 6888	11.35	8.45	8.54	6.35	8.01			2
	11.27	8.15	8.45				.30	6
NGC 2359	10.94	8.20	7.52		7.64			4
	11.00	8.25	7.27				−.98	8
RCW 78	11.04	8.92	8.07				−.85	8
MR 100	11.06	8.52	7.64				−.88	5
MR 26	10.95	8.55	7.55				−1.0	5
RCW 104	11.18	8.54	7.85				−.69	5
M1–67	11.34	8.10	8.70				0.60	7
Orion Neb	11.02	8.52	7.57	7.19	7.66	6.60		3

References.

1. Kwitter (1984)
2. Kwitter (1981)
3. Peimbert and Torres-Peimbert (1977)
4. Peimbert et al. (1984)
5. Rosa and Mathis (1990)
6. Esteban and Vilchez (1990)
7. Esteban et al. (1990b)
8. Esteban et al. (1990a)

shell-like H II, H I, or CO cavities swept out by the WR progenitor, or by the stellar wind of its parent OB association.

Apart from the Galactic ring-like shell around WR stars, some dozens of such objects have also been found in other galaxies of the Local Group (see Chu (1983), Georgelin et al. (1983), Braunsfurth and Feitzinger (1983), Rosado (1986); further references may be found in all of these papers). Ten WR rings have been identified in the LMC. No WR stars in the SMC have ring nebula except for the large supershell containing the WO4+O2 V binary Sand 1 and its parent cluster NGC 602c (Lozinskaya, 1990). Drissen et al. (1990) have recently identified 11 probable WR ring nebulae and eight less likely cases in M33.

Immediately after ring nebulae were identified in other galaxies, it was noticed that the shells around extragalactic WR stars were several times the size of those in the Milky Way. In all likelihood, this is a selection effect: within the Galaxy, it is easier to identify compact nebulae than large complexes, while in distant systems the small nebulae fall below the limit of resolution and large complexes are easily distinguished. In fact, after the foregoing classification scheme was introduced, it was found that most of the objects in the LMC were extended type-R_s shells. A comparison of objects belonging to the same class divulges no significant differences in size between those in our own Galaxy and those in external systems.

Without exception, galactic ring nebulae associated with WR stars are thermal radio sources. The thermal nature of their radio emission is revealed by the spectrum ($\alpha = 0 - 0.2$), the lack of linear polarization, and the fact that the emission measures derived from the radio and Hα flux agree to within the measurement errors. We emphasize that a thermal radio spectrum is the most important observational clue distinguishing bubbles swept out by stellar wind from decelerated supernova remnants. In point of fact, the two types of nebulae have similar filamentary shell structure; the expansion velocities of the bright filaments are almost the same (in IC 443 and NGC 6888, they are practically equal); and the x-ray emission from gas behind the shock front differs only quantitatively (it is a hundred times stronger in supernova remnants).

To conclude, we dwell upon one further question that is crucial to an understanding of the nature of ring-like shells around WR stars. The spectra of ten nebulae have been thoroughly analyzed in order to determine their chemical composition. The following nebulae showed relative abundances that differed, for a number of elements, from the normal cosmic values: NGC 6888 (Kwitter, 1981; Esteban and Vilchez, 1990), NGC 2359 (Peimbert et al., 1978; Esteban et al., 1990a), M 1–67 (Barker, 1978; Esteban et al., 1990b), Sh 308, NGC 3199, RCW 58, and the shell surrounding HD 191765 (MR 100) (Kwitter, 1984; Roza and Mathis, 1990), as well as others. In Table 21, we present estimates of the relative abundance (by number of atoms) of the most important elements in these objects, and for comparison, in the Orion nebula. In the earlier investigations, the chemical composition of the shells around WR stars was obtained by assuming that the spectrum is due to photoionization; interstellar absorption was estimated from the Balmer decrement, electron density according to the [S II] or [O II] line ratios, and in the absence of specific measurements, the temperature was obtained by equating the $O^+/H^+ + O^{++}/H^+$ relative abundance to that of the Orion nebula. In determining the total abundance of an element based on the proximity of ionization potentials, it has been assumed that $N/N^+ = S/S^+ = O/O^+$, $Ne/Ne^+ = Ar/Ar^+ = O/O^{++}$, and that He/He^+ ranges between O/O^+ and $4(O/O^+)$. This approach presupposes a normal oxygen abundance, but as can be seen from Table 21, the relative nitrogen and helium abundances are elevated in certain nebulae. The sulfur and argon abundances are lower limits, since in the spectral band investigated there were no high-ionization lines of those elements. The marked over-abundance of neon may be a systematic error, as the same anomaly also appears in planetary nebulae at a different ionization level; see Kwitter (1984).

Recent observations have encompassed many fainter nebulae, and more importantly, they have covered a wider spectral range (at higher spectral resolution); see Chu (1990), Vilchez and Esteban (1990), Dufour (1989), and references therein. Efforts have been made to identify distinct kinematic and chemical components in each nebula (e.g., ejected and swept-up interstellar material, different knots and filaments), and to account for the local interstellar abundance around WR rings. Since the newer investigations deal mainly with He, O, and N, we summarize both the older and more recent data in Table 21.

It is clear from Table 21 that nebulae consisting principally of ejecta all have a higher He/H abundance ratio than the standard value of 0.1, and a higher than normal N/O ratio. In contrast, N/O in the wind-blown bubble NGC 2359 is nearly normal (Dufour, 1989). According to Chu (1990), the increase in N/O in WR ring nebulae is due both to enhanced N abundance and reduced O abundance.

The anomalous abundances are easily accounted for if ring nebulae consist ejected material. On the other hand, the observations might also be explained in terms of enrichment of the interstellar medium by a stellar wind. This would not be difficult to ascertain, knowing the mass of the nebula (Table 20), its age as given by its size and the expansion velocity of the shell, the mass-loss rate of the star (Table 18), and the chemical composition of the outflowing matter, based on the spectrum of the central star (see Kwitter (1984)). The chemical composition of the surface of a Wolf–Rayet star is rather uncertain: estimates made by different authors often differ by an order of magnitude (see, e.g., Nugis (1982) and Willis (1982a)). Nevertheless, if we take $N/He = 10^{-2}$ (by number of atoms) for the star HD 192163, based on measurements by Nugis (1982) and Willis and Wilson (1979), the value observed in NGC 6888, $N/He = 1.55 \times 10^{-3}$, can be accounted for by the mixing of approximately five solar masses of interstellar matter with about $0.5 - 0.6\, M_\odot$ of matter from the stellar wind. As we have seen above, just that amount of mass loss as stellar wind over the lifetime of a shell is provided by the observed rate of outflow from the central star.

We therefore conclude by noting that although the recent data suggest chemical abundance anomalies, they provide no definitive clues that might enable us to distinguish between ejection of the products of nucleosynthesis and wind-impacted or wind-enriched interstellar gas.

The errors incurred in determining the chemical composition of a shell spectroscopically are large. This is a consequence of the inherent uncertainty in going from ion abundances to total abundances, the assumption of uniform temperature throughout the emitting gas, differences in local interstellar abundance, which are subject to the Galactic abundance gradient, and actual variations in the spectrum from filament to filament. Moreover, all calculations have been carried out assuming excitation via photoionization, while in fact we know that the expansion velocity of a number of nebulae — NGC 6888 in particular — may reach ~100 km/sec, and collisional ionization and excitation of the gas at a shock front can definitely make a contribution.

Future x-ray observations are bound to provide more reliable quantitative information, enabling us better to assess the degree to which the interstellar medium has been enriched in heavy elements by mass loss from WR stars.

15. Ring nebulae: stellar wind or stellar ejecta?

The nature of the shells surrounding Wolf–Rayet stars involves a puzzling situation. As early as the pioneering paper by Johnson and Hogg (1965), ring nebulae were attributed to the sweeping up of interstellar gas by a strong stellar wind; the theory of the interaction of stellar wind with the interstellar medium developed in parallel with observations of these nebulae, and forms the basis of our current understanding. Furthermore, all investigators concerned with this type of nebula or with stellar wind, be they observers or theoreticians, have subscribed wholeheartedly to that interpretation.

Masevich et al. (1975), however, attributed the very same nebulae to shell ejection in the course of the evolution of a massive close binary system. That interpretation also came into wide use by experts in the evolution of such systems (see, e.g., van den Heuvel (1976) and Moffat and Seggewiss (1979b)). It was also suggested that isolated Wolf–Rayet stars might produce ring nebulae via shell ejection during the red or blue supergiant stage (Bisnovatyi-Kogan and Nadezhin, 1972; Conti, 1976).

In this section we survey some of the observational consequences of these mechanisms, and having compared them with observations of nebulae and Wolf–Rayet stars, we will choose the most likely scenario.

The data summarized in Table 20 make it clear that ring nebulae may be anywhere from 2 to 100 pc in size, with expansion velocities of 10–100 km/sec. To judge by the best-studied shells, we know that the power and duration of outflow from the central star are sufficient to produce the nebula. It is also simple to demonstrate that the remaining shells in Table 20 could also have been produced by circumstellar gas having been swept up by stellar wind. Furthermore, even apart from the origin of the ambient gas (whether it was matter ejected from the star or interstellar gas), there are a number of observational factors that cannot be accounted for without invoking a strong stellar wind. For example, in the absence of ram pressure due to a hot wind, the bright compact knots and filaments observed in NGC 6888, which are 0.01–0.1 pc across and have a density $n_e \approx 500 \, cm^{-3}$ and temperature $T \approx 15,000 \, K$, would dissipate in 100 to 1000 years. But the kinematic age of the condensations, as implied by their velocity and distance from the center, are an order of magnitude greater, approximately $(1-5) \times 10^4$ yr. Without a strong stellar wind, it would also be hard to understand the very existence of a thin shell with $\Delta R/R \lesssim 0.1$ approximately 5 pc from the central star, since the thickness of a freely expanding shell is proportional to its radius.

The pressure due to hot gas — the shocked wind — precludes the dissipation of dense condensations, and it provides an explanation for a thin shell far from the star. Finally, the most compelling argument is that the thin optical shell of NGC 6888 is filled with the hot plasma of the decelerated wind, which shows up in x-ray observations of the nebula.

For the time being, then, we deal solely with the physics of nebulae, without broaching the subject of the central stars, as it lies completely within the confines of the model considered in Section 13, which incorporates the interaction of stellar wind with interstellar gas. The only observational evidence suggesting shell ejection comes from the kinematics of the nebulae M 1–67 and RCW 58, and the elemental abundances of certain objects.

However, if we turn to the observations of the central WR stars of ring nebulae and compare them with a complete sample of Galactic population I WR stars, we come to the conclusion that stellar wind alone does not suffice to explain the entire corpus of observational data.

Let us first introduce the basic properties of Wolf–Rayet stars, which determine their physics and evolutionary status (see IAU Symposium No. 99; Khaliullin and Cherepashchuk, 1982; van der Hucht et al., 1981, 1988; Torres et al., 1986; Breysacher et al., 1986; Doom, 1987; Smith, 1988c; Abbott and Conti, 1987; Conti and Underhill, 1988; further references are contained in these papers). The most recent data are presented in the proceedings of IAU Symposium No. 143 (van der Hucht and Hidayat (eds.), 1990).

Phenomenologically, WR stars may be classified according to whether the strong, broad lines of He I and He II are present, as well as lines of oxygen, carbon, and nitrogen from the ionization states O II–O VI, C II–C IV, and N II–N V. Physically, these are the products of the evolution of a massive star that has lost its outer hydrogen layers, i.e., a helium core with a hydrogen–helium shell enriched in C, N, and O. Two sequences of WR stars can be distinguished: the nitrogen (WN) stars, in which lines of N dominate the spectrum, and the carbon (WC) stars, with strong lines of C and O; both types have lines of He I and He II. A total of 161 WR stars have been identified in our Galaxy, distributed as follows over the spectral classes: WN2 (1 star), WN3 (3), WN4 (11), WN4.5 (6), WN5 (6), WN6 (21), WN7 (19), WN8 (10), WN9 (1), WC4 (5), WC5 (11), WC6 (14), WC7 (12), WC8 (8), WC8.5 (6), WC9 (18), WC10 (1), WO1 (1), and WO2 (1). The later spectral subtypes along each sequence display a lower degree of excitation.

According to the classification scheme presently in use, the four known WC4 pec and WC5 pec stars (two of them in the Magellanic Clouds) should be considered a separate WO sequence, defined by the relative strengths of the lines of O IV, O V, and O VI (Barlow and Hummer, 1982). This sequence represents the evolutionary stage following the WC stage. The small number of WO stars relative to WN and WC can be accounted for if the WO stage corresponds to the end of helium core burning, or if the star has gone on to core carbon burning. The WO sequence progresses from WO4 through WO3, WO2, and WO1 in order of increasing excitation.

There is a correlation between the distribution of WR stars of different spectral types and galactocentric distance. In addition, stars of the nitrogen sequence are most prevalent in the Magellanic Clouds. These two facts are probably related to differences in heavy-element abundances, and agree with evolutionary model predictions at the corresponding metallicities.

These stars have a typical photospheric radius of $3-4\,R_\odot$; the radius of the outflowing atmosphere may reach $\sim 40\,R_\odot$. The mass, which has been reliably determined for WR stars in binary systems, falls in the range $5-70\,M_\odot$, with $\sim 10\,M_\odot$ being the most likely value. The absolute luminosity, which has been determined from stars in the LMC or stars that are members of Galactic OB associations, varies quite markedly as a function of spectral type along the nitrogen sequence, but only weakly — or scarcely at all — along the carbon sequence: the spread in M_V may be $\Delta M_V = 0.5-0.7$ mag for a single type. The fact that M_V is correlated with spectral type means that late WN stars are high-mass objects and are more luminous, and that if stars evolve along a spectral sequence as they lose matter, then that evolution progresses from later types to earlier.

The bolometric luminosity of WR stars is highly uncertain, as the appropriate bolometric correction is poorly known: the spectrum is most certainly not blackbody radiation, it is difficult to distinguish radiation coming from the helium core from that originating in the extended atmosphere, and the line emission mechanism is somewhat unclear. Estimates of T_{eff} and M_{bol} are therefore uncertain. The location of individual WR stars in the Hertzsprung–Russell diagram is thus subject to significant error. Nevertheless, it seems to have been firmly established that WR stars exhibit excess luminosity for their mass, as compared with main sequence

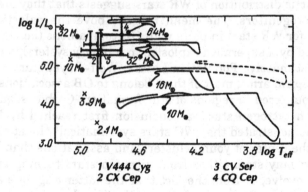

Fig. 86. Evolutionary tracks on the Hertzsprung–Russell diagram of high-mass Wolf–Rayet stars that belong to close binary systems (Tutukov and Yungel'son, 1973): 1) V 444 Cyg; 2) CX Cep; 3) CV Ser; 4) CQ Cep (Cherepashchuk, 1982). On the left is the sequence of homogeneous helium stars, and on the right is the zero-age main sequence.

Table 22. Effective Temperature and Relative Abundance of Chemical Elements in the Atmospheres of WN and WC Stars

Type	T_e, K	C/N	C/He	N/He
WN	50,000	4×10^{-3}	9×10^{-5}	2.5×10^{-2}
WC	30,000	3	9×10^{-3}	3×10^{-3}

stars, and that they lie between the latter and the evolutionary sequence of homogeneous helium stars (Fig. 86). The most probable evolutionary status of WR stars, as determined from their location in the Hertzsprung–Russell diagram, is that they are in the helium-burning stage of the convective core of a massive star that has lost its outer hydrogen shell either through mass exchange in a binary system or via outflow from a massive single star.

Wolf–Rayet stars are the strongest sources of stellar wind that we know of among the steady-state stars (see Section 12). Recent observations of their UV and optical spectra coupled with radio and IR free–free emission yield an average terminal wind velocity of $v_w = 2000 - 3000$ km/sec, and a mass-loss rate in the range $10^{-5} - 10^{-4}$ $M_\odot$/yr, with no significant differences among the mean mass-loss rates of (a) single and binary WR stars, (b) WN and WC stars, or (c) WN and WC subclasses (see Conti and Underhill (1988), Willis (1990), and references cited in those papers). The ratio between mechanical energy lost in the form of wind and luminosity is $L_w/L = 1/5 - 1/10$ (compared with $L_w/L \approx 1/100$ for Of stars and $L_w/L < 1/100$ for main-sequence O stars).

The Galactic distribution of WR stars suggests that they are produced by massive progenitors. The mean height above the galactic plane is $z = 80 - 90$ pc for WR stars in pairs with OB companions (i.e., for stars not yet accelerated by a supernova explosion in a binary system), a value typical of the most massive Population I stars. Wolf–Rayet stars are concentrated in the spiral arms, most of them belong to OB associations, and they are often encountered in regions of star formation. Carbon-sequence stars are older than nitrogen stars, a conclusion first reached by Mikulášek (1969), who demonstrated that WC stars systematically lie at greater distances from the center of young clusters and associations than WN stars. This situation may stem either from nitrogen stars turning into carbon stars as they evolve, or from the fact that the latter originate as stars of lower mass and therefore evolve less rapidly. The chemical composition of the surface of Wolf–Rayet stars lends support to the first interpretation.

In Table 22, we present the relative C, N, and He abundances in the atmospheres of the Wolf–Rayet stars of the two sequences, based on ultraviolet observations of five isolated WN stars and one WC star (Willis and Wilson, 1979). The hydrogen abundance determined from optical spectra of dozens of WR stars is $N_H/N_{He} \approx 0.1 - 0.2$. Table 22 makes it clear that the carbon and nitrogen abundances are markedly different in the two sequences; in fact, they differ exactly as predicted by the course of the CNO cycle if WN stars turn into WC (see Tutukov and Yungel'son (1983) and Maeder (1983)). There is a well-defined trend in the H/He ratio along the nitrogen sequence: in the later WN stars, H/He ≈ 1, and in the early ones, H/He ranges from 1 to 0. This is suggestive of evolution from later to earlier types, if the low hydrogen abundance is due to outflow from the outer layers of the star.

The most up-to-date outlook on this problem has been summarized by Maeder (1987c), Conti and Underhill (1988), and van der Hucht and Hidayat (1990). For the purpose of establishing the initial mass and evolutionary status of WR stars of various spectral types, studies of such stars in rich, young clusters whose age has been reliably determined by the turnoff point from the main sequence seem to be the most promising. In principle, given a copious selection of clusters of known age that contain WR stars of different spectral types, the problem ought to have a unique solution. At the moment, however, the sample is meager, and the membership of WR stars in a particular group is not always unambiguous. Nevertheless, a fair number of such studies have appeared in recent years, and on the whole, the conclusions they reach are mutually consistent (e.g., see Schild and Maeder (1984, 1987c), Lundstrom and Stenholm (1984), Langer (1987), van der Hucht and Hidayat (1990), and references therein). The lower limit on the initial mass of isolated WN stars is $M_i \gtrsim 18 M_\odot$, and it is $M_i \gtrsim 35 M_\odot$ for isolated WC stars. Stahl (1987) came up with approximately the same figures, having compared the distribution of WR stars with O stars of known mass that are no more than 2.5 kpc from the sun: WR stars have an initial mass of $M_i > 22 M_\odot$; WN stars are produced primarily by O stars with M_i ranging from 25 to 37 $M_\odot$, and from very massive stars with $M_i \gtrsim 50 M_\odot$; WC stars come from intermediate masses, $M_i \sim 35 - 40 M_\odot$; and single stars come preferentially from more massive progenitors than binaries. Using WR stars in OB associations, Humphreys et al. (1984) found $M_i \gtrsim 50 M_\odot$ but they did not allow for ongoing star formation in associations.

The evolutionary path followed by more massive stars with $M_i \gtrsim 60 M_\odot$ is $O \rightarrow Of \rightarrow WN7-8 \rightarrow WC4-8 \rightarrow SN$. In the range $35 \lesssim M_i \lesssim 60 M_\odot$, we have $RSG \rightarrow (WN6?) \rightarrow WC8.5-9 \rightarrow WO \rightarrow SN$. Stars with $M_i \lesssim 35 M_\odot$ probably evolve through the progression $RSG \rightarrow WN3-5 \rightarrow SN$ (Schild and Maeder, 1984; Langer, 1987). Qualitatively similar evolutionary scenarios were calculated by Doom et al. (1986), who took into account mass loss and mixing. Their three evolutionary tracks for stars of different initial mass were as follows:

$M_i \gtrsim 130 M_\odot$: η Car - like $\rightarrow$ Hubble – Sandage variable (HSV)
$\rightarrow Of \rightarrow$ transitional object $\rightarrow WN \rightarrow WC \rightarrow WO \rightarrow SN$;

$33 \lesssim M_i \lesssim 130\,M_\odot$: $O \to Of \to$ transitional object
$$\to WN \to WC \to WO \to SN;$$

$22 < M_i < 33\,M_\odot$: $O \to Of \to RSG \to HSV$
$$\to WN \to WC \to SN.$$

The duration of all successive WN and WC stages is approximately the same, and is about $(1-2) \times 10^5$ yr for stars with an initial mass of $\sim 30\,M_\odot$ (Tutukov and Yungel'son, 1973; Vanbeveren and Packet, 1979; Maeder, 1983).

Wolf–Rayet stars — the central stars of ring nebulae — are not much different from stars lacking shells, so long as one is concerned with individual stellar parameters having to do with their spectrum, luminosity, rate of outflow, and so forth. But upon comparing the statistical characteristics of the two types of WR stars — with and without ring nebulae — we find that the former is possessed of two distinctive features that are fundamentally important to an understanding of the "WR + shell nebula" phenomenon (Lozinskaya and Tutukov, 1981):

1. A complete sample of WR stars with heliocentric distance $r \leq 2.5$ kpc contains 47 members. Among those, there are ten that are paired with a massive OB companion, and six with a low-mass companion, possibly a neutron star (Hidayat et al., 1984). Meanwhile, only one of the ten nebulae classified as W or E in Table 20 are associated with a WR + OB system, and five are associated with a WR + compact companion system.

If we suppose that ring nebulae form only as a consequence of stellar wind, then the predominance of shells around WR stars with an invisible compact companion does not make sense. Cavities produced by stellar wind should, with equal probability, be observed around pairs with a high-mass companion, those with a low-mass companion, and isolated WR stars. In actual fact, an O or B star paired with a WR can only enhance, rather than diminish, the power of the stellar wind, and we know of no observational selection effects that can explain the paucity of windswept shells around WR + OB binary systems. On the contrary, the detection of a low-mass relativistic companion is an exceedingly difficult observational undertaking: the periodic variations in the brightness of the WR primary are $\sim 2-3\%$, while radial-velocity variations are at most $\sim 5\%$ of the linewidth. It is therefore possible that as-yet undetected binaries with low-mass companions lurk among the "single" WR stars associated with ring nebulae.

2. The second fact that is inexplicable from the standpoint of a "pure" stellar wind is the preponderance of nitrogen stars among the central stars of ring nebulae: of the 19 nebulae in Table 20, only three are associated with WC stars. In the solar neighborhood of the Galaxy ($r \leq 2.5$ kpc), there are 15 type WN stars and 29 type WC (van der Hucht et al., 1988). The two types of stars lose mass at the same rate; the mass outflow stage in carbon-sequence stars lasts twice as long as the wind-dominated stage in nitrogen

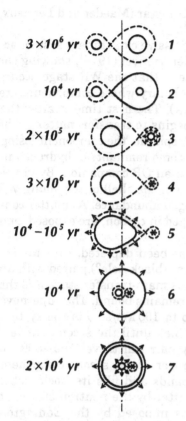

$3 \times 10^6 \, yr$ 1

$10^4 \, yr$ 2

$2 \times 10^5 \, yr$ 3

$3 \times 10^6 \, yr$ 4

$10^4 - 10^5 \, yr$ 5

$10^4 \, yr$ 6

$2 \times 10^4 \, yr$ 7

Fig. 87. Evolution of a massive close binary system, leading to the formation of a WR star and shells (Massevich, 1975). The various stages are as follows: 1) two main-sequence stars; 2) the more massive, helium-core star fills its Roche lobe; 3) Wolf–Rayet star paired with an OB star; 4) neutron star paired with an OB star following a supernova explosion; 5) second star almost fills its Roched lobe; 6) helium core with a compact relativistic companion inside a common envelope; 7) WR star with a compact companion and an expanding ring nebula. The approximate duration of each stage is indicated on the left.

nitrogen stars, if the former come from the latter. If ring nebulae resulted solely from the interaction of stellar wind with interstellar gas, the older stars of the carbon sequence would be the ones to have swept up the more massive shells. The observations clearly indicate otherwise.

Both difficulties can be avoided if one assumes that rather than sweeping up interstellar gas, the wind from a Wolf–Rayet star sweeps up the shell ejected in a previous stage of development of the massive system. Ejection of the outer hydrogen layers can take place either during the evolution of a single star or in a binary system. In either case, a ring nebula similar to those actually observed can form as a result of the subsequent action of the wind from a Wolf–Rayet star on the ejected gas (Lozinskaya and Tutukov, 1981).

The ejection of a shell by an isolated massive star ($M_{init} \approx 30 \, M_\odot$) is to be expected when hydrogen burning is complete and helium burning commences, i.e., after the star has moved to the red supergiant region (Bisnovatyi-Kogan and Nadezhin, 1972). At that point, there is a strong inversion of the density gradient in the outer layers of the star, which stimulates a powerful ($M \approx 0.5 \, M_\odot/yr$) outflow of limited duration — the essentially instantaneous ejection of a shell comprising a large fraction of the star's mass (about $15 - 20 \, M_\odot$).

A popular recent evolutionary scenario for a massive single star is $O \rightarrow RSG \rightarrow WN \rightarrow WC$, with mass outflow and convective mixing, which carries C, N, and O to the surface of the central star (Maeder and Lequeux, 1982; Maeder, 1982, 1983; Doom et al., 1986).

Figure 87 illustrates the evolution of a massive close binary system according to Masevich et al. (1975) and van den Heuvel (1976), showing the typical duration of each stage. It can be seen that the WR stage occurs twice in the development of a massive close binary whose components are of approximately equal mass ($M_{init} \gtrsim 20\,M_\odot$). The first time is after the more massive star fills its Roche lobe and begins to transfer matter to the second. As a result of this brief ($10^3 - 10^4$ yr) episode, the component losing matter — a virtually pure helium star with some remnants of hydrogen in the shell — turns into a WR star paired with an OB companion. By virtue of the CNO reactions, the carbon abundance in the shell of the young WR star is not too high, compared with the nitrogen abundance. As matter continues to leave the star, deeper layers, enriched in carbon, are exposed, and the WN star turns into a WC.

Subsequently, after its nuclear fuel has been depleted, the star explodes as a supernova, forming a neutron star (black hole?) paired with an OB companion. Since by virtue of the previous mass-transfer stage it is the less massive star that explodes, the system remains bound. The supernova explosion imparts a high space velocity (up to 100 km/sec) to the system (OB + neutron star), and over its lifetime (i.e., until the second star explodes), perhaps $\sim 3 \times 10^6$ yr, this runaway pair can travel 100–500 pc from its original location. After all the hydrogen in the second component — the OB star — has been burned, it expands and fills its Roche lobe. Since the rate of accretion of outflowing matter by the neutron star is at most $10^{-8} - 10^{-7}\,M_\odot$/yr, due to constraints imposed by the Eddington limit, the compact companion and the helium core of the OB star end up immersed in a common hydrogen envelope. That envelope probably lasts for no longer than the thermal time scale, i.e., $10^3 - 10^4$ yr. Afterwards, dynamical slowdown of the binary nucleus — the helium star and its relativistic companion — leads to the ejection of a massive ($10–30\,M_\odot$) common envelope. In the present scenario, the expansion velocity of that shell can be of the order of the escape velocity at its surface, i.e., 10–300 km/sec. Thus, as a massive close binary system evolves, it gives rise to a young Wolf–Rayet star (possibly a runaway) paired with a compact relativistic companion, surrounded by an expanding gaseous nebula.

The lifetime of the shell ejected by a massive isolated or binary star depends on how the emission measure of the expanding nebula falls off. A freely expanding shell with a mass $\lesssim 10–20\,M_\odot$ becomes difficult to detect against the Galactic background emission at a radius of about 2 pc (we have taken the limit of detectability to be at $E_m \geq 100\,\mathrm{cm}^{-6} \cdot \mathrm{pc}$; see below). Assuming a wind that sweeps up the ejecta into a thin layer, a ring nebula can be observed when it is as much as 10–20 pc across. Accordingly, the lifetime of a shell in this scenario is $10^4 - 10^5$ yr, which is consistent with the kinematic age of the nebulae listed in Table 20. As it expands further, and assuming only that the density of the ambient medium is not anomalously high, $n_0 \lesssim 10 \, \text{cm}^{-3}$, the expanding shell ceases to be observable when it reaches the level of the Galactic Hα background. The typical lifetime of $10^4 - 10^5$ yr explains why ring nebulae are seen primarily around stars of the nitrogen sequence — they become unobservable prior to the transition from the WN stage to the WC stage. In terms of this model, the wind-blown bubble G2.4+1.4 around the most evolved WO star known may have been observable because of its location at the edge of a dense cloud, and because of the star's extremely powerful but short-lived stellar wind.

A second feature of the nuclei of ring nebulae — the small number of WR + OB pairs — can be accounted for by the ejection of a common envelope following the supernova explosion and the formation of a neutron star in the binary system, or more trivially, when a shell is ejected by an isolated supergiant.

Is it possible to distinguish the two paths whereby Wolf–Rayet stars form ring nebulae — through the ejection of a shell by a single massive star, or as a process taking place in a binary system? There are three criteria to judge by, and they are all observational consequences of the supernova explosion that precedes the ejection of a common envelope in a binary system:

1. The presence of a compact companion — a neutron star — in a tight orbit. The detection of such a star, incidentally, is fraught with extreme difficulty, due to the broad emission lines and high luminosity of Wolf–Rayet stars.

2. A peculiar velocity $\sim 100 \, \text{km/sec}$ of the system as a whole (prior to a supernova explosion, population I Wolf–Rayet stars have typical velocities $\sim 10 \, \text{km/sec}$).

3. If the peculiar velocity of a runaway star is directed perpendicular to the galactic plane, the z component may be anomalously high for population of WR stars, which have a very flat distribution (for binary WR + OB systems, i.e., prior to a supernova explosion, the mean height above the plane is $z = 80$ pc).

The foregoing are the predicted hallmarks of Wolf–Rayet stars in ring nebulae for the scenario of common-envelope ejection in a binary system at the WR + neutron star stage of evolution. Turning now to the observational data summarized in Table 20, we see that there are eight objects that conform to this picture:

HD 50896: weak, periodic variability in brightness and periodic variations in the emission-line profiles with $P = 3\overset{d}{.}76$ have been detected. The observations may be interpreted as evidence for the presence of a compact companion, a neutron star of approximate mass $M = 1.3 \, M_\odot$ lying within the extended atmosphere of the WR star (Firmani et al., 1980). The star is high above the galactic plane at $z = -300$ pc. Rapid x-ray variation (on a time scale of 30 min) has been detected in HD 50896, which may possibly signal the presence of a black hole or white dwarf paired with the WR star (White and Long, 1986). The nature of the compact companion of the WR star, like the nature of the variability of HD 50896, thus to a certain extent remains open; see also Vreux (1985).

HD 96548: has a compact companion with period $P = 4\overset{d}{.}8$ and mass approximately $0.3\,M_\odot$ (Moffat and Isserstedt, 1980). The system is high above the galactic plane, $z = -300\,\mathrm{pc}$. (Recently Smith et al. (1985) obtained a period $P = 5\overset{d}{.}8$ and examined another possible interpretation for these periodic variations, unrelated to the presence of a compact companion. If their doubts are confirmed, then they should apply equally well to other objects, and the whole idea may have to be reexamined.)

HD 117688: high peculiar velocity, $v_{pec} = -90\,\mathrm{km/sec}$.

HD 187282: compact companion (neutron star), period $P = 3\overset{d}{.}85$, mass approximately $1\,M_\odot$ (Antokhin et al., 1982). High above the galactic plane, $z = -360\,\mathrm{pc}$.

HD 192163: compact companion; period $P = 4\overset{d}{.}5$, mass approximately $0.5\,M_\odot$ (Aslanov and Cherepashchuk, 1981). High peculiar velocity, $v_{pec} = 100\,\mathrm{km/sec}$. Observations by Vreux et al. (1985) have, however, cast the previously detected periodicity into doubt. Based on a large number of spectrograms, with results obtained at a high confidence level, these workers have detected periodicity, but with a more probable period of $0\overset{d}{.}45$ or $0\overset{d}{.}31$. In their opinion, the variability may be associated either with nonradial stellar pulsations or with a neutron star companion — but one with a very compact orbit, $a \sim 7\,R_\odot$. Observations of the periodic variability of HD 192163 must therefore be continued.

209 BAC: compact companion, period $P = 2\overset{d}{.}36$, mass $1-1.6\,M_\odot$ (Moffat et al., 1982a). High above the plane, $z = 260\,\mathrm{pc}$, and high peculiar velocity, $v_{pec} = 180\,\mathrm{km/sec}$.

HD 191765: compact companion, period $P = 7\overset{d}{.}4$, mass approximately $1.5\,M_\odot$ (Antokhin and Cherepashchuk, 1984).

HD 115473: high above the galactic plane, $z = 240\,\mathrm{pc}$.

For at least these eight objects, the model of a common envelope ejected in the WR + neutron star stage, with ejecta subsequently being swept up by stellar wind, seems to be justified.

The applicability of the scenario in question to those objects in Table 20 for which none of the appropriate hallmarks of a supernova explosion have been detected is presently uncertain. It is conceivable that some were produced through the expulsion of a shell by a single star, and that the ejecta were subsequently swept up by stellar wind. High-mass ring nebulae ($M \gtrsim 50-100\,M_\odot$) consist mainly of swept-up interstellar gas, and wind probably plays a more important role than instantaneous expulsion of matter in producing them.

Most likely, then, the answer to the question posed in the heading of this section is *ejected shell plus stellar wind*. These were the arguments upon which we based our WR ring creation scenario ten years ago (Lozinskaya and Tutukov, 1981). Modern high-resolution observations of certain ring nebulae show direct evidence of two distinct components,

Table 23. Statistics of Wolf–Rayet Stars with Nebulae

Sample population	Early WN $m_V \leq 11.1$ $A_V \leq 2.^m 5$	Late WN $m_V \leq 12$ $A_V \leq 2.^m 5$	WC $m_V \leq 11.1$ $A_V \leq 2.^m 5$	All WR
Number of stars	19	12	29	51
Bright background	2	2	2	6
Stars with H II regions	14 82%	9 90%	16 89%	39 87%
Stars with ring nebulae	7(+3?) 41% (60%?)	3(+2?) 30% (50%?)	3(+2?) 17% (28%?)	13(+7?) 29% (44%)

namely ejected clumps and a fine filamentary bubble blown by stellar wind; see Section 14.

There is yet another independent argument favoring the idea of wind sweeping up stellar, rather than interstellar, matter, with the former having been ejected after the explosion of a supernova in the system. Runaway WR stars, of which there are many among the central stars of ring nebulae, should sweep asymmetric shells with a characteristic bowed shape out of the quasistationary interstellar gas. The symmetric ring structure of the nebulosity around a runaway star may result from stellar wind sweeping up ejected gas that is comoving with the star. We point out, however, that the shape of a shell must be examined in each case individually, taking into consideration the direction of stellar motion, since it depends on the density distribution of the ambient interstellar gas and on the orientation of the ambient magnetic field.

It is possible to approach the problem of ring nebula formation from still another standpoint, by comparing their prevalence with the predictions based on shell ejection *vs.* wind-blown interstellar gas (see Lozinskaya (1983)).

If the sample of WR stars and ring nebulae is complete — i.e., it is drawn from a volume of space in which all stars and nebulae are above the detection threshold — then the ratio of the number of stars with shells, $N(\text{WR} + \text{shell})$, to the total number of stars, $N(\text{WR})$, is obviously

$$\frac{N(\text{WR} + \text{shell})}{N(\text{WR})} = \frac{t(\text{shell})}{t(\text{WR})}. \tag{15.1}$$

where $t(\text{WR})$ is the duration of the WR stage (approximately $(2-3) \times 10^5$ yr) and $t(\text{shell})$ is the lifetime of a shell. If shell formation is dominated by the sweeping up of interstellar gas, we have $N(\text{WR} + \text{shell})/N(\text{WR}) \approx 1$, since the duration of the adiabatic expansion stage is negligible, and the age of a nebula is essentially equal to the duration of the outflow stage (see Section 13).

If "instantaneous" expulsion of matter plays the major role, t(shell) will be determined by the falloff in emission measure as the ejected shell expands. We find the maximum radius R_{lim} at which the nebula is still observable ($E_m \geq 100 \, \text{cm}^{-6} \cdot \text{pc}$) and the corresponding age by noting that the typical mass ejected is $M_{ej} \leq 10 M_\odot$, the thickness of the shell is $\Delta R/R \lesssim 0.2 - 0.3$, and the expansion velocity is $\gtrsim 30 \, \text{km/sec}$.

The limiting magnitude for an extended object on the Palomar Sky Survey red plates corresponds to $E_m = 50 \, \text{cm}^{-6} \cdot \text{pc}$; special deep plates taken through narrow-band filters enable one to go much fainter. We have adopted a more stringent criterion of detectability because Wolf–Rayet stars belong to the flattest component of Galactic Population I, i.e., they must be observed against the Hα emission from the Galactic disk. Under these conditions, we obtain $R_{lim} \leq 2 - 3 \, \text{pc}$, $t_{ej} \leq 5 \times 10^4 - 10^5 \, \text{yr}$, and $N(\text{WR} + \text{shell})/N(\text{WR}) \leq 0.2 - 0.5$.

Either estimate of the expected number of stars with nebulae is crude at best, but they differ by a factor of two to three, and it is worthwhile to compare them with the observations. A count of the total number of stars brighter than some given magnitude shows that the sample of early Galactic WN stars and WC stars can be assumed to be complete to $m_V = 11$, and a sample of late WN stars should be complete to $m_V = 12$. Comparing the lowest detectable emission measure, $E_m = 100 \, \text{cm}^{-6} \cdot \text{pc}$, with the emission measure of ring nebulae (e.g., NGC 6888), we obtain a stringent bound on interstellar absorption, $A_V \leq 2.^m 5$. For $m_V \leq 11$ (or $m_V \leq 12$ for late WN stars) and $A_V \leq 2.^m 5$, both the sample of WR stars and the sample of ring nebulae are complete.

In Table 23, we present the statistics of Wolf–Rayet stars with ring nebulae (Lozinskaya, 1983), giving the total size of the sample, the number (and percentage) of WR stars associated with H II regions, and the number (and percentage) of stars with ring nebulae. The data are listed separately for early WN (WN3 to WN6), late WN (WN7 to WN9), and WC stars.

The data in Table 23 imply that approximately 80–90% of all WR stars are associated with diffuse H II regions; this is consistent the classical picture of how a Strömgren sphere is produced around a hot star. Diffuse H II regions form when the density of the ambient gas is high enough — in other words, in the cold or warm components of the gaseous disk of the Galaxy (see Section 17). The percentage of WR stars with diffuse nebulae is higher than $f_{c+w} \approx 50\%$, the fraction by volume of the Galactic disk occupied by cold or warm gas. This makes sense, inasmuch as WR stars tend to be associated with OB associations and gas and dust complexes.

A comparison of the observed number of WR stars that have shells (29%, or at most 44% when uncertain identifications are included) with the predicted number suggests that they may be formed in some other way than purely by stellar wind. In all likelihood, the instantaneous ejection of matter plays an important role in the formation of ring nebulae. What that role might be is still unclear., but it is not inconceivable that the expulsion of a shell produces an initial density distribution in the ambient gas that interacts more stably with the stellar wind.

In the opposite case, a cold wind can penetrate finger-like through the inhomogeneous interstellar medium without forming a regular structure. This penetration of a cold wind may be one reason why ring nebulae do not always form around strong sources of stellar wind, the theoretical prediction notwithstanding. One trivial explanation — that the density of the ambient interstellar medium is low — does not hold water, since the large number of WR stars that excite H II regions bespeaks a fairly high density. It might also be suggested that the scarcity of wind-swept shells results from the high velocity of the star relative to the circumstellar gas — but that too is not so. Out of 18 actual ring nebulae, the central stars of seven — Sh 308, RCW 78, NGC 6888, M 1-67, RCW 58, and the nebulae surrounding HD 187282 and HD 115473 — have high peculiar velocity or are anomalously high above the galactic plane, both evidence of rapid stellar motion.

It is also possible that the brief life span of a ring nebula, in contrast to the extended period over which outflow takes place from a Wolf–Rayet star, may stem from the fast leakage of hot wind through breaks in the regular structure of a shell (Lozinskaya, 1983). The inhomogeneity of the shell may result from multiple blast and reverse shock waves that come into being at the time of shell collapse (see Section 9). For a break in the shell to form, a density fluctuation $\Delta n_0/n_0 \approx 30\%$ in the ambient gas is sufficient, and the presence of a magnetic field facilitates fragmentation of the shell (Falle, 1975b). Evidence for the penetration of hot plasma includes the thin, faint radial jets and very faint tangential [O III] filaments outside NGC 6888 (see Section 14).

If the wind in fact flows rapidly through a porous shell, the layer of hot wind (layer b in Fig. 76) should be much less significant than in the scheme discussed in Section 13. The most promising observations to test this proposition would be of x-ray emission from the bubble formed by the stellar wind. In Table 24, we have set out the 0.1–4 keV luminosity predicted by the standard model of a bubble for a number of ring nebulae around WR and Of stars, taking interstellar absorption into account (Bochkarev and Lozinskaya, 1985). The calculation was carried out for plasma having a normal chemical composition, in ionization equilibrium; we used the shell parameters summarized in Tables 20 and 26. The columns of Table 24 contain

1) the name of the nebula;
2) the color excess of the central star assumed in estimating the interstellar absorption;
3) the predicted plasma temperature in layer b;
4) the predicted x-ray emission at 0.1–4 keV;

Table 24. Predicted Temperature and 0.1–4 keV X-ray Flux from Ring Nebulae Associated with Wr and Of Stars

Nebula	$E(B-V)$	T_B, 10^6 K	$F(0.1-4\,\text{keV})$ erg·cm^{-2}·sec	Limiting $F(0.1-4\,\text{keV})$
NGC 6888	0.6–0.55	9	$(1-2)\times10^{-11}$	1.2×10^{-12}
Sh 308	0.12–0.22	9	$\leq 1.7\times10^{-12}$	3.2×10^{-12}
NGC 2359	0.56–0.66	6–10	$(0.1-4)\times10^{-12}$	4×10^{-13}
NGC 6164–5* (second shell)	0.7	2	5×10^{-12}	1.1×10^{-12}
NGC 7635	0.73	2.5–4.5	$(0.1-6)\times10^{-14}$	2×10^{-13}
RCW 58	0.54	2–6	$\leq 0.14\times10^{-12}$	0.6×10^{-12}
NGC 3199	0.92	3–7	$(0.1-1)\times10^{-12}$	1.4×10^{-12}

*The calculations were carried out for the second shell of NGC 6164–5 from the center, since it was probably the only one produced by stellar wind (see Section 16).

5) the limiting detectable flux, at 10^4 sec integration time, in a scan with the Imaging Proportional Counter (IPC) of the Einstein observatory.

A number of ring nebulae around Wolf–Rayet stars were included in the Einstein observing program. According to Moffat et al. (1982b), the extended sources associated with the nebulae Sh 308 and RCW 58 were undetectable with a signal integration time of 10^4 sec. We thereby obtain upper limits on the flux from Sh 308,

$$F(0.1-4\,\text{keV}) = 3.2\times10^{-12}\ \text{erg·cm}^{-2}\cdot\text{sec}^{-1},$$

and RCW 58,

$$F(0.1-4\,\text{keV}) = 0.6\times10^{-12}\ \text{erg·cm}^{-2}\cdot\text{sec}^{-1};$$

both values are above the predicted flux.

To test the theoretical model of wind interaction with the interstellar medium — particularly the hypothesized outward penetration of the wind — observations of NGC 6888 would seem to be the most promising (Fig. 80 contains the predicted x-ray spectrum).

An extended x-ray source associated with NGC 6888 has recently been detected (Bochkarev et al., 1988; Bochkarev, 1988), but at less than the predicted flux: $F(0.1-4\,\text{keV}) \sim 10^{-12}\ \text{erg·cm}^{-2}\cdot\text{sec}^{-1}$. EXOSAT x-ray observations of NGC 6888 also yield a conservative upper limit of

10^{-12} erg·sec^{-1}·cm^{-2} in the 0.05–2 keV range, which is an order of magnitude less than predicted (Kahler et al., 1987). This object may provide some support for the leakage of hot wind through breaks in the regular structure of a shell, but we must also point out that calculations of the x-ray luminosity expected from wind-swept bubbles are still rather uncertain. Specifically, we have not taken into consideration the rapid dissipation of wind energy that accompanies thermal transport in a layer with an abrupt temperature gradient (see Dorland and Montmerle (1987)), and the departure of plasma from ionization equilibrium.

We shall return to some of the as-yet unresolved problems of the origin of ring nebulae in Section 16, following a discussion of Of rings.

16. Nebulae associated with Of stars; the interstellar medium around supernova precursors

It ought to be possible to shed some light on the questions posed in the preceding section by investigating objects with similar traits surrounding other strong sources of stellar wind. Foremost among these are the Of stars, with mass outflow rates that approach those of the Wolf–Rayet stars. The Of stars are hot, massive O stars which, along with absorption lines, display broad, bright emission lines due to He II (4686 Å), N III (4634, 4640, 4641 Å), and fainter lines of Hα, C III (5694 Å), and others. Certain absorption lines are faint or undetectable, being overwhelmed by the emission; the emission lines themselves frequently have P Cyg profiles.

The Of stars lose mass just as copiously as WR stars: $\dot{M} = 10^{-5} - 10^{-6}\ M_\odot/\text{yr}$, $v_w = (2-3) \times 10^3$ km/sec (see Table 18 and the data of Barlow and Cohen (1977), Andriesse (1980), and references included in Section 12). In contrast to WR stars, however, Of stars retain an outer hydrogen shell; the mass, based on Of stars in binary systems, can reach $\sim 60\ M_\odot$ (Conti, 1979).

The Of stars are progenitors of Wolf–Rayet stars — in any event, of the late-type WN7–9 (Conti, 1979; Willis, 1982b; Moffat and Seggewiss, 1979a, b; Maeder, 1982, 1983). WN7–9 stars differ from WC and late WN, and are similar to Of in a number of ways. They typically have high luminosity and a high effective temperature, narrower spectral lines, and — of particular importance — higher hydrogen abundance than early-type stars of the nitrogen sequence. WN7–9 are encountered in the very youngest star clusters. From the Hertzsprung–Russell diagram of the corresponding clusters, the age of WN7–9 stars is found to be 3×10^6 yr, while for early-type WN stars it is 5×10^6 yr (Moffat and Seggewiss, 1979c). In terms of the foregoing parameters, the WN7–9 stars resemble Of stars more than they resemble the earlier WR stars. The only practical difference between them consists of the narrower emission lines in Of spectra, suggesting stronger outflow from the atmospheres of the WN7–9. All of these properties can be explained by assuming that WN7–9 stars are formed as a result of mass loss from massive Of stars; see also Section 15. A recent analysis of all Of stars in young clusters (Feinstein et al., 1986) has confirmed their high luminosity (ranging from $M_V = -4.3$ to $M_V = -8.0$) and extreme youth, and has shown that they lie above the zero-age main sequence in the Hertzsprung–Russell diagram. Additional new data may be found in van der Hucht and Hidayat (1990).

In order to clarify the evolutionary status of Of stars, it would also be of interest to find out whether any are surrounded by shell nebulae, and if so, what the latter are like. The author began to address this problem in

Table 25. Stars of Spectral Type Of with Diffuse or Ring Nebulae

Luminosity class	V	III	I + II	Total
Number of Of stars	12	8	9	29
Number of Of stars associated with H II regions	11	6	7	23
	90%	75%	80%	79%
Number of Of stars associated with a ring nebula (including those above)	3	4	5	12
	25%	50%	55%	41%

1982, searching for and investigating ring nebulae around Of stars (see Lozinskaya and Lomovskii (1982), Lozinskaya (1982), and Lozinskaya et al. (1983)).

A number of nebulae with ring-like morphology, excited by Of stars, have long been known. We were interested, however, in finding out whether there is a well-defined class of such objects, whether it is well-populated, and whether all Of stars are associated with ring nebulae.

The complete list of Of and O(f) stars reviewed, taken from existing catalogs of Galactic stars with known spectral classification, consists of 108 objects. Since the spectroscopic markers identifying Of stars may be time-varying, we accepted stars identified as Of even if they appeared in only one catalog. The environment of stars with $\delta \geq -43°$ was inspected on Palomar Sky Survey prints and on deep photographs from the atlas of Parker et al. (1979). The use of deep photographs enabled us, first, to identify weak nebulosity, with emission measure $E_m \geq 20\,\mathrm{cm^{-6}} \cdot \mathrm{pc}$; second, it facilitated the recognition of ring morphology in the brightest nebulae on photographs taken in the [O II], [O III], and [S II] lines, in addition to Hα + [N II].

The results were as follows. Emission or strong absorption precluding the recognition of any nebulosity were observed in the vicinity of more than 95% of the stars inspected. Without further investigation of kinematics, it is difficult to establish a physical connection between faint H II that were identified and the stars. But we can state that only 3–5% of the Of stars in the region for which the sample of stars and nebulae was complete (see below) show no evidence of faint emission nebulosity, at least in the line of sight.

Of the 100 or so Of stars inspected, 46 are the sources of excitation in 42 H II regions, including 13 stars surrounded by ring nebulae. In order to understand whether this is a lot or a little, in light of the fact that Of stars 1) ionize gas, 2) give rise to stellar wind that sweeps up gas, and 3) can "instantaneously" expel matter, we consider a region free of selection effects. A count of the total number of stars brighter than a certain magnitude shows that the list of Of stars is complete to $m_V = 8$, or in other words, out to a distance of 1.5–2.5 kpc for $M_V = -5$ to -6. The maximum tolerable absorption, in terms of not precluding the detection of a ring nebula, is

$A_V = 2.^m 5$, assuming that the shells around Of and WR stars are identical (see Section 15).

A complete sample of stars satisfying these conditions ($m_V \leq 8$, $A_V \leq 2.^m 5$) numbers 36 objects; of those 36 stars, 25 are associated with H II regions, including 12 with ring nebulae. In Table 25, we give the distribution of Of stars with diffuse or ring nebulae, organized by luminosity class (except for those whose luminosity class is unknown).

Clearly, the percentage of Of stars associated with amorphous and ring nebulae is close to the number of WR stars with nebulae: approximately 70–80% of Of stars seem to be associated with H II regions, and among those, 30–40% of the stars are surrounded by ring nebulae. The differences between the various luminosity classes are not statistically significant, due to the small size of the sample.

The fraction of Of stars associated with nebulae is greater than the fraction of the volume of the Galactic disk occupied by cold or warm high-density gas. As in the case of the WR stars, the explanation is that the stars tend to be located in regions with dense molecular complexes and OB associations. In fact, of 29 stars belonging to clusters or associations, 22 are associated with nebulae. Out of six isolated Of stars, only two are associated with nebulae (Lozinskaya, 1982).

We have thus demonstrated that there exists a class of ring emission nebulae around Of stars, and those nebulae are as plentiful as the shells around WR stars. The basic parameters of the ring nebulae around Of stars are given in Table 26 as follows:

1) designation of nebula and morphological type;

2) Of star (other exciting stars, if they exist, are shown in parentheses);

3) angular and linear diameter;

4) expansion velocity;

5) mass of the shell;

6) remarks.

To estimate the linear size for stars in associations, we utilized the distance given by Humphreys (1978); for the rest, we assumed $M_V = -5$ to -6 and used the data of Neckel and Clare (1980) to take account of absorption, or we used the kinematic distance.

The nebulae associated with Of stars may be divided among the same four types discussed in Section 14. As in the case of the WR stars, most of the nebulae around Of stars are of type R_a — amorphous H II regions, which are listed in Lozinskaya (1982). Nebulae of this type are of a size consistent with the Strömgren sphere around O5–O9 stars, namely 60–200 pc in a medium with a uniform density of $n_0 = 1 \, cm^{-3}$. Stellar wind in these nebulae can show up as faint, high-velocity wings on lines in the optical spectrum. It is possible that deep plates taken in [O II], [O III], [N II], and [S II] with high angular resolution will reveal delicate filamentary shells near stars even in those objects which thus far have seemed only to be amorphous H II regions.

Table 26. Ring Nebulae around Of Stars

Nebula Morphological type	Of Star (other exciting stars)	Diameter (angular and linear)	v_{exp} km/sec	Mass $M_\odot$	Remarks
1	2	3	4	5	6
Sh 310 R_s	HD 57060 – UW CMa (HD 57061)	4° 70–100 pc			
IC 1396, Sh 131 R_s	HD 206267 (HD 204827, 205196, 205794, 205948, 206773)	3° 40–45 pc			
IC 1805, Sh 190 R_s	HD 15570, 15558, 15629 (cluster)	1.5° 60 pc			
Sh 264, SG 64 R_s	HD 36861 — λ Ori (HD 36881, 36895, 37232)	3.5° 30 pc	8		Expansion velocity of outer H I shell
RCW 113–116 R_s	HD 151804, 152248, 152408	3–4° 100–130 pc			
IC 1848, Sh 199 R_s	HD 17505 (HD 237015, 17505, 17520, and others)	2° 75 pc			

1	2	3	4	5	6
Rosette Nebula $R_s + W$	HD 46056 (NGC 2244)	60' 30 pc	20		Central bubble 20 pc across
Sh 134, SG 248 R_s	HD 210839 — λ Cep	3–4° 30–50 pc			Individual filaments, knots, dust clouds
Ring Nebula R_s	HD 153919	2° 30–35 pc			Dust shell
Sh 119, SG 240 R_s / W	HD 203064 — 68 Cyg	2–1.5° 34 pc	15	100–500	
NGC 7653 $W + R_s$	BD +60°2522	2.5' 2 pc	20	4–5	The shell NGC 7635 is embedded in the H II region Sh 162
NGC 6164–5 $E + W + R_s$	HD 148937	6' 2.5 pc	32	2	NGC 6164–5 is surrounded by two extensive shells; see text
Sh 22, SG 123 R_s / W	HD 162978	45 – 50' 10–18 pc			Surrounded by an outer, fainter shell
Anonymous	HD 313864		20		
Sh 60, SG 154 R_s	HD 172175	10 – 15'	16		Sct OB2 association
Sh 108, SG 196 R_s	HD 193514	8 – 10' ~5 pc			Cyg OB1 association

The second type of nebula, type R_s, is a ring-like H II region, and these are the most numerous in Table 26. In most nebulae of this type, the Of star is not the sole source of ionizing radiation, but it is the dominant source of stellar wind. Sometimes a ring nebula surrounds a compact, young cluster containing an Of star.

The Rosette Nebula surrounding the cluster NGC 2244, which contains the Of star HD 46056, is a good example of this phenomenon. The nebula is a bright, shell-like ionization-bound H II region with a mean electron density of $5-15\,\mathrm{cm}^{-3}$ (for various shell models); see Celnik (1985). Here the central cavity resulted from an intense period of star formation at the center of a dense cloud, and what remained of the cloud was then subjected to ionizing radiation and wind from the hot stars of the compact cluster. Several concentric layers can be discerned in the Rosette; the bright outermost shell shows no high-velocity, large-scale motions, and the inner cavity is surrounded by a shell of ionized gas expanding at 20 km/sec (Smith, 1973; Fountain et al., 1979). The strongest sources of wind in the cluster are two stars, of types O4 V and O5 V ($\dot{M} = (1-3) \times 10^{-6}\ M_\odot/\mathrm{yr}$, $v_w = 3100\,\mathrm{km/sec}$). The overall wind from OB stars in the cluster has $\dot{M} \approx 4 \times 10^{-6}\ M_\odot/\mathrm{yr}$ and $L_w = 10^{37}$ erg/sec (Dorland et al., 1986), and clearly the outflow from the O8f star HD 46054 ($\dot{M} = 2 \times 10^{-8}\ M_\odot/\mathrm{yr}$, $v_w = 1700\,\mathrm{km/sec}$) does not play a particularly important part in the formation of the cavity.

The nebula Sh 264, which surrounds the O8f star λ Ori, is an older type-R_s object. Three more O stars are located within the central cavity: λ Ori is a member of a cluster belonging to the Ori OB1 association, but it is associated with a ring nebula that is isolated from the extended shell that includes Barnard's Loop (see Section 18). At a distance $r \approx 500\,\mathrm{pc}$, the size of the nebula is $64 \times 48\,\mathrm{pc}$, with a density of $n_e \approx 3\,\mathrm{cm}^{-3}$; the radius of the Strömgren sphere around an O8 III star in a medium of that density is $R_{St} \approx 30\,\mathrm{pc}$. The ring nebula is surrounded by a shell of neutral hydrogen and dust clouds, and it is evidently interacting with the nearest molecular cloud (see Heiles et al. (1981) and references therein). The H I shell is expanding at approximately 8 km/sec. Motions within the ionized shell are subsonic; the expansion velocity is at most 5 km/sec. Clearly, then, the nebula around λ Ori can be considered a classic evolved shell-like H II region. The line ratios $I_{[\mathrm{N\,II}]}/I_{\mathrm{H}\alpha} = 0.2$ and $I_{[\mathrm{O\,III}]}/I_{\mathrm{H}\beta} = 0.02$ (Reynolds and Ogden, 1982) are also typical of H II regions.

At the same time, stellar wind cannot be ruled out as the formative agent for the expanding shell. The wind power required to produce an H I shell ~70 pc across, with an expansion velocity of ~8 km/sec in a medium having an ambient density of $n_0 \approx 1\,\mathrm{cm}^{-3}$ is $L_w \approx (2-3) \times 10^{35}$ erg/sec, and it must act for $t \approx 2 \times 10^6$ yr, which is consistent with the mass-loss rate and lifetime of an Of star. In addition, surrounding the ring-like H II region, there is a ring of molecular clouds, expanding at about 14 km/sec, as if a cavity has been swept out of a planar molecular layer (Maddalena and Morris, 1987).

Another two of the nebulae in Table 26, Sh 119 and Sh 22, are likely to be type R_s around individual stars, although it is possible that their ring-like morphology is partly due to stellar wind effects.

The nebula Sh 119 around 68 Cygni is a bright, elongated shell, possibly with double structure (Fig. 88). The author has participated in spectroscopic and interferometric observations of the nebula, radio observations at 105 MHz, and UBV photometry of the central star. The results may be summarized as follows (Esipov et al., 1982). Optically, the nebula shows a pure recombination spectrum; the radio spectrum is thermal, with $\alpha = -0.1$. There are no large-scale systematic motions discernable in the shell, and random motions fall in the velocity range $\Delta v = 20-30$ km/sec. The mean temperature of the nebula is $T_e = 12000 \pm 5000$ K, the density is $n_e = 40-150$ cm^{-3}, and the mass comes to $100-150\,M_\odot$. The star 68 Cyg is the source of ionization, and the wind from the central star may also play a certain role in forming the shell, which has a radius of 17 pc, and is expanding at no more than 15 km/sec in a medium with a density of $n_0 = 0.7$ cm^{-3}. The requisite wind power is $L_w \sim 3 \times 10^{35}$ erg/sec, which is a small fraction of the observed mass loss from 68 Cyg (see Table 18), and the age ($\sim 5 \times 10^5$ yr) is in accord with the lifetime of an Of star. Recent more detailed observations of the nebula Sh 119 in the radio, optical, and infrared have been obtained by Wisotzki and Wendker (1989), who give $T_e = 7500 \pm 1000$ K, $n_e \sim 2$ cm^{-3}, and indicate that the nebula represents a shell-like H II region. Due to the high peculiar velocity of 68 Cyg, the wind front is located about 1 pc from the star, and a bow shock has been created in the visible and IR instead of a classic wind-blown bubble.

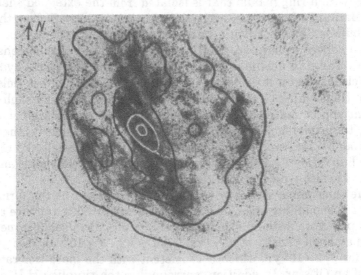

Fig. 88. Photograph of Sh 119, the nebula surrounding the Of star 68 Cyg; the image is overlaid with 1400-MHz radio isophotes (Wendker, 1971).

By analogy with WR stars — the nuclei of ring nebulae — we have proposed that the runaway star 68 Cyg may be a member of a close binary system with a relativistic companion, and to search for weak, periodic variability, we have carried out UBV photometry. Slight variations in brightness were in fact detected, with a period $P = 3.^d3$; this variability was confirmed by Alduseva et al. (1982), but at $P = 5^d$. However, subsequent more detailed measurements have failed to confirm either period, even though variability was detected at $P = 0.^d51$, a result that is difficult to interpret in terms of the spectroscopic duplicity of 68 Cyg (see Khyanni and Pel't (1986) and Zeinalov et al. (1987)).

The ring nebula Sh 22 (RCW 144; object number 123 in the catalog of Gaze and Shain (1955)), which is associated with the star HD 162978 (O8f II–III), is a twin of Sh 119 (Lozinskaya et al., 1983). Here we again see a double shell: a central, bright ring nebula 45–50′ in size and 15–20′ thick enveloped in thinner, diffuse filaments 65–80′ from the central star. The photometric distance to the star is 1.4 kpc, and the mean radial velocity of the nebula, $v_{LSR} = +7$ km/sec, yields a kinematic distance of 2.7 ± 1.6 kpc. Given the uncertainty of the distance, the radius of the inner bright shell is 9–18 pc, and that of the outer, 14–30 pc. Particularly noteworthy on the Palomar Sky Survey plates is the ring morphology of a large region outside the boundaries of Sh 22, with an extensive gas and dust complex 4–5° in size that brightens at the rim with the star HD 162978, and with the nebula Sh 22 at the center of symmetry. To prove that there is a genuine relationship between this outer shell plus Of star and Sh 22, it is necessary to investigate the velocity field of the whole complex. If the gas

Fig. 89. Photograph of NGC 7635, the nebula surrounding the star BD +60°2522 (Barlow et al., 1976).

and dust shell is at the same distance, its radius is 45–85 pc. It may conceivably have been created by ionizing radiation and the wind from stars in the Sgr OB1 association, which is 1.4 kpc away.

It is not possible to say anything more definitive about the mechanism whereby Sh 22 and its surrounding shells were produced without studying their kinematics. The wind from HD 162978 could form a 20-pc shell in a medium with a typical density $n_0 \sim 1 \, cm^{-3}$ in $(1-5) \times 10^5$ yr, if its rate of mass outflow is close to the average for Of stars.

Two of the nebulae in Table 26 are without question members of the class comprising bubbles blown by a strong stellar wind, the evidence being their characteristic fine filamentary morphology.

The nebula NGC 7635 around BD +60°2522 is a filamentary spherical shell approximately $\sim 150''$ across within an extended (close to $30'$) amorphous H II region, Sh 162 (Fig. 89). Distance estimates are discordant, ranging from 500 pc to 3.5 kpc; the most reliable seem to be the photometric distance to the star, $r_{phot} = 3.5$ kpc (Doroshenko, 1972; Israel et al., 1973) and the kinematic distance to the nebula, $r_{kin} = 3.3$ kpc (Deharveng-Baudel, 1973). At a distance of 1 kpc, the mass and density of the nebula implied by its optical brightness (allowing for a filling factor) are $n_e \sim 400-500 \, cm^{-3}$ and $M = 4-5 \, M_\odot$ (see Lynds and O'Neil (1983), Lomovskii and Klement'eva (1986), and references therein). The nebula has a thermal radio spectrum; its flux density corresponds to a mean electron density $n_e \sim 100 \, cm^{-3}$ and mass $M = 13 \, M_\odot$ (Israel et al., 1973). Near the central star there are two or three bright, compact knots about 0.03 pc in size, with a density of $(6-9) \times 10^3 \, cm^{-3}$ (Glushkov and Karyagina, 1972). The expansion velocity of the filamentary shell is approximately $15-25$ km/sec (Deharveng-Baudel, 1973); Lynds and O'Neil (1983).

Internal motions in the nebula have also been investigated by Maucherat and Vuillemin (1973), and by Pismis et al. (1983). The H II regions Sh 162 and NGC 7635 have been shown to represent a combined complex at a distance of about 3.4 kpc, with an expansion velocity of only 4 km/sec for NGC 7635.

The thin shell shares a common origin with the extended diffuse region Sh 162; the total mass of gas in the diffuse nebula is $4 \times 10^3 \, M_\odot$, and its mean density is $10-50 \, cm^{-3}$ (Israel et al., 1973). If we accept that value to be the density of the ambient gas swept up by stellar wind, then we find from (13.5) that the required wind energy input rate is $L_w = (1-2) \times 10^{35}$ erg/sec, and the age of the shell is approximately 2×10^4 yr. The observed rate of mass loss from the star BD +60°2522 (see Table 18) affords the necessary mechanical input power.

[O III] and Hβ images and spectrophotometry (see Dufour (1989) and papers cited) show that the ionization level within the wind-blown bubble NGC 7635 is higher than that of the nearby diffuse H II nebula, except for the cones shadowed by the brightest compact knots. The [O III]/Hβ intensity ratio in the bubble is as high as about 7 (when surrounding diffuse

H II emission is subtracted), convincingly indicating the presence of emission from shock-heated gas.

NGC 6164–5 is one of the most striking nebulae in the heavens, with four concentric shells surrounding the Of star HD 148937 that simultaneously provide an example of type-E stellar ejecta, a type-W wind-blown bubble, and an H II region with ring morphology (type R_s). A star is located at the center of symmetry of the bright s-shaped nebula NGC 6164–5, and the nebula, in turn, is surrounded by a network of thin, faint filaments embedded in a spherically symmetric H II region; the latter is bounded by a thin dust ring (Fig. 90). The most reliable estimate of the distance to the system is 1.4 kpc, yielding radii for the corresponding shells of 1.3 pc (NGC 6164–5), 5.7 pc (the faint filamentary shell), approximately 25 pc (the ring-like H II region), and about 30 pc (the dust shell). Beyond any doubt, the central location of the star within a system of concentric shells implies that they are all intimately related.

Studies of the kinematics of NGC 6164–5 have demonstrated that the bright central nebula was ejected by the star: two bright outer clumps are collinear, and are moving in opposite directions at the same speed, 32 km/sec; two inner faint filaments are moving more slowly, and were probably expelled later. Catchpole and Feast (1970) and Pismis (1974) have found the age of the nebula to be $(1-4)\times10^3$ yr, with an initial velocity for the ejecta of 180 km/sec.

The particular filamentary morphology of the second shell is typical of a bubble blown by stellar wind. Its age ($t = 5\times10^4 - 10^5$ yr) can be determined using Eqs. (13.5) if one assumes the ambient density of the surrounding H II region to be $n_0 \approx 10$ cm^{-3}, and takes the parameters of the wind to be $\dot{M} = 10^{-7}$ $M_\odot$/yr and $v_w = 2600$ km/sec (Bruweiler et al., 1981).

The structure of the enveloping diffuse nebula is that of a thick shell with strong [S II] emission, and in all likelihood it is a classic fully evolved H II region. It is also conceivable, however, that the ring morphology of the outer H II region is also a consequence of wind from the central star; in any event, the high brightness in the [S II] line is typical of the emission from gas behind a shock front.

More recent detailed studies of the kinematics and spectrum of NGC 6164–5 have now been carried out (Dufour et al., 1988; Leitherer and Chavarria, 1987). They have confirmed previous radial velocity measurements on the expanding clumps of ejected material, $v = 30-40$ km/sec, but by taking projection in the plane of the sky into consideration, Leitherer and Chavarria have ended up with a higher true expansion velocity, $v_{exp} = 350$ km/sec (the expulsion axis is inclined 5° to the plane of the sky). The presence of bipolar ejecta seems likely to be connected with rapid rotation of the star. The ejecta are estimated to have a mass $M_{ej} \sim 2\,M_\odot$; the mean density $n_e \sim 10^3$ cm^{-3} (up to 10^4 cm^{-3} in dense knots) and the temperature $T_e = 6400-8800$ K are typical of a "normal" H II region.

The ejecta have also been found to have a highly anomalous chemical composition. In the brighter knots of the ejected shell of NGC 6164–5, N/H

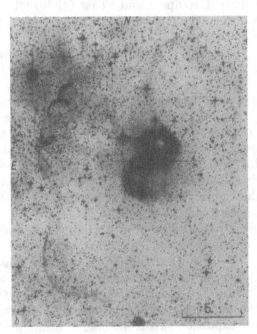

Fig. 90. The system of concentric ring nebulae around the Of star HD 148937. **a**) Overall view of the region in red light (Westerlund, 1960); **b**) the bright central shell NGC 6164–5, and the thin filamentary bubble swept out by stellar wind (Bruchweiler et al., 1981).

and He/H are enhanced and O/H is significantly depleted by a factor of about 3 as compared with "normal" H II regions. The diffuse faint inner filaments have a normal N/H abundance, while He/H and O/H are similar to the brightest knots. These data suggest that the bright s-shaped nebula consists of material ejected by the Of star at an advanced stage of evolution. Leitherer and Chavarria have observed extraordinarily high He I λ 6678 line intensity in two regions on the eastern edge of the second wind-driven shell, suggesting that it may be He-enriched by the wind material.

If we follow up on the conclusion reached in the preceding section that the ejecta of a shell facilitate the formation of a wispy filamentary cavity filled with hot wind, we can also speculate that the formation of the thin second shell from the center was stimulated by the virtually instantaneous expulsion of matter, as was the case with NGC 6164–5 (but preceding the latter). Assuming that the ejecta in both shells have essentially the same mass and are expanding with constant shell thickness ($E_m \propto R^{-4}$), and estimating the emission measure of NGC 6164–5 from the Hα intensity, $I_{Hα} = (3.5 - 18) \times 10^{-10}$ erg·cm^{-2}·sec (Johnson, 1972), we obtain $E_m \approx 130$ cm^{-6}·pc corresponding to a radius of 5.7 pc, which is consistent with the observed brightness of the second thin filamentary nebula. The preservation of the thickness of the expanding ejected shell is a consequence of hot wind pressure. If stellar wind sweeps up previously ejected gas, as in NGC 6164–5, then the overall similarity and identical orientation of the central and filamentary shells are understandable.

However, this set of circumstances may also result from there being the same asymmetry in the instantaneous and sustained ejection of matter, which is governed by the properties of the central star — its rotation, for example. According to Avedisova (1977), the ring-like shells may have acquired their ellipsoidal shape by virtue of different rates of mass loss at the poles and equator of a rapidly rotating star, through stellar wind or the expulsion of matter. It is clear in any event that HD 148937 is an evolved star that formed a large H II region and thin filamentary wind-blown bubble, and then underwent ejection (perhaps twice) of about $2 M_\odot$ of N-rich and probably O-poor processed stellar material. The outer H II region and the dust ring may have been formed either by the ionization front alone or by the ionization front plus stellar wind. The confirmation of that scenario would require, first and foremost, an investigation of the kinematics of the entire complex.

It is, in addition, not entirely clear what direct relationship might exist between the Of star with its system of shells and the compact association Ara OB I, the cluster NGC 6193, the extensive H I shell swept up by the wind from these stellar aggregates, and molecular CO clouds located in the vicinity (see Arnal et al. (1987), which also contains further references to the literature).

Miranda and Rosado (1987) have suggested three additional possible ring nebulae tentatively associated with Of stars by Lozinskaya (1982):

Sh 60 (SG 154) around HD 172175, Sh 108 (SG 196) around HD 193514 and an anonymous nebula around HD 313864. They obtained Hα images and Fabry–Perot interferograms of the three nebulae in order to check the association. In two of the nebulae, Sh 60 and Sh 108, they noted a limb brightening effect that is consistent with the idea that the Of stars create the shells. Preliminary results of these observations indicate Sh 108 to be at the same distance as the Of star, but there is no systematic expansion velocity of the shell. On the other hand, they estimated expansion velocities of about 16–20 km/sec for the two other nebulae. It will be necessary to obtain more interferograms of the nebulae in order to evaluate the kinematic distances and expansion velocities of the shells.

The new class of ring nebulae around Of stars requires every bit as much attention from observers as WR stars with shells, as only a comparative analysis of the two classes will elucidate the mechanism by which they are produced. Based upon data already in existence, we can draw certain conclusions. The Of stars tend to be connected with the same four kinds of nebulae as Wolf–Rayet stars, and the percentages of the two classes associated with ring and diffuse nebulae are approximately the same. The Of stars lose matter not only as stellar wind, but "instantaneously" as well, in the form of a slow-moving shell. The physical conditions within ring nebulae related to the two types of stars also tend to be similar, both in wind-swept cavities and within stellar ejecta. These stars can also shed matter more than once at an early evolutionary stage; the ejecta have a typical mass of $2 M_\odot$. The shell around the star HD 148937 provides convincing evidence that the effects of both processes — shell ejection and stellar wind — are intertwined. At this point, the problems that arise promise to yield the next approximation to be applied to the study of the interaction between the stellar and gaseous components of galaxies, including the effects of a strong stellar wind on an ejected shell, and the expansion of ejected material into a cavity produced by stellar wind (the latter may well be important to studies of young supernova remnants).

Among the Of stars with ring nebulae, there is one binary with a compact relativistic companion, namely the well-known massive x-ray binary HD 153919, an Of star paired with a neutron star. Two more objects in Table 26 show some hints of a supernova explosion having taken place in a binary system: the star HD 57060 in the nebula Sh 310, at $z = -150$ pc, is rather high above the galactic plane for a Population I object, and the star HD 152408 in the RCW 113–116 complex is noteworthy for its high peculiar velocity, $v_{pec} = -116$ km/sec. There are also massive pairs among the central stars, such as λ Cep and BD +60°2522 (Doroshenko, 1972; Cherepashchuk and Aslanov, 1984).

Searches for low-level periodic variability in HD 57060, HD 152408, and other Of stars with shells comprise an extremely interesting observational problem, since many of the links in the chain of events *supernova explosion in a binary system* → *ejection of a common shell* → *stellar wind*, which we considered in the previous section, are still uncertain.

In concluding this inventory of interesting problems, we should be remiss not to mention the obvious one — a search for ring nebulae around Of stars in nearby galaxies, the most promising of which is the Large Magellanic Cloud. The first step in this direction has already been taken: Walborn (1982) has detected the bright, doubled lines of [O II] 3727–3729 Å, He I 5876 Å, and [N II] 6584 Å in the spectra of four Of stars in the LMC. The line splitting corresponds to a velocity $\sim 30 - 50$ km/sec, which may be associated with a compact, nitrogen-enriched expanding shell like that of NGC 6164–5.

Thus, having selected the WR and Of stars as the most powerful sources of stellar wind in the hope that they might provide some insight into the "pure" interaction of stellar wind with interstellar gas, the observations have convinced us that in all likelihood, no such pure interaction exists in nature. The idealized scenario considered in Section 13 for the formation of a filamentary shell filled with hot wind was developed against the background of the "classical" evolution of an H II region. A strong, short-lived stellar wind at the WR stage foretells a sustained ($t \sim 2 \times 10^6$ yr) episode of lower-level wind from a main-sequence O star, which may then terminate with the "instantaneous" expulsion of an outer shell. In any event, the shells observed around WR stars seem most likely to have resulted from the effects of a strong wind on gas expelled by the stars in an earlier stage of development; this picture, however, remains somewhat uncertain.

The great variety observed among the nebulae associated with sources of strong stellar wind reflects different modes of interaction of those stars with the interstellar medium; the dominant influence for a particular source at any given time turns out to be either radiation, stellar wind, or shell ejection. The theory of that interaction has also developed by fits and starts — for one or two processes at a time — but not for all of them as a coherent whole. We are now capable of building a theoretical foundation to underlie the structure, kinematics, and physical conditions in any of the types of nebulae enumerated above that are observed around strong sources of stellar wind. There are still a number of questions, however, to which there are as yet no definitive answers. Let us summarize once again the principal difficulties that one encounters in interpreting the observations of ring nebulae around WR and Of stars, and some possible ways to overcome them.

Why is it — contrary to theoretical predictions — that not all stars with strong mass outflows are surrounded by filamentary ring nebulae swept out by stellar wind? According to the statistics, no more than 45% of the WR and Of star are surrounded by ring-like H II regions, and there are even fewer wind-blown bubbles and ejected shells among them. In Section 15, we discussed several possible explanations for the apparent fact that wind-swept shells are either short-lived, or never form around certain WR and Of stars. Firstly, if the wind sweeps up ejecta rather than interstellar gas, the lifetime will be governed by the reduction in the emission measure of the expanding ejecta, and will turn out to be shorter than the outflow stage. This would seem to be a reasonable explanation, in light of the fact that wind and radiation from the main-sequence progenitor most likely "evacuate" the interstellar gas from the immediate vicinity of the WR or Of star. Massive, extensive H II and H I shells, probably swept out by the progenitor or the OB association to which the WR/Of belongs, have actually been observed around many W- and E-type shells (see Lozinskaya (1988c)).

Secondly, the short lifetime of wind-swept shells may be associated with the rapid dissipation of the energy of the hot wind. The new energy dissipation mechanism examined by Dorland et al. (1986) and Dorland and Montmerle (1987) had essentially a minor effect on the bubble lifetime. But a hot plasma comprised of shocked wind may simply flow right through the gaps in a cold, swept-up shell. Here the evidence would seem to consist of the outer thin shells and high-speed motions that have been detected about several ring nebulae, and possibly the fact that the x-ray emission from NGC 6888 has been found to be an order of magnitude below the predicted level.

It may also be that in a highly inhomogeneous medium, "cold," unshocked wind might penetrate through tenuous passageways between dense clumps without giving rise to a distinct ring-like structure. In that case, incidentally, the expulsion of a "slow" shell could in fact trigger a redistribution of the circumstellar gas, facilitating the formation of a wind-swept bubble.

The second fact that was somewhat difficult to interpret was that the observed energy conversion factor ($\varepsilon = M_s v_s^2 / M v_w^2 t \sim 0.01 - 0.05$) and momentum conversion factor ($\pi = M_s v_s / M v_w t \sim 0.5$) were found by Treffers and Chu (1982) and Chu (1982) to deviate from those expected from the classical theory of an energy-driven bubble ($\varepsilon = 0.2$, $\pi > 1$); see Section 13. This may signify that most ring nebulae are already in a momentum-conserving stage of expansion, for which the theory gives $\varepsilon \ll 0.2$, $\pi = 1$. The reason for the rapid transition to momentum-conserving motion may either be the rapid dissipation of energy from the hot wind alluded to above. Another possible explanation of the discrepancy may be a large mass of neutral gas in the exterior layer of the shell that remains unaccounted for.

On the other hand, since we have concluded that some wind-blown bubbles actually do contain shells of stellar ejecta, the discrepancy between the predicted and observed momentum and energy conversion factors is not meaningful. Indeed, the shell kinetic energy may actually be determined by stellar ejecta instead of the total energy input due to stellar wind. Even WR and Of ring nebulae with no confirmed stellar ejecta may contain an unknown amount of stellar material, and their energetics and evolution cannot be described by the classical model of wind-blown bubbles.

There is one more aspect of ring nebulae that requires explanation. We have demonstrated that a powerful stellar wind affects interstellar gas in much the same way as a supernova explosion. The resemblance stems from the fact that a strong spherical shock wave, which propagates in both cases into the circumstellar gas, conveys to the latter the kinetic energy of the explosion or the stellar wind.

There is, however, one fundamental distinction: the thermal nature of the radio emission from the shells produced by the wind, as opposed to the synchrotron radiation emitted by old supernova remnants, still awaits explanation. In fact, if we were to correctly link the radio synchrotron emission of old supernova remnants with an enhanced interstellar magnetic field containing relativistic particles that is frozen into the compressed gas behind the radiative shock front (see Section 10), then the same mechanism ought to be operating in the shells produced by stellar wind, particularly in NGC 6888. The expansion velocity of the latter and the degree of compression of the shocked gas, as determined by the thickness and density of the filamentary shell are identical to those in the old remnant IC 443, but NGC 6888 has a purely thermal radio spectrum. Thus far, two reasonable explanations have suggested themselves. The first is that the radio synchrotron emission from old supernova remnants is due to relativistic electrons injected by a pulsar or accelerated at the shock front during the initial development of the remnant, instead of being produced by the interstellar field and particles. The second is that rather than sweeping up interstellar gas, the wind from WR and Of stars compresses previously ejected stellar gas, in which case the frozen-in magnetic field will be much weaker than the mean Galactic field (Lozinskaya, 1980c).

Let us now summarize. Despite the uncertainties emphasized here, the observations of ring nebula surrounding stars with copious mass loss are, on the whole, in agreement with the theory of the interaction of radiation and stellar wind with the interstellar medium. Furthermore, it is precisely the observations of shells around WR and Of stars that provide the key to an understanding of how the massive progenitors of supernovae "prepare" the surrounding interstellar medium prior to the final explosion.

We showed in Section 1 that the optical, ultraviolet, x-ray, infrared, and radio emission from supernovae are completely determined by the interaction of the supernova shell with the wind of the presupernova. The peculiarity of SN 1987A (Section 2) is a direct consequence of the dense matter in the slow wind of the red supergiant being swept out of the stellar environment by the high-velocity wind of the blue supergiant that eventually exploded. If our concern is with those properties of supernova remnants attributable to progenitor effects on the ambient medium, then the following considerations must first be borne in mind. In Cas A, we directly observe the interaction of the two-component supernova ejecta with the two-component matter lost by the progenitor — probably a Wolf–Rayet star — in the form of wind and a slow shell. The toroidal geometry of the oxygen-rich remnants may result not from the toroidal expulsion of matter in the supernova explosion, but instead, and equally likely, from asymmetric outflow (enhanced at the equator, for example) from the supernova progenitor.

The anomalously low kinetic energy and the lack of a fast-moving outer shell around the Crab Nebula may be a result of the wind from the progenitor having swept out the surrounding gas (this hypothesis has recently been confirmed by IRAS data; see Section 4). The existence of plerions and composite supernova remnants probably reflects not so much a difference between supernova explosions as a difference in the density (and structure?) of the circumstellar medium, which was formed under the direct influence of the radiation and stellar wind of the progenitor (or the stars of its parent association).

The thin outer radio rim of Tycho's remnant (SN Ia) can be accounted for by the interaction of the supernova with a planetary nebula previously expelled by the progenitor.

It would seem that the barrel-shaped structure of both young and old supernova remnants finds its most natural explanation in a regular asymmetry of the circumstellar gas and its frozen-in magnetic field, which is also produced under the influence of the radiation and wind of the progenitor.

A theoretical analysis of the interaction of stellar wind with interstellar gas essentially turns out to be even more complex than the theory of supernovae interacting with the interstellar medium. The complications get even more serious when the following are taken into account: the velocity of the wind and the rate of mass loss vary as the star evolves; a low-velocity shell can be thrown off; there is ionizing radiation and its accompanying photoionization evaporation of circumstellar cloudlets, which are then evacuated to the periphery of the H II region.

It would therefore be useful to summarize certain purely empirical conclusions about the structure of the interstellar medium around massive supernova precursors, relying upon the observations of ring nebulae.

The interstellar medium around massive presupernovae — Wolf–Rayet and Of stars — is characterized by regular, quasisymmetric shell structure. Shell-like formations are observed around something like 40–45% of all stars of these two types. Radiation and mass loss from high-mass stars produce different kinds of ring nebulae: massive, extended H I and H II shells (up to 100 pc in size, with a mass of order $10^3 M_\odot$ and $v_{exp} \lesssim 10$ km/sec); thin filamentary nebulae that have been swept up by a strong stellar wind (up to 10–30 pc across, with a mass of $10 - 10^2 M_\odot$ and $v_{exp} \sim 30 - 100$ km/sec); and stellar ejecta (patchy morphology, size $\sim 1-5$ pc, mass $\sim 2-3 M_\odot$, $v_{ej} \sim 30 - 300$ km/sec).

The interstellar medium near massive hot stars is typically a multi-shell, layered environment: within massive, extended shells of H I and H II, one sees small-scale ejecta and wind-swept cavities. Both wind-swept and ejected shells often occur within "supershells" that have been swept out by stars belonging to the parent association of a WR or Of (see Chapter 4). In one instance, what is seen is an object with four concentric shells: an Of star with ejecta surrounded by a filamentary, swept-up shell, and two extensive outer shells — a ring-like H II region, and a layer of dust enveloping it. What this structure suggests is that stellar wind and the ejection of a slow shell can characterize the same star, and the two processes need not be well-separated in time.

Both wind-swept and ejected shells are often elongated. The elongated shape may be due either to asymmetric mass loss or to strong regular interstellar magnetic fields hampering the shell's expansion across magnetic field lines. The hot wind in the energy-driven bubble is isobaric, and the bubble has to be symmetrized even in the case of asymmetric wind. In a momentum-driven bubble, the dynamical pressure of the cold wind may produce an asymmetric shell: either prolate via bipolar mass loss, or oblate via equatorial mass loss (prolate if seen edge-on). In at least one case there is clear-cut observational proof that a wind-swept shell is filled with hot plasma — shocked stellar wind.

The kinematic age of ejected and wind-swept shells is $(1-5)\times10^4$ yr. The nebula surrounding a massive star can therefore dissipate by the time the star explodes. In light of this it is of interest to consider the environments of WO stars. The four Sanduleak WO stars are the most advanced massive stars, and probably have only a few thousand years to go before a supernova explosion. Indeed, our observations of WR 102 (Sand 4) (Dopita et al., 1990; Dopita and Lozinskaya, 1990) indicate the absence of He in the star's envelope, implying that it is absent from the interior as well; the star's position on the HR diagram also supports our interpretation that it is the CO core of a massive star. Similarly, Sand 1 (M(WR) $= 14\,M_\odot$ and M(O4) $= 50\,M_\odot$) seems to be the most highly evolved WR star known in a close binary (Moffat et al., 1985). WO stars thus provide a unique opportunity to investigate the environment of a supernova shortly before the explosion.

Since the WO stage is very brief (the lifetime of the CO core of a massive star is only about 1% of the core He burning duration), we cannot expect more than about 1% of the approximately 300 WR stars known in the Galaxy and Magellanic clouds to be WO stars like WR 102, and the four objects are a quite representative sample.

WO stars are characterized by extremely fast "superwind", with $v_w \approx 4500-7400$ km/sec (Barlow and Hummer, 1982; Torres et al., 1986; Dopita et al., 1990), and the ambient gas is influenced by the previous "normal" outflow of the progenitor WR star. Both facts point to new avenues of research on the interaction of stellar wind with interstellar gas.

We have shown that the four WO stars appear to be associated with optical and/or IR shell-like structures, although the short-lived WO superwind is not the dominant influence in producing the shell (Lozinskaya, 1990).

The nebula G 2.4+1.4 around WR 102 is the only object out of the four which has been shown to be a classical wind-blown bubble (see Section 14, Dopita et al. (1990), and Dopita and Lozinskaya(1990)). We have shown that the shell's dynamical age, $t \approx 10^5$ yr, is much greater than the duration of the WO superwind. We propose that the progenitor star forms a ring nebula before reaching the WO stage, and the subsequent superwind produces the characteristic thin filamentary "scalloped" structure as a result of the shell's impact, probably accompanied by leakage of the superwind through holes punched in the shell. IRAS observations of the region reveal a large-scale, incomplete ring, with the optical nebula located at its edge.

The environments of the second oxygen star in the Galaxy, WR 142 (a member of the young open cluster Be 87) and the WO4+O2V binary Sand 1 in the SMC (the brightest star in NGC 602c) look similar. Both WO stars and their clusters are located within large-scale shells or supershells that are easily seen at both 60 and 100 μm (and in the visual in the SMC).

Despite the small number of WO stars and the very preliminary nature of our examination, we have reached a number of conclusions. The four WO stars are located inside optical and/or IR shell-like structures. Only the nebula G 2.4+1.4 around WR 102 is a classical wind-blown bubble, but even here, the short-lived "superwind" of the oxygen sequence stage seems not to dominate the creation of the shell. As for the stars inside large shells or supershells, their genetic relationship is not entirely clear. On the one hand, the winds of WO stars and their progenitors are capable of creating the corresponding shells. In the tenuous interior of a supershell around a young cluster, the short-lived WO superwind acts as a supplementary energy source that reheats the superbubble.

On the other hand, the propagation of a supershell in a dense molecular cloud may trigger the formation of supermassive stars, prerequisite to reaching a final oxygen-sequence stage. Nevertheless, observations of these nebulae yield a splendid model of how a massive progenitor modifies the structure of the circumstellar medium out to tens of parsecs, which in turn determines the physics and evolution of the remnants of massive SN II and SN Ib.

Rapidly expanding ejected shells are well known to exist around luminous blue variables, the best examples being AG Car and η Car; P Cyg is surrounded by extended (about 60″ in size) radio emission (Baars and Wendker, 1987), although no optical counterpart has been found (Stahl, 1989).

Observations by Stahl and Wolf (1986) have shown that most (all, according to the authors!) early-type supergiants with emission lines are also associated with circumstellar shells. Based on observations of [N II], [S II], and [O I], they deduced the presence of low-excitation circumstellar shells with a density of $10^4 - 10^5 \, cm^{-3}$ and an expansion velocity of 15 – 20 km/sec. These shells may possibly have been thrown off by their central stars in an S Dor type of explosion.

A dozen fast-moving OB stars have been found to create bow shocks easily seen on IRAS images (van Buren and McCray, 1988); several optical nebulae around B stars are listed by Chu (1990).

It is important to stress that similar phenomena are also observed in planetary nebulae. Double-shell nebulae can result from multiple ejections, or can be produced by stellar wind from the central star (as we have seen, the occurrence of central stars with stellar wind is widespread). Recent investigations (Chu, 1987; Chu et al., 1987; Jewitt et al., 1986) have shown that more than half of all planetary nebulae have either double-shell morphology or a faint outer halo (a faint shell) surrounding the bright central nebula. We have also known now for several decades that bipolar flows are a common phenomenon in planetary nebulae.

What this means is that ejecta and stellar wind from the low-mass progenitors of SN Ia also produce multishell quasisymmetric structure in the circumstellar medium, and that circumstellar shells are often asymmetric.

In short, then, the circumstellar medium around a progenitor turns out to have been prepared ahead of time for the eventual supernova explosion, and a meaningful analysis of the nature and evolution of supernova remnants — of practically any age — requires that a multilayer, shell-structured ambient medium with strong density gradients be taken into consideration as a possibility.

Chapter 4. The Effects of Supernovae and Stellar Wind on Gas and Dust in the Galaxy

17. Physical state of the interstellar medium: supernovae as a regulating influence

We have seen above that supernova explosions and strong stellar wind can abruptly change the temperature, density, velocity, and ionization state of the surrounding gas. Is this an important factor in the physics of the interstellar medium on a galactic scale? We already have enough information in hand to answer that question. In fact, we know the size of the region within which gas is accelerated and heated by the shock waves produced by an outburst or by a strong outflow of matter; we know how long that hot plasma bubble, surrounded by a dense, expanding shell, will endure; and we know how often supernovae explode and what kinds of stars possess a strong wind.

Supernova remnants live long enough that it is possible for one supernova to go off in the immediate vicinity of another, and for the young remnant to merge with the old. Cox and Smith (1974) first drew attention to this fact, pointing out that the confluence of hot plasma bubbles will lead to the formation of hot tunnels occupying a substantial fraction of the Galactic disk. Rayleigh–Taylor instability arising in the collision region between the young and old remnants disrupts the shells and results in the merging of the remnants. Making allowance for the inhomogeneity of the interstellar medium lends further credence to Cox and Smith's conclusions, since the hot gas behind the shock front propagates rapidly in the medium between dense clouds. This idea was developed in detail by McKee and Ostriker (1977), who came to a radical conclusion: supernovae play a preeminent role in determining the structure of the interstellar medium, since hot, tenuous gas (merging remnants) occupies the bulk of the Galactic disk. The observational stimulus that launched these investigations was the measurement of the soft x-ray background, which showed hot ($T_e \approx 6 \times 10^5$ K) gas with a density of $n_e \approx 0.007$ cm^{-3} to be widely distributed (Burstein et al., 1976).

To estimate the probability of coalescence, let us take the volume of an old remnant to be $V \propto R^3 \propto t^{3\eta}$, where η is the exponent that governs the rate of expansion of the shell (see Section 9). $Q(R)$, the predicted number of young remnants of radius $R \le R(t)$ contained within that volume, and f_{SNR}, the fraction of the total volume occupied by supernova remnants, are (McKee and Ostriker, 1977)

$$Q(R) = (1+3\eta)^{-1} SVt, \quad f_{SNR} = 1 - e^{-Q}. \tag{17.1}$$

Here S is the supernova rate, $S_{-13} \equiv S/10^{-13} \text{ pc}^{-3}\text{yr}^{-1}$. For the oldest remnants, whose radius and age are given by Eqs. (9.7) and (9.8), one can take $\eta = 0$, since at that stage the shell is no longer expanding (for young remnants, $\eta = 0.3 - 0.4$). Taking $t = t_{max}$, $R = R_{max}$, and $\eta = 0$, we find

$$Q_{SNR} = 10^{-0.29} E_{51}^{1.28} S_{-13} n_0^{-0.14} \tilde{P}_{04}^{-1.30}, \quad \tilde{P}_{04} \equiv 10^{-4} P_0/k. \tag{17.2}$$

If the external pressure is low ($T_0 \le 10^4$ K, $n_0 \le 0.3 \text{ cm}^{-3}$), we have $Q \ge 1$, i.e., supernova remnants can merge before they have completely dissipated in the interstellar medium. Inasmuch as a hot, tenuous plasma at a temperature of $5 \times 10^6 - 10^7$ K cools slowly, over a time $t \gtrsim 5 \times 10^6$ yr, hot bubbles coalesce, and the new explosions reheat the merged cavities, since a shock wave will propagate rapidly through tenuous gas, reestablishing a high temperature there. Here McKee and Ostriker drew a fundamental inference: the standard two-phase model of the interstellar medium, consisting of cold clouds and "warm" intercloud gas ($T_e \approx 5 \times 10^3$ K, $n_e \approx 0.1 \text{ cm}^{-3}$) is untenable. In addition to those two components, there must be a third, "hot," component, and hot gas occupies a substantial fraction of the gaseous disk of the Galaxy. All three phases are in a state close to dynamic equilibrium, and transitions from one state to another take place quickly, in a characteristic time $t \lesssim 10^7$ yr.

This means that the state of the interstellar medium, which as we saw in Section 9 governs the evolution of supernova remnants, is itself regulated by the entire prior history of outbursts. The equilibrium state of the interstellar medium can therefore only be ascertained within the broader context of a complete recycling of matter and energy over a large fraction of the Galaxy (and ultimately, when spiral density waves enter the picture, throughout the Galaxy as a whole).

The following basic components must partake of this recycling, where for the time being we consider only the physical state of the medium (Ikeuchi et al., (1984):

1. Molecular clouds and giant molecular clouds. The former have a mass of $10^3 - 10^4 \, M_\odot$, the latter, $10^5 - 10^6 \, M_\odot$.
2. Diffuse clouds, either neutral or ionized, with a mass of $10 - 10^3 \, M_\odot$. The diffuse clouds consist of a dense core (cold component) and a tenuous ionized corona (warm component).
3. A warm intercloud medium with a temperature of $\sim 10^4$ K and a density close to 0.1 cm^{-3}. (This component is actually linked to the warm corona of the clouds. In general, there are two more in-

termediate phases that can be taken into account: warm clouds with a low degree of ionization and low temperature (in the range 10^2 to 10^4 K), and extremely low-density H II regions that have been ionized by OB stars outside those H II regions. For our present purposes, however, these phases are unimportant.)

4. Stars formed in associations and clusters on the outskirts of giant molecular clouds.
5. Supernovae and strong sources of stellar wind.
6. Hot tenuous gas inside the cavities produced by supernovae and stellar wind.

A multitude of interactions takes place among these components, and they are regulated by the following balanced processes.

I. The influx of energy carried by ultraviolet radiation from the stars and supernovae, as well as the kinetic energy of the stellar wind and supernova shells, ionize, heat and accelerate the interstellar gas in the clouds and the intercloud medium. This is how interstellar gas makes the transition from the "cold" phase to the "warm" or "hot" phase. The heated gas loses energy radiatively, and energy is also dissipated in inelastic collisions between clouds.

II. Supernovae and stellar wind sweep up gas into dense shells that cool radiatively, and the fragmentation of cold shells leads to the formation of diffuse clouds; this is how the transition from hot to warm to cold takes place. There is simultaneously evaporation of cold clouds in contact with hot gas, and fragmentation and destruction of clouds by the shock waves induced by supernova explosions.

III. The coalescence upon collision of diffuse clouds in random motion results in the formation of massive molecular clouds. The large-scale gravitational instability of an ensemble of molecular clouds spanning a large region of the galactic disk produces giant molecular clouds.

IV. At the same time, there is continuous destruction (erosion, socalled) of molecular complexes by ionizing radiation, stellar wind, and supernova explosions.

V. Gravitational instability on the outskirts of giant molecular clouds, induced by the shock waves associated with supernovae, stellar wind, and the expansion of H II regions, initiates a new round of star formation.

VI. There is an exchange of matter between the disk and corona of the Galaxy, including an "upwelling" of hot cavities or supersonic gas-dynamic outflow ("Galactic fountains"), if the hot gas is not immersed in a thick layer of cold, dense gas of the disk. On the other hand, radiative cooling also takes place, and coronal gas falls into the Galactic disk in the guise of cold clouds.

VII. Matter is exchanged between stars and the interstellar medium: the gas in molecular clouds collapses into stars, and stars lose matter as stellar wind, in the ejecta of "slow" shells, and in supernova explosions.

The typical time scale for such processes as the merging of clouds, the formation of giant molecular complexes, their erosion, and bursts of star formation seems to be at least $10^7 - 10^8$ yr. Gas exchange between the halo and disk and between stars and the interstellar medium takes place even more slowly, with $t \gtrsim 10^9$ yr (Ikeuchi et al., 1984). Fastest of all (in a time $10^6 - 10^7$ yr) is the interaction of supernova- and wind-induced shock waves, along with ionizing radiation from the stars, with the gas of the interstellar medium, thereby regulating the equilibrium of the cold, warm, and hot gas. Therefore, in a region that does not encompass any giant molecular clouds or rich foci of star formation, and over a time span of at most 10^8 yr, the equilibrium state of the diffuse clouds and intercloud medium is determined solely by supernovae, wind, and ionizing radiation. Furthermore, the most efficient processes are the sweeping up of gas into shells, the fragmentation of highly evolved cold shells into randomly moving diffuse clouds, the heating of gas by shock waves and cooling by radiative losses, the ionization and recombination of clouds and the intercloud medium, and the evaporation of cold clouds embedded in hot gas. The equilibrium state of the medium, as determined by these factors, was first investigated analytically by McKee and Ostriker (1977). The problem was numerically addressed over a wider range of parameters of the interstellar gas by Habe et al. (1981) and Ikeuchi et al. (1984). These authors devoted special attention to the transition to the equilibrium state.

In order to obtain quantitative estimates of the parameters of the hot, cold, and warm media, McKee and Ostriker formally tracked the evolution of a supernova remnant via a "$Q(R)$" approach; specifically, they utilized (17.1) and the corresponding equations in Section 9 to find the relationship between Q and the remnant parameters of interest. Taking $Q = 1$, which corresponds to coalescence, one can determine state of the gas inside the merging remnants. Thus, specifying the variation of the shell radius in general form via (9.11), (17.1) yields

$$Q = \frac{10^{-14.64}}{1+3\eta}\left(\frac{\alpha}{\eta}\right)^{6/5} S_{-13}\left(\frac{E_{51}}{n_h}\right)^{3/5} t^{11/5}. \tag{17.3}$$

As an example, we give here the relations characterizing the moment at which radiative cooling begins (McKee and Ostriker, 1977):

$$R_{cool} = 10^{2.21} E_{51}^{0.04} \alpha^{0.19} \beta^{0.04} \left(\frac{Q_{cool}}{S_{-13}} \right)^{0.23} \approx 180 \, \text{pc},$$

$$T_{cool} = 10^{5.47} E_{51}^{0.30} \alpha^{-0.45} \beta^{0.30} \left(\frac{Q_{cool}}{S_{-13}} \right)^{-0.15} \approx 4.4 \times 10^5 \, \text{K}, \qquad (17.4)$$

$$Q_{cool} = 10^{-2.09} E_{51}^{-0.09} \alpha^{118} \beta^{-2.19} S_{-13} \Sigma^{2.10} \approx 0.5 .$$

In these equations, we have kept the notation used in Section 9; β is a factor that allows for the enhancement of radiative cooling due to temperature and density variations within the cloud: $L = \beta n_g^2 \Lambda(T_g) V$. For an adiabatic solution $\beta = 2.3$, and $\beta = 10$ when thermal conductivity and cloud evaporation are taken into account. The numerical values correspond to the standard model, with $E_{51} = 1$, $n_0 = 0.01 \, \text{cm}^{-3}$, and $\Sigma = 48$.

The mean parameters of the hot component found by McKee and Ostriker, making allowance for the fact that different remnants will be at different evolutionary stages, i.e., integrating over the full applicable range,

$$\langle f \rangle = \int_0^1 f(Q) \, dQ,$$

are

$$\langle n_h \rangle = 4.6 \times 10^{-3} \, \text{cm}^{-3},$$

$$\tilde{P}_h = 7.9 \times 10^3 \, \text{cm}^{-3} \cdot \text{K},$$

$$\langle T_h \rangle = 4.6 \times 10^5 \, \text{K},$$

or integrating only over the restricted range $0.25 \leq Q \leq 0.75$, which yields "typical" values,

$$n_h(\text{typ}) = 3.5 \times 10^3 \, \text{cm}^{-3},$$

$$\tilde{P}_h(\text{typ}) = 3.6 \times 10^3 \, \text{cm}^{-3} \cdot \text{K},$$

$$T_h(\text{typ}) = 4.5 \times 10^5 \, \text{K}.$$

At that point, knowing the characteristics of the hot gas, one can find the parameters of the cold clouds and warm corona, bearing in mind that the three components of the gas are at the same pressure and specifying a size distribution for the clouds. McKee and Ostriker adopt $n(a) \propto a^{-4}$, where a is the radius of a cloud, but their results are virtually insensitive to the choice of spectrum. Low-mass clouds most often lack a nucleus, and are ionized throughout. They evaporate quickly, even in the relatively cold medium that prevails between successive shock waves, and they are easily swept up by a collision with the expanding supernova shell. Clouds without nuclei can therefore be left out of considerations of the overall balance of the interstellar medium, and the minimum size in the cloud spectrum, $a_{0\,min}$, is given by the requirement that a cloud have a core. The upper limit on cloud size, $a_{0\,max}$, comes from gravitational instability. In this idealized picture, then, the fraction of space occupied by cold (f_c) or warm (f_w) gas is

$$f_c = 4\pi a_{cr}^3 N_{cl} K\left(\frac{n_w}{n_c}\right)\ln\frac{a_{0\,max}}{a_{0\,min}},$$

(17.5)

$$f_w = \frac{4}{3}\pi a_{cr}^3 N_{cl},$$

and typical intercloud distances $\lambda^{-1} \equiv \int N(a_0)\pi a^2\, da$ are

$$\lambda_c = \left[2.2K\pi N_{cl}a_{cr}^2\left(\frac{n_w}{n_c}\right)^{2/3}\right]^{-1},$$

(17.6)

$$\lambda_w = \left[\pi N_{cl}a_{cr}^2\right]^{-1}.$$

Here a_{cr} is the critical radius of a corona, which is determined by the requirement that there be a dense core, n_c and n_w are the density of the cold core and warm corona, K is the core-to-corona mass ratio in the lowest-mass clouds, and N_{cl} is the number of clouds per unit volume (pc^{-3}).

We may write the mean density and the equal-pressure condition in the form

$$\bar{n} = n_c f_c + n_w f_w \sim n_c f_c,$$

(17.7)

$$n_c T_c \approx (1 + X_w)n_w T_w \approx 9 \times 10^3 \tilde{P}_4,$$

where $X_w = n_e/n_w$ is the degree of ionization of the corona. The condition on the ratio between the minimum cloud size $a_{0\,min}$ and the critical radius a_{cr} is then

$$a_{0\,min} = \left[\frac{KT_c}{(1 + X_w)T_w}\right]^{1/3} a_{cr}.$$

(17.8)

The maximum size of a gravitationally stable cloud, in the absence of a magnetic field, is given by Spitzer (1969):

$$a_{0\,max} \approx 2.3(T_c/(80\ \text{K}))\tilde{P}_{04}^{-1/2}\ \text{pc}.$$

In the presence of a magnetic field $H \approx 3 \times 10^{-6}$ G,

$$a_{0\,max} \approx 11.5\tilde{P}_{04}^{-5/6} \approx 10\ \text{pc}$$

(McKee and Ostriker, 1977). From (17.5) and (17.6), we have

$$\bar{n} = \frac{11.5Ka_{cr}^3 N_{cl}\tilde{P}_4}{(1+X_w)T_w}\ln\frac{a_{0\,max}}{a_{0\,min}}. \tag{17.9}$$

The values of X_w and a_{cr} are given by the ionization balance condition:

$$a_{cr} = \frac{3}{2}\big[\sigma_H n_w (1-X_w)\big]^{-1},$$

$$\alpha_{rec}n_e^2\frac{4}{3}\pi N_{cl}a_{cr}^3 = \alpha_{rec}n_e^2 f_w = \epsilon_{UV}, \tag{17.10}$$

where σ_H is the effective ionization cross section, α_{rec} is the hydrogen recombination coefficient, and ϵ_{UV} is the number of ionizing photons in $\text{cm}^{-3}\cdot\text{sec}^{-1}$.

Equations (17.7)–(17.10) yield $a_{0\,min}$, a_{cr}, X_w, and Q_{cool} for given K, T_c, T_w, $a_{0\,max}$, $\bar{n}$, and ϵ_{UV}, and for a "typical" value of the hot gas pressure of the coalescing remnants. Observations of diffuse Galactic clouds give $T_c \approx 80$ K, $T_w \approx 8000$ K, and $\bar{n} \approx 1\,\text{cm}^{-3}$ (Spitzer, 1981). Since clouds with no core break up rapidly, one can take $K \approx 2$. Now ionizing radiation from supernovae in the energy range 13.6–40 eV constitutes approximately 30% of the initial energy E_0 (Chevalier, 1974; see also Section 9); upon averaging over the Galactic disk, this yields $\epsilon_{UV} = 1.2 \times 10^{-15}$ photons·cm^{-3}·sec^{-1}. The background ionizing radiation from B stars is $\sim 0.7 \times 10^{-15}$ photons·cm^{-3}·sec^{-1}, and O stars outside H II regions contribute an average of 2×10^{-15} photons·cm^{-3}·sec^{-1} (Mezger, 1978). Taking the total incident flux of ionizing radiation to be $\epsilon_{UV} = (2-3) \times 10^{-15}$ photons·cm^{-3}·sec^{-1} to within a factor of two or three, and using the cloud temperature and mean density cited above, McKee and Ostriker (1977) have obtained the cloud parameters summarized in Table 27. It is possible that the 40–120 eV x-ray background from supernovae (approximately $10^{-16}S_{-13}E_{51}$ (Chevalier, 1974)) traverses the warm, ionized corona and produces an intermediate weakly ionized warm layer with a density of $n \approx 0.16\,\text{cm}^{-3}$, degree of ionization $x \approx 0.15$, and temperature $10^2 - 10^4$ K around the core that occupies a

Table 27. Parameters of Interstellar Clouds

Parameter	Cold core	Warm ionized corona
Hydrogen density, cm^{-3}	42	0.25
Degree of ionization	0.001	0.68
Adopted temperature, K	80	8000
Volume relative to Galactic disk	0.024	0.23
Intercloud distance, pc	88	12
Cloud radius, pc:		
maximum	10	10.8
mean	1.6	2.1
minimum	0.38	2.1
Surface density, 10^{19} cm^{-2}:		
maximum	173	0.22
mean	27	0.22
minimum	0.5	0.22

fractional volume $f \approx 0.1 - 0.15$, but these figures are quite uncertain. For completeness, mention should also be made of low-density H II regions ionized by O stars lying outside dense clouds.

Figure 91 illustrates the results obtained by McKee and Ostriker, giving a schematic cross section through a typical cloud with a core, corona, and intermediate layer that is immersed in hot gas, as well as a representation of the three-component structure of the interstellar medium on different scales.

The parameters of the three components of the interstellar gas obtained by assuming that they are regulated by supernova explosions are consistent with the observational data. We emphasize that here we are not concerned with individual H I and H II regions or hot intercloud gas, which reflect the local conditions of stellar interaction with the gaseous medium, but average conditions typical of the solar neighborhood of the Galaxy. This stipulation is important, if only because the supernova rate S_{-13} and mean gas density $\bar{n}$ depend on galactocentric distance. A complete compilation of observations of the various components of the gaseous medium of the Galaxy has been published by Myers (1978), summarizing survey results in seven radio and UV spectral lines. Figure 92 contains a plot of temperature, density, and pressure estimates, from which it is clear that the above "theoretical" values are in agreement with the observations. Noting that the hot intercloud gas and the warm and cold clouds are at the same pressure $\tilde{P} = 3700 \, \text{cm}^{-3} \cdot \text{K}$, Myers obtained $f_g = 0.3 - 0.8$ from the

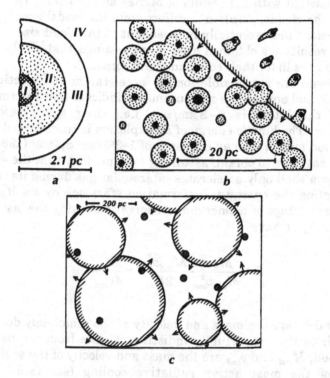

Fig. 91. Structure of the interstellar medium as regulated by supernova explosions (McKee and Ostriker, 1977).

a) An "average" cloud embedded in hot gas. For the core (I), the intermediate warm, weakly ionized layer (II), the warm corona (III), and the hot medium (IV), the temperature, density, and degree of ionization are, respectively

I) 80 K, 42 cm^{-3}, 10^{-3};

II) 8000 K, 0.37 cm^{-3}, 0.15;

III) 8000 K, 0.25 cm^{-3}, 0.68;

IV) 4.5×10^5 K, 3.5×10^{-3} cm^{-3}, 1.

b) Large-scale picture. Cloud sizes reflect the actual porosity of the interstellar medium. The supernova shell is moving away from the upper right-hand corner.

c) The scale here has been reduce by another factor of 20. We now see only relatively young supernova remnants with radii $R < R_{cool}$ ($R_{cool} \approx 180$ pc in a hot tenuous medium), and the largest clouds, with radii $a \geq 7$ pc.

observations plotted here, a mean value $f_h = 0.5$, and $f_w = 0.2-0.4$. If one then takes this pressure for all 45 observed diffuse H I clouds, they are found to range in size from 0.2 to 12 pc, and in density from 7 to 61 cm^{-3}, which is consistent with the results of McKee and Ostriker. The size distribution of the clouds, $N(a) \propto a^{-1}$, differs from the model above, but the basic conclusions are essentially independent of the cloud size spectrum, and similar results are obtained even if one assumes that all clouds have the same mean radii for their core and warm corona.

Supernovae are responsible for the acceleration of Galactic diffuse clouds; 21-cm and optical line measurements indicate that the rms radial velocity of the latter is 6-8 km/sec, i.e., their space velocity is 10-13 km/sec. The kinetic energy of an explosion is transformed into the energy of cloud motion with an efficiency of 1-3% (see Spitzer (1981)). This result is an easy one to obtain, assuming for simplicity that the explosion of a supernova shell only accelerates interstellar gas during its radiative phase. Denoting the energy transformation efficiency by $k \equiv M_{cl}v_{cl}^2/2E_0$, and taking advantage of momentum conservation during the last stage of shell expansion, we have

$$k = \frac{M_{cl}v_{cl}^2}{M_{cool}v_{cool}^2} \frac{M_{cool}v_{cool}^2}{2E_0} \approx \frac{v_{cl}}{4v_{cool}}. \qquad (17.11)$$

Here M_{cl} and v_{cl} are the mass and velocity of the completely decelerated shell, which by then is virtually indistinguishable from a conventional Galactic cloud; M_{cool} and v_{cool} are the mass and velocity of the shell at t_{cool}, the time of the most active radiative cooling (see (9.3)). Taking $v_{cl} \approx 10$ km/sec and $v_{cool} \approx 100-200$ km/sec, we find $k \approx 2-3\%$, assuming that by that evolutionary stage, approximately 50% of the initial energy E_0 has been radiated away, and that the thermal and kinetic energy of the shell are about equal.

The more rigorous analysis of McKee and Ostriker (1977) for completely inelastic collisions yields, for the late stage $t > t_{cool}$, an energy influx, averaged over shell velocities and angles between clouds, of

$$\Delta E_+ = 2\pi R^2 \, dR \, M_{cl}(a_0) N(a_0) \, da_0 \frac{M_{sh}}{M_{cl}} \left(\frac{M_{sh}}{M_{cl}} v_{sh}^2 - \frac{1}{3} v_{cl}^2 \right) \qquad (17.12)$$

per supernova with radius between R and $R + dR$, over cloud radii ranging from a_0 to $a_0 + da_0$. We have assumed here that the mass per unit surface area of a cloud exceeds that of the shell — $M_{cl} \equiv M_{cl}/a^2 \gg M_{sh} \equiv M_{sh}/R^2$ — and that the shell velocity is greater than that of the clouds.

Integrating (17.12) over the cloud spectrum and the supernova remnant's radius, from the time of formation of the shell (R_{cool}) to coalescence (R_{co}) with $\eta' = 0.39$ and $Q = 0.5$ yields

$$\langle\langle \Delta E_+ \rangle\rangle = \frac{3\pi^2 N_{cl}}{16\rho_w} a_{cr}[M(R_{cool})]^2 R_{cool}^3 v_{cool}^3 \approx 5\times 10^{49} \text{ erg/SN}, \quad (17.13)$$

which is about 5% of the initial supernova energy $E_{51} = 1$ assumed for the calculations.

The numerical results obtained by Cowie et al. (1981a) gave a similar value of the supernova energy injected into the cloud component of the

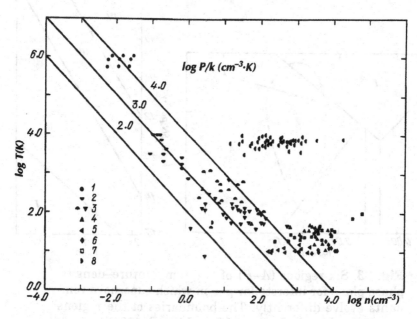

Fig. 92. A summary of temperature and density observations for various components of the interstellar gas (Myers, 1978).

1) Hot gas observed in the O VI line;
2) intercloud gas observed at 21 cm;
3) diffuse clouds (observed at 21 cm by a number of workers);
4) warm clouds (21 cm);
5) warm clouds (2.6 mm CO line);
6) globules (2.6 mm CO);
7) moderate-size molecular clouds associated with emission nebulae (2.6 mm CO);
8) H II regions (observed in H109α at 6 cm).

medium (see Fig. 57). Averaging the energy loss of a supernova over the whole Galactic disk, with a supernova rate $S_{-13} = 1$, we find an influx of kinetic energy

$$\frac{d\langle\langle E_+\rangle\rangle}{dV} = 6 \times 10^{-27} \text{ erg} \cdot \text{cm}^{-3} \cdot \text{sec}^{-1}.$$

The influx of energy from supernovae balances the energy dissipated in inelastic collisions between clouds. Taking the cloud collision cross sec-

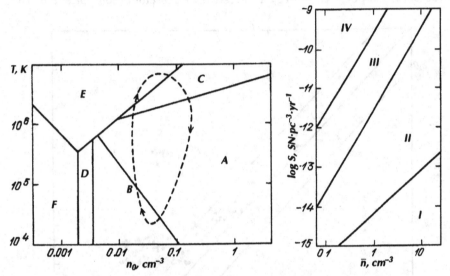

Fig. 93. Six regions (A–F) of the temperature–density phase plane for interstellar gas in which supernova remnants evolve differently. The boundaries of the regions come from the calculations of Ikeuchi et al. 1984), assuming a supernova rate $S_{-13} = 1$ and cloud density $N_{cl} = 10^{-4}$ pc^{-3}. The dashed curved show the cyclical variation of interstellar gas temperature and density under the influence of supernovae; see text.

Fig. 94. Four states of the interstellar medium determined by the supernova rate (S) and mean gas density(n).

I) Equilibrium between warm and cold gas;
II) equilibrium between warm, hot, and cold gas (model of McKee and Ostriker (1977));
III) cyclical state variations;
IV) hot gas leaving the Galaxy dominates (Ikeuchi et al., 1984).

tion to be half its geometrical value and averaging over a Gaussian velocity distribution and the previously assumed cloud-size distribution, McKee and Ostriker found an energy loss density

$$\frac{d\langle\langle E_-\rangle\rangle}{dV} \approx -1.3 \times 10^{-44} \langle v\rangle^3 \; \text{erg} \cdot \text{cm}^{-3} \cdot \text{sec}^{-1}.$$

Equating the influx of supernova energy to the energy losses of clouds, they obtained an rms cloud velocity $\langle v\rangle = 8$ km/sec, close to the observed value.

As noted earlier, the equilibrium state of the three-phase interstellar medium considered here, controlled by supernova explosions, pertains to the average solar neighborhood of the Galaxy, which is devoid of giant molecular clouds. McKee and Ostriker, having ascertained that remnants can coalesce and that hot, tenuous matter takes up a large fraction of the Galactic disk, went on to analyze the situation for a low initial density, $n_0 = 0.01 \, \text{cm}^{-3}$. But the equilibrium state determined by the supernova rate and initial pressure (temperature and density of the ambient medium) could also be otherwise — in particular, it could be different in the corona and the disk, in the spiral arms and interarm space, at the Galactic center and on the periphery. Defining the demise of a remnant as either the time at which it merges (t_{co}) or the time at which the shell dissipates after the internal and external pressures become equal (t_p), one can distinguish six different regions for the equilibrium state of the interstellar gas, as regulated by supernovae, in the density–temperature phase plane (Ikeuchi et al., 1984). To establish a notation for these zones, Ikeuchi et al. identify three transition times in the evolution of a remnant (see Section 9): t_1, the

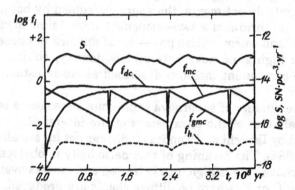

Fig. 95. Cyclical variations in the supernova rate (S) and the relative mass in molecular clouds (f_{mc}), giant molecular clouds (f_{gmc}), diffuse clouds (f_{dc}), and hot bubbles or merging supernova remnants (f_h) (Ikeuchi et al., 1984). Calculations for the solar neighborhood of the Galaxy, with $\bar{n} = 1.7 \, \text{cm}^{-3}$. Here $f_i \equiv M_i / \Sigma M_i$.

transition from the adiabatic stage dominated by evaporation to the standard adiabatic stage; $t_2 \equiv t_{cool}$, the transition from the adiabatic to the radiative stage; and t_3, the onset of shell dissipation.

The six regions in the density–temperature phase plane are then the following:

A) pressures are equalized in the radiative stage: $t_p > t_2$, $t_{co} > t_3$.
B) coalescence of remnants takes place in the radiative stage: $t_2 < t_{co} < t_3$ (this is the case considered by McKee and Ostriker).
C) pressures are equalized in the adiabatic expansion stage: $t_1 < t_p < t_2$ (this is the case analyzed by Cox (1979)).
D) coalescence is the outcome of the adiabatic stage: $t_1 < t_{co} < t_2$.
E) pressures are equalized in the evaporation stage: $t_p < t_1$.
F) coalescence takes place in the evaporation stage: $t_{co} < t_1$.

These six possibilities are shown in Fig. 93; estimates of the temperature and density of the ambient gas have been made by Ikeuchi et al. (1984) for a supernova rate $S_{-13} = 1$ and a density of "average" diffuse clouds (mass ~40 $M_\odot$, size 2 pc) $N_{cl} = 10^{-4}$ pc^{-3}. The calculations of Habe et al. (1981) have shown that the equilibrium state of a medium regulated by supernova explosions sets in after $(1-3) \times 10^7$ yr for $S_{-13} = 1$, and an order of magnitude later for $S_{-13} = 0.1$, given reasonable variations in the initial temperature, the mean gas density $\bar{n}$, and the flux of ionizing radiation. Under those circumstances, at least four equilibrium states of the interstellar gas are possible, and are plotted in the $\langle \bar{n}, S_{-13} \rangle$ phase plane in Fig. 94; see Ikeuchi et al. (1984).

I)   For a high mean gas density and low supernova rate, supernova remnants do not merge, the volume occupied by hot gas is small, and equilibrium of a two-component interstellar medium — cold clouds and warm ionized gas — is established in phase region A.
II) For a high supernova rate or low density, equilibrium of the three-component medium described above is established in region B.
III) As the number of supernova explosions increases, a certain cyclical behavior of the parameters of the interstellar medium predicted by Ikeuchi and Tomita (1983) can set in; see also Ikeuchi et al. (1984). The meaning of this periodicity is obvious. When the supernova rate is high, gas is heated and crosses over into phase region C or E, where no diffuse clouds are produced, but instead evaporate readily. Hot gas gradually fills the whole region, and its density rises due to the evaporation of clouds. The increase in hot gas density can lead to a dominance of radiative cooling over shock-induced heating, and the medium will move into region A, where diffuse clouds will be formed anew upon fragmentation of the shells swept up by supernovae. The appearance of small,

dense clouds reduces the mean density of the ambient gas, and carries the system into phase B, where efficient heating by supernovae begins once again. This cyclical variation of the state of the interstellar gas is depicted by the dashed curve in Fig. 93.

IV) When the supernova rate is very high, radiative cooling fails to take precedence over heating in regions C–E, and the entire volume becomes occupied by hot, tenuous gas.

The straight lines bounding these four equilibrium states in the $\langle \bar{n}, S_{-13} \rangle$ phase plane in Fig. 94 come from the following conditions:

Transition between states I and II: $t_{sw} \sim t_i$, where $t_{sw} \sim (R_{cool}^3 S)^{-1}$ is the characteristic time for gas to be swept up, and $t_i \sim \zeta^{-1}$ is the characteristic ionization time.

Transition between states II and III: $t_{sw} \sim t_{ev} \sim t_{he}$, where $t_{ev} \sim (R_{ev}^3 S)^{-1}$ is the evaporation time, R_{ev} is the radius of the remnant upon leaving the evaporative stage for the adiabatic (see Section 9), and $t_{he} \sim 3\bar{n}kT/f_g E_0 S$ is the characteristic time for gas heating by supernovae.

Transition between states III and IV: $t_{he} \sim t_{cool}$.

The possibility of cyclical variations in the state of the interstellar medium thus far remains purely speculative, since it is completely unconfirmed by observations. Complicating matters by allowing for the creation and destruction of giant molecular clouds, and star formation therein, would clearly bring one closer to the actual observational situation. The state of the interstellar medium, incorporating all of the components enumerated at the beginning of this section (apart from gas exchange between the disk and halo), is also subject to cyclicalvariations (Ikeuchi et al., 1984) that conform to the following main branches. Gravitational instability of an ensemble of molecular clouds leads to the coalescence of dozens of such clouds at a time, each with a mass of $M \sim 10^4\ M_\odot$, into giant molecular clouds with a total mass of $M \sim 10^5 - 10^6\ M_\odot$ (Cowie, 1981; Elmegreen, 1987a; Gorbatskii and Usovich, 1986). The largest gravitationally bound cells in galaxies such as our own are the "superclouds" ($M \sim 10^6 - 5 \times 10^7\ M_\odot$), which contain individual, denser molecular fragments (Elmegreen and Elmegreen, 1987). The appearance of this sort of molecular complex on the scene induces a burst of star formation. Starting on the outskirts of the giant molecular clouds, star formation carries a portion of a cloud into a warm phase, and after the onset of supernovae explosions, into a hot phase. A wave of star formation and supernova outbursts eats into the giant molecular cloud and then abates, having depleted its source of nourishment. There is a competing chain of processes: warm and hot gas are swept up and cool radiately; cold, dense shells form and fragment; diffuse clouds form and coalesce into molecular clouds, again augmenting the mass of molecular clouds per unit surface of the Galactic disk.

When that density rises above a certain critical threshold, gravitational instability develops once again in the ensemble of molecular clouds (see, e.g., Cowie (1981) and Elmegreen (1987a, b)).

Figure 95 shows the results obtained numerically by Ikeuchi et al. (1984) for the solar neighborhood, assuming $\bar{n} = 1.7 \, \text{cm}^{-3}$, Salpeter's initial mass function $N(M) \propto M^{-2.35}$ over the mass range $5 \leq M_{init} \leq 60 \, M_\odot$, and stellar lifetime as a function of mass given by

$$\log t(M) = 10.02 - 3.57 \log M + 0.9 (\log M)^2. \qquad (17.14)$$

The curves clearly display periodic variations in the supernova rate and the fractional mass of gas contained in hot cavities, molecular clouds, and giant molecular clouds, the periodicity being 1.2×10^8 yr. Depending on the mean gas density, epicyclic frequency, and velocity dispersion of the molecular clouds, i.e., at different galactocentric distances, the calculations yield variations of different duration and amplitude. At $r_{gal} \sim 5 \, \text{kpc}$, $\bar{n} \sim 10 \, \text{cm}^{-3}$, corresponding to the ring of molecular clouds in our Galaxy, the period drops to 4×10^7 yr, and the amplitude of the variation in the supernova rate falls off. In outlying regions with $r_{gal} \sim 12 \, \text{kpc}$ and $\bar{n} = 0.7 \, \text{cm}^{-3}$, long-period ($t = 2.4 \times 10^8$ yr), large-amplitude variations are to be expected.

These cyclical variations in the state of the interstellar medium bear a strong resemblance to oscillatory (Belousov–Zhabotinskii) chemical reactions, or to the periodic population swings observed in simple ecological systems — for example, the conspicuous maxima and minima observed in the populations of hares, caribou, and lynx approximately every 10 years in the closed ecosystem of Newfoundland (Bergerud, 1983). In the present case, the "predators" are the stars and supernovae that destroy the giant molecular clouds where they were born. We may hope that the actual variations in star-forming activity and its various manifestations (specifically, the number of giant H II regions and the supernova rate), being observable from afar, will be fairly easily detected via statistical analysis of a sufficiently large number of galaxies. In particular, the enhanced rate of star formation observed in certain dwarf galaxies (Klein, 1984) may conceivably result from just this sort of active phase (Ikeuchi et al., 1984). In spiral galaxies, such large-scale, long-period cycles are disrupted by the passage of spiral density waves. The foregoing picture is therefore most likely to hold in corotating regions, on the outskirts of spirals, and in irregular galaxies.

The density gradient in the Galaxy and the buoyancy of hot plasma bubbles — supernova remnants — gives rise to the penetration of the corona by hot gas. If hot gas occupies a significant fraction of the disk, i.e., it does not have to percolate through large masses of cold, dense gas, then its ascent to the corona may turn out to be the main source of coronal gas, with a rate of influx of order $1 \, M_\odot/\text{yr}$. Type Ia supernova outbursts high above the Galactic plane also heat the coronal gas, partly compensating for radiative cooling.

The mean temperature of coronal gas is at most 10^6 K, and a typical density is 10^{-3} cm^{-3}. If the energy influx from type Ia supernovae heats the coronal gas reasonably efficiently, the gas may reach a scale height of 7–8 kpc (Chevalier and Oegerle, 1979). If more hot gas is produced than the gravitational field of the Galaxy can retain, coronal gas will leave the Galaxy. The upshot is that the galactic wind will also be regulated by supernova explosions.

Radiative cooling of coronal gas may turn out to be significant at a z-height of 2–5 kpc. Cooling is accompanied by thermal instability, and leads to the condensation of neutral clouds. Cold, condensed gas falls back into the galactic disk at a typical velocity of 100 km/sec. It is possible that high-velocity, high-latitude H I clouds may be formed as a result of this process, which has been termed a *galactic fountain* (Shapiro and Field, 1976; Cox, 1981). In actuality, the overwhelming majority of high-latitude clouds have negative velocities of the right order of magnitude. If coronal gas cools rapidly, then gas exchange between the disk and corona can reach several solar masses per year. It should be pointed here that investigations of the process whereby a hot corona is formed are still in their infancy, and these estimates are highly speculative.

The role played by supernovae in producing the corona of spiral galaxies and galactic wind was recently reexamined by Heiles (1987) in a detailed paper based on our current understanding of the supernova rate, supernova energetics, and the evolution of remnants in the various components of the interstellar medium. Heiles pointed out that the most substantial contribution comes from the explosion of high-mass stars, which are concentrated in the neighborhood of OB associations. The hot bubbles swept out by multiple supernovae around rich, young stellar aggregates can easily break out into the halo, due to the instability of the polar regions, when they become comparable in size to the thickness of the gaseous disk of the Galaxy; see also Section 18.

Note that although the role of supernova as regulators of the physical state of the interstellar medium is beyond doubt, quantitative estimates must be treated cautiously, bearing in mind the highly idealized foundation upon which the edifice has been built.

More to the point, in all of the studies of the equilibrium state of the multiphase interstellar medium cited above, supernova were assumed to be uniformly distributed throughout the galactic disk. Meanwhile, the massive progenitors of SN II and SN Ib favor the vicinity of clusters and associations, and to a lesser degree, they may determine the physics of the gaseous disk as a whole.

In early calculations, no allowance was made for the pressure exerted by the interstellar magnetic field and relativistic particles, although it can dominate the thermal gas pressure, at least in certain phases.

Table 28. Mass Loss in Stellar Wind, Influx of Wind Energy, Dust, and Ionizing Radiation from Massive Stars

Component	Stellar contribution by spectral type, %			Total contribution per kpc^2 in the Galactic plane
	WR	O	B, A	
$\dot{M}$, wind	56	36	8	$10^{-4}(M_\odot/\mathrm{yr})\cdot\mathrm{kpc}^{-2}$
$L_w = \frac{1}{2}\dot{M}v_w^2$	50	47	3	$(2-3)\times10^{38}\,\mathrm{erg\cdot sec^{-1}\cdot kpc^{-2}}$
Radiation	6	66	28	$(3-4)\times10^{40}\,\mathrm{erg\cdot sec^{-1}\cdot kpc^{-2}}$
$\dot{M}$, dust	100	0	0	$5\times10^{-8}(M_\odot/\mathrm{yr})\cdot\mathrm{kpc}^{-2}$
N_* ($M_{init} \geq 15\,M_\odot$)	4	49	47	$40.3\,\mathrm{stars\cdot kpc}^{-2}$

Assumed initial conditions — particularly the pressure of the different phases of the interstellar medium, the supernova rate, and the evolutionary path taken by supernova remnants — may also not be entirely realistic. Cox (1988) has reviewed both this situation and a number of others in the theory of the multiphase interstellar medium controlled by supernovae, and has come to the discouraging conclusion that our fundamental framework for dealing with the multiphase interstellar medium, which to a considerable extent owes its existence to the pioneering work of Cox and Smith (1974) on the coalescence of supernova remnants, may be wrong.

According to Cox (1988), we cannot rule out the possibility that the warm component occupies more of the Galactic disk than the hot component — in any case, much more than $f_w = 0.2$, the value adopted by McKee and Ostriker (1977). The most realistic estimate, according to Cox (1988), might be $f_w \approx 0.35$, although under certain circumstances it might even be $f_w \approx 1$. The cloud and intercloud components of the gaseous medium are not necessarily at the same thermal pressure: pressure balance between the various phases must also include the magnetic field. When the latter is taken into consideration, the interstellar medium is found to be more "resilient" in the face of the destructive action of supernovae. Supernova explosions unquestionably lead to the formation of hot bubbles; some, like Loop I, are quite large. There is also no question but that hot superbubbles, which are associated with stellar wind and multiple supernova explosions in rich stellar aggregates, also exist; see Section 18.

Cox is somewhat doubtful, however, that there is a large volume within the disk of the Galaxy filled with hot, merged bubbles — i.e., that the combination of the warm and cold components is dominated by the hot. So far, direct observational evidence that a large fraction of the Galactic disk is occupied by a hot, tenuous component is also lacking. In fact, on the contrary, the observations of the last few years seem to suggest that warm H I regions are distributed more evenly and occupy more space than implied by the model of McKee and Ostriker (1977); see the references in McCray and Kafatos (1987).

Does stellar wind affect the large-scale structure of the gaseous medium of the Galaxy to the same extent as supernova explosions? In particular, what is the contribution of the energy lost by stars in the form of wind compared with the energy imparted by explosions? We are now able to answer that question definitively, since we know what kinds of star lose mass, and at what rate; we also know the number of stars of various spectral types in the solar neighborhood. This knowledge is sufficient to literally count up all sources of stellar wind, taking each individually into account, as has been done by Abbott (1982). More recently, similar values of injected mass, the mechanical energy input due to stellar wind, radiation, and the mass of dust supplied to the interstellar medium by stars of various types have also been obtained by van der Hucht et al. (1987). It can be seen from Figs. 74 and 75 that strong outflow is a property inherent to stars with $M_{init} \gtrsim 15 M_\odot$, and that value can be taken as a limit distinguishing stars with wind from those without. The most recent catalogs of massive early-type stars are complete to $r \leq 3$ kpc. Those stars constitute the main source of kinetic wind energy (see Table 28). Early-type dwarfs with stellar wind are visible out to 500 pc, and A and B supergiants out to 1.5 kpc, but the wind from those stars typically makes but a small contribution as compared with the massive early-type stars. Also negligible is the wind energy input due to cold supergiants (whose total wind energy is approximately 10^4 times lower than that of hot OB stars), cold dwarfs (10^3 times lower), and the central stars of planetary nebulae (10^2 times lower), so the counts cited yield a reasonably complete picture out to 3 kpc. Estimates of the ionizing radiation from stars and the mechanical energy input due to wind are presented in Tables 28 and 29. The overall mean mechanical wind-energy input rate of all stars with $r \leq 3$ kpc in the solar neighborhood is 1.9×10^{38} erg $\cdot$ sec$^{-1} \cdot$ kpc^{-1}; the most important energy contribution comes from the stellar wind issuing from the relatively rare WR stars and those of early spectral type, and the same is true of the flux of ionizing radiation. The combined wind power of 50 WR stars, whose outflow stage is an order of magnitude shorter than the hydrogen-burning stage in O stars, exceeds the power of the roughly 500 remaining O stars in the same region. The wind power of roughly 500 B and A stars can be all but neglected when it comes to the large-scale dynamics of the interstellar medium.

We have seen in Section 13 that the kinetic energy imparted by stellar wind is transmitted to the ambient interstellar gas, and goes into heating the gas via shock waves and accelerating the swept-up shell. It is reradiated by the gas of the shell and by the hot layer of wind. In the standard wind-blown bubble considered in Section 13, which was produced by stellar wind with power $L_w = 10^{36}$ erg/sec in a medium of density $n_0 = 1$ cm^{-3}, approximately 30% of the wind energy is transformed into thermal energy of the hot layer in a 10^6-year lifetime. In addition, about 10% goes into the

Table 29. Stellar Wind Power of Associations in the Solar Neighborhood ($r_\circ \le 3$ kpc)

Association	L_w, erg·sec^{-1}	Association	L_w, erg·sec^{-1}
Cyg OB1, OB2, OB3	10^{39} *	Cru OB1	2.4×10^{37}
Car OB1	8.3×10^{38} *	Mon OB2	1.7×10^{37}
Sco OB1	1.7×10^{38}	Cep OB2	1.3×10^{37}
Sgr OB1	1.3×10^{38}	Ori OB1	1.3×10^{37}
Cas OB6	1.1×10^{38}	Cas OB5	1.3×10^{37}
Gum Nebula	7×10^{37}	Pup OB1	1.2×10^{37}
Ser OB1, OB2	9×10^{37}	Cen OB1	10^{37}
Per OB1	3.2×10^{37}	Vul OB1	10^{37}
Ara OB1a	2.7×10^{37}		

*These values are probably somewhat of an overestimate.

kinetic energy of expansion of the cold, dense shell, about 27% is radiated by hot gas, and about 33% by swept-up interstellar gas (Weaver et al., 1977). The energy radiated within the bubble in turn goes into heating of H I and H II regions.

The efficiency of these processes, in which the energy of stellar wind is transmitted to the various components of the interstellar medium, is summarized in Table 30. For comparison, the table includes the influx of energy from supernovae and ionizing stellar radiation in the same solar neighborhood, $r_0 \le 3$ kpc. The estimated contribution from supernovae is more uncertain, as it incorporates not just the initial energy of supernova types Ia, Ib, and II, but also the actual number of supernovae within the prescribed volume. The latter cannot simply be counted, as can stellar sources of radiation and wind. The number of supernovae has been estimated by assuming that every star of mass $M_{init} \ge 8\,M_\odot$ eventually explodes as a supernova, and that the number of stars with $8 \le M_{init} \le 15\,M_\odot$ can be derived from the initial mass function.

A similar count based on a large number of measurements of $\dot{M}$ and v_w, and making use of a slightly different initial mass function to take supernovae into consideration, has been carried out by van Buren (1985). His conclusions about the relative contribution made by stars of different mass to the mass, energy, and momentum flux into the interstellar medium are in qualitative agreement with Abbott's estimates, but the actual values of the overall contribution differ somewhat. Specifically, within 2 kpc of the sun, van Buren obtains the following values for the total amount of mass, energy, and momentum injected by stellar wind and supernova explosions (per unit area of the Galactic disk):

Table 30. Ionizing Radiation, Stellar Wind, and Supernovae: Power Input to the Interstellar Medium (Abbott, 1982)

Energy input $erg \cdot sec^{-1} \cdot kpc^{-2}$	Stellar radiation	Stellar wind	Supernovae	Total
Shock heating of hot gas	—	6×10^{37}	2×10^{38}	4×10^{38}
Kinetic energy of cloud motion	2×10^{37}	2×10^{37}	6×10^{37}	9×10^{37}
Heating of H II regions by ionizing radiation	2×10^{39}	3×10^{37}	2×10^{38}	2×10^{39}
Heating of H I regions by shock waves and radiation	3×10^{38}	10^{36}	3×10^{38}	6×10^{38}
Total energy output	3×10^{40}	2×10^{38}	10^{39}	

$$\dot{M} = 10^{-3} \, M_\odot yr^{-1} \cdot kpc^{-2},$$

$$\dot{P} = 10^{31} \, dyn \cdot kpc^2,$$

$$\dot{E} = 2.7 \times 10^{39} \, erg \cdot sec^{-1} \cdot kpc^{-2}.$$

To summarize the counts carried out by Abbott, van Buren, and van der Hucht et al., we can state that stars lose energy principally in the form of radiation: the total energy in stellar wind is at most 1% (5% for Wolf–Rayet stars), and the total amount of energy released in explosions is no more than 2% of the energy in photon radiation. Nevertheless, due to the high efficiency with which the kinetic energy of stellar wind and supernovae is transmitted to interstellar gas, all three sources turn out to have comparable large-scale effects on the interstellar medium. In fact, supernovae tend to dominate in heating the hot component of the intercloud gas, radiation dominates the heating of H II regions, and the two processes play approximately equivalent roles in the heating of neutral clouds. On average, the mechanical energy of stellar wind is about 20% of the energy released in supernova explosions. For that reason, the three-phase structure of the interstellar medium discussed at the beginning of this section is actually attributable mainly to supernovae. In the vicinity of OB associations, however, where stellar wind is typically two orders of magnitude stronger than average, it will exert much more of an influence on interstellar gas than will supernovae.

18. Shells and supershells produced by supernovae and by the stellar wind from OB associations

Nowhere are the effects of supernovae and stellar wind upon interstellar gas felt as strongly as in the vicinity of OB associations. Fully 50–90% of all early-type massive stars belong to associations and clusters, which implies that most supernova explosions — in any event, SN II and SN Ib — take place in associations. The intimate relationship between a number of Galactic supernova remnants and OB associations is revealed by their telltale morphology and kinematics (see Lozinskaya (1980b)). The Large Magellanic Cloud presents an even clearer picture of that relationship, since distance estimates to its remnants are unencumbered by the usual errors. The centers of ~40% of the supernova remnants in the LMC lie within OB associations; the total area covered by the latter is at most about 4% of the visible area of the LMC, so the probability of the coincidence being random is negligible (Cowie et al., 1979).

We can estimate the supernova rate in OB associations by making use of the total number of OB association in the Galaxy and the mean interval between outbursts, $\tau = 25 - 35$ yr (Section 11); we also note that 50–90% of all massive stars belong to associations, and that at least ~50% of all supernovae are SN II and SN Ib. This yields a rate in an "average" OB association of approximately one supernova every $(2-3) \times 10^5$ yr. A different estimate, based on the number and lifetime of OB stars in an association, yields the same characteristic time between outbursts, $\tau \approx 10^5$ yr (Cowie et al., 1979).

In the LMC, the supernova rate in associations can be determined even more accurately from the total number of associations and their associated remnants, knowing the typical time spent by a remnant above the limit of detectability, $t \lesssim 10^4$ yr. Such counts yields a mean interval of $\tau \approx 5 \times 10^5$ yr in any one association (Cowie et al., 1979).

For an association lifetime $\tau \approx 5 \times 10^7$ yr, the total kinetic energy injected by supernovae accordingly comes to about 10^{53} erg. It can be seen from Table 30 that the energy pumped in by stellar wind is 3×10^{52} erg for an "average" association, and an order of magnitude higher in giant stellar aggregates like those in Carina and Cygnus. Having mentioned the overall effect of wind and supernovae in associations, we must further note that the density of the unperturbed gas there may be tens to hundreds of times higher than the average density in the galactic disk.

The interaction of ionizing stellar radiation, wind, and multiple supernova explosions leads to the formation of shells and supershells associated with OB associations and young open clusters. The proliferation of terminology to be found in the literature — *bubble, cavity, hole, shell, supershell, ring nebula* — reflects the variety of observable manifestations of that interaction. These include neutral hydrogen shells, H II regions with brightening at their boundaries, thin filamentary nebulae, shell-like gas and dust complexes, thermal and synchrotron radio shell sources, and extended soft x-ray sources — structures with more or less clear-cut shell features and sizes ranging from tens of parsecs to several kiloparsecs.

Supershells hundreds or thousands of parsecs across are observed primarily in the nearest galaxies. This is in part a selection effect: observing as we do from the inside out, we can only poorly discriminate such structures against the background of galactic emission within our own stellar system. But there are also physical reasons, to be discussed below, why the formation of giant shells is favored in irregular galaxies.

High angular resolution 21-cm observations of M31 (Brinks and Bajaja, 1986) have revealed 140 cavities and shells $10^2 - 10^3$ pc across in the neutral hydrogen distribution, expanding at $\sim 10 - 15$ km/sec. The corresponding kinematic age of those supershells is $(3 - 30) \times 10^6$ yr, the mass of H I removed is from 10^3 to 10^7 $M_\odot$, and the kinetic energy lies between 10^{49} and 10^{53} erg. Shells no more than 300 pc across are related to OB associations, while for larger ones the relationship is uncertain, since the probability of a random coincidence increases.

Photographs of M33 taken in Hα with the 6-m telescope at the Special Astrophysical Observatory of the USSR (Courtes et al., 1987) have revealed an unbroken "foam" of cavities, shells, supershells, and arcs from 40 to 300 pc in size, both within the spiral arms of the galaxy and between them.

The most favorable circumstances for searching for interstellar shells may be found in our nearest face-on galaxy, the LMC. Shells and supershells have been widely observed in the LMC both in the optical and radio. Recently, a multitude of giant shells and "layers" of neutral hydrogen have been identified, disclosing high-velocity motion at up to ±100 km/sec (Meaburn et al., 1987a; see also the references in that paper). Such high velocities provide indisputable evidence of the activity of supernovae and/or stellar wind.

In Fig. 96 we have reproduced a photograph of the LMC taken by Davis et al. (1976) that shows the full diversity of emission ring nebulae. Most of them can be identified with OB associations or single stars with strong stellar wind, or they are supernova remnants. In all, out of 122 OB associations in the LMC, 50 exhibit ring-like morphology, to a greater or lesser extent, in the surrounding H II regions (see Braunsfurth and Feitzinger (1983) and references therein).

By way of example, let us consider a number of the best-studied shells surrounding OB associations in the Milky Way and LMC, as we attempt to elucidate the mechanism by which they were formed. We must immediately point out that this one-at-a-time approach is currently the most promising, as there still exists no general, complete theory of how ionizing radiation, stellar wind, and supernova remnants in an association interact with the interstellar medium. Moreover, a detailed theoretical analysis is probably even impossible in view of the marked inhomogeneities in the density distribution of the interstellar medium and the distribution of energy sources on scales encompassed by perturbations, not to mention the serial nature of the process whereby stars and associations are formed, and gas and dust complexes are destroyed.

Figure 97 shows the supershell LMC 2. Its overall size is ~900 pc; it abuts the extremely bright H II region 30 Dor (N 157) on the northwest, where a number of extensive molecular clouds are to be found. Within LMC 2 there are five young stellar associations, which ionize the H II regions 30 Dor, N 158C, N 160A, N 159, and N 164. Two more associations are unrelated to any dense local clouds, and may be an ionizing source for gas in the supershell: direct measurements of their ultraviolet luminosity

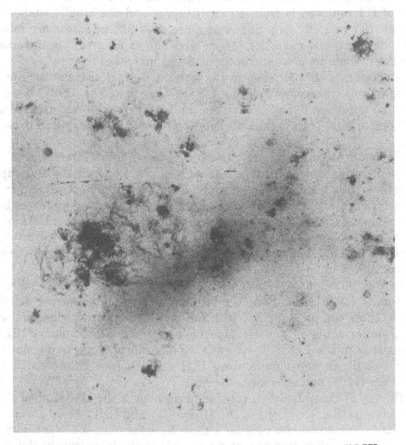

Fig. 96a. The Large Magellanic Cloud in Hα + [N II] (Davies et al., 1976).

$(L_{UV} = (3-4) \times 10^{51}$ photons·sec^{-1}) are consistent with the thermal radio brightness of the shell (Caulet et al., 1982). The kinematic age given by the size and expansion velocity (30 km/sec) is $5 \times 10^6 - 10^7$ yr. Using (13.5), the wind power required to produce the supershell is $L_w \sim 10^{39}$ erg/sec, and it must last for $t \sim 10^7$ yr if the ambient gas in the region has a density $n_0 \sim 0.4 - 0.5$ cm^{-3}.

Caulet et al. (1982) estimate the total power from all sources of stellar wind inside associations belonging to LMC 2 — five WR stars, 26 supergiants, and approximately 400 massive early-type stars — to be $L_w \sim (0.2 - 2.5) \times 10^{39}$ erg/sec. If the rate of outflow remains constant, the total wind energy delivered over the lifetime of the supershell will reach $(2-3) \times 10^{53}$ erg. That level is also supportable by supernova explosions if star formation within the complex continues over $\sim 10^7$ yr, and explosions occur at a constant rate of one supernova every 5×10^5 yr. The observed

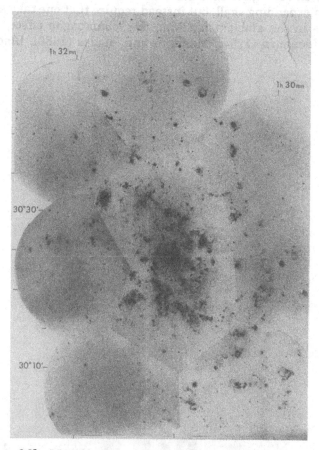

Fig. 96b. M33 in Hα, photographed with the 6-m telescope of the Special Astrophysical Observatory, USSR Academy of Sciences (Courtes et al., 1987).

expansion velocity and the mass obtained from the optical and radio brightness yield a shell kinetic energy of about 10^{53} erg, so the combined influence of stellar wind and supernovae within the extensive stellar aggregate could have produced the giant shell LMC 2.

We shall not touch upon any other extragalactic supershells; in just one galaxy, the LMC, they number in the dozens (see Georgelin et al. (1983), Meaburn (1980), and references therein; see also new data in Proc. IAU Symp. No. 148, *The Magellanic Clouds and Their Dynamical Interaction with the Milky Way*, Sydney (1990)). High-velocity gas motions and a high [S II]/Hα ratio in those nebulae attest to the influence of a shock wave engendered by stellar wind or supernovae. The effects of ionizing radiation, stellar wind, and multiple supernova explosions do a presentable job of explaining the observed morphology and kinematics of supershells. The complicated filamentary structure and kinematics of certain objects — N 59A, N 57, and N 70, for example — fail to conform to a simple scheme describing an expanding shell, but instead require that one take account of irregularities in the ambient medium, the "champagne effect," etc. (see Goudis and Meaburn (1984), Meaburn and Blades (1980), Blades et al.

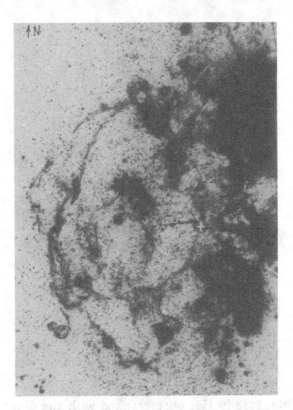

Fig. 97. The giant shell LMC 2. The small white cross marks the center of expansion (Caulet et al., 1982).

(1980), and Meaburn et al. (1982)).

Inspection of Einstein maps reveals shell-like morphology in the diffuse x-ray emission from the vicinity of seven OB associations and H II complexes in the LMC (Chu and MacLow, 1990). Assuming that these are supershells blown by the wind of OB associations, these authors find the x-ray luminosity expected from hot plasma to be an order of magnitude lower than actually observed. According to their calculations, the addition of an off-center supernova explosion, which heats the ionized supershell from the inside, may increase the predicted x-ray emission to the observed value.

In the Milky Way there are 13 known giant H II shells (Georgelin et al., 1979) and approximately 50 H I shells (Heiles, 1979; Heiles, 1984). One of the best-studied of the shell complexes is the Gum Nebula, the largest H II region in the Galaxy. At a distance of 400 pc, the size of the central bright shell is 260 pc; deep plates taken with narrow-band filters in Hα show that the bright shell is surrounded by a faint, filamentary halo, and possibly a second shell more than 300 pc across (Chanot and Sivan, 1983).

It has been proposed that this giant H II region is

a) the relic H II region ionized by a star that exploded, leaving behind the supernova remnant Vela XYZ;

b) a normal, evolved H II region, the ionization source of which is ζ Pup and γ^2 Vel;

c) a supernova remnant older than Vela XYZ, ionized by ζ Pup and γ^2 Vel;

d) a cavity produced by stellar wind from ζ Pup.

The optical spectrum of the nebula is typical of H II regions: the relative line intensities are $I_{[NII]}/I_{H\alpha} = 0.4$ and $I_{[OIII]}/I_{H\beta} = 0.2-0.3$, with the latter increasing to $0.6-1.3$ near the exciting stars. The intensity ratio of the components of [S II] yields a density of $n_e \lesssim 100\,cm^{-3}$ (Chanot and Sivan, 1983). The bright central shell is expanding at $20-30\,km/sec$ (Reynolds, 1976a).

The radiation from ζ Pup and γ^2 Vel is sufficient to ionize the nebula,

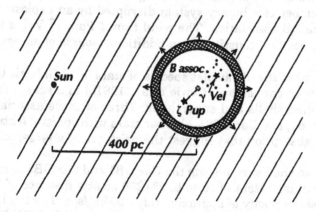

Fig. 98. The Gum Nebula (Reynolds, 1976b).

no matter what the mechanism that produced the expanding shell. The emission measure of the nebula actually ranges from 50 to 600 $cm^{-6} \cdot pc$, and the corresponding Hα luminosity, $L_{H\alpha} \approx 2.2 \times 10^{49}$ photons·sec^{-1}, requires an ionizing luminosity of $L_{UV} \approx 7.3 \times 10^{49}$ photons·sec^{-1}. The two stars supply at least $L_{UV} \approx (7-8) \times 10^{49}$ photons·sec^{-1} (only radiation from the O9 I component of γ^2 Vel has been taken into account, as the effective temperature of the WC8 star is highly uncertain) (see Chanot and Sivan, 1983).

Two of the possible explanations for the Gum Nebula — that it is a relic H II region or a fully evolved Strömgren sphere — do not fit the measured expansion velocity. The third possibility, that it was produced by stellar wind, seems more likely. Using the relations (13.5), we find that the wind power from ζ Pup alone ($v_w = 2700$ km/sec, $\dot{M} = 10^{-5}$ $M_\odot$/yr) would be sufficient to produce a ring-like shell 250 pc in diameter and expanding at ~30 km/sec in a medium with density $n_0 = 0.3$ cm^{-3}, if the outflow were to last for about 10^6 yr. In addition, γ^2 Vel and the rest of the stars in the association are a further source of wind. The wind hypothesis, however, does not explain the markedly off-center position of the most powerful source of wind in the symmetric shell (Fig. 98). Furthermore, the star ζ Pup has a high peculiar velocity, close to 60 km/sec in the plane of the sky, which ought to have deformed the otherwise perfect ring morphology of the wind-blown bubble.

An alternative proposed by Reynolds (1976b) holds that the nebula is a very old supernova remnant, dating back 10^6 yr — i.e., it was created long before the explosion that gave rise to Vela XYZ. The presence of the ionizing sources ζ Pup and γ^2 Vel explains why the optical supernova remnant has remained visible so long. In this model, the interior of the shell should be filled with hot plasma ($T_e \gtrsim 10^6$ K, $n_e \approx 3 \times 10^{-3}$ cm^{-3}) and should be visible in x rays. Interestingly enough, if we look at the proper motion of the runaway star ζ Pup over the course of 10^6 yr, we find that it would have started right at the center of the expanding shell and inside Vela OB2, the association to which γ^2 Vel belongs (see Fig. 98). It would seem that ζ Pup was once a member of a binary system disrupted by an explosion, which belonged to the OB association. The wind from ζ Pup, γ^2 Vel, and other stars in the association constitutes an additional source of energy for the shell.

Observations of the ultraviolet spectra of stars in 13 Galactic OB associations have been carried out (Cowie et al., 1981b) in a search for large, rapidly expanding shells and supershells. Three of those associations — Ori OB1, Car OB1, and Car OB2 — harbor stars with spectral features due to interstellar absorption that suggest the existence of a large, expanding gaseous shell.

The common shell surrounding the Car OB1 and Car OB2 associations has a mean angular extent of close to 3° (200 pc at a distance of 2.3 kpc), and an expansion velocity of approximately 100 km/sec. The total mass of gas in the rapidly expanding shell is 100 $M_\odot$ (Cowie et al., 1981b), and the age deduced from the size and velocity is 4×10^5 yr. It is conceivable that this low-density, high-velocity shell was produced by a supernova that exploded 4×10^5 years ago, and by a powerful wind. Here also we see denser, slow-moving features: faint optical filaments and neutral hydrogen clouds that border the shell. These dense clouds, expanding at about 10 km/sec, have most likely been swept up in previous supernova explosions and by the wind from the association. The main source of that powerful stellar wind would seem to be the peculiar massive star η Car and three WR stars; 12 Of stars and 10 O3–O9 stars have served to sustain the effects of the wind (Cowie et al., 1981b).

An extensive region near η Car is a source of soft x rays (Seward and Chlebowski, 1982). The x-ray source is as much as 50 pc across, and compact bright spots about 1 pc across can be made out in the diffuse background. The x-ray spectrum yields a temperature $T_e = 5 \times 10^6 - 10^7$ K; the plasma density in the extended source is about $0.1 \mathrm{cm}^{-3}$, and in the bright spots, it is $1 \mathrm{cm}^{-3}$. The luminosity of the brightest region near η Car, $L_{0.2-4\,\mathrm{keV}} \sim 3 \times 10^{32}$ erg/sec, may be attributable to wind from nearby excitatory stars (invoking radiation from the layer of hot wind labeled b in Fig. 76; the hot coronas of these stars are also observable as x-ray point sources with a harder spectrum). The total luminosity of the diffuse region, $L_{0.2-4\,\mathrm{keV}} \sim 2 \times 10^{35}$ erg/sec, yields a mass of x-ray emitting plasma of $3-30\, M_\odot$; what we are observing here may in part be radiation associated with the remnant of the recent supernova explosion G 287.8–0.5 (see Section 7).

The array of optical, radio, and x-ray observations of the region paint the following picture (Seward and Chlebowski, 1982). Two associations, Car OB1 and Car OB2, are associated with a dense cloud of gas and dust. Ionizing radiation, wind, and supernova explosions within these associations are in the process of destroying the cold, dense cloud. Cold gas is found to be irregularly mixed in with warm and hot gas. The warm gas ($T_e \sim 10^4$ K, $n_e \sim 200-300\,\mathrm{cm}^{-3}$) constitutes most of the mass in the optical nebulae. These H II regions also house hot zones ($T_e \sim 5 \times 10^6 - 10^7$ K, $n_e \sim 0.1\,\mathrm{cm}^{-3}$) and individual high-density hot clouds ($T_e \sim 10^7$ K, $n_e \sim 1\,\mathrm{cm}^{-3}$) radiating x rays, but they are observable only on the near side of the cloud.

X rays seem to provide the most promise in searching for supershells produced by wind and supernova explosions around nearby Galactic associations. Such observations began relatively recently, but already they have yielded a great deal of information.

A supershell has been discovered in Cygnus — a shell-like source of soft x rays approximately 20° across (400–500 pc at a distance of 1–2 kpc), with total luminosity $L_{0.5-1\,\mathrm{keV}} = 5 \times 10^{36}$ erg/sec and temperature $T_e = (1-2) \times 10^6$ K (Cash et al., 1980). In all likelihood, the Cygnus supershell is not an isolated entity. According to Bochkarev and Sitnik (1985), we are observing in that portion of the sky the combined effect, seen in projection, of bubbles produced by the wind from eight associations containing a total of close to 100 O stars, including 20 WR stars and eight Of stars, plus about a dozen isolated supernova remnants. When we look in the direction of Cygnus, we are sighting along a spiral arm, and that accounts for the anomalously high density of OB associations, high-luminosity stars, and supernova remnants.

Another supershell detected in x rays has been linked to the Mon OB1 and Mon OB2 associations (Nousek et al., 1981). In that object we observe a shell-like source of soft x rays 20° across (linear diameter 200 pc); the mean gas density in the shell is about $0.01\,\mathrm{cm}^{-3}$, and the temperature is $T_e \approx 3 \times 10^6$ K. Like the Carina supershell, the hot Monoceros–Gemini shell is bordered by clouds of neutral hydrogen, and here as well we encounter a large-scale loop — a nonthermal radio source (supernova remnant) (see Section 11).

The extended x-ray source related to Cep OB3 and the H II region Sh 155 possesses similar properties (Fabian and Stewart, 1983). Its luminosity, $L_{0.2-4\,\mathrm{keV}} \approx 2 \times 10^{32}$ erg/sec, may be entirely due to emission from the hot wind of stars in the association.

Finally, we come to the giant Orion–Eridanus shell of gas and dust, in which the interactions taking place between stars in the association and interstellar gas are most clear-cut. This region, a rich center of gas, dust, young star clusters, and star-forming regions, has long held the attention of observers. A comprehensive review of investigations of this complex may be found in Goudis (1982); here we simply touch upon the large-scale structure and kinematics due to the effects of wind and supernovae. The region, which includes the Ori OB1 and λ Ori associations, is diagrammed in Fig. 99. A multitude of both diffuse and thin filaments of ionized and neutral gas and dust form a unique shell approximately 280 pc in size (Reynolds and Ogden, 1979). The shell is expanding at 15–25 km/sec and the mass of ionized gas contained within it is $M_{\mathrm{HII}} \approx 8 \times 10^4\, M_\odot$, but that comprises only about 10% of the total mass of neutral gas. Barnard's Loop

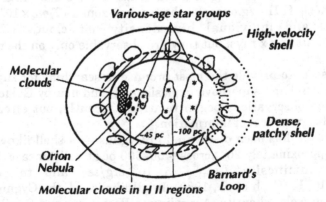

Fig. 99. Structure of the giant Orion–Eridanus shell (Cowie et al., 1979; Reynolds and Ogden, 1979); see text.

(approximately 140 pc diameter) and a number of isolated nebulae lie within the H I shell. The λ Ori association extends beyond Barnard's Loop, and is surrounded by a ring-like H II region of its own (see Section 16), but it does belong to the same giant shell complex. The total kinetic energy of the supershell is somewhat greater than 10^{52} erg, and the source of ionization is made up of the stars in Ori OB1 ($L_{UV} \gtrsim 4 \times 10^{49}$ photons·sec^{-1}) and λ Ori.

Two sets of absorption lines show up in UV spectra of the stars of Ori OB1 and λ Ori (Cowie et al., 1979). One, the low-velocity system ($v \approx 10-20$ km/sec), corresponds to high density and a high degree of ionization, and it varies markedly from star to star. This absorption is associated with ionized gas in Barnard's Loop, in the ring nebula around λ Ori, and in the other clouds making up the overall patchwork shell. The other, which is common to all stars investigated, is associated with absorption in a uniform, low-density shell that is expanding at 100–120 km/sec. The high-velocity shell is made up of ionized gas (mass approximately 100 $M_\odot$) and is probably the layer of radiatively cooling gas trailing the shock front that is expanding into a low-density medium with $n_0 \approx 3 \times 10^{-3}$ cm^{-3} (Cowie et al., 1979).

The age of the rapidly expanding shell given by its size and velocity is $t \approx 3 \times 10^5$ yr. The radiative shock wave that is sweeping up gas was most probably formed by the wind from the association and by supernova explosions. The most important contribution is probably due to the most recent outburst, which took place about $t \approx 3 \times 10^5$ yr years ago. Gas in the high-velocity outer shell is heated by the shock wave to a temperature $T_e \approx (1-2) \times 10^5$ K, and its column density reaches $N_H \approx (2-3) \times 10^{17}$ cm^{-2}. The centers of expansion of the gas behind the high-velocity radiative shock front and the clouds of ionized and neutral gas in the slow-moving shell coincide. The dense, slow-moving clouds seem likely to have been accelerated by the passage of shock waves induced by previous supernova explosions and the wind from the association. Within the slow shell are concentrated dense molecular clouds, bright H II regions (including the Orion Nebula), and regions of star formation. The star formation process in this complex has progressed from right to left through the molecular clouds depicted in Fig. 99.

All of the shells and supershells chosen as examples are observed around young OB associations; their characteristic age ranges from several million to several tens of millions of years. We can now attempt to trace out the evolutionary sequence followed by supershells.

Initially, the shell structure around a young unevolved cluster or association will be produced solely by stellar wind, since the most massive stars only start to explode after $\sim 2 \times 10^6$ yr. The most obvious example of a shell in the earliest stage of development is the Rosette Nebula, whose central cavity has been produced by wind from the compact cluster NGC 2244 (including an Of star; see Section 16), which is the youngest subgroup of stars in the Mon OB2 association. The ring-like emission nebula around the λ Ori association is a similar object. The latter is a more extensive shell formed by the wind of the association, but still lacking any visible signs of supernova explosions.

In all likelihood, the gas and dust shell surrounding the association Cep OB4 (with the open cluster Be 59) and the bright H II region W 1 (G 118.6+4.8,G 118.1+5.0) is a young object of the type in question. Lozinskaya and Sitnik (1977) revealed that the nebula is expanding at 35–40 km/sec, and the shell radius is 25–30 pc. The shell is spatially coincident with Cep OB4, their centers lie in the vicinity of the bright nebula W 1, their angular dimensions are similar, and they are at the same distance. This object is not yet a supershell, but more likely a supershell in the very earliest stage of development. Blanco and Williams (1959), having discovered the association, detected anomalous reddening of high-luminosity stars, which they attributed to the extreme youth of the cluster — the early-type stars are still enshrouded in an envelope of embryonic matter. Forty-four OB stars and a number of protostellar objects indicative of ongoing star formation have been identified in the association, including 24 faint stars with Hα emission, most of which are variable, T Tauri stars, peculiar nebulous starlike objects (Cohen and Kuhi, 1976), and an OH maser source (Rudnitskii, 1978).

Cohen and Kuhi (1976) maintain that star formation did not take place all at once in this object. They have identified different age subgroups of stars and starlike objects, spanning the range 10^5 to 2×10^6 yr.

A second, inner shell has been revealed by analyzing optical, radio, and infrared observations (Lozinskaya et al., 1987). It has been demonstrated that both large gas and dust shells are most likely to have been formed by stellar wind coming from Cep OB4 and the cluster Be 59. Expansion at 30–40 km/sec may be related to a powerful but short-term wind (either a short-lived or invisible source of stellar wind) or the explosion of a supernova $\sim 3 \times 10^5$ years ago. Synchrotron emission from that remnant ought to be much weaker than the thermal radiation coming from the ionized gas of the shell. According to the IR measurements, the mass of gas swept out of the inner shell is $\sim 10^3 \, M_\odot$, and the mass of the entire gas and dust complex associated with Cep OB4 is approximately $2 \times 10^4 \, M_\odot$ (Lozinskaya et al., 1987).

As the process of star formation continues, as new sources of stellar wind wind turn on, and as supernovae subsequently explode, the size of a shell will increase and the amount of both neutral and ionized gas that has been swept up will grow. The more than 40 OB stars and aforementioned protostellar and stellar objects constitute a reservoir of ionizing radiation, wind energy, and supernova explosions sufficient to produce a supershell several hundred parsecs in size during the process of evolution.

Heiles (1979, 1984) has quite possibly detected the final stage in the evolution of a supershell produced in an association by wind and supernova explosions: H I shells $10^2 - 10^3$ pc across, some of which are expanding at 10–20 km/sec. Incorporating corrections due to Bruchweiler et al. (1980),

Table 31. Supershells around OB Associations: Predicted Expansion Velocity and Radius Parallel and Perpendicular to the Galactic Plane (Bruchweiler et al., 1980)

Parameter	Phase I	Phase II	Phase III
$t, 10^6$ yr	3	11	19
v, km/sec	20	5	5
$R_\parallel$, pc	106	185	250
$R_\perp$, pc	105	207	500

the kinetic energy of one of the highest-energy shells known, the object G 139.03–69, is at most 5×10^{53} erg. As can be seen from Table 29, the wind energy of the two richest groups of stars in the Cygnus and Carina regions reaches $\sim 10^{53}$ erg over a lifetime spanning 10^7 yr.

No more than 10–20% of all H I supershells — specifically, the smallest ones — have been identified with OB associations. The H II shells and supershells referred to above, and associated with OB associations, are also smaller than the average H I supershells identified by Heiles. This makes sense if we note that H II shells and supershells are ionized by main-sequence O and B stars over the course of the first $\sim 10^7$ years, and then, as massive stars finish evolving, the flux of ionizing radiation diminishes, but the size and mass of the swept-up shell continue to increase by virtue of the energy injected by supernovae (see below). Furthermore, an association within which massive OB stars were already fully evolved would be difficult to discriminate within an extended H I supershell.

The scale of the phenomenon — the size and mass of the shells and giant supershells — is entirely determined by the flux of ionizing radiation, the power and duration of the wind, and the number of supernovae, i.e., ultimately, by the richness of the cluster. Three giant supershells have been identified in the local spiral arm of the Milky Way: in Cygnus (Cash et al., 1980), in Scorpio–Centaurus (Weaver, 1979), and, as described above, in Orion–Eridanus. The centers of all three lie quite high above the Galactic plane, at $z \approx 100$ pc. That fact, combined with the reconstructed large-scale magnetic field structure of the associations and gas and dust clouds suggest that star formation in these regions has been influenced by Rayleigh–Taylor instability (Bochkarev and Sitnik, 1985). According to Bochkarev (1984), the lack of spiral density waves in the local arm may possibly be due to its location near the corotation radius, and that does not preclude the development of Rayleigh–Taylor instability right through the stage at which gravitationally coupled massive star complexes are being formed. Under those circumstances, rich star clusters are formed, and around them giant shells typically ~ 1 kpc across. It is for this very reason that giant shell complexes are primarily observed in irregular galaxies like the Large Magellanic Cloud, and on the outskirts (or in favorable zones near the corotation radius) of spiral galaxies. Star formation initiated by a spiral density wave will lead to the formation of somewhat sparser associations with shells 200–500 pc in size.

We have thus shown, with a purely empirical set of examples based on Galactic and extragalactic objects, that wind energy and the explosion of massive stars in associations suffice to produce supershells. The stars in the associations are the source of ionizing radiation; in the late stages, after evolved stars have exploded, supermassive shells consist mainly of neutral gas. Is it at all possible to construct a theoretical model of the combined interaction with interstellar gas of the UV radiation, stellar wind, and supernovae in an association? Bruchweiler et al. (1980) have attempted to do so for a "standard" association consisting of about three dozen stars earlier than B0, with the following idealizations. The shell is produced in an interstellar medium having no small-scale fluctuations, and with a density gradient perpendicular to the Galactic plane, $n = n_0 \exp(-z/z_0)$. The first $\sim 3 \times 10^6$ yr of supershell evolution are governed solely by wind, after which all stars with initial mass $M_{init} \geq 15 M_\odot$ simultaneously explode. The second phase of supershell expansion is determined by the energy of these supernovae and the wind of as-yet unevolved stars. In the third phase, we have the simultaneous explosion of all stars with $M_{init} \geq 8 M_\odot$.

The first phase of expansion is described by Eqs. (13.5), with subsequent allowance for the energy influx from supernovae in the snowplow stage (see Section 9); gravitation is also taken into consideration, retarding the expansion of sufficiently large shells in the direction normal to the plane of the Galaxy.

Table 31 contains the computed results: the duration of the three stages, the expansion velocity, and the radius (parallel and perpendicular to the plane of the Galaxy) of a supershell at a galactocentric distance of 10 kpc (for $z_0 = 150$ pc and $n_0 = 1$ cm^{-3}). At 5 kpc ($z_0 = 70$ pc, $n_0 = 3$ cm^{-3}), the final shell radius reaches 180 pc, while at 20 kpc ($z_0 = 500$ pc, $n_0 = 0.1$ cm^{-3}), it reaches 500–700 pc. Clearly, the predicted size and expansion velocity of the "theoretical" supershell are close to the observed values.

A different and perhaps more realistic model has been computed by Tomisaka et al. (1981). An association with 20–100 early-type stars injects energy for the first $\sim 3 \times 10^6$ yr only in the form of stellar wind, at a rate $L_w = 1.3 \times 10^{36}$ erg/sec. (The authors have explicitly adopted a low value; this is the energy imparted by the wind of a single WR star or a sparse association containing no WR or Of stars or early supergiants; see Table 30.) The next stage (wind + supernovae) is characterized by a constant supernova rate with either $\tau = 2 \times 10^5$ yr or $\tau = 10^6$ yr (each explosion yields $E_0 = 10^{51}$ erg, $\Delta M = 3 M_\odot$). That stage continues for $(20 - 100)\tau$ yr. The unperturbed gas density is assumed constant and fairly low: $n_0 = 1$, 0.1, or 0.01 cm^{-3}. The calculations, which terminate when the shell has slowed to $v = 8$ km/sec, show that wind and some dozens of supernovae induce the formation of a 200–1000-pc shell that expands as $R \propto t^\eta$, where

$\eta = 0.25 - 0.50$, depending on the density of the ambient medium and the time τ between outbursts.

We now give the relations obtained by McCray and Kafatos (1987) for the variation of radius, expansion velocity, and kinetic energy of a supershell around an association containing N_* stars of initial mass $7 \lesssim M_{init} \lesssim 30\,M_\odot$. These workers assumed that supernovae are the dominant source energizing the supershell between $t \sim 5 \times 10^6$ yr (at which time stars with $M_{init} \gtrsim 30\,M_\odot$ leave the main sequence) and $t \sim 5 \times 10^7$ yr (when stars with $M_{init} \sim 7\,M_\odot$ have finished evolving). Prior to the onset of efficient cooling of the supershell, its evolution is given by

$$R_s = 97(N_* E_{51}/n_0)^{1/5} t_7^{3/5} \text{ pc,} \tag{18.1}$$

$$v_s = 5.7(N_* E_{51}/n_0)^{1/5} t_7^{-2/5} \text{ km/sec.} \tag{18.2}$$

The kinetic energy reaches

$$E_s = 4 \times 10^{49}(N_* E_{51}) t_7 \text{ erg.} \tag{18.3}$$

The mean gas density is

$$n_s = n_0 (v_s/a_s)^2 = 32(N_* E_{51})^{2/5} n_0^{3/5} a_s^{-2} t_7^{-4/7} \text{ cm}^{-3}, \tag{18.4}$$

where $a_s = (kT_s/\mu + H_s^2/4\pi\rho_s)^{1/2}$ km/sec is the magnetoacoustic velocity in the shell.

Adiabatic expansion of the supershell gives way to the radiative cooling stage, in which hot-plasma energy losses become comparable to the influx of energy from stellar wind and supernovae. In that phase, the motion of the shell is governed by the equations

$$R(t) \propto R_{cool}(t/t_{cool})^{0.25},$$

$$R_{cool} \approx 50(N_* E_{51})^{0.4} \zeta^{-0.9} n_0^{-0.6} \text{ pc,} \tag{18.5}$$

$$t_{cool} \approx 4 \times 10^6 (N_* E_{51})^{0.3} \zeta^{-1.5} n_0^{-0.7} \text{ yr.}$$

Here ζ is the metallicity of the interstellar medium, normalized to $\zeta = 1$ for the solar plasma.

We see, then, that old shells are comparable in size to the thickness of the galactic disk, and the expansion velocity is close to the random velocities of gas clouds. The "polar" regions of a supershell are Rayleigh–Taylor unstable, a situation that can lead to the leaking of hot plasma into the corona of spiral galaxies (Tomisaka and Ikeuchi, 1986; McCray, 1988; Igumentshchev et al., 1990).

In irregular galaxies, on the one hand, there are no constraints on the size of a supershell engendered by this sort of leakage from the disk. On the other, the low metallicity and uniform density of the gas ($\zeta = 0.3$ and $n_0 \sim 0.35\,\mathrm{cm}^{-3}$ in the LMC) increase R_{cool} and t_{cool}; see (18.5). Both factors in concert facilitate the creation of giant supershells in irregular galaxies like the LMC, and in the outer reaches of spirals. In irregulars, moreover, the process is not curtailed by the passage of spiral density waves. The observations of Courtes et al. (1987) have actually shown that with increasing galactocentric distance in the spiral galaxy M33, the many shells and supershells become larger, and they also come closer to being perfectly spherical.

The agreement between the numerical models and observational results should not be overestimated; an elaboration of this sort of theoretical analysis would seem to be rather unpromising. In fact, the supershells around an average association turn out to be comparable in size with the thickness of the Galactic disk. In no way can the interstellar medium be considered uniform on such scales, and as we discussed in Chapters 2 and 3, the density of the ambient gas itself directly affects the structure and kinematics of supernova remnants and wind-formed shells. In addition, young OB associations are intimately related to dense and extremely inhomogeneous molecular complexes, and the process whereby associations are created — and consequently the way in which ionizing radiation, wind, and supernovae are turned on in these complexes — is played out serially, which further complicates the modeling. Even within the confines of a single association, stars of differing mass are probably also of different ages (see Doom et al. (1985)).

Finally, the most serious assumption is that wind and supernovae act upon homogeneous unperturbed interstellar gas, whereas only ionizing radiation, prior to the turn-on of wind and supernovae, can produce ring-like H I and H II regions (see the series of calculations by Beltrametti et al. (1982), Tenorio-Tagle et al. (1982), and Bodenheimer et al. (1983); all of these papers contain further references to the literature). In a rich cluster containing hundreds of OB stars of various masses, both their departure from the main sequence and a slight reduction in the effective temperature during the hydrogen-burning stage take place at different times. There are two competing processes: the number of ionizing photons falls off as the more massive stars leave the main sequence, and there is a decrease in the density of ionized gas as an H II region expands, thereby reducing the recombination rate; under certain conditions, ring structures in the surrounding H I and H II regions can result. According to the calculations of Beltrametti et al. (1982), the total amount of ionizing radiation from a rich cluster containing, for example, 2×10^5 stars with an initial mass function specified by an exponent $\gamma = 2.45$ can be assumed constant and equal to $L_{UV} = 2 \times 10^{51}$ photons·sec^{-1} for the first 2×10^6 yr (with the further assumption that all stars with $0.5 \leq M_{init} \leq 30\,M_\odot$ came into being at the same time). Then, when at $t_{crit} \approx (4-6) \times 10^6$ yr the most massive stars with $M_{init} \approx 30\,M_\odot$ begin to leave the main sequence, the ionizing radiation is gradually supplied by the less massive stars, and varies as $L_{UV} \propto t^{-m}$, $m = 5$ until $t = 5 \times 10^7 - 10^8$ yr. Beltrametti et al. have identified four stages in the evolution of an H II region associated with such a cluster. The first $(t < t_{crit})$ is the classic expansion of an H II region (Spitzer, 1981; Kaplan and Pikel'ner, 1979); the radius of the ionization front is given by

$$R_i(t) = R_{St}\left(1 + \frac{7}{4}c_{II}\frac{t}{R_{St}}\right)^{4/7}, \qquad (18.6)$$

where R_{St} is the Strömgren radius, and c_{II} is the speed of sound in the ionized gas. The expansion of an H II region in this phase is accompanied by a shock wave that compresses the neutral gas, and a rarefaction wave in the ionized gas.

The second stage (recombination and formation of an H I shell) relates to the decrease in ionizing flux after t_{crit}. If the decrease in the number of ionizations due to the falloff in L_{UV} is greater than the decrease in the number of recombinations due to the falloff in density within the expanding H II region, the radius of the ionization front will decrease as well:

$$R_i(t) = R_{crit}\left(\frac{t_{crit}}{t}\right)^{m/3}, \quad R_{crit} \equiv R(t_{crit}). \qquad (18.7)$$

This decrease continues as long as the velocity $\dot{R}_i$ is comparable to the speed of sound. At the end of the second phase, a shell expanding according to (18.6) will be neutral.

In the third stage, the numerical calculations (Beltrametti et al., 1982) show that the drop in density, $n_{HII} \propto t^{-8.5}$, dominates the decrease in L_{UV}, the radius varies as $R_i \propto t^4$, and the ionization front propagates through the neutral shell until such time as the latter is fully ionized. The ionization front propagates somewhat further into the ambient gas, until in the fourth and last stage, full recombination of the entire ionized region takes place, due to the increasing size of the ionized region and the continuing decrease in the flux of ionizing radiation.

An H II region may not pass through stage II — i.e., a shell of neutral gas may not form — if the density of the unperturbed gas is high, $n_0 \gtrsim 400\,\mathrm{cm}^{-3}$, or the ionizing flux falls off slowly. In that event, the ionization front continues to expand despite the dropoff in ionizing flux.

It is clear, then, that even prior to supernova explosions, and without allowing for the effects of stellar wind, there may be either an H II region with a more or less uniform density distribution (phase I), an extensive, massive, expanding H I shell (phase II), or a giant ring nebula or expanding H II shell (phase III) around an OB association. For an association with $L_{UV} = 2 \times 10^{51}$ photons·sec^{-1}, $t_{crit} = 5 \times 10^{6}$ yr, and $n_0 = 100\,\mathrm{cm}^{-3}$, the shell can reach hundreds of parsecs in size, with an expansion velocity of about 10 km/sec.

If a cluster and H II region are located at the boundary of a dense molecular complex (which is in fact where they usually occur), this symmetric picture of a multilayer H II region will be complicated by the "champagne effect" (Tenorio-Tagle et al., 1982).

In actual molecular and stellar complexes, ionizing radiation, wind, and supernovae act in concert. Wind accelerates the formation of shell-like structures around associations, and produces isolated, small-scale cavities around the most powerful sources — WR, Of, and other stars — within supershells, as we observe, for example, in the extensive Cygnus and Carina complexes. Explosions of the most massive stars, beginning after some millions of years, interact with the gas of H II regions that are already fully evolved, and which have been perturbed by ionization fronts and stellar wind. At any given stage, the dominant mechanism in a particular supershell can be identified, as noted above, via the presence of soft x rays (a sign of stellar wind or a recent supernova explosion), radio synchrotron emission (signifying a supernova), or the intensity ratios [N II]/Hα, [S II]/Hα, and so on in the spectrum of the optical nebula (photoionization or collisional excitation).

Building a theory that takes all three factors into account is a task for the future. At this point, two-dimensional numerical calculations have been carried out for a model supershell around an association in a galactic disk with a density gradient, taking into account the indicated evolutionary process for an H II region, and two subsequent sets of explosions, the first with 17 stars of initial mass $M_{init} \geq 15\,M_\odot$, and the second of 180 stars of initial mass $8 < M_{init} < 15\,M_\odot$ (Bodenheimer et al., 1983). In this sort of model, it is easy to obtain a supershell of size ~100–200 pc, expanding at $\lesssim 100$ km/sec. But to explain the giant structures that are observed to be ~500–1000 pc across, it is necessary to sustain the expansion of the shell for more than 10^7 years. In that long a time, tidal destruction of the shell due to differential rotation of the Galaxy becomes a serious issue, as does the serial nature of star and association creation within the boundaries of a giant star complex and giant molecular cloud. It is conceivable that the very largest galactic supershells are formed as a result of other large-scale perturbations — those associated, for example, with the fallout of high-latitude clouds that have cooled from the corona into the gaseous disk (Tenorio-Tagle, 1981; Tenorio-Tagle et al., 1987). There is no need for such "exotic" mechanisms for producing giant shells, however, if the large-scale structure of galactic molecular clouds and star complexes is taken into consideration.

According to our present understanding, massive giant molecular clouds, where star clusters and associations form, are fragments of coherent, giant superclouds that are hundreds to thousands of parsecs across. One consequence of this is the hierarchical clustering of stellar aggregates: a few (two to five) clusters and associations form a group, two or three groups comprise a stellar complex, and complexes combine into giant "beads," which form the spiral arms of galaxies (Efremov, 1984; Efremov and Sitnik, 1988). On the scale of a rich aggregate or stellar complex, giant shells can be explained naturally in terms of the effects of radiation, stellar wind, and supernovae.

The real picture — a multifaceted one — is of course more complicated. Since most of the parent cloud is still present when a cluster is born, stars of the next generation continue to be formed, a process initiated by the propagation of ionization fronts and shock waves. This active phase in the evolution of a molecular complex continues for $10^7 - 10^8$ years, a period comparable to the lifetime of massive stars and star associations, and it exceeds the time for stars to condense out of the protostellar cloud, as well as the time required for dissipation of supernova remnants and the shells swept out by stellar wind from an isolated star. One therefore observes genetically related groups of clusters and associations, giant molecular clouds and H II regions, giant expanding shells produced by stellar wind and

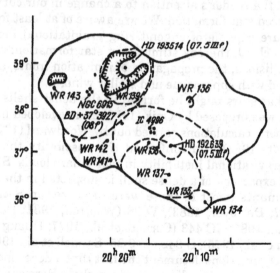

Fig. 100. The "hierarchical" multishell structure in the interstellar medium produced by radiation and stellar wind in the vicinity of Cyg OB1. It consists of a shell encompassing Cyg OB1, two components surrounding NGC 6913 and IC 4996, and small-scale interior shells and bubbles around the Of and WR stars (Lozinskaya and Sitnik, 1988).

multiple supernova explosions, regions of star formation at their boundaries, and in addition small-scale shells — supernova remnants and cavities swept out by the wind from individual stars. Good examples of such star and gas complexes are the supershell LMC 2, which we mentioned at the beginning of this section, and the giant complex of shells and supershells in the Perseus arm, which is associated with Cas OB2, NGC 7510, and Ba 3 (see Lozinskaya et al. (1986)).

Many supershells display a clear-cut hierarchical structure, possibly reflecting the sequence of star formation. This sort of nested hierarchy of different-size shells is evident, in particular, around the Cyg OB1 association and its constituent young clusters NGC 6913 and IC 4996 (Lozinskaya and Sitnik, 1988).

Figure 100 contains a diagram of the region, with an extensive common shell around Cyg OB1, two of its components around NGC 6913 and IC 4996, and small-scale shells around the WR and Of stars belonging to Cyg OB1. The northwestern part of the common shell (around IC 4996) is most clearly visible in optical and radio emission from ionized gas; the eastern part (around NGC 6913) is best seen in the 60 and 100 μm IRAS bands via the infrared emission from heated dust (Lozinskaya and Repin, 1990).

To conclude this discussion of supershells around young groups of stars, we direct the reader's attention to a change in our current ideas on supernova-induced star formation. We are aware of at least four sources of external pressure capable of engendering gravitational instability in a giant molecular cloud, thereby stimulating star formation: spiral density waves, cloud collisions, the propagation of ionization fronts, and the shock waves associated with supernovae and stellar wind.

The idea that stars might be formed in the dense shells of old supernova remnants was proposed by Öpik (1953) and developed in a number of subsequent papers; calculations carried out by Woodward (1976) and Krebs and Hillebrandt (1983) also confirmed the possibility that supernovae might induce gravitational instability in molecular clouds. Star formation induced by the expansion of a dense shell is suspected in the following old supernova remnants (most of these were discussed in Section 7): W 44 (Wootten, 1977; DeNoyer, 1983), W 28 (Wootten, 1981; DeNoyer, 1983; Odenwald et al., 1983), IC 443 (Cornett et al., 1977; Huang et al., 1986), S 147 (Wootten et al., 1975), G 342.01+0.25 (Sandell et al., 1983). For these remnants, the observations document the fact that a dense molecular cloud has been shocked by an expanding supernova shell. Here one finds compact IR sources and OH or H_2O masers, phenomena that are often associated with star-forming regions.

Searches for CO, OH, and H_2CO clouds impacted by the shock waves induced by shell expansion have been carried out in dozens more supernova remnants, none with unambiguously positive results (Slysh et al., 1979). Even for the objects listed above, however, there is no hard evidence of a connection between the maser sources and embryonic stars; in addition, those supernova remnants are located in regions populated by young star groups, so that regions of star formation may be flagged not by an individual supernova outburst, but rather by the combined activity of radiation, wind, and a number of supernovae. Two such examples are IC 443 (a compact infrared source is located 12 pc from the boundary of the remnant, and its age is $\sim 10^5$ yr (Odenwald and Shivanandan, 1985), much greater than that of the remnant; see Section 7) and HB 3 (the known star-forming region W 3, although it is located at the boundary of the supernova remnant, was probably not directly triggered by it; see Lozinskaya and Sitnik (1980) and Landecker et al. (1987)).

All of the Galactic star-forming regions that have been investigated, including CMa R1, Ori A, W 3, Cep OB3, Cep OB4, Sco OB1, and Per OB1, are associated with young star clusters and associations. The clear implication is that rather than isolated supernovae being the external pressure source inducing gravitational instability (which culminates in star formation), as was once believed, the source consists of expanding shells and supershells produced by the radiation, wind, and multiple supernovae in OB associations and clusters. The expansion of such shells and supershells in dense clouds or "superclouds" with randomly distributed cold, dense molecular clouds stimulates the formation of massive second-generation stars. A general idea of the stage at which gravitational instability leads to the fragmentation of a supershell, leading to a new round of star formation, is given by the following relations (McCray and Kafatos, 1987). For a shell expanding according to (18.1) and (18.2), we have

$$t_{frag} \sim 3.2 \times 10^7 (N_* E_{51})^{-1/8} n_0^{-1/2} a_s^{5/8} \text{ yr,}$$

$$R_{frag} \sim 200 (N_* E_{51})^{1/8} n_0^{-1/2} a_s^{3/8} \text{ pc,} \qquad (18.8)$$

$$M \sim 5 \times 10^4 (N_* E_{51})^{-1/8} n_0^{-1/2} a_s^{29/8} M_\odot .$$

If the hot plasma pressure within a supershell has already fallen, due to radiative cooling ($t > t_{cool}$) or breakthrough of the shell, then

$$t_{frag} \sim 1.2 \times 10^7 (N_* E_{51})^{-1/15} n_0^{-11/15} a_s^{4/5} R_{100}^{-7/15} \text{ yr,}$$

$$R_{frag} \sim 100 (N_* E_{51})^{1/15} n_0^{-4/15} a_s^{1/5} R_{100}^{7/15} \text{ pc.} \qquad (18.9)$$

where $R_{100} \equiv R_{cool}/100 \text{ pc}$.

The dependence of t_{frag} and R_{frag} on the magnetoacoustic velocity means that a regular interstellar magnetic field inhibits the development of instability.

Stars that form in expanding shells, i.e., that are initiated by shock waves, should be more massive than the central stars of the first generation (Elmegreen and Lada, 1977).

These massive second-generation stars turn out to be located on the outskirts of the large common shell; they are relatively short-lived ($t \lesssim 10^7$ yr), and according to calculations of Kolesnik and Silich (1988), their explosion in the late stages of supershell evolution comprises an important source of energy. Figure 101 diagrams the scenario for induced star formation in a supershell expanding within a giant supercloud containing dense molecular clouds.

Naturally, the sizes and ages of supershells in which induced fragmentation and second-generation star formation are taking place, as given by Eqs. (18.8) and (18.9) and shown in Fig. 101, are crude approximations. Nevertheless, observations of LMC-class irregular galaxies, where the development of giant shells is favored by low metallicity, a homogeneous interstellar medium, and the absence of spiral density waves, actually confirm the existence of a "secondary" cascade of star formation induced by the gravitational instability of supershells (Dopita, 1987; Feitzinger, 1987; Smith et al., 1987; McCray, 1988; references cited in these papers).

In the LMC, a number of giant ring-like structures have been observed that contain OB associations, H II regions, and supernova remnants, and these bound giant cavities in the neutral hydrogen distribution. The most impressive examples are the giant ring complex (stellar aggregate) Shapley III (more than 600 pc in diameter), and the ring-like chain of dozens of young OB associations around Loop IV (approximately 1500 pc across) (Dopita et al., 1985). Clear-cut annular regions of induced star formation have also been revealed by UV observations of the LMC (Smith et al.,

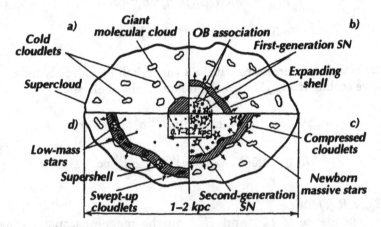

Fig. 101. A cascade of induced star formation in a supershell surrounding an OB association. The latter is expanding inside a giant cloud containing smaller dense, cold molecular clouds (Kolesnik and Silich, 1987).

1987). A perfect example of a giant ring of young stars and stellar objects has been discovered in the irregular galaxy NGC 4449 (Bothun, 1986). Such structures most probably bear witness to the gravitational fragmentation of giant shells, and a new round of star formation.

In our own Galaxy, spiral density waves inhibit the formation of such giant structures. Furthermore, shells "leak," due to the instability of the "polar" regions, when they become comparable in size to the thickness of the gaseous disk of the Galaxy. But, within the Milky Way, we have indeed observed somewhat less grandiose instances of serial star formation induced by the expansion of shells produced by radiation, wind, and supernovae in rich OB associations. The aforementioned hierarchy of nested shells may conceivably reflect a process of step-by-step star formation within an association. In the vicinity of Cas OB2, we see approximately 20 young stars and protostellar objects, and they are all located on the periphery of a shell expanding into the molecular complex. In the Cyg OB1 region, the youngest members of the stellar aggregate are massive WR and Of stars located preferentially on the outskirts of the common shell, forming secondary small-scale cavities about themselves (Lozinskaya and Sitnik, 1988).

IRAS observations have ushered in a new era in the search for star-forming regions engendered by the expansion of shells and supershells. Objects found include IR sources in dense, shocked knots of molecular clouds within IC 443 (Huang et al., 1986), and the so-called IRAS stellar shells (Schwartz, 1987). The IRAS observations have shown that discrete Galactic IR sources — young stars, pre-main sequence stars, compact H II regions, and the very youngest objects of related types — form well-defined shell structures typically 40 to 300 pc in size. That sort of structure could well result from star formation induced by the expansion of a shell or supershell.

This has been far from a complete list of examples demonstrating serial or induced star formation. A detailed review of the most recent achievements in this field can be found in the volumes *Physical Processes in Interstellar Clouds* (1987), *Star-Forming Regions* (1987), and *Star Formation in Galaxies* (1987).

Conclusion

In closing, it would be well to repeat with complete conviction a remark
made at the outset: the nature of supernovae and their interaction with the
interstellar medium is far from being well-understood. This derives not
only and not so much from a paucity of observational data or theoretical
research as from the complexity of the phenomenon itself. Consideration of
stellar wind and the ejection of a slow-moving shell prior to the explosion of
a supernova opens a new avenue of research relating to the interaction between stellar and interstellar matter. But in essence, we have only just
realized that the actual issue involves the need to study stellar wind and
supernovae simultaneously, piecing together a more or less clear-cut set of
individual building blocks: the interaction of a supernova shell with the
gas of the wind, the interaction of an old supernova remnant with the interstellar medium, and the interaction of stellar wind with interstellar gas.
We have thus far not managed to construct an orderly edifice out of these
blocks, although both the path to a solution and the difficulties involved
are quite apparent.

Here we present just one final example. The expansion of a supernova
shell, which is governed by the propagation of the blast wave into the circumstellar gas and the back propagation of the reverse shock in the ejecta
(see Sections 8 and 9) presents a complex picture. That picture is further
complicated if one takes into account the multilayer inhomogeneity of the
ambient medium (with strong density gradients), which is due to mass
losses from the progenitor. Even in the simplest models of the density distribution of stellar wind, ejecta, and interstellar gas, the physical processes
in young supernova remnants ($R \sim 1-2$ pc, $t \approx 100-500$ yr) are determined
by the interplay of at least four shock waves (see, for example, the calculations of Itoh and Fabian (1984)). At the boundary between the stellar and
interstellar gas, a reverse rarefaction wave first appears; this is followed by
a compression wave and a strong, inward-reflected shock that heats the
gas of the wind anew. The collision of the latter with the expanding ejecta
initiates a second shock wave in the ejecta, propagating inward, and a
third shock in the matter of the wind, which is reflected outward. And this
description applies only to a "two-layer" circumstellar gas, comprised of the
slow wind from a red supergiant and the homogeneous interstellar gas beyond! In fact, it is really necessary to consider any pre-existing fast wind
from a main-sequence O star as well, which perturbs the interstellar
medium out to a distance of perhaps 10 pc, and a possible short-lived phase
involving an extremely powerful wind immediately preceding the supernova explosion.

The low-mass progenitors of type Ia lose matter in the form of a slow-
moving shell — a planetary nebula; the progenitors of WR stars probably

expel a massive exterior hydrogen shell (see Chapter 3). The high-velocity shell thrown off by the explosion of a supernova overtakes the matter in the slow-moving ejecta; the wind from the progenitor (which in WR and Of stars is actually observable) also interacts with the slow-moving shell... The complexity of the problem is staggering — what more could one possibly say?

The interaction of stellar wind with the interstellar medium is also still far from clear, even in what would seem to be the simple case of an isolated strong source of wind — a Wolf–Rayet or Of star. The large-scale influence of supernovae and stellar wind on the gaseous media of galaxies must be viewed as one branch in a complex chain of events determining the evolution of stellar and interstellar matter. In Chapter 4, we have only sketched in the major links in this chain. Each one is in fact a complex problem for which, in general, we have only just formulated the questions and indicated a path to their solution.

References

A

Abbott, D. C., Astrophys. J. **263**, 723 (1982).

Abbott, D. C. and Conti, P. S., Ann. Rev. Astron. Astrophys. **25**, 113 (1987).

Abbott, D. C., Bieging, J. H., Churchwell, E., and Casinelli, J. P., Astrophys. J. **238**, 196 (1980).

Abbott, D. C., Telesco, C. M., and Wolff, S. C., Astrophys. J. **279**, 225 (1984).

Abbott, D. C., Bieging, J. H., Churchwell, E., and Torres, A. V., Astrophys. J. **303**, 239 (1986).

Abranin, É. P., Bazelyan, L. L., and Goncharov, N. Yu., Astron. Zh. **54**, 781 (1977) [Sov. Astron. **21**, 441 (1977)].

Agafonov, M. I., Aslanyan, A. M., Gulyan, A. G., Ivanov, V. N., Martirosyan, R. M., Stankevich, K. S., and Stolyarov, S. P., Pis'ma Astron. Zh. **12**, 275 (1986) [Sov. Astron. Lett. **12**, 112 (1986)].

Agafonov, M. I., Aslanyan, A. M., Gulyan, A. G., Ivanov, V. N., Martirosyan, R. M., Stankevich, K. S., and Stolyarov, S. P., Astron. Zh. **64**, 60 (1987) [Sov. Astron. **31**, 31 (1987)].

Aglietta, M. et al., in: *ESO Workshop on the SN 1987A; ESO Conference and Workshop Proceedings No. 26*, I. J.Danziger (ed.) (1987), p. 207.

Albinson, J. S. and Gull, S. F., in: *Regions of Recent Star Formation*, R. Roger and P. Dewdney (eds.), Reidel, Dordrecht (1982), pp. 193–199.

Albinson, J. S., Tuffs, R. J., Swinbank, E., and Gull, S. F., Mon. Not. R. Astron. Soc. **219**, 427 (1986).

Alduseva, V. Y, Aslanov, A. A., Kolotilov, E. A., and Cherepashchuk, A. M., Pis'ma Astron. Zh. **8**, 717 (1982) [Sov. Astron. Lett. **8**, 386 (1982)].

Alekseev (Alexeyev), E. N., Alekseeva (Alexeyeva), L. M., Krivosheina, I. V., and Volchenko, V. I., in: *ESO Workshop on the SN 1987A; ESO Conference and Workshop Proceedings No. 26*, I. J.Danziger (ed.) (1987), p. 237.

Allakhverdiev (Allakhverdiyev), A. O., Guseinov, O. H., Kasumov, F. K., and Iusifov (Jusifov), I. M., Astrophys. Space Sci. **97**, 287 (1983).

Allen, D. (ed.), *AAO Newsletter*, No. 45 (1988), cover.

Allen, R. J., Goss, W. H., Kerr, R., and de Bruyn, A. G., Astron. Astrophys. **48**, 253 (1976).

Aller, H. D. and Reynolds, S. P., Astrophys. J. Lett. **293**, L73 (1985).

Aller, H. D. and Reynolds, S. P., in: *The Crab Nebula and Related Supernova Remnants*, M. C. Kafatos and R. B. C. Henry (eds.), Cambridge University Press, Cambridge (1985), p. 75.

Andrews, M. D., Basart, J. P., Lamb, R. C., and Becker, R. H., Astrophys. J. **266**, 684 (1983).

Andriesse, C. D., Mon. Not. R. Astron. Soc. **192**, 95 (1980).

Angerhofer, P. E., Becker, R. H., and Kundu, M. R., Astron. Astrophys. **55**, 11 (1977).

Angerhofer, P. E., Wilson, A. S., and Mould, J. R., Astrophys. J. **236**, 143 (1980).

Angerhofer, P. E., Strom, R. G., Velusamy, T., and Kundu, M. R., Astron. Astrophys. **94**, 313 (1981).

Antokhin, I. I., Aslanov, A. A., and Cherepashchuk, A. M., Pis'ma Astron. Zh. **8**, 734 (1982) [Sov. Astron. Lett. **8**, 395 (1982)].

Antokhin, A. A. and Cherepashchuk, A. M., Pis'ma Astron. Zh. **10**, 370 (1984) [Sov. Astron. Lett. **10**, 155 (1984)].

Antokhin, A. A. and Cherepashchuk, A. M., Pis'ma Astron. Zh. **11**, 10 (1985) [Sov. Astron. Lett. **11**, 4 (1985)].

Antokhina, É, A. and Cherepashchuk, A. M., Pis'ma Astron. Zh. **11**, 10 (1985) [Sov. Astron. Lett. **11**, 4 (1985)].

Ardelyan, N. V., Bisnovatyi-Kogan, G. S., and Popov, Yu. P., Astron. Zh. **56**, 1244 (1979) [Sov. Astron. **23**, 705 (1979)].

Arendt, R. G., in: IAU Colloq. No. 101, *Supernova Remnants and the Interstellar Medium*, R. Roger and T. Landecker (eds.), Cambridge University Press, Cambridge (1988).

Arkhipova, V. P. and Lozinskaya, T. A., Astron. Zh. **55**, 1320 (1978a) [Sov. Astron. **22**, 751 (1978)].

Arkhipova, V. P. and Lozinskaya, T. A., Pis'ma Astron. Zh. **4**, 16 (1978b) [Sov. Astron. Lett. **4**, 7 (1978)].

Arnal, E. M., Cersosimo, J. C., May, J., and Bronfman, L., Astron. Astrophys. **174**, 78 (1987).

Arnett, W. D., Astrophys. J. **319**, 136 (1987).

Arnett, W. D., Astrophys. J. **331**, 377 (1988).

Arnett, W. D. and Fu, A., Astrophys. J. **340**, 396 (1989).

Aschenbach, B., Space Sci. Rev. **40**, 447 (1985).

Aschenbach, B., in: IAU Colloq. No. 101, *Supernova Remnants and the Interstellar Medium*, R. Roger and T. Landecker (eds.), Cambridge University Press, Cambridge (1988), p. 99.

Aslanov, A. A. and Cherepashchuk, A. M., Pis'ma Astron. Zh. **7**, 482 (1981) [Sov. Astron. Lett. **7**, 265 (1981)].

Aslanyan, A. M., Gulyan, A. G., Ivanov, V. N., Martirosyan, R. M., Stankevich, K. S., and Stolyarov, S. P., *Twelfth All-Union Conference on Radio Astronomy, Abstracts* [in Russian], Tallinn (1987), p. 154.

Assousa, G. E. and Erkes, J. W., Astron. J. **78**, 885 (1973).

Assousa, G. E., Balick, B., and Erkes, J. W., Bull. Am. Astron. Soc. **7**, 35 (1975).

Avedisova, V. S., Astron. Zh. **48**, 894 (1971) [Sov. Astron. **15**, 708 (1972)].

Avedisova, V. S., Pis'ma Astron. Zh. **3**, 405 (1977) [Sov. Astron. Lett. **3**, 217 (1977)].

Axelrod, T. S., in: *Type I Supernovae*. Proc. Texas Workshop on Type I Supernovae, J. C. Wheeler (ed.), University of Texas Press, Austin (1980), pp. 80–95.

B

Baade, W. and F. Zwicky, Phys. Rev. **46**, 76 (1934).

Baade, W., Astrophys. J. **97**, 119 (1943).

Baade, W. and Minkowski, R., Astrophys. J. **119**, 206 (1954).

Baars, J. W. M. and Wendker, H. J., Astron. Astrophys. **181**, 210 (1987).

Baars, J. W. M., Genzel, R., Pauliny-Toth, I. I. K., and Witzel, A., Astron. Astrophys. **61**, 99 (1977).

Baars, J. W. M., Dickel, H. R., and Wendker, H. J., Astron. Astrophys. **62**, 13 (1978).

Baker, J. R., Preuss, E., and Whiteoak, J. B., Astrophys. Lett. **14**, 123 (1973).

Balick, B. and Preston, H. L., Astron. J. **94**, 958 (1987).

Balinskaya, I. S. and Bychkov, K. V., Soobshch. Spets. Astrofiz. Observ. No. 26, 51 (1979).

Balinskaya, I. S. and Bychkov, K. V., Soobshch. Spets. Astrofiz. Observ. No. 31, 49 (1981).

Ballet, J., Arnaud, M., Chieze, J. P., Magne, B., and Rothenflug, R., in: IAU Colloq. No. 101, *Supernova Remnants and the Interstellar Medium*, R. Roger and T. Landecker (eds.), Cambridge University Press, Cambridge (1988a), p. 141.

Ballet, J., Caplan, J., Rothenflug, R., and Soutol, A., in: IAU Colloq. No. 101, *Supernova Remnants and the Interstellar Medium*, R. Roger and T. Landecker (eds.), Cambridge University Press, Cambridge (1988), p. 411.

Ballet, J., Caplan, J., Rothenflug, R., Dubreuil, D., and Soutol, A., Astron. Astrophys. **211**, 217 (1989).

Band, D. L. and Liang, E. P., Astrophys. J. **334**, 266 (1988).

Bandiera, R., Pacini, F., and Salvati, M., Astron. Astrophys. **126**, 7 (1983).

Bandiera, R., Pacini, F., and Salvati, M., Astrophys. J. **285**, 134 (1984).

Bandiera, R., Astrophys. J. **319**, 885 (1987).

Barbon, R., in: *Type I Supernovae*. Proc. Texas Workshop on Type I Supernovae, J. C. Wheeler (ed.), University of Texas Press, Austin (1980), pp. 16-19.

Barbon, R., Ciatti, F., and Rosino, L., Astron. Astrophys. **25**, 241 (1973).

Barbon, R., Ciatti, F., and Rosino, L., Astron. Astrophys. **72**, 287 (1979).

Barbon, R., Cappellaro, E., Ciatti, F., Turatto, M., and Kowal, C. T., Astron. Astrophys. Suppl. Ser. **58**, 735 (1984).

Barbon, R., Cappellaro, E., and Turatto, M., *Asiago Supernova Catalogue, Version of July 26, 1988* (1988).

Barker, T., Astrophys. J. **219**, 914 (1978).

Barlow, M. J. and M. Cohen, Astrophys. J. **213**, 737 (1977).

Barlow, M. J. and Cohen, M., in: IAU Symp. No. 99, *WR Stars: Observations, Physics, Evolution*, C. W. De Loore and A. J. Willis (eds.), Reidel, Dordrecht (1982), p. 387.

Barlow, M. J., Cohen, M., and Gull, T. R., Mon. Not. R. Astron. Soc. **176**, 359 (1976).

Barlow, M. J., Smith, L. J., and Willis, A. J., Mon. Not. R. Astron. Soc. **196**, 101 (1981).

Barrett, P., in: *ESO Workshop on the SN 1987A; ESO Conference and Workshop Proceedings No. 26*, I. J.Danziger (ed.) (1987), p. 173.

Bartunov, O. S., Blinnikov, S. I., Levakhina, L. V., and Nadezhin, D. K., Pis'ma Astron. Zh. **13**, 744 (1987) [Sov. Astron. Lett. **13**, 313 (1987)].

Bartunov, O. S. and D. Yu. Tsvetkov, Astrophys. Space Sci. **122**, 343 (1986).

Bash, F. N., Green, E., and Peter, Willis, A. J.. L., Astrophys. J. **217**, 464 (1977).

Bates, B., Brown-Kerr, W., Giaretta, D. L., and Keenan, F. P., Astron. Astrophys. **122**, 64 (1983).

Becker, R. H., in: IAU Symp. No. 101, *Supernova Remnants and their X-Ray Emission*, J. Danziger and P. Gorenstein (eds.), Reidel, Dordrecht (1983), p. 321.

Becker, R. H. and Fesen, R. A., Astrophys. J. **334**, L35 (1988).

Becker, R. H. and Helfand, D. J., Astrophys. J. **283**, 154 (1984).

Becker, R. H. and Helfand, D. J., Nature **313**, 115 (1985a).

Becker, R. H. and Helfand, D. J., Astrophys. J. **297**, L25 (1985b).

Becker, R. H., Boldt, B. A., Holt, S. S., Pravdo, S. H., Rothschild, R. E., Serlemitsos, P. J., and Swank, J. H., Astrophys. J. **209**, 65 (1976).

Becker, R. H., Holt, S. S., Smith, B. W., White, N. E., Boldt, E. A., Mushotzky, R. F., and Serlemitsos, P. J., Astrophys. J. Lett. **234**, L73 (1979).

Becker, R. H., Boldt, E. A., Holt, S. S., Serlemitsos, P. J., and White, N. E., Astrophys. J. Lett. **237**, L77 (1980a).

Becker, R. H., Holt, S. S., Smith, B. W., White, N. E., Boldt, E. A., Mushotzky, R. F., and Serlemitsos, P. J., Astrophys. J. Lett. **235**, L5 (1980b).

Becker, R. H., Helfand, D. J., and Szymkowiak, A. E., Astrophys. J. **255**, 557 (1982).

Becker, R. H., Helfand, D. J., and Szymkowiak, A. E., Astrophys. J. Lett. **268**, L93 (1983).

Becker, R. H. and Szymkowiak, A. E., Astrophys. J. **248**, L23 (1981).

Begelman, M. C. and Sarazin, C. L., Astrophys. J. Lett. **302**, L59 (1986).

Bell, A. R., Mon. Not. R. Astron. Soc. **179**, 573 (1977).

Bell, A. R., Mon. Not. R. Astron. Soc. **182**, 147 (1978a).

Bell, A. R., Mon. Not. R. Astron. Soc. **182**, 443 (1978b).

Beltrametti, M., Tenorio-Tagle, G., and Yorke, U. W., Astron. Astrophys. **112**, 1 (1982).

Benvenuti, P., Dopita, M. A., and D'Odorico, S., in: *Proc. Third European IUE Conf.*, Madrid, May 1982 (1982), p. 434.

Berezhko, E. G. and Krymskii, G. F., Usp. Fiz. Nauk **154**, 49 (1988) [Sov. Phys. Usp. **31**, 27 (1988)].

Berezinskii, V. S. and Ginzburg, V. L., Pis'ma Astron. Zh. **13**, 931 (1987) [Sov. Astron. Lett. **13**, 391 (1987)].

Berezinskii, V. S., Ginzburg, V. L., and Prilutskii, O. F., Pis'ma Astron. Zh. **10**, 98 (1984) [Sov. Astron. Lett. **10**, 38 (1984)].

Bergerud, A. T., Sci. Am. **249**, No. 6, 130 (1983).

Berkhuijsen, E. M., Astron. Astrophys. **14**, 359 (1971).

Berkhuijsen, E. M., Astron. Astrophys. **24** 143 (1973).

Berkhuijsen, E. M., Astron. Astrophys. **120**, 147 (1983).

Berkhuijsen, E. M., Astron. Astrophys. **166**, 257 (1986).

Berkhuijsen, E. M., Astron. Astrophys. **181**, 398 (1987).

Berkhuijsen, E. M., in: IAU Colloq. No. 101, *Supernova Remnants and the Interstellar Medium*, R. Roger and T. Landecker (eds.), Cambridge University Press, Cambridge (1988), p. 285.

Bertola, F., A. Mammano, and M. Perinotto, Contr. Osservat. Astrophis. Univ. Padova Asiago, No. 174 (1965), pp. 51–61.

Bieging, J. H., Abbott, D. C., and Churchwell, E. B., Astrophys. J. **340**, 518 (1989).

Bignami, G. F. and Caraveo, P. A., Astrophys. J. **325**, L5 (1988).

Bignell, R. C. and Seaquist, E. R., Astrophys. J. **270**, 140 (1983).

Binette, L., Dopita, M. A., and Tuohy, I. R., Astrophys. J. **297**, 476 (1985).

Bisnovatyi-Kogan, G. S., Astron. Zh. **47**, 813 (1970) [Sov. Astron. **14**, 1012 (1971)].

Bisnovatyi-Kogan (Bisnovaty-Kogan), G. S. and Nadezhin (Nadyozhin), D. K., Astrophys. Space Sci. **15**, 353 (1972).

Bisnovatyi-Kogan, G. S., Illarionov, A., and Slysh, V. I., Preprint Inst. Kosm. Issled. Akad. Nauk SSSR, Moscow (1988).

Bisnovatyi-Kogan, G. S., Lozinskaya, T. A., and Silich, S. A., Astrophys. Space Sci. **166**, 277 (1990).

Blades, J. C., Elliott, K. H., and Meaburn, J., Mon. Not. R. Astron. Soc. **192**, 101 (1980).

Blair, W. P. and Kirshner, R. P., Astrophys. J. **289**, 582 (1985).

Blair, W. P., Kirshner, R. P., Fesen, R. A., and Gull, T. R., Astrophys. J. **282**, 161 (1984b).

Blair, W. P., Raymond, J. C., Fesen, R. A., and Gull, T. R., Astrophys. J. **279**, 708 (1984a).

Blair, W. P., Raymond, J. C., Danziger, I. J., and Matteucci, F., in: IAU Colloq. No. 101, *Supernova Remnants and the Interstellar Medium*, R. Roger and T. Landecker (eds.), Cambridge University Press, Cambridge (1988), p. 187.

Blair, W. P., Kirshner, R. P., and Winkler, P. F., Astrophys. J. **272**, 84 (1983).

Blair, W. P. and Schild, R. E., Astrophys. Lett. **24**, 189 (1985).

Blanco, V. M. and Williams, A. D., Astrophys. J. **130**, 482 (1959).

Blanco, V. M., Kunkel, W., Hiltner, W. A., Lynga, G., Bradt, H., Clark, G., Naranan, S., Rappaport, S., and Spada, G., Astrophys. J. **152**, 1015 (1968).

Blanco, V. M. et al., Astrophys. J. **320**, 589 (1987).

Blandford, R. D., in: IAU Colloq. No. 101, *Supernova Remnants and the Interstellar Medium*, R. Roger and T. Landecker (eds.), Cambridge University Press, Cambridge (1988), p. 309.

Blandford, R. and Eichler, D., Phys. Rep. **154**, 3 (1987).

Blandford, R. D., Kennel, C. F., and McKee, C. F., Nature **301**, 586 (1983).

Blinnikov, S. I., Chugai, N. N., and Lozinskaya, T. A., Sov. Sci. Rev. Ser. E: Astrophys. Space Phys. **6**, 195 (1988).

Blinnikov, S. I., Imshennik, V. S., and Utrobin, V. P., Pis'ma Astron. Zh. **8**, 671 (1982) [Sov. Astron. Lett. **8**, 361 (1982)].

Blomme, R. and Van Rensbergen, W., Astron. Astrophys. **207**, 70 (1988).

Borken, R. J. and Iwan, D., Astrophys. J. **218**, 511 (1977).

Bochkarev, N. G., Pis'ma Astron. Zh. **10**, 184 (1984) [Sov. Astron. Lett. **10**, 76 (1984)].

Bochkarev, N. G., Astron. Zh. **62**, 875 (1985) [Sov. Astron. **29**, 509 (1985)].

Bochkarev, N. G., Astron. Zh. **64**, 38 (1987) [Sov. Astron. **31**, 20 (1987)].

Bochkarev, N. G., Astrophys. Space Sci. **138**, 229 (1987).

Bochkarev, N. G., Nature **332**, 518 (1988).

Bochkarev, N. G. and Lozinskaya, T. A., Astron. Zh. **62**, 103 (1985) [Sov. Astron. **29**, 60 (1985)].

Bochkarev, N. G., Lozinskaya, T. A., Piskunov, N., Pravdikova, V. I., and Sitnik, T. G., in: *Wolf–Rayet Stars and Related Objects* [in Russian], T. Nugis and I. Pustil'nik (eds.), Estonian Academy of Sciences, Tallinn (1988), p. 168.

Bochkarev, N. G. and Sitnik, T. G., Astrophys. Space Sci. **108**, 237 (1985).

Bodenheimer, P. and Woosley, S. E., Astrophys. J. **269**, 281 (1983).

Bodenheimer, P., Yorke, H. W., Tenorio-Tagle, G., and Beltrametti, M., in: IAU Symp. No. 101, *Supernova Remnants and their X-Ray Emission*, J. Danziger and P. Gorenstein (eds.), Reidel, Dordrecht (1983), p. 399.

Bodo, G., Ferrari, A., Massaglia, S., and Tsinganos, K., Astron. Astrophys. **149**, 246 (1985).

Bothun, G. D., Astron. J. **91**, 507 (1986).

Bouchet, P. and Danziger, I. J., IAU Circ. No. 4575 (1988).

Bouchet, P., Stanga, R., Moneti, A., Le Bertre, Th., Manfroid, J., Silvestro, G., and Slezak, E., in: *ESO Workshop on the SN 1987A; ESO Conference and Workshop Proceedings No. 26*, I. J.Danziger (ed.) (1987a), p. 79.

Bouchet, P., Stanga, R., Le Bertre, T., Epchtein, N., Hamann, W. R., and Lorenzetti, D., Astron. Astrophys. **177**, L9 (1987b).

Boyarchuk, A. A., Gershberg, R. E., Zvereva, A. M., Petrov, P. P., Severnyi, A. B., Terebizh, A. V., Hua, Ch.-T., and Sheikhet, A. I., Pis'ma Astron. Zh. **13**, 739 (1987) [Sov. Astron. Lett. **13**, 311 (1987)].

Branch, D. and F. Greenstein, Astrophys. J. **167**, 89 (1971).

Branch, D., Astrophys. J. **248**, 1076 (1981).

Branch, D., Astrophys. J. **300**, L51 (1986).

Branch, D., in: *Supernovae: a Survey of Current Research*, M. J. Rees and R. J. Stoneham (eds.), Reidel, Dordrecht (1982), pp. 267–279.

Branch, D. et al., Astrophys. J. Lett. **244**, L61 (1982).

Branch, D., Falk, S. W., McCall, M. L., Rybski, P., Uomoto, A. K., and Willis, B. J., Astrophys. J. **244**, 780 (1981).

Branch, D. and Greenstein, F., Astrophys. J. **167**, 89 (1971).

Branch, D., Lucy, C. H., McCall, M. L., Sutherland, P. G., Uomoto, A., Wheeler, J. C., and Wills, B. J., Astrophys. J. **270**, 123 (1983).

Branch, D. and Nomoto, K., Astron. Astrophys. **164**, L13 (1986).

Brandt, J. C., Roosen, R. G., Thompson, J., and Ludden, D. J., Astrophys. J. **208**, 109 (1976).

Braun, R., Astron. Astrophys. **171**, 233 (1987).

Braun, R. and Strom, R. G., Astron. Astrophys. **164**, 208 (1986a).

Braun, R. and Strom, R. G., Astron. Astrophys. **164**, 193 (1986b).

Braun, R. and Strom, R. G., Astron. Astrophys. Suppl. Ser. **63**, 345 (1986c).

Braun, R., Goss, W. M., Danziger, I. J., and Boksenberg, A., in: IAU Symp. No. 101, *Supernova Remnants and their X-Ray Emission*, J. Danziger and P. Gorenstein (eds.), Reidel, Dordrecht (1983), p. 159.

Braun, R., Goss, W. M., Caswell, J. L., and Roger, R. S., Astron. Astrophys. **162**, 259 (1986).

Braun, R., Gull, S. F., and Perley, R. A., Nature **327**, 395 (1987).

Braun, R., Goss, W. M., and Lyne, A. G., Astrophys. J. **340**, 355 (1989).

Braunsfurth, E. and Feitzinger, J. V., Astron. Astrophys. **127**, 113 (1983).

Brecher, K. and Wasserman, I., Astrophys. J. Lett. **240**, L105 (1980).

Breysacher, J., Astron. Astrophys. **160**, 185 (1986).

Brinkmann, W., Aschenbach, B., et al., Nature **313**, 662 (1985).

Brinks, E. and Bajaja, E., Astron. Astrophys. **169**, 14 (1986).

Brown, L. W., Woodgate, B. E., and Petre, R., Astrophys. J. **334**, 852 (1988).

Bruchweiler, F. C., Gull, T. R., Kafatos, M., and Sofia, S., Astrophys. J. Lett. **238**, L27 (1980).

Bruchweiler, F. C., Gull, T. R., Henize, K. G., and Cannon, R. D., Astrophys. J. **251**, 126 (1981).

Bunner, A. N., Astrophys. J. **220**, 261 (1978).

Burkert, W., Zimmermann, H. U., Aschenback, B., Bräuninger, H., and Williamson, F., Astron. Astrophys. **115**, 167 (1982).

Burstein, D., Borken, R. J., Kraushaar, W. L., and Sanders, W. T., Astrophys. J. **213**, 405 (1976).

Burton, M. G., in: IAU Colloq. No. 101, *Supernova Remnants and the Interstellar Medium*, R. Roger and T. Landecker (eds.), Cambridge University Press, Cambridge (1988), p. 399.

Bychkov, K. V., Astron. Zh. **50**, 907 (1973) [Sov. Astron. **17**, 577 (1973)].

Bychkov, K. V., Astron. Zh. **51**, 317 (1974a) [Sov. Astron. **18**, 184 (1974)].

Bychkov, K. V., Astron. Zh. **51**, 712 (1974b) [Sov. Astron. **18**, 420 (1974)].

Bychkov, K. V., Astron. Zh. **55**, 1222 (1978a) [Sov. Astron. **22**, 695 (1978)].

Bychkov, K. V., Astron. Zh. **55**, 755 (1978b) [Sov. Astron. **22**, 431 (1978)].

Bychkov, K. V., Astron. Zh. **56**, 781 (1979) [Sov. Astron. **23**, 438 (1979)].
Bychkov, K. V., Izv. Spets. Astrofiz. Observ. **21**, 58 (1986).
Bychkov, K. V. and Fedorova, O. V., Astron. Zh. **64**, 509 (1987) [Sov. Astron. **31**, 268 (1987)].
Bychkov, K. V. and Lebedev, V. S., Astron. Astrophys. **80**, 167 (1979).
Bychkov, K. V. and Pikel'ner, S. B., Pis'ma Astron. Zh. **1**, 29 (1975) [Sov. Astron. Lett. **1**, 37 (1975)].
Bychkov, K. V., Sitnik, T. G., and Fedorova, O. V., Astron. Zh. **63**, 939 (1986) [Sov. Astron. **30**, 555 (1986)].

C

Caldwell, C. N. and Oemler, A., Astron. J. **86**, 1424 (1981).
Cameron, A. G. W. and Iben, I., Astrophys. J. **305**, 228 (1986).
Canizares, C. R., Kriss, G. A., and Feigelson, F. D., Astrophys. J. Lett. **253**, L17 (1982).
Canizares, C. R., Winkler, P. F., Markert, T. H., and Berg, C., in: IAU Symp. No. 101, *Supernova Remnants and their X-Ray Emission*, J. Danziger and P. Gorenstein (eds.), Reidel, Dordrecht (1983), p. 205.
Canto, J., Astron. Astrophys. **61**, 641 (1977).
Canto, J., Johnson, P. G., Meaburn, J., Mikhail, J. S., Terrett, D. L., and White, N. J., Mon. Not. R. Astron. Soc. **187**, 673 (1979).
Cappa de Nicolau, C. C. and Niemela, V. S., Astron. J. **89**, 1398 (1984).
Cappa de Nicolau, C. C., Niemela, V. S., and Arnal, E. M., Astron. J. **92**, 1414 (1986).
Cappa de Nicolau, C. C., Niemela, V. S., Dubner, G., and Arnal, E. M., Rev. Mex. Astron. Astrophys. **14**, 611 (1987).
Cappellaro, E. and Turatto, M., Astron. Astrophys. **190**, 10 (1988).
Carlini, A. and Treves, A., Astron. Astrophys. **215**, 283 (1989).
Cash, W., Charles, P., Bowyer, S., Walter, F., Garmire, G., and Riegler, G., Astrophys. J. Lett. **238**, L71 (1980).
Casinelli, J. P., Castor, J. L., and Lamers, H., Publ. Astron. Soc. Pac. **90**, 496 (1978a).
Casinelli, J. P., Olson, G., and Salio, R., Astrophys. J. **220**, 573 (1978b).
Casinelli, J. P., Waldron, W. L., Sanders, W. T., Harnden, F. R., Rosner, R., and Vaiana, G. S., Astrophys. J. **250**, 677 (1981).
Cassatella, A., in: *ESO Workshop on the SN 1987A; ESO Conference and Workshop Proceedings No. 26*, I. J.Danziger (ed.) (1987), p. 101.
Cassinelli, J. P. and van der Hucht, K. A., in: *Instabilities in Luminous Early-Type Stars*, Proceedings of a Workshop in Honor of C. de Jager, Lunteren (Netherlands), H. Lamers and C. de Loore (eds.) (1987).
Castor, J., McCray, R., and Weaver, R., Astrophys. J. Lett. **200**, L107 (1975).
Caswell, J. L., Mon. Not. R. Astron. Soc. **187**, 431 (1979).

Caswell, J. L., in: IAU Colloq. No. 101, *Supernova Remnants and the Interstellar Medium*, R. Roger and T. Landecker (eds.), Cambridge University Press, Cambridge (1988), p. 269.

Caswell, J. L., Haynes, R. F., Milne, D. K., and Wellington, K. J., Mon. Not. R. Astron. Soc. **190**, 881 (1980).

Caswell, J. L., Kesteven, M. J., Komesaroff, M. M., Haynes, R. F., Milne, D. K., Stewart, R. T., and Wilson, S. G., Mon. Not. R. Astron. Soc. **225**, 329 (1987).

Caswell, J. L. and Lerche, I., Mon. Not. R. Astron. Soc. **187**, 201 (1979).

Caswell, J. L., Milne, D. K., and Wellington, K. J., Mon. Not. R. Astron. Soc. **195**, 89 (1981).

Caswell, J. L., Murray, J. D., Roger, R. S., Cole, D. J., and Cooke, D. J., Astron. Astrophys. **45**, 239 (1975).

Catchpole, R. M. and Feast, M. W., Observatory **90**, 136 (1970).

Caulet, A., Deharveng, L., Georgelin, J. M., and Georgelin, J. P., Astron. Astrophys. **110**, 185 (1982).

Celnik, W. E., Astron. Astrophys. **144**, 171 (1985).

Chalabaev, A., Perrier, C., and Mariotti, J. M., IAU Circ. No. 4481 (1987).

Chalabaev, A., Perrier, C., and Mariotti, J. M., Preprint (1988).

Chanan, G. A., Helfand, D. J., and Reynolds, S. P., Astrophys. J. Lett. **287**, L23 (1984).

Chanot, A. and Sivan, J. P., Astron. Astrophys. **121**, 19 (1983).

Charles, P. A., Kahn, S. M., and McKee, C. F., Astrophys. J. **295**, 456 (1985).

Chechetkin, V. M., Denisov, A. A., Koldoba, A. V., Poveschenko, Yu. A., and Popov, Yu. P., in: IAU Colloq. No. 101, *Supernova Remnants and the Interstellar Medium*, R. Roger and T. Landecker (eds.), Cambridge University Press, Cambridge (1988), p. 27.

Cherepashchuk, A. M., Astrophys. Space Sci. **86**, 299 (1982).

Cherepashchuk, A. M. and Aslanov, A. A., Astrophys. Space Sci. **102**, 97 (1984).

Chevalier, R. A., Astrophys. J. **188**, 501 (1974).

Chevalier, R. A., Astrophys. J. **208**, 826 (1976).

Chevalier, R. A., in: *Supernovae* (Astrophys. Space Sci. Libr. No. 66), 3rd ed., D. N. Schramm (ed.), Reidel, Dordrecht (1977a), p. 53.

Chevalier, R. A., Nature **266**, 701 (1977b).

Chevalier, R. A., Astrophys. J. **251**, 259 (1981).

Chevalier, R. A., Astrophys. J. **259**, 302 (1982a).

Chevalier, R. A., Astrophys. J. **258**, 790 (1982b).

Chevalier, R. A., in: *Eleventh Texas Symposium on Relativistic Astrophysics*, D. S. Evans (ed.), (Annals of the New York Academy of Sciences Vol. 422) (1984a), p. 215.

Chevalier, R. A., Astrophys. J. **280**, 797 (1984b).

Chevalier, R. A., Astrophys. J. Lett. **285**, L63 (1984c).

Chevalier, R. A., in: *The Crab Nebula and Related Supernova Remnants*, M. C. Kafatos and R. B. C. Henry (eds.), Cambridge University Press, Cambridge (1985), p. 63.

Chevalier, R. A., in: *ESO Workshop on the SN 1987A; ESO Conference and Workshop Proceedings No. 26*, I. J.Danziger (ed.) (1987), p. 481.

Chevalier, R. A., in: IAU Colloq. No. 101, *Supernova Remnants and the Interstellar Medium*, R. Roger and T. Landecker (eds.), Cambridge University Press, Cambridge (1988), p. 31.

Chevalier, R. A. and Fransson, C., in: *Supernovae as a Distance Indicator*, N. Bartel (ed.), Springer-Verlag, Berlin (1985), p. 123.

Chevalier, R. A. and Fransson, C., Nature **328**, 44 (1987).

Chevalier, R. A. and Kirshner, R. P., Astrophys. J. **219**, 931 (1978).

Chevalier, R. A. and Kirshner, R. P., Astrophys. J. **233**, 154 (1979).

Chevalier, R. A. and Oegerle, W. R., Astrophys. J. **227**, 398 (1979).

Chevalier, R. A. and Raymond, J. C., Astrophys. J. Lett. **225**, L27 (1978).

Chevalier, R. A., Roberetson, J. W., and Scott, J. S., Astrophys. J. **207**, 450 (1976).

Chevalier, R. A., Kirshner, R. P., and Raymond, J. C., Astrophys. J. **235**, 186 (1980).

Chieze, J. P. and Lazareff, B., Astron. Astrophys. **95**, 194 (1981).

Chiosi, C., Astron. Astrophys. **93**, 163 (1981).

Chu, You-Hua, Astrophys. J. **249**, 195 (1981).

Chu, You-Hua, Astrophys. J. **269**, 202 (1983).

Chu, You-Hua, in: IAU Symp. No. 131, *Planetary Nebulae*, S. Torres-Peimbert (ed.), Kluwer, Dordrecht (1987).

Chu, You-Hua, Publ. Astron. Soc. Pac. **100**, 986 (1988).

Chu, You-Hua, in: IAU Symp. No. 143, *Wolf–Rayet Stars and Interrelations with other Massive Stars in Galaxies*, K. A. van der Hucht and B. Hidayat (eds.), Kluwer, Dordrecht (1990).

Chu, You-Hua and MacLow, M.-M., in: IAU Symp. No. 148, *The Magellanic Clouds*, R. Haynes and D. Milne (eds.), Kluwer, Dordrecht (1991), p. 99.

Chu, You-Hua, Jacoby, G. H., and Arendt, R., Astrophys. J. Suppl. Ser. **64**, 529 (1987).

Chu, You-Hua and Treffers, R. R., Astrophys. J. **249**, 586 (1981a).

Chu, You-Hua and Treffers, R. R., Astrophys. J. **250**, 615 (1981b).

Chu, You-Hua, Treffers, R. R., and Kwitter, K. B., Astrophys. J. Suppl. Ser. **53**, 937 (1983).

Chugai, N. N., Astron. Tsirk. No. 1105, 1 (1980).

Chugai, N. N., Astron. Zh. **59**, 1134 (1982) [Sov. Astron. **26**, 683 (1982)].

Chugai, N. N., Pis'ma Astron. Zh. **12**, 461 (1986a) [Sov. Astron. Lett. **12**, 192 (1986)].

Chugai, N. N., Astron. Tsirk. No. 1465, 3 (1986b).

Chugai, N. N., Pis'ma Astron. Zh. **13**, 671 (1987) [Sov. Astron. Lett. **13**, 282 (1987)].

Chugai, N. N., Astrophys. Space Sci. **146**, 375 (1988).

Cioffi, D. F., McKee, C. F., and Bertschinger, E., Astrophys. J. **334**, 252 (1988).

Clark, D. H., Mon. Not. R. Astron. Soc. **175**, 77P (1976).

Clark, D. H., Andrews, P. J., and Smith, R. C., Observatory **101**, 203 (1981).

Clark, D. H. and Caswell, J. L., Mon. Not. R. Astron. Soc. **174**, 267 (1976).

Clark, D. H., Murdin, P., Wood, R., Gilmozzi, R., Danziger, I. J., and Furr, A. W., Mon. Not. R. Astron. Soc. **204**, 415 (1983).

Clark, D. H., Murdin, P., Zarnecki, J. C., and Culhane, J. L., Mon. Not. R. Astron. Soc. **188**, 11P (1979).

Clark, D. H. and Stephenson, F. R., *The Historical Supernovae*, Pergamon Press, Oxford (1977), p. 83.

Clark, D. H. and Stephenson, F. R., in: *Supernovae: A Survey of Current Research*, M. J. Rees and R. J. Stoneham (eds.), Reidel, Dordrecht (1982), p. 355.

Clark, D. H., Tuohy, I. R., Long, K. S., Szymkowiak, A. E., Dopita, M. A., Mathewson, D. S., and Culhane, J. L., Astrophys. J. **255**, 440 (1982).

Clayton, C. A., Mon. Not. R. Astron. Soc. **226**, 493 (1987).

Clear, J., Bennett, K., Buccheri, R., Grenier, I. A., Hermsen, W., Mayer-Hasselwander, H. A., and Sacco, B., Astron. Astrophys. **174**, 85 (1987).

Clifton, T. R., Backer, D. C., Foster, R. S., Fruchter, A. S., Kulkarni, S., and Taylor, J. H., IAU Circ. No. 4422 (1987).

Clocchiatti, A. and Marraco, H. G., Astron. Astrophys. **197**, L1 (1988).

Cohen, J. G., Publ. Astron. Soc. Pac. **89**, 626 (1977).

Cohen, M. and Barlow, M. J., Astrophys. Lett. **16**, 165 (1975).

Cohen, M. and Kuhi, L. V., Astrophys. J. **210**, 365 (1976).

Colgate, S. A. and McKee, C. F., Astrophys. J. **157**, 623 (1969).

Collins, G. W., Mon. Not. R. Astron. Soc. **213**, 279 (1985).

Colomb, F. R. and Dubner, G., Astron. Astrophys. **112**, 141 (1982).

Conti, P. S., Mem. Soc. R. Sci. Liège, 6ᵉ Sér. **9**, 193 (1976).

Conti, P. S., Ann. Rev. Astron. Astrophys. **16**, 371 (1978).

Conti, P. S., in: IAU Symp. No. 83, *Mass Loss and Evolution of O-Type Stars*, P. S. Conti and C. W. H. De Loore (eds.), Reidel, Dordrecht (1980), p. 431.

Conti, P. S. and Underhill, A. B. (eds.), "O-type stars and Wolf–Rayet stars," Monograph Series on Nonthermal Phenomena in Stellar Atmospheres, NASA SP-497 (1988).

Contini, M., Astron. Astrophys. **183**, 53 (1987).

Contini, M., Kozlovsky, B. Z., and Shaviv, G., Astron. Astrophys. **92**, 273 (1980).

Cook, W. R., Palmer, D., Prince, T., Schindler, S., Starr, C., and Stone, E., IAU Circ. No. 4527 (1988a).

Cook, W. R., Palmer, D. M., Prince, T. A., Schindler, S. M., Starr, C. H., and Stone, E. C., Astrophys. J. **334**, L87 (1988b).

Cornett, R. H. and Hardee, P. E., Astron. Astrophys. **38**, 157 (1975).

Cornett, R. H., Chin, G., and Knapp, G. R., Astron. Astrophys. **54**, 889 (1977).

Courtes, G., Ann. Astrophys. **23**, 115 (1960).
Courtes, G., Petit, H., Sivan, J.-P., Dodonov, S., and Petit, M., Astron. Astrophys. **174**, 28 (1987).
Cowan, J. J. and Branch, D., Astrophys. J. **258**, 31 (1982).
Cowan, J. J. and Branch, D., Bull. Amer. Astron. Soc. **16**, 541 (1984).
Cowan, J. J. and Branch, D., Astrophys. J. **293**, 400 (1985).
Cowie, L. L., Astrophys. J. **245**, 66 (1981).
Cowie, L. L., in: *Interstellar Processes*, D. Hollenbach and H. Thronson (eds.), Reidel, Dordrecht (1987).
Cowie, L. L. and McKee, C. F., Astrophys. J. **211**, 135 (1977).
Cowie, L. L., Songaila, A., and York, D. G., Astrophys. J. **230**, 469 (1979).
Cowie, L. L., McKee, C. F., and Ostriker, J. P., Astrophys. J. **247**, 908 (1981a).
Cowie, L. L., Hu, E. M., Taylor, W., and York, D. G., Astrophys. J. Lett. **250**, L25 (1981b).
Cox, D. P., Astrophys. J. **178**, 143 (1972a).
Cox, D. P., Astrophys. J. **178**, 159 (1972b).
Cox, D. P., Astrophys. J. **234**, 863 (1979).
Cox, D. P., Astrophys. J. **245**, 534 (1981).
Cox, D. P., in: IAU Colloq. No. 101, *Supernova Remnants and the Interstellar Medium*, R. Roger and T. Landecker (eds.), Cambridge University Press, Cambridge (1988), p. 73.
Cox, D. P. and Raymond, J. C., Astrophys. J. **298**, 651 (1985).
Cox, D. P. and Smith, B. W., Astrophys. J. Lett. **189**, L105 (1974).
Cristiani, S., Babel, J., Barwig, H., et al., Astron. Astrophys. **177**, L5 (1987a).
Cristiani, S., Bouchet, P., Gouiffes, C., Sauvageot, J. L., Arsenault, R., François, P., in: *ESO Workshop on the SN 1987A; ESO Conference and Workshop Proceedings No. 26*, I. J.Danziger (ed.) (1987), p. 65.
Cropper, M., Bailey, J., McCowage, J., Cannon, R. D., Couch, W. J., Walsh, J. R., Strade, J. O., and Freeman, F., Mon. Not. R. Astron. Soc. **231**, 695 (1988).
Crotts, A., IAU Circ. No. 4561 (1988a).
Crotts, A. P. S., Astrophys. J. **333**, L51 (1988b).

D

Danziger, I. J., in: IAU Symp. No. 101, *Supernova Remnants and their X-Ray Emission*, J. Danziger and P. Gorenstein (eds.), Reidel, Dordrecht (1983), p. 193.
Danziger, I. J., Fosbury, R. A. E., Alloin, D., Cristiani, S., Dachs, J., Gouiffes, C., Jarvis, B., and Sahu, K. C., Astron. Astrophys. **177**, L13 (1987).
Danziger, I. J. and Goss, W. M., Mon. Not. R. Astron. Soc. **190**, 47 (1980).
Danziger, I. J., Clark, D. H., and Murdin, P., Mem. Soc. Astron. Ital. **49**, 559 (1978).

Danziger, I. J., Murdin, P., Clark, D. H., and D'Odorico, S., Mon. Not. R. Astron. Soc. **186**, 555 (1978).

Davelaar, J., Bleeker, J., Deerenberg, A., Tanaka, Y., Hayakawa, S., and Yamashina, K., Astrophys. J. **230**, 428 (1979).

Davelaar, J., Bleeker, J., and Deerenberg, A., Astron. Astrophys. **92**, 231 (1980).

Davelaar, J., Smith, A., and Becker, R. H., Astrophys. J. **300**, L59 (1986).

Davidsen, A. F., Henry, R. C., Snyder, W. A., Friedman, H., Fritz, G., Naranan, S., Shulman, S., and Yentis, D., Astrophys. J. **215**, 541 (1977).

Davidson, K., Astrophys. J. **228**, 179 (1979).

Davidson, K. and Fesen, R. A., Ann. Rev. Astron. Astrophys. **23**, 119 (1985).

Davidson, K., Gull, T. R., Maran, S. P., Stecher, T. P., Fesen, R. A., Parise, R. A., Harvel, C. A., Kafatos, M., and Trimble, V. L., Astrophys. J. **253**, 698 (1982).

Davies, R. D., Elliott, K. H., and Meaburn, J., Mem. R. Astron. Soc. **81**, 89 (1976).

Davies, R. D., Elliott, K. H., Goudis, C., Meaburn, J., and Terbutt, N. J., Astron. Astrophys. Suppl. Ser. **31**, 271 (1978).

de Bruyn, A. G., Astron. Astrophys. **119**, 301 (1983).

Deharveng-Baudel, L., Mem. Soc. R. Sci. Liège, Coll. 8, 6e Sér. **5** 357 (1973).

de Jager, C., Nieuwenhuijzen, H., and van der Hucht, K. A., Astron. Astrophys. **177**, 217 (1987).

Dennefeld, M., Astron. Astrophys. **112**, 215 (1982).

Dennefeld, M. and Pequignot, D., Astron. Astrophys. **127**, 42 (1983).

DeNoyer, L. K., Mon. Not. R. Astron. Soc. **183**, 187 (1978).

DeNoyer, L. K., Astrophys. J. Lett. **232**, L165 (1979).

DeNoyer, L. K., Astrophys. J. **264**, 141 (1983).

DeNoyer, L. K. and Frerking, M. A., Astrophys. J. **246**, L37 (1981).

de Vaucouleurs, G., de Vaucouleurs, A., and Corwin, H. G., *Second Reference Catalog of Bright Galaxies*, University of Texas, Austin (1976).

Dickel, H. R., Habing, H. J., and Isaacman, R., Astrophys. J. Lett. **238**, L39 (1980).

Dickel, J. R. and D'Odorico, S., Mon. Not. R. Astron. Soc. **206**, 351 (1984).

Dickel, J. R. and Greisen, E. W., Astron. Astrophys. **75**, 44 (1979).

Dickel, J. R. and Jones, E. M., Astrophys. J. **288**, 707 (1985).

Dickel, J. R. and Spangler, S. R., Astron. Astrophys. **79**, 243 (1979).

Dickel, J. R., Dickel, H. R., and Crutcher, R. M., Publ. Astron. Soc. Pac. **88**, 840 (1976).

Dickel, J. R., D'Odorico, S., Felli, M., and Dopita, M. A., Astrophys. J. **252**, 582 (1982).

Dickel, J. R., Sault, R., Arend, R. G., Matsui, Y., and Korista, K. T., Astrophys. J. **330**, 254 (1988).

Dinerstein, H. L., Lester, D. F., Rank, D. M., Werner, W., and Wooden, D. H., Astrophys. J. **312**, 314 (1987).

D'Odorico, S., The Messenger, No. 49, 34 (1987).

D'Odorico, S. and Sabbadin, F., Astron. Astrophys. Suppl. Ser. **28**, 439 (1977).

D'Odorico, S., Dopita, M. A., and Benvenuti, P., Astron. Astrophys. Suppl. Ser. **40**, 67 (1982).

Doom, C., Astron. Astrophys. **182**, L43 (1987).

Doom, C., Astron. Astrophys. **192**, 170 (1988).

Doom, C., de Greve, J. P., and de Loore, C., Astrophys. J. **290**, 185 (1985).

Doom, C., de Greve, J. P., and de Loore, C., Astrophys. J. **303**, 136 (1986).

Dopita, M. A., Astrophys. J. Suppl. Ser. **33**, 437 (1977).

Dopita, M. A., in: IAU Symp. No. 115, *Star Forming Regions*, M. Peimbert and J. Jugaku (eds.), Reidel, Dordrecht (1987), p. 501.

Dopita, M. A., in: IAU Symp. No. 108, *Atmospheric Diagnostics of Stellar Evolution*, Tokyo, 1987, K. Nomoto (ed.), Springer-Verlag, New York (1988a).

Dopita, M. A. and Mathewson, D. S., Astrophys. J. Lett. **231**, L147 (1979).

Dopita, M. A. and Tuohy, I. R., Astrophys. J. **282**, 135 (1984c).

Dopita, M. A. and Lozinskaya, T. A., Astrophys. J. **359**, 419 (1990).

Dopita, M. A., Mathewson, D. S., and Ford, V. L., Astrophys. J. **214**, 179 (1977).

Dopita, M. A., Ford, V. L., McGregor, P. J., Mathewson, D. S., and Wilson, I. R., Astrophys. J. **250**, 103 (1978b).

Dopita, M. A., Tuohy, I. R., and Mathewson, D. S., Astrophys. J. Lett. **248**, L105 (1981).

Dopita, M. A., Evans, R., Cohen, M., and Schwartz, R., Astrophys. J. Lett. **287**, L69 (1984a).

Dopita, M. A., Binette, L., and Tuohy, I. R., Astrophys. J. **282**, 142 (1984b).

Dopita, M. A., Mathewson, D. S., and Ford, V. L., Astrophys. J. **297**, 599 (1985).

Dopita, M. A., Achilleos, N., Dawe, J. A., Flynn, C., Meatheringham, S. J., and McNaught, R. D., Proc. Astron. Soc. Austr. **7**, 141 (1987a).

Dopita, M. A., Meatheringham, S. J., Nulsen, P., and Wood, P. R., Astrophys. J. **322**, L85 (1987b).

Dopita, M. A., Dawe, J. A., Achilleos, N., Brissenden, R. J. V., Flynn, C., Meatherington, S. J., Rawlings, S., Tuohy, I. R., McNaught, R. D., Coates, D. W., Hancy, S., Thompson, K., and Shobbrock, R. R., Astron. J. **95**, 1717 (1988).

Dopita, M. A., Lozinskaya, T. A., McGregor, P. J., and Rawlings, S. J., Astrophys. J. **351**, 563 (1990).

Dorland, H. and Montmerle, T., Astron. Astrophys. **177**, 243 (1987).

Dorland, H., Montmerle, T., and Doom, C., Astron. Astrophys. **160**, 1 (1986).

Doroshenko, V. T., Astron. Zh. **47**, 292 (1970) [Sov. Astron. **14**, 237 (1970)].

Doroshenko, V. T., Astron. Zh. **49**, 494 (1972) [Sov. Astron. **16**, 402 (1972)].

Doroshenko, V. T. and Lozinskaya, T. A., Pis'ma Astron. Zh. **3**, 541 (1977) [Sov. Astron. Lett. **3**, 295 (1977)].

Doroshkevich, A. G. and Zel'dovich, Ya. B., Zh. Teor. Éksp. Fiz. **80**, 801 (1981) [Sov. Phys. JETP **53**, 405 (1981)].

Dotani, T., Hayashita, K., Inoue, H., et al. (37 authors), Nature **330**, 230 (1987).

Downes, A., Mon. Not. R. Astron. Soc. **203**, 695 (1983).

Downes, A., Pauls, T., and Salter, C. J., Astron. Astrophys. **103**, 277 (1981).

Downes, A., Pauls, T., and Salter, C. J., Mon. Not. R. Astron. Soc. **218**, 393 (1986).

Downes, D., Astron. J. **76**, 305 (1971).

Draine, B. T. and Salpeter, E. E., Astrophys. J. **231**, 77 (1979a).

Draine, B. T. and Salpeter, E. E., Astrophys. J. **231**, 438 (1979b).

Drake, S. A. and Linsky, J. L., Astrophys. J. Lett. **274**, L77 (1983).

Drissen, L., Shara, M. M., and Moffat, A. F. J., in: IAU Symp. No. 143, *Wolf–Rayet Stars and Interrelations with other Massive Stars in Galaxies*, K. A. van der Hucht and B. Hidayat (eds.), Kluwer, Dordrecht (1990).

Drury, L. O. C., Rep. Progr. Phys. **46**, 973 (1983).

Dubner, G. and Arnal, M., in: IAU Colloq. No. 101, *The Interaction of Supernova Remnants with the Interstellar Medium*, R. Roger and T. Landecker (eds.), Cambridge University Press, Cambridge (1988), p. 249.

Dubner, G. M., Niemela, V. S., and Purton, C. R., Astron. J. **99**, 857 (1990).

Dufour, R. J., Parker, R. A. R., and Henize, K. G., Astrophys. J. **327**, 859 (1988).

Dufour, R. J., Rev. Mex. Astron. Astrofiz. **18**, 87 (1989).

Duin, R. M. and van der Laan, H., Astron. Astrophys. **40**, 111 (1975).

Duric, N. and Seaquist, E. R., Astrophys. J. **301**, 308 (1986).

Dwek, E., Astrophys. J. **247**, 614 (1981).

Dwek, E., Astrophys. J. **274**, 175 (1983).

Dwek, E., in: IAU Colloq. No. 101, *Supernova Remnants and the Interstellar Medium*, R. Roger and T. Landecker (eds.), Cambridge University Press, Cambridge (1988), p. 363.

Dwek, E., Astrophys. J. **322**, 812 (1987).

Dwek, E., Petre, R., Szymkowiak, A., and Rice, W. L., Astrophys. J. Lett. **320**, L27 (1987).

Dyson, J. E., Astron. Astrophys. **23**, 381 (1973).

Dyson, J. E., Astron. Astrophys. **59**, 161 (1977).

Dyson, J. E., Astron. Astrophys. **62**, 269 (1978).

Dyson, J. E. and De Vries, J., Astron. Astrophys. **20**, 223 (1972).

Dyson, J. E. and Ghanbari, J., Astron. Astrophys. **226**, 270 (1989).

E

Ebisuzaki, T. and Shibazaki, T., Astrophys. J. **327**, L5 (1988).

Efremov, Yu. N., Vestn. Akad. Nauk SSSR **12**, 56 (1984).

Efremov, Yu. N. and Sitnik, T. G., Pis'ma Astron. Zh. **14**, 817 (1988) [Sov. Astron. Lett. **14**, 347 (1988)].

Elias, J. H., K. Matthews, G. Neugebauer, and S. E. Persson, Astrophys. J. **296**, 379 (1985).

Elliott, K. H., Nature **226**, 1236 (1970).

Elliott, K. H., Mem. Soc. Astron. Ital. **49**, 477 (1978).

Elliott, K. H., Mon. Not. R. Astron. Soc. **186**, 9 (1979).

Elliott, K. H., Goudis, C., and Meaburn, J., Mon. Not. R. Astron. Soc. **175**, 605 (1976).

Elliott, K. H. and Malin, D. F., Mon. Not. R. Astron. Soc. **186**, 45P (1979).

Elliott, K. H. and Meaburn, J., Mon. Not. R. Astron. Soc. **170**, 237 (1975a).

Elliott, K. H. and Meaburn, J., Mon. Not. R. Astron. Soc. **172**, 427 (1975b).

Elliott, K. H., Meaburn, J., and Terrett, D. L., Mon. Not. R. Astron. Soc. **184**, 527 (1978).

Elmegreen, B. G., Astrophys. J. **205**, 405 (1976).

Elmegreen, B. G., Astrophys. J. **312**, 626 (1987a).

Elmegreen, B. G., in: *Interstellar Processes*, D. Hollenbach and H. Thronson (eds.), Reidel, Dordrecht (1987b).

Elmegreen, B. G. and Elmegreen, D. M., Astrophys. J. **320**, 182 (1987).

Elmegreen, B. G. and Lada, C. J., Astrophys. J. **214**, 725 (1977).

Emmering, R. T. and Chevalier, R. A., Astrophys. J. **338**, 388 (1989).

Erickson, W. C. and Mahoney, M. J., Astrophys. J. **290**, 596 (1985).

Esipov, V. F. and Lozinskaya, T. A., Astron. Zh. **45**, 1153 (1968) [Sov. Astron. **12**, 913 (1969)].

Esipov, V. F. and Lozinskaya, T. A., Astron. Zh. **48**, 449 (1971) [Sov. Astron. **15**, 353 (1971)].

Esipov, V. F., Klement'eva, A. Yu., Kovalenko, A. V., Lozinskaya, T. A., Lyutyi, V. M., Sitnik, T. G., and Udal'tsov, V. A., Astron. Zh. **59**, 965 (1982) [Sov. Astron. **26**, 582 (1982)].

Esteban, C. and Vílchez, J. M., in: IAU Symp. No. 143, *Wolf–Rayet Stars and Interrelations with other Massive Stars in Galaxies*, K. A. van der Hucht and B. Hidayat (eds.), Kluwer, Dordrecht (1990).

Esteban, C., Vílchez, J. M., Mancando, A., and Edmunds, M. G., Astron. Astrophys. **227**, 515 (1990a).

Esteban, C., Vílchez, J. M., Smith, L. J., and Manchado, A., (1990b), in press.

F

Fabbiano, G., Doxsey, R. E., Griffiths, R. E., and Johnston, M. D., Astrophys. J. Lett. **235**, L163 (1980).

Fabian, A. C., Nature **314**, 130 (1985).

Fabian, A. C. and Stewart, G. C., Mon. Not. R. Astron. Soc. **202**, 697 (1983).

Fabian, A. C., Willingale, R., Pye, J. P., Murray, S. S., and Fabbiano, G., Mon. Not. R. Astron. Soc. **193**, 175 (1980).

Fabian, A. C., Eggleton, P. P., Hut, P., and Pringle, P. E., Astrophys. J. **305**, 333 (1986).

Fahlman, G. G. and Gregory, P. C., in: IAU Symp. No. 101, *Supernova Remnants and their X-Ray Emission*, J. Danziger and P. Gorenstein (eds.), Reidel, Dordrecht (1983), p. 445.

Falk, S. W. and Arnett, W. D., Astrophys. J. Suppl. Ser. **33**, 515 (1977).

Falle, S. A. E. G., Mon. Not. R. Astron. Soc. **172**, 55 (1975a).

Falle, S. A. E. G., Astron. Astrophys. **43**, 323 (1975b).

Falle, S. A. E. G., Mon. Not. R. Astron. Soc. **195**, 1011 (1981).

Falle, S. A. E. G., in: IAU Colloq. No. 101, *Supernova Remnants and the Interstellar Medium*, R. Roger and T. Landecker (eds.), Cambridge University Press, Cambridge (1988), p. 419.

Falle, S., Garlick, A. R., and Pidsley, P. H., Mon. Not. R. Astron. Soc. **208**, 925 (1984).

Fedorenko, V. N., Astron. Zh. **58**, 790 (1981) [Sov. Astron. **25**, 451 (1981)].

Fedorenko, V. N., Pis'ma Astron. Zh. **10**, 214 (1984) [Sov. Astron. Lett. **10**, 89 (1984)].

Fedorenko, V. N., Preprint No. 1195, Fiz.–Tekhn. Inst., Leningrad (1987).

Feinstein, A., Vázquez, R. A., and Benvenuto, O. G., Astron. Astrophys. **159**, 223 (1986).

Feitzinger, J. V., in: IAU Symp. No. 115, *Star Forming Regions*, M. Peimbert and J. Jugaku (eds.), Reidel, Dordrecht (1987), p. 521.

Fejes, I., Astron. Astrophys. **168**, 69 (1986).

Felli, M. and Panagia, N., Astrophys. J. **262**, 650 (1982).

Felli, M. and Perinotto, M., Astron. Astrophys. **76**, 69 (1979).

Fesen, R. A., Astrophys. J. Lett. **270**, L53 (1983).

Fesen, R. A., Astrophys. J. **281**, 658 (1984).

Fesen, R. A., in: *The Crab Nebula and Related Supernova Remnants*, M. Kafatos and R. B. C. Henry (eds.), Cambridge University Press, Cambridge (1985).

Fesen, R. A. and Gull, T. R., Publ. Astron. Soc. Pac. **95**, 196 (1983).

Fesen, R. A. and Gull, T. R., Astrophys. Lett. **24**, 197 (1985).

Fesen, R. A. and Gull, T. R., Astrophys. J. **306**, 259 (1986).

Fesen, R. A. and Itoh, H., Astrophys. J. **295**, 43 (1985).

Fesen, R. A. and Ketelsen, D. A., in: *The Crab Nebula and Related Supernova Remnants*, M. C. Kafatos and R. B. C. Henry (eds.), Cambridge University Press, Cambridge (1985), p. 89.

Fesen, R. A. and Kirshner, R. P., Astrophys. J. **242**, 1023 (1980).

Fesen, R. A. and Kirshner, R. P., Astrophys. J. **258**, 1 (1982).

Fesen, R. A., Blair, W. P., and Kirshner, R. P., Astrophys. J. **262**, 171 (1982).

Fesen, R. A., Blair, W. P., and Kirshner, R. P., Astrophys. J. **292**, 29 (1985).

Fesen, R. A., Becker, R. H., and Blair, R. H., Astrophys. J. **313**, 378 (1987).

Fesen, R. A., Wu, C. C., Leventhal, M., and Hamilton, A. J. S., Astrophys. J. **327**, 164 (1988a).

Fesen, R. A., Kirshner, R. P., and Becker, R. H., in: IAU Colloq. No. 101, *Supernova Remnants and the Interstellar Medium*, R. Roger and T. Landecker (eds.), Cambridge University Press, Cambridge (1988b), p. 55.

Fesen, R. A., Becker, R. H., and Goodrich, R. W., Astrophys. J. Lett. **329**, L89 (1988c).

Fesen, R. A., Shull, J. M., and Saken, J. M., Nature **334**, 229 (1988d).

Fesen, R. A., Becker, R. H., Blair, W. P., and Long, K. S., Astrophys. J. **338**, L13 (1989).

Fesenkov, V. G., Kazachevskii, V. M., and Tulenkova, L. N., Astron. Zh. **31**, 224 (1954).

Filippenko, A. V. and Sargent, W. L. W., Nature **316**, 407 (1985).

Filippenko, A. V., Proc. Astron. Soc. Austr. **7**, 540 (1988).

Filippenko, A. V., *Atlas of Supernova Spectra*, in preparation.

Firmani, C., Koenigsberger, G., Bissiacchi, G., Moffat, A. F. J., and Isserstedt, J., Astrophys. J. **239**, 607 (1980).

Fischbach, K. F., Canizares, C. R., Markert, T. H., and Coyne, J. M., in: IAU Colloq. No. 101, *Supernova Remnants and the Interstellar Medium*, R. Roger and T. Landecker (eds.), Cambridge University Press, Cambridge (1988), p. 153.

Fosbury, R. A. F., Danziger, I. J., Lucy, L. B., Gouiffes, C., and Cristiani, S., in: *ESO Workshop on the SN 1987A; ESO Conference and Workshop Proceedings No. 26*, I. J.Danziger (ed.) (1987), p. 139.

Fountain, W. F., Gary, G. A., and O'Dell, C. R., Astrophys. J. **229**, 971 (1979).

Fransson, C., Astron. Astrophys. **133**, 264 (1984).

Fransson, C. and Chevalier, R. A., Astrophys. J. Lett. **322**, L15 (1987).

Fransson, C., Benvenuti, P., Gordon, C., Hempe, K., Palumbo, G. G. C., Panagia, N., Reimers, D., and Wamsteker, W., Astron. Astrophys. **132**, 1 (1984).

Fransson, C., Grewing, M., Cassatella, A., Panagia, N., and Wamsteker, W., Astron. Astrophys. **177**, L33 (1987).

Freeman, K. C., Rodgers, A. W., and Lynga, G., Nature **219**, 251 (1968).

Fruchter, A. S., Taylor, J. H., Backer, D. C., Clifton, T. R., Foster, R. S., and Wolszczan, A., Nature **331**, 53 (1988).

Fu, A. and Arnett, W. D., Astrophys. J. **340**, 414 (1989).

Fukui, Y. and Tatematsu, K., in: IAU Colloq. No. 101, *Supernova Remnants and the Interstellar Medium*, R. Roger and T. Landecker (eds.), Cambridge University Press, Cambridge (1988), p. 261.

Fürst, E. and Reich, W., Astron. Astrophys. **154**, 303 (1986).

Fürst, E., Reich, W., Reich, P., Sofue, Y., and Handa, T., Nature **314**, 720 (1985).

Fürst, E., Reich, W., and Sofue, Y., Astron. Astrophys. Suppl. Ser. **71**, 63 (1987).

Fusco-Femiano, R. and Preite-Martinez, A., Astrophys. J. **281**, 593 (1984).

G

Galas, C. M. F., Tuohy, I. R., and Garmire, Garmire, G. P., Astrophys. J. Lett. **236**, L13 (1980).

Galas, C. M. F., Venkatesan, D., and Garmire, G. P., Astrophys. Lett. **22**, 103 (1982).

Garmany, C. D. and Conti, P. S., Astrophys. J. **284**, 705 (1984).

Garmany, C. D. and Conti, P. S., Astrophys. J. **293**, 407 (1985).

Garmany, C. D., Olson, G. L., Conti, P. S., and Van Steenberg, M. E., Astrophys. J. **250**, 660 (1981).

Gaskell, C. M., Cappellaro, E., Dinerstein, H. L., Garnett, D. R., Harkness, R. P., and Wheeler, J. C., Astrophys. J. Lett. **306**, L77 (1986).

Gathier, R., Lamers, H., and Snow, T. P., Astrophys. J. **247**, 173 (1981).

Gaze, V. F. and Shain, G. A., Izv. Krymsk. Astrofiz. Observ. **15**, 11 (1955).

Gehrels, N., MacCallum, C. J., and Leventhal, M., Astrophys. J. Lett. **320**, L19 (1987).

Geldzahler, B. J. and Shaffer, D. B., Astrophys. J. Lett. **260**, L69 (1982).

Geldzahler, B. J., Pauls, T., and Salter, C. J., Astron. Astrophys. **84**, 237 (1980).

Geldzahler, B. J., Shaffer, D. B., and Kühr, H., Astrophys. J. **286**, 284 (1984).

Georgelin, J. M., Georgelin, J. P., and Sivan, J. P., in: IAU Symp. No. 84, *Large-Scale Characteristics of the Galaxy*, W. B. Burton (ed.), Reidel, Dordrecht (1979), p. 65.

Georgelin, J. M., Georgelin, J. P., Laval, A., Monnet, G., and Rosado, M., Astron. Astrophys. Suppl. Ser. **54**, 459 (1983).

Georgelin, J. M., Lortet, M. C., and Testor, G., Astron. Astrophys. **162**, 265 (1986).

Gershberg, R. E. and Shcheglov, P. V., Astron. Zh. **41**, 425 (1964) [Sov. Astron. **8**, 337 (1964)].

Gilmozzi, R., in: *ESO Workshop on the SN 1987A; ESO Conference and Workshop Proceedings No. 26*, I. J.Danziger (ed.) (1987), p. 19.

Gilmozzi, R., Proc. Astron. Soc. Austr. **7**, No. 4, 397 (1988).

Gilmozzi, R., Cassatella, A., Clavel, J., Fransson, C., González, R., Gry, C., Panagia, N., Talavera, A., and Wamsteker, W., Nature **328**, 318 (1987).

Ginzburg, V. L. (ed.), *Cosmic-Ray Astrophysics* [in Russian], Nauka, Moscow (1984).

Ginzburg, V. L. and Syrovatskii, S. I., *The Origin of Cosmic Rays* [in Russian], USSR Academy of Sciences, Moscow (1963).

Ginzburg, V. L. and Syrovatskii, S. I., Usp. Fiz. Nauk **87**, 65 (1965) [Sov. Phys. Usp. **8**, 674 (1966)].

Giovanelli, R. and Haynes, M. P., Astrophys. J. **230**, 404 (1979).

Girard, T. and Van Altena, W., IAU Circ. No. 4378 (1987).

Gladyshev, S. A., Goranskii, V. P., and Cherepashchuk, A. M., Astron. Zh. **64**, 1037 (1987) [Sov. Astron. **31**, 541 (1987)].

Glushkov, Yu. I. and Karyagina, Z. V., Astron. Tsirk. No. 711, 4 (1972).

Glushkov, Yu. I., Denisyuk, É. K., Karyagina, Z. V., and Vil'koviskii, É. A., Astron. Tsirk. No. 1078, 4 (1979).

Golovatyi, V. V. and Pronik, V. I., Tsirk. Astron. Observ. L'vovsk. Univ. No. 52, 3 (1977).

Golovatyi, V. V. and Novosyadlyi, B. S., Pis'ma Astron. Zh. **12**, 440 (1986) [Sov. Astron. Lett. **12**, 184 (1986)].

Golovatyi, V. V. and Pronik, V. I., Astrofiz. **25**, 57 (1986a).

Golovatyi, V. V. and Pronik, V. I., Astrofiz. **25**, 329 (1986b).

Goncharskii, A. V., Metlitskaya, Z. Yu., and Cherepashchuk, A. M., Astron. Zh. **61**, 124 (1984) [Sov. Astron. **28**, 74 (1984)].

González, R., Wamsteker, W., Gilmozzi, R., Walborn, N., and Lauberts, A., in: *ESO Workshop on the SN 1987A; ESO Conference and Workshop Proceedings No. 26*, I. J.Danziger (ed.) (1987), p. 33.

González, J. and Rosado, M., Astron. Astrophys. **134**, L21 (1984).

Gorbatskii, V. G. and Usovich, K. I., Astrofiz. **25**, 124 (1986).

Gosachinskii, I. V. and Khersonskii, V. K., Astron. Zh. **59**, 237 (1982) [Sov. Astron. **26**, 146 (1982)].

Gosachinskii, I. V. and Khersonskii, V. K., Preprint No. 4, Spets. Astrofiz. Observ. Akad. Nauk SSSR, p. 1 (1983).

Goss, W. M. and Van Gorkom, J. H., J. Astrophys. Astron. **5**, 425 (1984).

Goss, W. M. and Viallefond, F., J. Astron. Astrophys. **6**, 145 (1985).

Goss, W. M., Caswell, J. L., and Robinson, B. J., Astron. Astrophys. **14**, 481 (1971).

Goss, W. M., Schwarz, U. J., and Wesselius, R. P., Astron. Astrophys. **28**, 305 (1973).

Goss, W. M., Shiver, P. A., Zealey, W. J., Murdin, P., and Clark, D. H., Mon. Not. R. Astron. Soc. **188**, 357 (1979).

Goss, W. M., Kalberla, P., and Schwarz, U. J., in: IAU Colloq. No. 101, *Supernova Remnants and the Interstellar Medium*, R. Roger and T. Landecker (eds.), Cambridge University Press, Cambridge (1988), p. 239.

Goudis, C., Astrophys. and Space Sci. Libr. Vol. 90, Reidel, Dordrecht (1982), p. 1.

Goudis, C. and Meaburn, J., Astron. Astrophys. **51**, 401 (1976).

Goudis, C. and Meaburn, J., Astron. Astrophys. **62**, 283 (1978).

Goudis, C. and Meaburn, J., Astron. Astrophys. **138**, 57 (1984).

Goudis, C., Hippelein, H., and Münch, G., Astron. Astrophys. **117**, 127 (1983).

Goudis, C., Hippelein, H., Meaburn, J., and Songsathaporn, R., Astron. Astrophys. **137**, 245 (1984).

Goudis, C., Meaburn, J., and Whitehead, M. J., Astron. Astrophys. **191**, 341 (1988).

Gouiffes, C., Rosa, M., Melnick, J., Danziger, I. J., Remy, M., Santini, C., Sauvageot, J. L., Jakobsen, P., and Ruiz, M. T., Astron. Astrophys. **198**, L9 (1988).

Graham, D. A., Haslam, C. G. T., Salter, C. J., and Wilson, W. E., Astron. Astrophys. **109**, 145 (1982).

Graham, J. R., Meikle, W. P. S., Allen, D. A., Longmore, A. J., Williams, P. M., Mon. Not. R. Astron. Soc. **218**, 93 (1986).

Graham, J. R., Wright, G., and Longmore, A. J., Astrophys. J. **313**, 847 (1987a).

Graham, J. R., Evans, A., Albinson, J. S., Bode, M. F., and Meikle, W. P. S., Astrophys. J. **319**, 126 (1987b).

Grasberg, É. K., and Nadezhin, D. K., Pis'ma Astron. Zh. **12**, 168 (1986) [Sov. Astron. Lett. **12**, 68 (1986)].

Grasberg, É. K., and Nadezhin, D. K., Astron. Zh. **64**, 1199 (1987) [Sov. Astron. **64**, 629 (1987)].

Grasberg, É. K., Imshennik, V. S., and Nadezhin, D. K., Astrophys. Space Sci. **10**, 3 (1971).

Grasberg, É. K., Imshennik, V. S., Nadezhin, D. K., and Utrobin, V. P., Pis'ma Astron. Zh. **13**, 547 (1987) [Sov. Astron. Lett. **13**, 227 (1987)].

Grebenev, S. A. and Syunyaev, R. A., Pis'ma Astron. Zh. **13**, 945 (1987a) [Sov. Astron. Lett. **13**, 397 (1987)].

Grebenev, S. A. and Syunyaev, R. A., Pis'ma Astron. Zh. **13**, 1042 (1987b) [Sov. Astron. Lett. **13**, 438 (1987)].

Green, D. A., Mon. Not. R. Astron. Soc. **211**, 433 (1984a).

Green, D. A., Mon. Not. R. Astron. Soc. **209**, 449 (1984b).

Green, D. A., Mon. Not. R. Astron. Soc. **218**, 533 (1986a).

Green, D. A., Mon. Not. R. Astron. Soc. **219**, 39P (1986b).

Green, D. A., Mon. Not. R. Astron. Soc. **225**, 11P (1987).

Green, D. A., Astrophys. Space Sci. **148**, 3 (1988a).

Green, D. A., in: IAU Colloq. No. 101, *Supernova Remnants and the Interstellar Medium*, R. Roger and T. Landecker (eds.), Cambridge University Press, Cambridge (1988b), p. 51.

Green, D. A. and Downes, A. J. B., Mon. Not. R. Astron. Soc. **227**, 221 (1987).

Green, D. A. and Gull, S. F., Nature **229**, 606 (1982).

Green, D. A. and Gull, S. F., Mon. Not. R. Astron. Soc. **237**, 555 (1989).

Gregory, P. C., Braun, R., and Gull, S. F., in: IAU Symp. No. 101, *Supernova Remnants and their X-Ray Emission*, J. Danziger and P. Gorenstein (eds.), Reidel, Dordrecht (1983), p. 437.

Gronenschild, E. and Mewe, R., Astron. Astrophys. Suppl. Ser. **48**, 305 (1982).

Gronenschild, E., Mewe, R., Heise, J., den Boggende, A. J. F., Schrijver, J., and Brinkman, A. C., Astron. Astrophys. **65**, L9 (1978).

Gull, T. R., Mon. Not. R. Astron. Soc. **161**, 47 (1973).

Gull, T. R., Mon. Not. R. Astron. Soc. **171**, 263 (1975).

Gull, T. R. and Fesen, R. A., Astrophys. J. Lett. **260**, L75 (1982).

Gulliford, P., Astrophys. Space Sci. **31**, 241 (1974).

Guseinov, O. H., Kasumov, F. K., and Kalinin, E. V., Astrophys. Space Sci. **68**, 385 (1980).

H

Habe, A., Ikeuchi, S., and Tanaka, Y. D., Publ. Astron. Soc. Jap. **33**, 23 (1981).

Hameury, J. M., Boclet, D., Durouchoux, Ph., Cline, T. L., Paciesas, W. S., Teegarden, B. J., Tueller, J., and Haymes, R. C., Astrophys. J. **270**, 144 (1983).

Hamilton, A. J. S. and Sarazin, C. L., Astrophys. J. **281**, 682 (1984a).

Hamilton, A. J. S. and Sarazin, C. L., Astrophys. J. **287**, 282 (1984b).

Hamilton, A. J. S. and Sarazin, C. L., Astrophys. J. **284**, 284 (1984c).

Hamilton, A. J. S. and Fesen, R. A., Astrophys. J. **327**, 178 (1988).

Hamilton, A. J. S., Sarazin, C. L., and Szymkowiak, A. E., Astrophys. J. **300**, 713 (1986a).

Hamilton, A. J. S., Sarazin, C. L., and Szymkowiak, A. E., Astrophys. J. **300**, 698 (1986b).

Hanami, H. and Sakashita, S., Astron. Astrophys. **181**, 343 (1987).

Hanbury Brown, R., Davies, R., and Hazard, C., Observatory **80**, 191 (1960).

Hanuschik, R. W. and Dachs, J., in: *ESO Workshop on the SN 1987A; ESO Conference and Workshop Proceedings No. 26*, I. J.Danziger (ed.) (1987a), p. 153.

Hanuschik, R. W. and Dachs, J., Astron. Astrophys. **182**, L29 (1987b).

Hanuschik, R. W., Thimm, G., and Dachs, J., Mon. Not. R. Astron. Soc. **234**, 41P (1988).

Harnden, F. R., in: IAU Symp. No. 101, *Supernova Remnants and their X-Ray Emission*, J. Danziger and P. Gorenstein (eds.), Reidel, Dordrecht (1983), p. 131.

Harnden, F. R. and Seward, F. D., Bull. Amer. Astron. Soc. **16**, 542 (1984).

Hayakawa, S., Kato, T., Nagase, F., Tanaka, Y., Yamashita, K., and Murakami, T., Astrophys. J. Lett. **213**, L109 (1977).

Hayakawa, S., Kato, T., Nagase, F., Yamashita, K., and Tanaka, Y., Astron. Astrophys. **62**, 21 (1978).

Heap, S. R., in: IAU Symp. No. 99, *WR Stars: Observations, Physics, Evolution*, C. W. De Loore and A. J. Willis (eds.), Reidel, Dordrecht (1982), p. 423.

Heap, S. R., in: IAU Symp. No. 103, *Planetary Nebulae*, D. R. Flower (ed.), Reidel, Dordrecht (1983), p. 375.

Hearn, D. R., Larsen, S. E., and Richardson, J. A., Astrophys. J. Lett. **235**, L67 (1980).

Heathcote, S. and Suntzeff, N., IAU Circ. No. 4567 (1988).

Heckathorn, J. N., Bruchweiler, F. C., and Gull, T. R., Astrophys. J. **252**, 230 (1982).

Heiles, C., Astrophys. J. **229**, 533 (1979).

Heiles, C.Astrophys. J. Suppl. Ser. **55**, 585 (1984).

Heiles, C., Astrophys. J. **315**, 555 (1987).

Heiles, C., Chu You-Hua, Reynolds, R. J., Yegingil, I., and Troland, T. H., Astrophys. J. **242**, 533 (1980).

Heiles, C., Chu You-Hua, and Troland, T. H., Astrophys. J. Lett. **247**, L77 (1981).

Helfand, D. J., in: IAU Symp. No. 101, *Supernova Remnants and their X-Ray Emission*, J. Danziger and P. Gorenstein (eds.), Reidel, Dordrecht (1983), p. 471.

Helfand, D. J. and Becker, R. H., Nature **307**, 215 (1984).

Helfand, D. J. and Becker, R. H., Nature **313**, 118 (1985).

Helfand, D. J. and Becker, R. H., Astrophys. J. **314**, 203 (1987).

Helfand, D. J., Chanan, G. A., and Novick, R., Nature **283**, 337 (1980).

Helfand, D. J., Velusamy, T., Becker, R. H., and Lockman, F. J., Astrophys. J. **341**, 151 (1989).

Helfer, H. L. and Savedoff, M. P., Astrophys. J. **304**, 581 (1986).

Henbest, S. N., Mon. Not. R. Astron. Soc. **190**, 833 (1980).

Henry, R. B. C., in: *Stellar Nucleosynthesis*, S. C. Chiosi and A. Renzini (eds.), Reidel, Dordrecht (1984), p. 43.

Henry, R. B. C., Publ. Astron. Soc. Pac. **98**, 1044 (1986).

Henry, R. B. C. and MacAlpine, G. M., Astrophys. J. **258**, 11 (1982).

Henry, R. B. C. and Fesen, R. A., in: IAU Colloq. No. 101, *Supernova Remnants and the Interstellar Medium*, R. Roger and T. Landecker (eds.), Cambridge University Press, Cambridge (1988a), p. 183.

Henry, R. B. C. and Fesen, R. A., Astrophys. J. **329**, 693 (1988b).

Henry, R. B. C., MacAlpine, G. M., and Kirshner, R. P., Bull. Amer. Astron. Soc. **14**, 887 (1982).

Henry, R. B. C., MacAlpine, G. M., and Kirshner, R. P., Astrophys. J. **278**, 619 (1984).

Hesser, J. E. and van den Bergh, S., Astrophys. J. **251**, 549 (1981).

Hester, J. J., Astrophys. J. **314**, 187 (1987).

Hester, J. J. and Cox, D. P., Astrophys. J. **300**, 675 (1986).

Hester, J. J. and Kulkarni, S., Astrophys. J. **340**, 362 (1989).

Hester, J. J., Parker, R. A. R., and Dufour, R. J., Astrophys. J. **273**, 219 (1983).

Hester, J. J., Raymond, J. C., and Danielson, G. E., Astrophys. J. **303**, L17 (1986).

Hidayat, B., Admiranto, A. G., and van der Hucht, K. A., Astrophys. Space Sci. **99**, 175 (1984).

Higgs, L. A., Landecker, T. L., and Roger, R. S., Astron. J. **82**, 718 (1977).

Higgs, L. A., Landecker, T. L., and Seward, F. D., in: IAU Symp. No. 101, *Supernova Remnants and their X-Ray Emission*, J. Danziger and P. Gorenstein (eds.), Reidel, Dordrecht (1983), p. 281.

Hill, I. E., Mon. Not. R. Astron. Soc. **169**, 59 (1974).

Hillebrandt, W., Höflich, P., Kafka, P., Müller, E., Schmidt, H. U., and Truran, J. W., Astron. Astrophys. **180**, L20 (1987a).

Hillebrandt, W., Höflich, P., Truran, J. W., and Weiss, A., Nature **327**, 597 (1987b).

Hillebrandt, W., Höflich, P., Schmidt, H. U., and Truran, J. W., Astron. Astrophys. **186**, L9 (1987c).

Holt, S. S., in: IAU Symp. No. 101, *Supernova Remnants and their X-Ray Emission*, J. Danziger and P. Gorenstein (eds.), Reidel, Dordrecht (1983), p. 17.

Howarth, I. D. and Phillips, A. P., Mon. Not. R. Astron. Soc. **222**, 809 (1988).

Huang, Y.-L. and Thaddeus, P., Astrophys. J. **295**, L13 (1985).

Huang, Y.-L., Dickman, R. L., and Snell, R. L., Astrophys. J. **302**, L63 (1986).

Huber, M. C. E., Nussbaumer, H., Smith, L., Willis, A. J., and Wilson, R., Nature **278**, 697 (1979).

Hughes, J. P., in: IAU Colloq. No. 101, *Supernova Remnants and the Interstellar Medium*, R. Roger and T. Landecker (eds.), Cambridge University Press, Cambridge (1988), p. 125.

Hughes, J. P., Astrophys. J. **314**, 103 (1987).

Hughes, J. P. and Helfand, D. J., Astrophys. J. **291**, 544 (1985).

Hughes, J. P., Helfand, D. J., and Kahn, S. M., Astrophys. J. Lett. **281**, L25 (1984a).

Hughes, J. P., Harten, R. H., Costain, C. H., Nelson, L. A., and Viner, M. R., Astrophys. J. **283**, 147 (1984b).

Humphreys, R. M., Astrophys. J. Suppl. Ser. **38**, 309 (1978).

Humphreys, R. M., Nichols, M., and Massey, R., Astron. J. **90**, 101 (1984).

Humphreys, R. M., Publ. Astron. Soc. Pac. **99**, 5 (1987).

I

IAU Symp. No. 34, *Planetary Nebulae*, D. E. Osterbrock and C. R. O'Dell (eds.), Reidel, Dordrecht (1960).

IAU Symp. No. 46, *The Crab Nebula*, R. G. Davoes and F. G. Smith (eds.), Reidel, Dordrecht (1971).

IAU Symp. No. 83, *Mass Loss and Evolution of O-Type Stars*, P. S. Conti and C. W. H. De Loore (eds.), Reidel, Dordrecht (1980).

IAU Symp. No. 99, *WR Stars: Observations, Physics, Evolution*, C. W. De Loore and A. J. Willis (eds.), Reidel, Dordrecht (1982).

IAU Symp. No. 101, *Supernova Remnants and their X-Ray Emission*, J. Danziger and P. Gorenstein (eds.), Reidel, Dordrecht (1983).

IAU Symp. No. 103, *Planetary Nebulae*, D. R. Flower (ed.), Reidel, Dordrecht (1983).

Iben, I., Astrophys. J. **304**, 201 (1986).

Iben, I. and Tutukov, A. V., in: *Stellar Nucleosynthesis*, S. C. Chiosi and A. Renzini (eds.), Reidel, Dordrecht (1984), p. 181.

Iben, I. and Tutukov, A. V., Astrophys. J. Suppl. Ser. **54**, 335 (1984b).

Igumentshchev, I. V., Shustov, B. M., and Tutukov, A. V., Astron. Astrophys. **234**, 396 (1990).

Ikeuchi, S. and Tomita, H., Publ. Astron. Soc. Jap. **35**, 77 (1983).

Ikeuchi, S., Habe, A., and Tanaka, J. D., Mon. Not. R. Astron. Soc. **207**, 909 (1984).

Ilovaisky, S. A. and Lequeux, J., Astron. Astrophys. **20**, 347 (1972).

Imshennik, V. S. and Nadezhin, D. K., Nauchn. Inform. Astrosoveta Akad. Nauk SSSR **29**, 27 (1970).

Imshennik, V. S. and Nadezhin, D. K., Sov. Sci. Rev. Ser. E, Astrophys. Space Phys. **2**, 75 (1985).

Imshennik, V. S. and Nadezhin, D. K., Usp. Fiz. Nauk **156**, 561 (1988).

Imshennik, V. S. and Nadezhin, D. K., Sov. Sci. Rev. Ser. E, Astrophys. Space Phys. **7** (1989).

Innes, D. E., Giddings, J. R., and Falle, S. A. E. G., Mon. Not. R. Astron. Soc. **227**, 1021 (1987).

Inoue, H., Koyama, K., Matsuoka, M., Ohashi, T., Tanaka, Y., and Tsunemi, H., Astrophys. J. **238**, 886 (1980).

Imshennik, V. S. and Nadezhin, D. K., in: *Itogi Nauki i Tekhniki. Astronomiya*, Vol. 21, VINITI (1982), p. 63.

Isenberg, P. A. Astrophys. J. **217**, 597 (1977).

Israel, F. P., Habing, H. J., and de Jong, T., Astron. Astrophys. **27**, 143 (1973).

Itoh, H., Publ. Astron. Soc. Jap. **31**, 541 (1979).

Itoh, H., Astrophys. J. **285**, 601 (1984).

Itoh, H. and Fabian, A. S., Mon. Not. R. Astron. Soc. **208**, 645 (1984).

Itoh, H., Kumagai, S., Shigeyama, T., Nomoto, K., and Nishimura, J., Nature **330**, 233 (1987).

Itoh, H., Masai, K., and Nomoto, K., Astrophys. J. **334**, 279 (1988).

Ivanov, V. P., Barabanov, A. P., Stankevich, K. S., and Stolyarov, S. P., Astron. Zh. **59**, 963 (1982a) [Sov. Astron. **26**, 581 (1982)].

Ivanov, V. P., Bubukin, I. T., and Stankevich, K. S., Pis'ma Astron. Zh. **8**, 83 (1982b) [Sov. Astron. Lett. **8**, 42 (1982)].

Iwan, D., Astrophys. J. **239**, 316 (1980).

Iyudin, A. F., Kirillov-Ugryumov, V. G., Kotov, Yu. D., Smirnov, Yu. V., Yurov, V. N., Kurnosova, L. V., Fradkin, M. I., Damle, S. V., Gokhale, G. S., Kunte, P. K., and Sreekantan, B. V., Pis'ma Astron. Zh. **10**, 104 (1984) [Sov. Astron. Lett. **10**, 41 (1984)].

J

Jansen, F. A., Smith, A., Bleeker, J. A. M., de Korte, P. A. J., Peacock, A., and White, N. E., Astrophys. J. **331**, 949 (1988).

Jenkins, E. B., Astrophys. J. **220**, 107 (1978).

Jenkins, E. B. and Meloy, D. A., Astrophys. J. Lett. **193**, L121 (1974).

Jenkins, E. B., Silk, J., and Wallerstein, G., Astrophys. J. Lett. **209**, L87 (1976).

Jenkins, E. B., Silk, J., Wallerstein, G., and Leep, E. M., Astrophys. J. **248**, 977 (1981).

Jenkins, E. B., Wallerstein, G., and Silk, J., Astrophys. J. **278**, 649 (1984).

Jewitt, D. C., Danielson, G. E., and Kupferman, P. N., Astrophys. J. **302**, 727 (1986).

Johnson, H. M., Astrophys. J. **176**, 645 (1972).

Johnson, H. M., Astrophys. J. **194**, 337 (1974).

Johnson, H. M., Astrophys. J. **198**, 111 (1975).

Johnson, H. M., Astrophys. J. **206**, 243 (1976).

Johnson, H. M., Mon. Not. R. Astron. Soc. **184**, 727 (1978).

Johnson, H. M., Astrophys. J. **235**, 66 (1980).

Johnson, H. M., Astrophys. J. **256**, 559 (1982a).

Johnson, H. M., Astrophys. J. Suppl. Ser. **50**, 551 (1982b).

Johnson, H. M. and Hogg, D. E., Astrophys. J. **142**, 1033 (1965).

Johnson, P. G. and Songsathaporn, R., Mon. Not. R. Astron. Soc. **195**, 51 (1981).

K

Kahler, H., Ule, T., and Wendker, H. J., Astrophys. Space Sci. **135**, 105 (1989).

Kahn, F. D., in: IAU Symp. No. 103, *Planetary Nebulae*, D. R. Flower (ed.), Reidel, Dordrecht (1983), p. 305.

Kahn, S. M., Bordie, J., Bowyer, S., and Charles, P. A., Astrophys. J. **269**, 212 (1983).

Kamper, K. and van den Bergh, S., Astrophys. J. Suppl. Ser. **32**, 351 (1976).

Kamper, K. and van den Bergh, S., Astrophys. J. **224**, 851 (1978).

Kaplan, S. A. and Pikel'ner, S. B., *The Interstellar Medium*, Fizmatgiz, Moscow (1963) [Harvard University Press, Cambridge (1970)].

Kaplan, S. A. and Pikel'ner, S. B., *Physics of the Interstellar Medium* [in Russian], Nauka, Moscow (1979).

Kardashev, N. S., Astron. Zh. **39**, 393 (1962) [Sov. Astron. **6**, 317 (1962)].

Kardashev, N. S., Astron. Zh. **41**, 807 (1964) [Sov. Astron. **8**, 643 (1964)].

Karovska, M., Nisenson, P., Noyes, R., and Papaliolios, C., IAU Circ. No. 4382 (1987).

Kellett, B. J., Branduardi-Raymont, G., Culhane, J. L., Mason, I. M., Mason, K. O., and Whitehouse, D. R., Mon. Not. R. Astron. Soc. **225**, 199 (1987).

Kemp, J. C., Henson, G. D., Kraus, D. J., Carroll, L. C., Beardsley, I. S., Takagashi, K., Jugaku, J., Matsuoka, M., Leibowitz, E. M., Mazeh, T., and Mendelson, H., Astrophys. J. **305**, 805 (1986).

Kenney, J. D. and Dent, W. A., Astrophys. J. **298**, 644 (1985).

Kennicutt, R. C., Astrophys. J. **277**, 361 (1984).

Kesteven, M. J. and Caswell, J. L., Astron. Astrophys. **183**, 118 (1987).

Kesteven, M. J. and Caswell, J. L., in: IAU Colloq. No. 101, *Supernova Remnants and the Interstellar Medium*, R. Roger and T. Landecker (eds.), Cambridge University Press, Cambridge (1988), p. 359.

Khaliullin, A. Kh. and Cherepashchuk, A. M., in: *Itogi Nauki i Tekhniki. Astronomiya*, Vol. 21, VINITI (1982), p. 5.

Khokhlov, A. M. and Érgma, É, V., Pis'ma Astron. Zh. **12**, 366 (1986) [Sov. Astron. Lett. **12**, 152 (1986)].

Khromov, G. S., *Planetary Nebulae* [in Russian], Nauka, Moscow (1985).

Khyanki, U. and Pel't, Ya., Pis'ma Astron. Zh. **12**, 541 (1986) [Sov. Astron. Lett. **12**, 228 (1986)].

Kirshner, R. P., in: *ESO Workshop on the SN 1987A; ESO Conference and Workshop Proceedings No. 26*, I. J.Danziger (ed.) (1987), p. 121.

Kirshner, R. P. and Kwan, J., Astrophys. J. **193**, 27 (1974).

Kirshner, R. P. and Kwan, J., Astrophys. J. **197**, 415 (1975).

Kirshner, R. P. and Taylor, K., Astrophys. J. Lett. **208**, L83 (1976).

Kirshner, R. P. and Chevalier, R. A., Astron. Astrophys. **67**, 267 (1978).

Kirshner, R. P. and Fesen, R. A., Astrophys. J. **224**, 59 (1978).

Kirshner, R. P. and Arnold, C. N., Astrophys. J. **229**, 147 (1979).

Kirshner, R. P. and Winkler, P. F., Astrophys. J. **227**, 853 (1979).

Kirshner, R. P. and Chevalier, R. A., Astrophys. J. Lett. **242**, L77 (1980).

Kirshner, R. P., Oke, J. B., Penston, M., and Searle, L., Astrophys. J. **185**, 303 (1973a).

Kirshner, R. P., Wilner, S. P., Becklin, E. E., Neugebauer, G., and Oke, J. B., Astrophys. J. Lett. **180**, L97 (1973b).

Kirshner, R. P., Gull, T. R., and Parker, R. A. R., Astron. Astrophys. Suppl. Ser. **31**, 261 (1978).

Kirshner, R. P., Sonneborn, G., Crenshaw, D. M., and Nassiopoulos, G. E., Astrophys. J. **320**, 602 (1987a).

Kirshner, R. P., Winkler, P. F., and Chevalier, R. A., Astrophys. J. Lett. **315**, L135 (1987b).

Kirshner, R. P., Morse, J. A., Winkler, P. F., and Blair, W. P., Astrophys. J. **342**, 260 (1989).

Klein, U., Emerson, D. T., Haslam, C. G., and Salter, C. J., Astron. Astrophys. **76**, 120 (1979).

Knapp, G. R. and Kerr, F. J., Astron. Astrophys. **33**, 463 (1974).

Knapp, G. R. and Morris, M., Astrophys. J. **292**, 640 (1985).

Knapp, G. R., Phillips, T. G., Leighton, R. B., Lo, K. Y., Wannier, P. G., and Wootten, H. A., Astrophys. J. **252**, 616 (1982).

Kolesnik, I. G. and Silich, S. A., in: IAU Colloq. No. 101, *Supernova Remnants and the Interstellar Medium*, R. Roger and T. Landecker (eds.), Cambridge University Press, Cambridge (1988), p. 469.

Kompaneets, D. A., Dokl. Akad. Nauk SSSR **130**, 1001 (1960).

Königl, A., Mon. Not. R. Astron. Soc. **205**, 471 (1983).

Koshiba, M., in: *ESO Workshop on the SN 1987A; ESO Conference and Workshop Proceedings No. 26*, I. J.Danziger (ed.) (1987), p. 207.

Koyama, K. et al., Publ. Astron. Soc. Jap. **39**, 801 (1987).

Krebs, J. and Hillebrandt, W., Astron. Astrophys. **128**, 411 (1983).

Kriss, G. A., Becker, R. H., Helfand, D. J., and Canizares, C. R., Astrophys. J. **288**, 703 (1985).

Ku, W. H. M., Kahn, C. M., and Pisarski, R., Astrophys. J. **278**, 615 (1984).

Kudritzki, R. P., Groth, H. G., Butler, K., Husfeld, D., Becker, S., Eber, F., and Fitzpatrick, E., in: *ESO Workshop on the SN 1987A; ESO Conference and Workshop Proceedings No. 26*, I. J.Danziger (ed.) (1987a), p. 39.

Kudritzki, R. P., Pauldrach, A., and Puls, J., Astron. Astrophys. **173**, 293 (1987b).

Kulkarni, S. and Heiles, C., in: *Galactic and Extragalactic Radio Astronomy*, K. I. Kellermann and G. L. Verschuur (eds.), Springer-Verlag, New York (1987).

Kulkarni, S., Clifton, T. C., Backer, D. C., Foster, R. S., Fruchter, A. S., and Taylor, J. H., Nature **331**, 50 (1988).

Kumagai, S., Itoh, M., Shigeyama, T., Nomoto, K., and Nishimura, J., Astron. Astrophys. **197**, L7 (1988).

Kumar, C. K., Kallman, T. R., and Thomas, R. J., Astrophys. J. **272**, 219 (1983).

Kundt, W., Nature **284**, 191 (1980).

Kundt, W., Astron. Astrophys. **121**, L15 (1983).

Kundt, W., Astron. Astrophys. **150**, 276 (1985).

Kundu, M. R., Angerhofer, P. E., Fürst, E., and Hirth, W., Astron. Astrophys. **92**, 225 (1980).

Kwitter, K. B., Astrophys. J. **245**, 154 (1981).

Kwitter, K. B., Astrophys. J. **287**, 840 (1984).

Kwok, S., in: IAU Symp. No. 103, *Planetary Nebulae*, D. R. Flower (ed.), Reidel, Dordrecht (1983), p. 293.

L

Lamers, H., Astrophys. J. **245**, 593 (1981).

Lamontagne, R. and Moffat, A. F. J., Astron. J. **94**, 1008 (1987).

Landecker, T. L. and Wielebinski, R., Austral. J. Phys. Astrophys. Suppl. No. 16, 1 (1970).

Landecker, T. L., Roger, R. S., and Higgs, L. A., Astron. Astrophys. Suppl. Ser. **39**, 133 (1980).

Landecker, T. L., Pineault, S., Routledge, D., and Vaneldik, J. F., Astrophys. J. Lett. **261**, L41 (1982).

Landecker, T. L., Vaneldik, J. F., Dewdney, P. E., and Routledge, D., Astron. J. **94**, 111 (987).

Landecker, T. L., Pineault, S., Routledge, D., and Vaneldik, J. F., Mon. Not. R. Astron. Soc. **237**, 277 (1989).

Langer, N., Astron. Astrophys. **171**, L1 (1987).

Langer, N. and Eid, M. F. E., Astron. Astrophys. **167**, 265 (1986a).

Langer, N. and Eid, M. F. E., Astron. Astrophys. **167**, 274 (1986b).

Lasker, B. M., Astrophys. J. **223**, 109 (1978).

Lasker, B. M., Publ. Astron. Soc. Pac. **91**, 153 (1979).

Lasker, B. M., Astrophys. J. **237**, 765 (1980).

Lasker, B. M., Astrophys. J. **244**, 517 (1981).

Leahy, D. A., Astrophys. J. **322**, 917 (1987a).

Leahy, D. A., Mon. Not. R. Astron. Soc. **228**, 907 (1987b).

Leahy, D. A. and Roger, R. S., in: IAU Colloq. No. 101, *Supernova Remnants and the Interstellar Medium*, R. Roger and T. Landecker (eds.), Cambridge University Press, Cambridge (1988), p. 301.

Leahy, D. A., Venkatesan, D., Long, K. S., and Naranan, S., Astrophys. J. **294**, 183 (1985a).

Leahy, D. A., Naranan, S., and Singh, K. P., Mon. Not. R. Astron. Soc. **213**, 15P (1985b).

Leibowitz, E. M. and Danziger, I. J., Mon. Not. R. Astron. Soc. **204**, 273 (1983).

Leitherer, C. and Chavarria, K. C., Astron. Astrophys. **175**, 208 (1987).

Lerche, I., Astron. Astrophys. **85**, 141 (1980).

Lerche, I. and Milne, D. K., Astron. Astrophys. **81**, 302 (1980).

Levine, A., Petre, R., Rappaport, S., Smith, G. C., Evans, K. D., and Rolf, D., Astrophys. J. Lett. **228**, L99 (1979).

Lipunov, V. M. and Postnov, K. A., Astron. Astrophys. **144**, L13 (1985).

Litvinova, I. Yu. and Nadezhin, D. K., Preprint No. 23, Inst. Teor. Éksp. Fiz. (1982).

Litvinova, I. Yu. and Nadezhin, D. K., Pis'ma Astron. Zh. **11**, 351 (1985) [Sov. Astron. Lett. **11**, 145 (1985)].

Lockhart, I. A., Goss, W. M., Caswell, J. L., and McAdam, W. B., Mon. Not. R. Astron. Soc. **179**, 147 (1977).

Lockman, F. J., Hobbs, L. M., and Shull, J. M., Astrophys. J. **301**, 380 (1986).

Lomovskii, A. I. and Klement'eva, A. Yu., Astron. Zh. **63**, 246 (1986) [Sov. Astron. **30**, 150 (1986)].

Long, K. S., Blair, W. P., Kirshner, R. P., and Winkler, P. F., in: IAU Colloq. No. 101, *Supernova Remnants and the Interstellar Medium*, R. Roger and T. Landecker (eds.), Cambridge University Press, Cambridge (1988a), p. 197.

Long, K. S., Blair, W. P., and van den Bergh, S., Astrophys. J. **333**, 749 (1988b).

Lortet, M. C., Georgelin, J. P., and Georgelin, J. M., Astron. Astrophys. **180**, 65 (1987).

Lozinskaya, T. A., Astron. Tsirk. No. 299, 1 (1964).

Lozinskaya, T. A., Astron. Zh. **47**, 122 (1970) [Sov. Astron. **14**, 98 (1970)].

Lozinskaya, T. A., Astron. Zh. **48**, 1145 (1971) [Sov. Astron. **15**, 910 (1972)].

Lozinskaya, T. A., Astron. Zh. **50**, 950 (1973a) [Sov. Astron. **17**, 603 (1973)].

Lozinskaya, T. A., Astron. Zh. **50**, 496 (1973b) [Sov. Astron. **17**, 317 (1973)].

Lozinskaya, T. A., Pis'ma Astron. Zh. **1**, 25 (1975a) [Sov. Astron. Lett. **1**, 35 (1975)].

Lozinskaya, T. A., Astron. Zh. **52**, 515 (1975b) [Sov. Astron. **19**, 315 (1975)].

Lozinskaya, T. A., Astron. Zh. **53**, 38 (1976) [Sov. Astron. **20**, 19 (1976)].

Lozinskaya, T. A., Pis'ma Astron. Zh. **3**, 306 (1977) [Sov. Astron. Lett. **3**, 163 (1977)].

Lozinskaya, T. A., Pis'ma Astron. Zh. **4**, 353 (1978a) [Sov. Astron. Lett. **4**, 190 (1978)].

Lozinskaya, T. A., Astron. Astrophys. **64**, 123 (1978b).

Lozinskaya, T. A., Astron. Zh. **56**, 900 (1979a) [Sov. Astron. **23**, 506 (1979)].

Lozinskaya, T. A., Astron. Astrophys. **71**, 29 (1979b).

Lozinskaya, T. A., Astron. Astrophys. **84**, 26 (1980a).

Lozinskaya, T. A., Astron. Zh. **57**, 707 (1980b) [Sov. Astron. **24**, 407 (1980)].

Lozinskaya, T. A., Pis'ma Astron. Zh. **6**, 350 (1980c) [Sov. Astron. Lett. **6**, 193 (1980)].

Lozinskaya, T. A., Astron. Zh. **57**, 1197 (1980d) [Sov. Astron. **24**, 691 (1980)].

Lozinskaya, T. A., Pis'ma Astron. Zh. **7**, 29 (1981) [Sov. Astron. Lett. **7**, 17 (1981)].

Lozinskaya, T. A., Astrophys. Space Sci. **87**, 313 (1982).

Lozinskaya, T. A., Pis'ma Astron. Zh. **9**, 469 (1983) [Sov. Astron. Lett. **9**, 247 (1983)].

Lozinskaya, T. A., Astron. Zh. **63**, 914 (1986) [Sov. Astron. **30**, 542 (1986)].

Lozinskaya, T. A., in: IAU Colloq. No. 101, *Supernova Remnants and the Interstellar Medium*, R. Roger and T. Landecker (eds.), Cambridge University Press, Cambridge (1988a), p. 95.

Lozinskaya, T. A., Proc. Astron. Soc. Australia **7**, 535 (1988b).

Lozinskaya, T. A., in: IAU Symp. No. 143, *Wolf–Rayet Stars and Interrelations with other Massive Stars in Galaxies*, K. A. van der Hucht and B. Hidayat (eds.), Kluwer, Dordrecht (1990).

Lozinskaya, T. A. and Lomovskii, A. I., Pis'ma Astron. Zh. **8**, 224 (1982) [Sov. Astron. Lett. **8**, 119 (1982)].

Lozinskaya, T. A. and Sitnik, T. G., Astron. Zh. **54**, 807 (1977) [Sov. Astron. **21**, 456 (1977)].

Lozinskaya, T. A. and Sitnik, T. G., Pis'ma Astron. Zh. **4**, 509 (1978) [Sov. Astron. Lett. **4**, 274 (1978)].

Lozinskaya, T. A. and Sitnik, T. G., Pis'ma Astron. Zh. **5**, 348 (1979) [Sov. Astron. Lett. **5**, 186 (1979)].

Lozinskaya, T. A. and Sitnik, T. G., Astron. Zh. **57**, 997 (1980) [Sov. Astron. **24**, 572 (1980)].

Lozinskaya, T. A. and Sitnik, T. G., Pis'ma Astron. Zh. **14**, 240 (1988) [Sov. Astron. Lett. **14**, 100 (1988)].

Lozinskaya, T. A. and Tutukov, A. V., Nauchn. Inform. Astrosoveta Akad. Nauk SSSR No. 49, 21 (1981).

Lozinskaya, T. A., Klement'eva, A. Yu., Zhukov, G. V., and Shenavrin, V. I., Astron. Zh. **52**, 682 (1975) [Sov. Astron. **19**, 416 (1975)].

Lozinskaya, T. A., Lar'kina, V. V., and Putilina, E. V., Pis'ma Astron. Zh. **9**, 662 (1983) [Sov. Astron. Lett. **9**, 344 (1983)].

Lozinskaya, T. A., Lomovskii, A. I., Pravdikova, V. V., and Surdin, V. G., Pis'ma Astron. Zh. **14**, 909 (1988) [Sov. Astron. Lett. **14**, 385 (1988)].

Lozinskaya, T. A., Sitnik, T. G., and Lomovskii, A. I., Astrophys. Space Sci. **121**, 357 (1986).

Lozinskaya, T. A., Sitnik, T. G., and Toropova, M. S., Astron. Zh. **64**, 939 (1987) [Sov. Astron. **31**, 493 (1987)].

Lucke, R. L., Woodgate, B. E., Gull, T. R., and Socker, D. G., Astrophys. J. **235**, 882 (1980).

Lucke, R. L., Zarnecki, J. C., Woodgate, B. E., Culhane, J. L., and Socker, D. G., Astrophys. J. **228**, 763 (1979).

Lucy, L. B., in: *ESO Workshop on the SN 1987A; ESO Conference and Workshop Proceedings No. 26*, I. J.Danziger (ed.) (1987), p. 417.

Lundqvist, P. and Fransson, C., Astron. Astrophys. **192**, 221 (1988a).

Lundqvist, P. and Fransson, C., in: IAU Colloq. No. 101, *Supernova Remnants and the Interstellar Medium*, R. Roger and T. Landecker (eds.), Cambridge University Press, Cambridge (1988), p. 15.

Lundqvist, P., Fransson, C., and Chevalier, R. A., Astron. Astrophys. **162**, L6 (1986).

Lundstron, I. and Stenholm, B., Astron. Astrophys. Suppl. Ser. **58**, 163 (1984).

Lynds, B. T. and O'Neil, E., Astrophys. J. **274**, 650 (1983).

Lyne, A. G., in: *Supernovae: A Survey of Current Research*, M. J. Rees and R. J. Stoneham (eds.), Reidel, Dordrecht (1982), p. 405.

Lyne, A. G., Anderson, B., and Salter, M. J., Mon. Not. R. Astron. Soc. **201**, 503 (1982).

M

MacAlpine, G. M., McGaugh, Mazzarella, J. M., and Uomoto, A., Astrophys. J. **342**, 364 (1989).

Maddalena, R. J. and Morris, M., Astrophys. J. **323**, 179 (1987).

Maeder, A., Astron. Astrophys. **105**, 149 (1982).

Maeder, A., Astron. Astrophys. **120**, 113 (1983).

Maeder, A., in: *ESO Workshop on the SN 1987A; ESO Conference and Workshop Proceedings No. 26*, I. J.Danziger (ed.) (1987), p. 251.

Maeder, A., Astron. Astrophys. **173**, 247 (1987b).

Maeder, A. and Lequeux, J., Astron. Astrophys. **114**, 409 (1982).

Mahoney, W. A., Varnell, L. S., Jacobson, A. S., Ling, J. C., Radocinski, R. G., and Wheaton, W. A., Astrophys. J. **334**, L81 (1988).

Malina, R., Lampton, M., and Bowyer, S., Astrophys. J. **207**, 894 (1976).

Manchanda, R. K., Bazzano, A., La Padula, C. D., Polcaro, V. F., and Ubertini, P., Astrophys. J. **252**, 172 (1982).

Manchester, R. N., Astron. Astrophys. **171**, 205 (1987).

Manchester, R. N. and Durdin, J. M., in: IAU Symp. No. 101, *Supernova Remnants and their X-Ray Emission*, J. Danziger and P. Gorenstein (eds.), Reidel, Dordrecht (1983), p. 421.

Manchester, R. N., Tuohy, I. R., and D'Amico, N., Astrophys. J. Lett. **262**, L31 (1982).

Manchester, R. N., Durdin, J. M., and Newton, L. M., Nature **313**, 374 (1985a).

Manchester, R. N., D'Amico, N., and Tuohy, I. R., Mon. Not. R. Astron. Soc. **212**, 975 (1985b).

Mansfield, V. N. and Salpeter, E. E., Astrophys. J. **190**, 305 (1974).

Marcher, S. J., Meikle, W. P. S., and Morgan, B. L., IAU Circ. No. 4391 (1987).

Margon, B., Science **215**, 247 (1982).

Margon, B., Ann. Rev. Astron. Astrophys. **22**, 507 (1984).

Markert, T. H., Canizares, C. R., Clark, G. W., and Winkler, P. F., Astrophys. J. **268**, 134 (1983).

Markert, T. H., Blizzard, P. L., Canizares, C. R., and Hughes, J. P., in: IAU Colloq. No. 101, *Supernova Remnants and the Interstellar Medium*, R. Roger and T. Landecker (eds.), Cambridge University Press, Cambridge (1988), p. 129.

Marsden, P. L., Gillett, F. C., Jennings, R. E., Emerson, J. P., De Jong, T., and Olnon, F. M., Astrophys. J. Lett. **278**, L29 (1984).

Marston, A. P. and Meaburn, J., Mon. Not. R. Astron. Soc. **235**, 391 (1988).

Masai, K, Hayakawa, S., Itoh, M., and Nomoto, K., Nature **330**, 235 (1987).

Mason, K. O., Kahn, S. M., Charles, P. A., Lampton, M. L., and Blissett, R., Astrophys. J. Lett. **230**, L163 (1979).

Massevich (Massevitch), A. G., Tutukov, A. V., and Yungelson, L. R., Astrophys. Space Sci. **40**, 115 (1975).

Mathewson, D. S., Dopita, M. A., Tuohy, I. R., and Ford, V. L., Astrophys. J. Lett. **242**, L73 (1980).

Mathewson, D. S., Ford, V. L., Dopita, M. A., Tuohy, I. R., Long, K. S., and Helfand, D. J., in: IAU Symp. No. 101, *Supernova Remnants and their X-Ray Emission*, J. Danziger and P. Gorenstein (eds.), Reidel, Dordrecht (1983), p. 541.

Mathewson, D. S., Ford, V. L., Dopita, M. A., Tuohy, I. R., Mills, B. Y., and Turtle, A. J., Astrophys. J. Suppl. Ser. **55**, 189 (1984).

Mathewson, D. S., Ford, V. L., Tuohy, I. R., Mills, B. Y., Turtle, A. J., and Helfand, D. J., Astrophys. J. Suppl. Ser. **58**, 197 (1985).

Matsui, Y. and Long, K. S., in: *The Crab Nebula and Related Supernova Remnants*, M. C. Kafatos and R. B. C. Henry (eds.), Cambridge University Press, Cambridge (1985), p. 211.

Matsui, Y., Long, K. S., Dickel, J. R., and Greisen, E. W., Astrophys. J. **287**, 295 (1985).

Matsui, Y., Long, K. S., and Tuohy, I. R., in: IAU Colloq. No. 101, *Supernova Remnants and the Interstellar Medium*, R. Roger and T. Landecker (eds.), Cambridge University Press, Cambridge (1988a), p. 157.

Matsui, Y., Long, K. S., and Tuohy, I. R., Astrophys. J. **329**, 838 (1988b).

Matveenko, L. I., Astron. Tsirk. No. 360, 3 (1966).

Matveenko, L. I., Pis'ma Astron. Zh. **10**, 111 (1984) [Sov. Astron. Lett. **10**, 44 (1984)].

Matz, S. M., Share, G. H., Leising, M. D., Chupp, E. L., and Vestrand, W. T., IAU Circ. No. 4510 (1987).

Matz, S. M., Share, G. H., Leising, M. D., Chupp, E. L., Vestrand, W. T., Purcell, W. R., Strickman,, M. S., and Reppin, C., Nature **331**, 416 (1988).

Mauche, C. W. and Gorenstein, P., in: *The Crab Nebula and Related Supernova Remnants*, M. C. Kafatos and R. B. C. Henry (eds.), Cambridge University Press, Cambridge (1985), p. 81.

Maucherat, A. et al. (28 authors), Astron. Astrophys. **23**, 147 (1973).

Mazeh, T., Aguilar, L. A., Treffers, R. R., Königl, A., and Sparke, L. S., Astrophys. J. **265**, 235 (1983).

McCray, R., in: IAU Colloq. No. 101, *Supernova Remnants and the Interstellar Medium*, R. Roger and T. Landecker (eds.), Cambridge University Press, Cambridge (1988), p. 447.

McCray, R. and Kafatos, M., Astrophys. J. **317**, 190 (1987).

McCollough, M. L. and Mufson, S. L., in: IAU Colloq. No. 101, *The Interaction of Supernova Remnants with the Interstellar Medium*, R. Roger and T. Landecker (eds.), Cambridge University Press, Cambridge (1988), p. 403.

McKee, C. F., in: IAU Colloq. No. 101, *Supernova Remnants and the Interstellar Medium*, R. Roger and T. Landecker (eds.), Cambridge University Press, Cambridge (1988), p. 205.

McKee, C. F. and Cowie, L. L., Astrophys. J. **195**, 715 (1975).

McKee, C. F. and Cowie, L. L., Astrophys. J. **215**, 213 (1977).

McKee, C. F. and Ostriker, J. P., Astrophys. J. **218**, 148 (1977).

McKee, C. F., Cowie, L. L., and Ostriker, J. P., Astrophys. J. Lett. **219**, L23 (1978).

McKee, C. F., Van Buren, D., and Lazareff, B., Astrophys. J. Lett. **278**, L115 (1984).

McKee, C. F., Hollenbach, D. J., Gregory, C., Seab, C., and Teelens, A. G. G. M., Astrophys. J. **318**, 674 (1987).

McLean, I. S., Aspin, C., and Reitsema, H., Nature **304**, 243 (1983).

Meaburn, J., in: *Topics in Interstellar Matter*, Astrophys. Space Sci. Libr. Vol. 70 (1977), p. 81.

Meaburn, J., Observatory **99**, 176 (1979).

Meaburn, J., Mon. Not. R. Astron. Soc. **192**, 365 (1980).

Meaburn, J., Mon. Not. R. Astron. Soc. **196**, 19P (1981).

Meaburn, J. and Blades, J. C., Mon. Not. R. Astron. Soc. **190**, 403 (1980).

Meaburn, J., Terrett, D. L., and Blades, J. C., Mon. Not. R. Astron. Soc. **197**, 19 (1981).

Meaburn, J., Marston, A. P., McGee, R. X., and Newton, L. M., Mon. Not. R. Astron. Soc. **225**, 591 (1987).

Meaburn, J., Wolstencroft, D. A., and Walsh, J. R., Astron. Astrophys. **181**, 333 (1987).

Menzies, J. W., in: *ESO Workshop on the SN 1987A; ESO Conference and Workshop Proceedings No. 26*, I. J.Danziger (ed.) (1987), p. 73.

Menzies, J. W., Catchpole, R. M., van Vuuren, G., Winkler, H., Lancy, C. D., Whitelock, P. A., Cousins, A. W. J., Carter, B. S., Marang, F., Lloyd Evans, T. H. H., Roberts, G., Kilkenny, D., Jones, J. S., Sekiguchi, K., Fairall, A. P., and Wolstencroft, R., Mon. Not. R. Astron. Soc. **227**, 39P (1987).

Merrill, P. W., Publ. Astron. Soc. Pac. **50**, 350 (1938).

Meyerott, R., Astrophys. J. **239**, 257 (1980).

Mezger, P. G., Astron. Astrophys. **70**, 565 (1978).

Mezger, P. G., Tuffs, R. J., Chini, R., Kreysa, E., and Gemünd, H.-P., Astron. Astrophys. **167**, 145 (1986).

Middleditch, J. and Pennypacker, C. R., Nature **313**, 659 (1985).

Middleditch, J., Pennypacker, C. R., and Burns, M. S., Astrophys. J. **274**, 313 (1983).

Middleditch, J., Pennypacker, C. R., and Burns, M. S., Astrophys. J. **315**, 142 (1987).

Mikulášek, Z., Bull. Astron. Inst. Czech. **20**, 215 (1969).

Miller, J. S., Astrophys. J. **189**, 239 (1974).

Mills, B. Y., Turtle, A. J., Little, A. G., and Durdin, J. M., Austral. J. Phys. **37**, 321 (1984).

Milne, D. K., Austral. J. Phys. **23**, 425 (1970).

Milne, D. K., Austral. J. Phys. **25**, 307 (1972).

Milne, D. K., Austral. J. Phys. **32**, 83 (1979a).

Milne, D. K., Proc. Astron. Soc. Austral. **3**, 341 (1979b).

Milne, D. K., Astron. Astrophys. **81**, 293 (1980).

Milne, D. K., Austr. J. Phys. **40**, 771 (1987).

Milne, D. K., in: IAU Colloq. No. 101, *Supernova Remnants and the Interstellar Medium*, R. Roger and T. Landecker (eds.), Cambridge University Press, Cambridge (1988), p. 351.

Milne, D. K. and Wilson, T. L., Astron. Astrophys. **10**, 220 (1971).

Milne, D. K. and Dickel, J. R., Austral. J. Phys. **28**, 209 (1975).

Milne, D. K. and Manchester, R. N., Astron. Astrophys. **167**, 117 (1986).

Minkowski, R., Publ. Astron. Soc. Pac. **53**, 224 (1941).

Minkowski, R., Rev. Mod. Phys. **30**, 1048 (1958).

Minkowski, R., in: IAU Symp. No. 9, *Radio Astronomy*, Paris (1959), p. 315.

Miranda, A. I., and Rosado, M., Rev. Mex. Astron. Astrofiz. **14**, 479 (1987).

Morrison, P. and Roberts, D., Nature **313**, 661 (1985).

Morton, D. C., Astrophys. J. **150**, 535 (1967).

Moffat, A. F. J. and Seggewiss, W., in: IAU Symp. No. 88, *Wolf–Rayet Binaries*, Toronto (1979a), p. 181.

Moffat, A. F. J. and Seggewiss, W., in: IAU Symp. No. 83, *Mass Loss and Evolution of O-Type Stars*, P. S. Conti and C. W. H. De Loore (eds.), Reidel, Dordrecht (1979b), p. 447.

Moffat, A. F. J. and Seggewiss, W., Astron. Astrophys. **77**, 128 (1979c).

Moffat, A. F. J. and Isserstedt, J., Astron. Astrophys. **91**, 147 (1980).

Moffat, A. F. J., Lamontagne, R., and Seggewiss, W., Astron. Astrophys. **114**, 135 (1982a).

Moffat, A. F. J., Firmani, C., McLean, I. S., and Seggewiss, W., in: IAU Symp. No. 99, *WR Stars: Observations, Physics, Evolution*, C. W. De Loore and A. J. Willis (eds.), Reidel, Dordrecht (1982b), p. 577.

Moffat, A. F. J., Breysacher, J., and Seggewiss, W., Astrophys. J. **292**, 511 (1985).

Moffat, P. H., Mon. Not. R. Astron. Soc. **153**, 401 (1971).

Morini, M., Robba, N. R., Smith, A., and van der Klis, M., Astrophys. J. **333**, 777 (1988).

Mufson, S. L., McCollough, M. L., Dickel, J. R., Petre, R., White, R., and Chevalier, R. A., Astron. J. **92**, 1349 (1987).

Murdin, P. and Clark, D. H., Mon. Not. R. Astron. Soc. **189**, 501 (1979).

Murdin, P. and Clark, D. H., Mon. Not. R. Astron. Soc. **190**, 65P (1980).

Murdin, P. and Clark, D. H., Nature **294**, 543 (1981).

Murdin, P., Clark, D. H., and Culhane, J. L., Mon. Not. R. Astron. Soc. **184**, 79 (1978).

Murray, S. S., Fabbiano, G., Fabian, A. C., Epstein, A., and Giacconi, R., Astrophys. J. Lett. **234**, L69 (1979).

Mustel', É. R., Izv. Krymsk. Astrofiz. Observ. **21**, 24 (1959a).

Mustel', É. R., Ann. Astrophys. Suppl. Ser. **8**, 125 (1959b).

Mustel', É. R., Astron. Zh. **48**, 665 (1971) [Sov. Astron. **15**, 527 (1972)].

Mustel', É. R., Astron. Zh. **49**, 15 (1972) [Sov. Astron. **16**, 10 (1972)].

Mustel', É. R., Astron. Zh. **50**, 1121 (1973) [Sov. Astron. **17**, 711 (1973)].

Mustel', É. R. and Chugai, N. N., Astrophys. Space Sci. **32**, 25 (1975).

Myers, P., Astrophys. J. Suppl. Ser. **26**, 83 (1973).

Myers, P., Astrophys. J. **225**, 380 (1978).

N

Nadezhin, D. K., Preprint No. 1, Inst. Teor. Éksp. Fiz. (1981).

Nadezhin, D. K. and Utrobin, V. P., Astron. Zh. **54**, 996 (1977) [Sov. Astron. **21**, 564 (1977)].

Neckel, Th. and Clare, G., Astron. Astrophys. Suppl. Ser. **42**, 251 (1982).

Neugebauer, G. et al., Astrophys. J. Lett. **278**, L1 (1984).

Nichols-Bohlin, J. and Fesen, R. A., Astron. J. **92**, 642 (1986).

Niemela, V. S. and Cappa de Nicolau, C. E., in: IAU Symp. No. 143, *Wolf–Rayet Stars and Interrelations with other Massive Stars in Galaxies*, K. A. van der Hucht and B. Hidayat (eds.), Kluwer, Dordrecht (1990).

Niemela, V. S., Ruiz, M. T., and Phillips, M. M., Astrophys. J. **289**, 52 (1985).

Nisenson, P., Papaliolios, C., Karovska, M., and Moyes, R., Astrophys. J. Lett. **320**, L15 (1987).

Nomoto, K., in: *The Crab Nebula and Related Supernova Remnants*, M. C. Kafatos and R. B. C. Henry (eds.), Cambridge University Press, Cambridge (1985), p. 97.

Nomoto, K. and Tsuruta, S., in: IAU Symp. No. 101, *Supernova Remnants and their X-Ray Emission*, J. Danziger and P. Gorenstein (eds.), Reidel, Dordrecht (1983), p. 509.

Nomoto, K. and Tsuruta, S., Astrophys. J. **305**, L19 (1986).

Nomoto, K. and Tsuruta, S., Astrophys. J. **312**, 711 (1987).

Nomoto, K., Thielemann, F. K., and Wheeler, J., Astrophys. J. Lett. **279**, L23 (1984).

Nomoto, K., Shigeyama, T., and Hashimoto, M., in: *ESO Workshop on the SN 1987A; ESO Conference and Workshop Proceedings No. 26*, I. J.Danziger (ed.) (1987), p. 325.

Nousek, J. A., Cowie, L. L., Hu, E., Lindblad, C. J., and Garmire, G. P., Astrophys. J. **248**, 152 (1981).

Nugent, J. J., Pravdo, S. H., Garmire, G. P., Becker, R. H., Tuohy, I. R., and Winkler, P. F., Astrophys. J. **284**, 612 (1984).

Nugis, T., in: IAU Symp. No. 99, *WR Stars: Observations, Physics, Evolution*, C. W. De Loore and A. J. Willis (eds.), Reidel, Dordrecht (1982), p. 127.

O

Odegard, N., Astrophys. J. **301**, 813 (1986).

Odenwald, S. F., Shivanandan, K., Fazio, G. G., Rengarajan, T. N., McBreen, B., Campbell, B., Campbell, M. F., and Moseley, H., Astrophys. J. **279**, 162 (1984).

Odenwald, S. F. and Shivanandan, K., Astrophys. J. **292**, 460 (1985).

Oemler, A. and Tinsley, B. M., Astron. J. **84**, 985 (1979).

Ögelman, H., Koch-Miramond, L., and Aurière, M., Astrophys. J. **342**, L83 (1989).

Olson, G. L. and Castor, J. I., Astrophys. J. **244**, 179 (1981).

Oort, J. H., Mon. Not. R. Astron. Soc. **106**, 159 (1946).

Oort, J. H. and Spitzer, L., Astrophys. J. **121**, 6 (1955).

Öpik, E. J., Irish Astron. J. **2**, 219 (1953).

Ostriker, J. P. and McKee, C. F., Rev. Mod. Phys. **60**, 1 (1988).

P

Pacini, F. and Salvati, M., Astrophys. J. **186**, 249 (1973).

Pacini, F. and Salvati, M., Astrophys. J. Lett. **245**, L107 (1981).

Panagia, N., in: *Proceedings of the Third European IUE Conference*, Madrid (1982), p. 31.

Panagia, N., Space Telescope Science Institute, Preprint No. 83 (1985a).

Panagia, N., in: *Supernovae as a Distance Indicator*, N. Bartel (ed.), Springer-Verlag, Berlin (1985b), p. 14.

Panagia, N., in: *ESO Workshop on the SN 1987A; ESO Conference and Workshop Proceedings No. 26*, I. J.Danziger (ed.) (1987), p. 55.

Panagia, N. et al. (27 authors), Mon. Not. R. Astron. Soc. **192**, 861 (1980).

Panagia, N., R. A. Sramek, and K. W. Weiler, Astrophys. J. Lett. **300**, L55 (1986).

Parker, R. A. R., Astrophys. J. **139**, 493 (1964).

Parker, R. A. R., Astrophys. J. **224**, 873 (1978).

Parker, R. A. R., Gull, T. R., and Kirshner, R. P., *An Emission-Line Survey of the Milky Way*, NASA, Washington, D.C. (1979).

Pashchenko, M. I. and Slysh, V. I., Astron. Astrophys. **35**, 153 (1974).

Patnaik, A. R., Velusamy, T., and Venugopal, V. R., Nature **332**, 136 (1988).

Pauldrach, A., Puls, J., Kudritzki, R. P., Mendez, R. H., and Heap, S. R., Astron. Astrophys. **207**, 123 (1988).

Peimbert, M., Astrophys. J. **170**, 261 (1971).

Peimbert, M. and van den Bergh, S., Astrophys. J. **167**, 223 (1971).

Peimbert, M. and Torres-Peimbert, S., Mon. Not. R. Astron. Soc. **179**, 217 (1977).

Peimbert, M., Torres-Peimbert, S., and Rayo, J. F., Astrophys. J. **220**, 516 (1978).

Pequignot, D. and Dennefeld, M., Astron. Astrophys. **120**, 249 (1983).

Perek, L. and Kohoutek, L., *Catalog of Galactic Planetary Nebulae*, Czech. Inst. Sci., Prague (1967).

Perinotto, M., in: IAU Symp. No. 103, *Planetary Nebulae*, D. R. Flower (ed.), Reidel, Dordrecht (1983), p. 323.

Petre, R., Canizares, C. R., Kriss, G. A., and Winkler, P. F., Astrophys. J. **258**, 22 (1982).

Petre, R., Canizares, C. R., Winkler, P. F., Seward, F. D., Willingale, R., Rolf, D., and Woods, N., in: IAU Symp. No. 101, *Supernova Remnants and their X-Ray Emission*, J. Danziger and P. Gorenstein (eds.), Reidel, Dordrecht (1983), p. 289.

Petre, R., Szymkowiak, A. E., Seward, F. D., and Willingale, R., Astrophys. J. **335**, 215 (1988).

Phillips, A. P. and Gondhalekar, P. M., Mon. Not. R. Astron. Soc. **202**, 483 (1983).

Physical Processes in Interstellar Clouds, Proc. NATO Summer School, G. Morfill and M. Scholer (eds.), Reidel, Dordrecht (1987).

Pikel'ner, S. B., Izv. Krymsk. Astrofiz. Observ. **12**, 93 (1954).

Pikel'ner, S. B., Astron. Zh. **33**, 785 (1956).

Pikel'ner, S. B., Astron. Zh. **38**, 21 (1961) [Sov. Astron. **5**, 14 (1961)].

Pikel'ner, S. B., Astrophys. Lett. **2**, 97 (1968).

Pikel'ner, S. B., Astrophys. Lett. **15**, 91 (1973).

Pikel'ner, S. B. and Shcheglov, P. V., Astron. Zh. **45**, 953 (1968) [Sov. Astron. **12**, 757 (1969)].

Pineault, S., Pritchet, C. J., Landecker, T. L., Routledge, D., and Vaneldik, J. F., Astron. Astrophys. **151**, 52 (1985).

Pineault, S., Landecker, T. L., and Routledge, D., Astrophys. J. **315**, 580 (1987).

Pinto, P. A. and Woosley, S. E., Astrophys. J. **329**, 820 (1988).

Pisarski, R. L., Helfand, D. J., and Kahn, S. M., Astrophys. J. **277**, 710 (1984).

Pismis, P., Rev. Mex. Astron. Astrofiz. **1**, 45 (1974).

Pismis, P. and Recillas-Cruz, E., Rev. Mex. Astron. Astrofiz. **4**, 271 (1979).

Pismis, P., Recillas-Cruz, E., and Hasse, I., Rev. Mex. Astron. Astrofiz. **2**, 209 (1977).

Pismis, P., Moreno, M. A., and Hasse, I., Rev. Mex. Astron. Astrophys. **8**, 51 (1983).

Pollock, A. M. T., Astron. Astrophys. **150**, 339 (1985).

Porter, A. C. and A. V. Filippenko, Astron. J. **93**, 1372 (1987).

Pottasch, S. R., Baud, B., Beintema, D., Emerson, J., Habing, H. J., harris, S., Houck, J., Jennings, R., and Marsden, P., Astron. Astrophys. **138**, 10 (1984).

Poveda, A. and Woltjer, L., Astron. J. **73**, 65 (1968).

Pravdo, S. H. and Serlemitsos, P. J., Astrophys. J. **246**, 484 (1981).

Pravdo, S. H., Smith, B. W., Charles, P. A., and Tuohy, I. R., Astrophys. J. Lett. **235**, L9 (1980).

Pronik, V. I., Izv. Krymsk. Astrofiz. Observ. **30**, 104 (1963).

Pronik, V. I., Chuvaev, K. K., and Chugai, N. N., Astron. Zh. **53**, 1182 (1976) [Sov. Astron. **20**, 666 (1976)].

Pskovskii, Yu. P., Astron. Zh. **45**, 945 (1968) [Sov. Astron. **12**, 750 (1969)].

Pskovskii, Yu. P.Pis'ma Astron. Zh. **3**, 403 (1977a) [Sov. Astron. Lett. **3**, 215 (1977)].

Pskovskii, Yu. P., Astron. Zh. **54**, 1188 (1977b) [Sov. Astron. **21**, 675 (1977)].

Pskovskii, Yu. P., Astron. Zh. **55**, 737 (1978a) [Sov. Astron. **22**, 420 (1978)].

Pskovskii, Yu. P., in: *Neutrinos 77*, Vol. 1, Nauka, Moscow (1978b), p. 145.

Pskovskii, Yu. P., Astron. Zh. **55**, 350 (1978c) [Sov. Astron. **22**, 201 (1978)].

Pskovskii, Yu. P., Astron. Zh. **61**, 1125 (1984) [Sov. Astron. **28**, 658 (1984)].

Pskovskii, Yu. P., Pis'ma Astron. Zh. **3**, 403 (1977a) [Sov. Astron. Lett. **3**, 215 (1977)].

Purvis, A., Mon. Not. R. Astron. Soc. **202**, 605 (1983).

Pye, J. P., Pounds, K. A., Rolf, D. P., Seward, F. D., Smith, A., and Willingale, R., Mon. Not. R. Astron. Soc. **194**, 569 (1981).

R

Rappaport, S., Doxsey, R., Solinger, A., and Borken, R., Astrophys. J. **194**, 329 (1974).

Rappaport, S., Petre, R., Kayat, M., Evans, K., Smith, G., and Levine, A., Astrophys. J. **227**, 285 (1979).

Raymond, J. C., Astrophys. J. Suppl. Ser. **39**, 1 (1979).

Raymond, J. C., Ann. Rev. Astron. Astrophys. **22**, 75 (1984).

Raymond, J. C. and Smith, B. W., Astrophys. J. Suppl. Ser. **35**, 419 (1977).

Raymond, J. C., Cox, D. P., and Smith, B. W., Astrophys. J. **204**, 290 (1976).

Raymond, J. C., Davis, M., Gull, T. R., and Parker, R. A. R., Astrophys. J. Lett. **238**, L21 (1980).

Raymond, J. C., Blair, W. P., Fesen, R. A., and Gull, T. R., Astrophys. J. **275**, 636 (1983).

Raymond, J. C., Hester, J. J., Cox, D., Blair, W. P., Fesen, R. A., and Gull, T. R., Astrophys. J. **324**, 869 (1988).

Read, P. L., Mon. Not. R. Astron. Soc. **194**, 863 (1981).

Reich, W. and Braunsfurth, E., Astron. Astrophys. **99**, 17 (1981).

Reich, W., Berkhuijsen, E. M., and Sofue, Y., Astron. Astrophys. **72**, 270 (1979).

Reich, W., Fürst, E., and Sieber, W., in: IAU Symp. No. 101, *Supernova Remnants and their X-Ray Emission*, J. Danziger and P. Gorenstein (eds.), Reidel, Dordrecht (1983), p. 377.

Reich, W., Fürst, E., and Sofue, Y., Astron. Astrophys. **133**, L4 (1984).

Reich, W., Fürst, E., Reich, P., and Junkes, N., in: IAU Colloq. No. 101, *Supernova Remnants and the Interstellar Medium*, R. Roger and T. Landecker (eds.), Cambridge University Press, Cambridge (1988), p. 293.

Reid, P. B., Becker, R. H., and Long, K. S., Astrophys. J. **261**, 485 (1982).

Renzini, A., Mem. Soc. Astron. Ital. **49**, 389 (1978).

Repin, S. V., Strel'nitskii, V. S., and Chugai, N. N., Astron. Tsirk. No. 1532, 23 (1988).

Rester, A. C., Eichhorn, G., and Coldwell, R. L., IAU Circ. No. 4526 (1987).

Reynolds, R. J., Astrophys. J. **203**, 151 (1976a).

Reynolds, R. J., Astrophys. J. **206**, 679 (1976b).

Reynolds, R. J., Astrophys. J. **291**, 152 (1985).

Reynolds, R. J. and Ogden, P. M., Astrophys. J. **229**, 942 (1979).

Reynolds, S. P., in: IAU Colloq. No. 101, *Supernova Remnants and the Interstellar Medium*, R. Roger and T. Landecker (eds.), Cambridge University Press, Cambridge (1988a), p. 331.

Reynolds, S. P., Astrophys. J. **327**, 853 (1988b).

Reynolds, S. P. and Aller, H. D., Astrophys. J. **327**, 845 (1988).

Reynolds, S. P. and Chanan, G. A., Astrophys. J. **281**, 673 (1984).

Reynolds, S. P. and Chevalier, R. A., Astrophys. J. **245**, 912 (1981).

Reynolds, S. P. and Chevalier, R. A., Astrophys. J. **278**, 630 (1984).

Reynolds, S. P. and Gilmore, D. M., Astron. J. **92**, 1138 (1986).

Reynolds, S. P. and Ogden, P. M., Astron. J. **87**, 306 (1982).

Roger, R. and Dewdney, P. (eds.), *Regions of Recent Star Formation*, Reidel, Dordrecht (1982).

Roger, R. S., Milne, D. K., Kesteven, M. J., Wellington, K. J., and Haynes, R. F., Astrophys. J. **332**, 940 (1988).

Rohlfs, K., Braunsfurth, E., and Hills, D. L., Astron. Astrophys. Suppl. Ser. **30**, 369 (1977).

Romani, R. W., Reach, W. T., Koo, B.-C., and Heiles, C., Astrophys. J. **349**, L51 (1990).

Rosa, M., IAU Circ. No. 4564 (1988).

Rosa, M. R. and Mathis, J. S., in: *Properties of Hot Luminous Stars*, Boulder–Münich workshop, C. D. Garmany (ed.), Publ. Astron. Soc. Pac. Conf. Ser. **7**, 135 (1990).

Rosado, M., Astron. Astrophys. **160**, 211 (1986).

Rosado, M. and González, J., Rev. Mex. Astron. Astrofiz. **5**, 93 (1981).

Rosado, M., Georgelin, J. M., Laval, A., and Monnet, G., in: IAU Symp. No. 101, *Supernova Remnants and their X-Ray Emission*, J. Danziger and P. Gorenstein (eds.), Reidel, Dordrecht (1983), p. 567.

Rousseau, J., Martin, N., Prevot, L., Rebeirot, A. R., and Brunet, J. P., Astron. Astrophys. Suppl. Ser. **31**, 243 (1978).

Rózyczka, M., Astron. Astrophys. **143**, 59 (1985).

Rózyczka, M. and Tenorio-Tagle, G., Astron. Astrophys. **147**, 202; 209; 220 (1985).

Rózyczka, M. and Tenorio-Tagle, G., Astron. Astrophys. **176**, 329 (1987).

Rudnitskii, G. M., Astron. Zh. **55**, 345 (1978) [Sov. Astron. **22**, 199 (1978)].

Ruiz, M. T., Astrophys. J. **243**, 814 (1981).

Ruiz, M. T., in: IAU Symp. No. 101, *Supernova Remnants and their X-Ray Emission*, J. Danziger and P. Gorenstein (eds.), Reidel, Dordrecht (1983a), p. 241.

Ruiz, M. T., Astron. J. **88**, 1210 (1983b).

Rupen, M. P., Van Gorkom, J. H., Knapp, G. R., Gunn, J. E., and Schneider, D. P., Astron. J. **94**, 61 (1987).

Rust, B. W., "The use of supernova light curves for testing the expansion hypothesis and their cosmological relations," Ph.D. Thesis, Univ. of Illinois (1974).

S

Sabbadin, F., Astron. Astrophys. **80**, 212 (1979).

Saio, H., Kato, M., and Nomoto, K., Astrophys. J. **331**, 388 (1988).

Sakhibov, F. Kh. and Smirnov, M. A., Pis'ma Astron. Zh. **8**, 281 (1982) [Sov. Astron. Lett. **8**, 150 (1982)].

Sakhibov, F. Kh. and Smirnov, M. A., Astron. Zh. **60**, 676 (1983) [Sov. Astron. **27**, 395 (1983)].

Sandell, G., Scalise, E., and Braz, M. N., Astron. Astrophys. **124**, 139 (1983).

Sandie, W., Nakano, G., Chase, L., Fishman, G., Meegan, C., Wilson, R., Paciesas, W., and Lasche, G., IAU Circ. No. 4526 (1988a).

Sandie, W., Nakano, G., Chase, L., Fishman, G., Meegan, C., Wilson, R., Paciesas, W., and Lasche, G., Astrophys. J. **334**, L91 (1988b).

Sanduleak, N., Contr. CTIO **89** (1970).

Sastry, C. V., Dwarakanath, K. S., and Shevgaonkar, R. K., J. Astrophys. Astron. **2**, 339 (1981).

Schaeffer, R., Cassé, M., Mochkovitch, R., and Cahen, S., Astron. Astrophys. **184**, L1 (1987).

Schaeffer, R. E., Astrophys. J. **323**, L51 (1987).

Schild, H. and Maeder, A., Astron. Astrophys. **136**, 237 (1984).

Schmidt, G. D. and Angel, J. R. P., Astrophys. J. **227**, 106 (1979).

Schmidt, M., in: *Galactic Structure; Stars and Stellar Systems*, Vol. 5, A. Blaauw and M. Schmidt (eds.), University of Chicago Press, Chicago (1965), p. 513.

Schneps, M. N., Haschick, A. D., Wright, E. L., and Barrett, A. H., Astrophys. J. **243**, 184 (1981).

Schneps, M. N. and Wright, E. L., Sky and Telescope **59**, 195 (1980).

Schnopper, H. W., Delvaille, J. P., Rocchia, R., Blondel, C., Cheron, C., et al., Astrophys. J. **253**, 131 (1982).

Schwartz, P. R., Astrophys. J. **320**, 258 (1987).

Schwarz, U. J., Arnal, E. M., and Goss, W. M., Mon. Not. R. Astron. Soc. **192**, 67P (1980).

Schweizer, F. and Lasker, B. M., Astrophys. J. **226**, 167 (1978).

Scoville, N. Z., Irvine, W. M., Wannier, P. G., and Predmore, C. R., Astrophys. J. **216**, 320 (1977).

Sedov, L. I., *Similarity and Dimensionality Methods in Mechanics* [in Russian], Gostekhizdat, Moscow (1957).

Sedov, L. I., *Similarity and Dimensionality Methods in Mechanics* [in Russian], Nauka, Moscow (1981).

Seward, F. D., in: IAU Symp. No. 101, *Supernova Remnants and their X-Ray Emission*, J. Danziger and P. Gorenstein (eds.), Reidel, Dordrecht (1983), p. 405.

Seward, F. D., Bull. Amer. Astron. Soc. **20**, 1049 (1989).

Seward, F. D. and Chlebowski, T., Astrophys. J. **256**, 530 (1982).

Seward, F. D. and Harnden, F. R., Astrophys. J. Lett. **287**, L19 (1984).

Seward, F. D., Gorenstein, P., and Tucker, W., Astrophys. J. **266**, 287 (1983a).

Seward, F. D., Harnden, F. R., Murdin, P., and Clark, D. H., Astrophys. J. **267**, 698 (1983b).

Seward, F. D., Harnden, F. R., Szymkowiak, A., and Swank, J., Astrophys. J. **281**, 650 (1984).

Sgro, A. C., Astrophys. J. **197**, 621 (1975).

Shapiro, P. R. and Moore, R. T., Astrophys. J. **207**, 460 (1976).

Shapiro, P. R. and Field, G. B., Astrophys. J. **205**, 762 (1976).

Sharpless, S., Astrophys. J. Suppl. Ser. 4, 257 (1959).

Shaver, P. A., Astron. Astrophys. **105**, 306 (1982).

Shaver, P. A., Salter, C. J., Patnaik, A. R., Van Gorkom, J. H., and Hunt, G. C., Nature **313**, 113 (1985).

Shcheglov, P. V., Astron. Tsirk. No. 266, 2 (1963).

Shcheglov, P. V., Astron. Tsirk. No. 395, 2 (1966).

Shibazaki, T. and Ebisuzaki, T., Astrophys. J. **327**, L9 (1988).

Shigeyama, T., Nomoto, K., Hashimoto, M., and Sugimoto, D., Nature **328**, 320 (1987).

Shigeyama, T., Nomoto, K., and Hashimoto, M., Astron. Astrophys. **196**, 141 (1988).

Shirkey, R. C., Astrophys. J. **224**, 477 (1978).

Shklovskii, I. S., Dokl. Akad. Nauk SSSR **90**, 983 (1953).

Shklovskii, I. S., Astron. Zh. **37**, 369 (1960a) [Sov. Astron. **4**, 355 (1960)].

Shklovskii, I. S., Astron. Zh. **37**, 256 (1960b) [Sov. Astron. **4**, 243 (1960)].

Shklovskii, I. S., Astron. Zh. **39**, 209 (1962) [Sov. Astron. **6**, 162 (1962)].

Shklovskii, I. S., *Supernovae and Related Problems* [in Russian], Nauka, Moscow (1976a).

Shklovskii, I. S., Pis'ma Astron. Zh. **2**, 244 (1976) [Sov. Astron. Lett. **2**, 95 (1976)].

Shklovskii, I. S., Astron. Zh. **55**, 726 (1978) [Sov. Astron. **22**, 413 (1978)].

Shklovskii, I. S., Nature **279**, 703 (1979).

Shklovskii, I. S., Publ. Astron. Soc. Pac. **92**, 125 (1980a).

Shklovskii, I. S., Astron. Zh. **57**, 673 (1980b) [Sov. Astron. **24**, 387 (1980)].

Shklovskii, I. S., Pis'ma Astron. Zh. **7**, 479 (1981a) [Sov. Astron. Lett. **7**, 263 (1981)].

Shklovskii, I. S., Astron. Zh. **58**, 554 (1981b) [Sov. Astron. **25**, 315 (1981)].

Shklovskii, I. S., Pis'ma Astron. Zh. **9**, 474 (1983) [Sov. Astron. Lett. **9**, 250 (1983)].

Shklovskii, I. S., Adv. Space Rev. **3**, 241 (1984a).

Shklovskii, I. S., Pis'ma Astron. Zh. **10**, 723 (1984b) [Sov. Astron. Lett. **10**, 302 (1984)].

Shklovskii, I. S. and Sheffer, E. K., Nature **231**, 173 (1971).

Shull, J. M., Astrophys. J. **237**, 769 (1980).

Shull, J. M., Astrophys. J. Suppl. Ser. **46**, 27 (1981).

Shull, J. M., Astrophys. J. **262**, 308 (1982).

Shull, P., Astrophys. J. **269**, 218 (1983).

Shull, J. M., in: *Interstellar Processes*, D. Hollenbach and H. Thronson (eds.), Reidel, Dordrecht (1987), panel discussion.

Shull, J. M. and McKee, C. F., Astrophys. J. **227**, 131 (1979).

Shull, P., Parker, R. A. R., Gull, T. R., and Dufour, R. J., Astrophys. J. **253**, 682 (1982).

Shull, P., Carsenty, U., Sarcander, M., and Neckel, T., Astrophys. J. Lett. **285**, L75 (1984).

Shull, M., Fesen, R. A., and Saken, J. N., **346**, 860 (1989).

Sitnik, T. G. and Toropova, M. S., Pis'ma Astron. Zh. **8**, 679 (1982) [Sov. Astron. Lett. **8**, 366 (1982)].

Slysh, V. I., Pis'ma Astron. Zh. **1**, No. 8, 12 (1975) [Sov. Astron. Lett. **1**, 161 (1975)].

Slysh, V. I., Wilson, T. L., Pauls, T., and C. Henkel, in: IAU Symp. No. 87, *Interstellar Molecules*, B. H. Andrew (ed.), (1979), p. 473.

Smith, A., in: IAU Colloq. No. 101, *Supernova Remnants and the Interstellar Medium*, R. Roger and T. Landecker (eds.), Cambridge University Press, Cambridge (1988a), p. 119.

Smith, A., Jones, L. R., Watson, M. G., Willingale, R., Wood, N., and Seward, F. D., Mon. Not. R. Astron. Soc. **217**, 99 (1985b).

Smith, A., Davelaar, J., Peacock, A., Taylor, B. G., Morini, M., and Robba, N. R., Astrophys. J. **325**, 288 (1988a).

Smith, A. M., Cornett, R. H., and Hill, R. S., Astrophys. J. **320**, 609 (1987).

Smith, B. W. and Jones, E. M., in: IAU Colloq. No. 101, *Supernova Remnants and the Interstellar Medium*, R. Roger and T. Landecker (eds.), Cambridge University Press, Cambridge (1988), p. 133.

Smith, L., in: *Wolf-Rayet Stars*, Proceeding of a symposium at Boulder, Colorado (1968), p. 23.

Smith, L. F., Astrophys. J. **327**, 128 (1988b).

Smith, L. F. and Batchelor, R. A., Austral. J. Phys. **23**, 203 (1970).

Smith, L. F. and Maeder, A., Astron. Astrophys. **211**, 71 (1989).

Smith, L. J., Pettini, M., Dyson, J. E., and Hartquist, T. W., Mon. Not. R. Astron. Soc. **211**, 679 (1984).

Smith, L. J., Lloyd, C., and Walker, E. N., Astron. Astrophys. **146**, 307 (1985).

Smith, L. J., Pettini, M., Dyson, J. E., and Hartquist, T. W., Mon. Not. R. Astron. Soc. **234**, 625 (1988b).

Smith, M. G., Astrophys. J. **182**, 111 (1973).

Snow, T. P., Astrophys. J. Lett. **253**, L39 (1982).

Snow, T. P. and Morton, D. C., Astrophys. J. Suppl. Ser. **32**, 429 (1976).

Snyder, W. A., Davidsen, A. F., Henry, R. C., Shulman, S., Fritz, G., and Friedman, H., Astrophys. J. Lett. **222**, L13 (1978).

Sofue, Y., Publ. Astron. Soc. Jap. **25**, 207 (1973).

Sofue, Y., Astron. Astrophys. **48**, 1 (1976).

Sofue, Y., Astron. Astrophys. **67**, 409 (1978).

Sofue, Y., Hamajima, K., and Fujimoto, M., Publ. Astron. Soc. Jap. **26**, 399 (1974).

Sofue, Y., Fürst, E., and Hirth, W., Publ. Astron. Soc. Jap. **32**, 1 (1980).

Sofue, Y., Takahara, F., and Hirabayashi, H., Publ. Astron. Soc. Jap. **35**, 447 (1983).

Solf, J. and Carsenty, U., Astron. Astrophys. **116**, 54 (1982).

Sonneborn, G., Altner, B., and Kirshner, R. P., Astrophys. J. Lett. **323**, L51 (1987).

Spitzer, L. Jr., *Diffuse Matter in Space*, Wiley, New York (1968).

Spitzer, L. Jr., *Physical Processes in the Interstellar Medium*, Wiley, New York (1978).

Spoelstra, T. A. T., Astron. Astrophys. **21**, 61 (1972).

Sramek, R. A., van der Hulst, J. M., and Weiler, K. W., IAU Circ. No. 3557 (1980).

Sramek, R. A., Panagia, N., and Weiler, K. W., Astrophys. J. Lett. **285**, L59 (1984).

Sramek, R. A., Weiler, K. W., van der Hulst, J. M., and Panagia, N., Bull. Amer. Astron. Soc. **17**, 566 (1985).

Stahl, O., Astron. Astrophys. **182**, 229 (1987).

Stahl, O., in: *Physics of Luminous Blue Variables*, K. Davidson, A. F. J. Moffat, and H. J. Lamers (eds.), Kluwer, Dordrecht (1990), p. 149.

Stahl, O. and Wolf, B., Astron. Astrophys. **158**, 371 (1986).

Star Formation in Galaxies, Proceedings of the Second IRAS Conference, N. Scoville (ed.), (1987).

IAU Symp. No. 115, *Star Forming Regions*, M. Peimbert and J. Jugaku (eds.), Reidel, Dordrecht (1987).

Steigman, G., Strittmatter, P. A., and Williams, R. E., Astrophys. J. **198**, 575 (1975).

Stephenson, F. R., Clark, D. H., and Crawford, D. F., Mon. Not. R. Astron. Soc. **180**, 567 (1977).

Stewart, G. C., Fabian, A. C., and Seward, F. D., in: IAU Symp. No. 101, *Supernova Remnants and their X-Ray Emission*, J. Danziger and P. Gorenstein (eds.), Reidel, Dordrecht (1983), p. 59.

Storey, M. C. and Manchester, R. N., Nature **329**, 421 (1987).

Straka, W. C., Dickel, J. R., Blair, W. P., and Fesen, R. A., Astrophys. J. **306**, 266 (1986).

Strom, R. G., Astrophys. J. **319**, L103 (1987).

Strom, R. G. and Blair, W. P., Astron. Astrophys. **149**, 259 (1985).

Strom, R. G. and Sutton, J., Astron. Astrophys. **42**, 299 (1975).

Strom, R. G., Angerhofer, P. E., and Dickel, J. R., Astron. Astrophys. **139**, 43 (1984).

Strom, R. G., Goss, W. M., and Shaver, P. A., Mon. Not. R. Astron. Soc. **200**, 473 (1982).

Suchkov, A. A. and Shchekinov, Yu. A., Pis'ma Astron. Zh. **10**, 35 (1984) [Sov. Astron. Lett. **10**, 13 (1984)].

Svoboda, K. et al. (20 authors), in: *ESO Workshop on the SN 1987A; ESO Conference and Workshop Proceedings No. 26*, I. J.Danziger (ed.) (1987), p. 229.

Swinbank, E., Mon. Not. R. Astron. Soc. **193**, 451 (1980).

Swinbank, E. and Pooley, G., Mon. Not. R. Astron. Soc. **186**, 775 (1979).

Syunyaev, R. A., Kapiovskii, A., Efremov, V., et al., Pis'ma Astron. Zh. **13**, 1027 (1987a) [Sov. Astron. Lett. **13**, 431 (1987)].

Syunyaev (Sunyaev), R. A., Kaniovskii (Kaniovsky), A., Efremov, V., et al. (34 authors), Nature **330**, 227 (1987b).

T

Talent, D. L. and Dufour, R. J., Astrophys. J. **233**, 888 (1979).

Tammann, G. A., *Supernovae and Supernova Remnants*, Reidel, Dordrecht (1977), pp. 155-185.

Tammann, G. A., in: *Supernovae* (Astrophys. Space Sci. Libr. No. 66), 3rd ed., D. N. Schramm (ed.), Reidel, Dordrecht (1977), p. 95.

Tammann, G. A., Mem. Soc. Astron. Ital. **49**, 315 (1978).

Tammann, G. A., in: *Supernovae: A Survey of Current Research*, M. J. Rees and R. J. Stoneham (eds.), Reidel, Dordrecht (1982), p. 371.

Tatematsu, K., Fukui, Y., Inata, T., Kogure, T., Ogawa, H., and Kawabata, K., Astron. Astrophys. **184**, 279 (1987).

Tatematsu, K., Fukui, Y., Inata, T., and Nakano, in: IAU Colloq. No. 101, *Supernova Remnants and the Interstellar Medium*, R. Roger and T. Landecker (eds.), Cambridge University Press, Cambridge (1988), p. 257.

Taylor, G. I., Proc. R. Soc. London A **101**, 159 (1950).

Taylor, K. and Münch, G., Astron. Astrophys. **70**, 359 (1978).

Tenorio-Tagle, G., Astron. Astrophys. **94**, 338 (1981).

Tenorio-Tagle, G. and Rózyczka, M., Astron. Astrophys. **155**, 120 (1986).

Tenorio-Tagle, G. and Rózyczka, M., Astron. Astrophys. **176**, 329 (1987).

Tenorio-Tagle, G., Beltrametti, M., Bodenheimer, P., and York, H. W., Astron. Astrophys. **112**, 104 (1982).

Tenorio-Tagle, G., Franco, J., Bodenheimer, P., and Rózyczka, M., Astron. Astrophys. **179**, 219 (1987).

Teske, R. G. and Kirshner, R. P., Astrophys. J. **292**, 22 (1985).

Teske, R. G. and Petre, R., Astrophys. J. **318**, 370 (1987).

Tinsley, B. M., Publ. Astron. Soc. Pac. **87**, 837 (1975).

Tinsley, B. M., in: *Supernovae*, D. N. Schramm (ed.), Reidel, Dordrecht (1977), p. 117.

Tinsley, B. M., Astrophys. J. **229**, 1046 (1979).

Tomisaka, K., and Ikeuchi, S., Publ. Astron. Soc. Jap. **38**, 697 (1986).

Tomisaka, K., Habe, A., and Ikeuchi, S., Astrophys. Space Sci. **78**, 273 (1981).

Toor, A., Astron. Astrophys. **85**, 184 (1980).

Toor, A., Palmeri, T. M., and Seward, F. D., Astrophys. J. **209**, 96 (1976).

Torres, A. V., Conti, P. S., and Massey, P., Astrophys. J. **300**, 379 (1986).

Treffers, R. R., Astrophys. J. **233**, L17 (1979).

Treffers, R. R., Astrophys. J. **250**, 213 (1981).

Treffers, R. R. and Chu, You-Hua, Astrophys. J. **254**, 569 (1982).

Trimble, V., Rev. Mod. Phys. **54**, 1183 (1982).

Trimble, V., Rev. Mod. Phys. **55**, 511 (1983).

Trimble, V., J. Astrophys. Astron **5**, 389 (1984).

Troland, T. H., Crutcher, R. M., and Heiles, C., Astrophys. J. **298**, 808 (1985).

Truran, J. W. and Weiss, A., in: *ESO Workshop on the SN 1987A; ESO Conference and Workshop Proceedings No. 26*, I. J.Danziger (ed.) (1987), p. 271.

Trushkin, S. A., Pis'ma Astron. Zh. **12**, 198 (1986a) [Sov. Astron. Lett. **12**, 81 (1986)].

Trushkin, S. A., Astron. Tsirk. No. 1453, 4 (1986b).

Tsunemi, H., Yamashita, R., Masai, K., Hayakawa, S., and Koyama, K., Astrophys. J. **306**, 248 (1986).

Tsvetkov, D. Yu., Astron. Zh. **60**, 37 (1983) [Sov. Astron. **27**, 22 (1983)].

Tsvetkov, D. Yu., Perem. Zvezdy **22**, 279 (1986).

Tsvetkov, D. Yu., Astron. Zh. **64**, 79 (1987a) [Sov. Astron. **31**, 39 (1987)].

Tsvetkov, D. Yu., Pis'ma Astron. Zh. **13**, 894 (1987b) [Sov. Astron. Lett. **13**, 376 (1987)].

Tsvetkov, D. Yu., private communication (1989).

Tuffs, R. J., in: IAU Symp. No. 101, *Supernova Remnants and their X-Ray Emission*, J. Danziger and P. Gorenstein (eds.), Reidel, Dordrecht (1983), p. 49.

Tuffs, R. J., Mon. Not. R. Astron. Soc. **219**, 13 (1986).

Tuohy, I. R., Clark, D. H., and Burton, W. M., Astrophys. J. Lett. **260**, L65 (1982).

Tuohy, I. R., Clark, D. H., and Garmire, G. P., Mon. Not. R. Astron. Soc. **189**, 59P (1979b).

Tuohy, I. R. and Dopita, M. A., Astrophys. J. Lett. **268**, L7 (1983).

Tuohy, I. R., Dopita, M. A., Mathewson, D. S., Long, K. S., and Helfand, D. J., in: IAU Symp. No. 101, *Supernova Remnants and their X-Ray Emission*, J. Danziger and P. Gorenstein (eds.), Reidel, Dordrecht (1983), p. 559.

Tuohy, I. R., Garmire, G. P., Manchester, R. N., and Dopita, M. A., Astrophys. J. **268**, 778 (1983a).

Tuohy, I. R., Mason, K. O., Clark, D. H., Cordova, F., Charles, P. A., Walter, F. M., and Garmire, G. P., Astrophys. J. Lett. **230**, L27 (1979d).

Tuohy, I. R., Nousek, J. A., and Garmire, G. P., Astrophys. J. **234**, 101 (1979c).

Tuohy, I. R., Nugent, J. J., Garmire, G. P., and Clark, D. H., Nature **279**, 139 ((1979a).

Turtle, A. J., Campbell-Wilson, D., Bunton, J. D., Jauncey, D. L., Kesteven, M. J., Manchester, R. N., Norris, R. P., Storey, M. C., and Reynolds, J. E., Nature **327**, 38 (1987).

Tutukov, A. V. and Yungel'son, L. R., Nauchn. Inform. Astrosoveta Akad. Nauk SSSR **27**, 58; 70 (1973).

Tutukov, A. V. and Yungel'son, L. R., Pis'ma Astron. Zh. **9**, 230 (1983) [Sov. Astron. Lett. **9**, 124 (1983)].

U

Udal'tsov, V. A., Pynzar', A. V., and Glushak, A. P., Usp. Fiz. Nauk **124**, 725 (1978) [Sov. Phys. Usp. **21**, 365, (1978)].

Ulmer, M., Crane, P. C., Brown, R. L., and van der Hulst, J. M., Nature **285**, 151 (1980).

Underhill, A. B., Astrophys. J. **276**, 583 (1984).

Uomoto, A., Astrophys. J. **310**, L35 (1986).

Uomoto, A. and Kirshner, R. P., Astrophys. J. **308**, 685 (1986).

Utrobin, V. P., Astrophys. Space Sci. **55**, 441 (1978).

Utrobin, V. P. and Chugai, N. N., Preprint No. 86, Inst. Prikl. Mat. Akad. Nauk SSSR (1979), p. 3.

Utrobin, V. P., Astrophys. Space Sci. **98**, 115 (1984).

V

Vanbeveren, D. and Packet, W., Astron. Astrophys. **80**, 242 (1979).

Van Buren, D., Astrophys. J. **294**, 567 (1985).

Van Buren, D., Astrophys. J. **306**, 538 (1986).

Van Buren, D. and McCray, R., Astrophys. J. **329**, L93 (1988).

van den Bergh, S., Astrophys. J. **165**, 457 (1971).

van den Bergh, S., Astrophys. Space Sci. **38**, 447 (1975).

van den Bergh, S., Astrophys. J. Lett. **208**, L17 (1976).

van den Bergh, S., Publ. Astron. Soc. Pac. **89**, 637 (1977).

van den Bergh, S., Astrophys. J. Lett. **220**, L9 (1978a).

van den Bergh, S., Astrophys. J. Suppl. Ser. **38**, 119 (1978b).

van den Bergh, S., Astron. Astrophys. **86**, 155 (1980a).

van den Bergh, S., Astrophys. J. Lett. **236**, L23 (1980b).

van den Bergh, S., in: IAU Symp. No. 101, *Supernova Remnants and their X-Ray Emission*, J. Danziger and P. Gorenstein (eds.), Reidel, Dordrecht (1983).

van den Bergh, S., in: *ESO Workshop on the SN 1987A; ESO Conference and Workshop Proceedings No. 26*, I. J.Danziger (ed.) (1987), p. 557.

van den Bergh, S., Astrophys. J. **327**, 156 (1988).

van den Bergh, S. and Kamper, K. W., Astrophys. J. **218**, 617 (1977).

van den Bergh, S. and Kamper, K. W., Astrophys. J. **268**, 129 (1983).

van den Bergh, S. and Kamper, K. W., Astrophys. J. Lett. **280**, L51 (1984).

van den Bergh, S. and Kamper, K. W., Astrophys. J. **293**, 537 (1985).

van den Bergh, S. and Maza, J., Astrophys. J. **204**, 519 (1976).

van den Bergh, S. and Pritchet, C. J., Publ. Astron. Soc. Pac. **98**, 448 (1986).

van den Bergh, S., Marscher, A. P., and Terzian, Y., Astrophys. J. Suppl. Ser. **26**, 19 (1973).

van den Bergh, S., McClure, R. D., and Evans, R., Astrophys. J. **323**, 44 (1987).

van den Heuvel, E. P. J., in: IAU Symp. No. 73, *Structure and Evolution of Close Binary Systems*, P. Eggleton, P. C. Mitton, and J. H. Whelan (eds.), Reidel, Dordrecht (1976), p. 35.

van der Hucht, K. A., in: IAU Symp. No. 105, *Observational Tests of the Stellar Evolution Theory*, A. Maeder and A. Renzini (eds.), Reidel, Dordrecht (1984), p. 273.

van der Hucht, K. A. and Hidayat, B. (eds.), IAU Symp. No. 143, *Wolf–Rayet Stars and Interrelations with other Massive Stars in Galaxies*, Kluwer, Dordrecht (1990).

van der Hucht, K. A., Cassinelli, J. P., and Williams, P. M., Astron. Astrophys. **168**, 111 (1986).

van der Hucht, K. A., Conti, P. S., Lundström, I., and Stenholm, B., Space Sci. Rev. **3**, 227 (1981).

van der Hucht, K. A., Jurriens, T. A., Olnon, F. M., Thé, P. S., Wesselius, R. P., and Williams, P. M., Astron. Astrophys. **145**, L13 (1985).

van der Hucht, K. A., Williams, P. M., and Thé, P. S., Quart. J. R. Astron. Soc. **28**, 254 (1987).

van der Laan, H., Mon. Not. R. Astron. Soc. **124**, 125 (1962).

Van Riper, K. A., in: IAU Symp. No. 101, *Supernova Remnants and their X-Ray Emission*, J. Danziger and P. Gorenstein (eds.), Reidel, Dordrecht (1983), p. 513.

Vartanian, M. N., Lum, R. S. R., and Ku, W. H.-M., Astrophys. J. Lett. **288**, L5 (1985).

Vedder, P. W., Canizares, C. R., Markert, T. H., and Pradhan, A. K., Astrophys. J. **307**, 269 (1986).

Velusamy, T., Nature **308**, 251 (1984).

Velusamy, T., Mon. Not. R. Astron. Soc. **212**, 359 (1985).

Velusamy, T., in: IAU Colloq. No. 101, *Supernova Remnants and the Interstellar Medium*, R. Roger and T. Landecker (eds.), Cambridge University Press, Cambridge (1988), p. 265.

Velusamy, T. and Becker, R. H., Astron. J. **95**, 1162 (1988).

Velusamy, T. and Kundu, M. R., Astron. Astrophys. **32**, 375 (1974).

Velusamy, T. and Sarma, N. V. G., Mon. Not. R. Astron. Soc. **181**, 455 (1977).

Venger, A. P., Gosachinskii, I. V., Grachev, V. G., Egorova, T. M., Ryzhkov, N. F., and Khersonskii, V. K., Astron. Zh. **58**, 1187 (1981) [Sov. Astron. **25**, 675 (1981)].

Venger, A. P., Gosachinskii, I. V., Grachev, V. G., Egorova, T. M., Ryzhkov, N. F., and Khersonskii, V. K., Astron. Zh. **59**, 20 (1982) [Sov. Astron. **26**, 20 (1982)].

Vid'machenko, A. P., Gnedin, Yu. N., Larionov, V. M., and Larionova, L. V., Pis'ma Astron. Zh. **14**, 387 (1988) [Sov. Astron. Lett. **14**, 163 (1988)].

Vílchez, J. M. and Esteban, C., in: IAU Symp. No. 143, *Wolf–Rayet Stars and Interrelations with other Massive Stars in Galaxies*, K. A. van der Hucht and B. Hidayat (eds.), Kluwer, Dordrecht (1990).

Vil'koviskii, É. Ya., Astrofiz. **17**, 309 (1981).

Vinyaikin, E. N. and Razin, V. A., Astron. Zh. **56**, 913 (1979) [Sov. Astron. **23**, 525 (1979)].

Vinyaikin, E. N., Razin, V. A., and Khrulev, V. V., Pis'ma Astron. Zh. **6**, 620 (1980) [Sov. Astron. Lett. **6**, 324 (1980)].

Vinyaikin, E. N., Volodin, Yu. V., Dagkesamanskii, R. D., and Sokolov, K. P., Astron. Zh. **64**, 271 (1987) [Sov. Astron. **31**, 141 (1987)].

Volk, K. and Kwok, S., Astron. Astrophys. **153**, 79 (1985).

Vreux, J.-M., Publ. Astron. Soc. Pac. **97**, 274 (1985).

Vreux, J.-M., Andrillat, J., and Gosset, E., Astron. Astrophys. **149**, 337 (1985).

W

Walborn, N. R., Astrophys. J. **256**, 452 (1982).

Walborn, N. R. and Hesser, J. E., Astrophys. J. **252**, 156 (1982).

Walborn, N. R., Lasker, B. M., Laidler, V. G., and Chu You-Hua, Astrophys. J. Lett. **321**, L41 (1987).

Wallerstein, G. and Jacobsen, T. S., Astrophys. J. **207**, 53 (1976).

Wamsteker, W., Gilmozzi, R., Cassatella, A., and Panagia, N., IAU Circ. No. 4410 (1987).

Wang, Z.-R., and Seward, F. D., Astrophys. J. **285**, 607 (1984).

Wang, Z.-R., Liu, J. Y., Gorenstein, P., and Zombeck, M. V., Highlights of Astronomy **7**, 583 (1986).

Wannier, P. G. and Sahai, R., Astrophys. J. **311**, 335 (1986).

Watson, M. G., Willingale, R., Pye, J. P., Rolf, D. P., Wood, N., Thomas, N., and Seward, F. D., in: IAU Symp. No. 101, *Supernova Remnants and their X-Ray Emission*, J. Danziger and P. Gorenstein (eds.), Reidel, Dordrecht (1983a), p. 273.

Watson, M. G., Willingale, R., Grindlay, J. E., and Seward, F. D., Astrophys. J. **273**, 688 (1983b).

Weaver, H., in: IAU Symp. No. 84, *Large-Scale Characteristics of the Galaxy*, W. B. Burton (ed.), Reidel, Dordrecht (1979), p. 295.

Weaver, R., McCray, R., Castor, J., Shapiro, P., and Moore, R., Astrophys. J. **218**, 377 (1977).

Weaver, T. A. and Woosley, S. E., Ann. N. Y. Acad. Sci. **336**, 335 (1980).

Westerlund, B. E., Archiv für Astron. Bd. 2 **44**, 467 (1960).

Weiler, K. W., Astron. Astrophys. **84**, 271 (1980).

Weiler, K. W., Observatory **103**, 85 (1983).

Weiler, K. W. and Panagia, N., Astron. Astrophys. **90**, 269 (1980).

Weiler, K. W., van der Hulst, J. M., Sramek, R., and Panagia, N., Astrophys. J. Lett. **243**, L151 (1981).

Weiler, K. W., Sramek, R., van der Hulst, J. M., and Panagia, N., in: IAU Symp. No. 101, *Supernova Remnants and their X-Ray Emission*, J. Danziger and P. Gorenstein (eds.), Reidel, Dordrecht (1983a), p. 171.

Weiler, K. W., Sramek, R. A., Panagia, N., van der Hulst, J. M., and Salvati, M., Astrophys. J. **301**, 790 (1986).

Weisskopf, M. C., Silver, E. H., Kestenbaum, H. L., Long, K. S., Novick, R., Astrophys. J. Lett. **220**, L117 (1978).

Wendker, H. J., Astron. Astrophys. **13**, 65 (1971).

Wendker, H. J., Smith, L. F., Israel, F., Habing, H. J., and Dickel, H. R., Astron. Astrophys. **42**, 173 (1975).

West, R. M., Lauberts, A., Jorgensen, H. E., and Schuster, H.-E., Astron. Astrophys. **177**, L1 (1987).

Westerlund, B. E., Arch. für Astron. Bd. 2 **44**, 467 (1960).

Wheeler, J. C. and Levreault, R., Astrophys. J. **294**, L17 (1985).

Wheeler, J. C., Harkness, R. P., Barker, E. S., Cochran, A. L., and Wills, D., Astrophys. J. Lett. **313**, L69 (1987).

Wheeler, J. C., Mem. Soc. Astron. Ital. **49**, 349 (1978).

Wheeler, J. C., and Bash, F. N., Nature **268**, 706 (1977).

Wheeler, J. C. and Levreault, R., Astrophys. J. Lett. **294**, L17 (1985).

White, G. J., Rainey, R., Hayashi, S. S., and Kaifu, N., Astron. Astrophys. **173**, 337 (1987).

White, R. L. and Long, K. S., Astrophys. J. **264**, 196 (1983).

White, R. L. and Long, K. S., Astrophys. J. **310**, 832 (1986).

Whitehead, M. J., Meaburn, J., and Goudis, C., Astron. Astrophys. **196**, 261 (1988).

Whitehead, M. J., Meaburn, J., and Clayton, C. A. Mon. Not. R. Astron. Soc. **237**, 1109 (1989).

Whitelock, P. A., Catchpole, R. M., Menzies, J. W., Feast, M. W., et al., Preprint AAO (1988).

Whittet, D. C. B., Somerville, W. B., McNally, D., Blades, J. C., Mon. Not. R. Astron. Soc. **189**, 519 (1979).

Williams, P. M., van der Hucht, K. A., and Thé, P. S., Quart. J. R. Astron. Soc. **28**, 249 (1987).

Williams, R. E., in: *ESO Workshop on the SN 1987A; ESO Conference and Workshop Proceedings No. 26*, I. J.Danziger (ed.) (1987).

Williams, R. E., in: IAU Symp. No. 108, *Atmospheric Diagnostics of Stellar Evolution*, Tokyo, 1987, K. Nomoto (ed.), Springer-Verlag, New York (1988).

Willis, A. G., Astron. Astrophys. **26**, 237 (1973).

Willis, A. G., in: IAU Symp. No. 99, *WR Stars: Observations, Physics, Evolution*, C. W. De Loore and A. J. Willis (eds.), Reidel, Dordrecht (1982a), p. 87.

Willis, A. G., Mon. Not. R. Astron. Soc. **198**, 897 (1982b).

Willis, A. G., in: IAU Symp. No. 143, *Wolf–Rayet Stars and Interrelations with other Massive Stars in Galaxies*, K. A. van der Hucht and B. Hidayat (eds.), Kluwer, Dordrecht (1990).

Willis, A. G. and Wilson, R., in: IAU Symp. No. 83, *Mass Loss and Evolution of O-Type Stars*, P. S. Conti and C. W. H. De Loore (eds.), Reidel, Dordrecht (1980), p. 461.

Wilson, A. S., Observatory **103**, 73 (1983).

Wilson, A. S., Astrophys. J. **302**, 718 (1986).

Wilson, A. S. and Weiler, K. W., Astron. Astrophys. **53**, 89 (1976).

Wilson, A. S. and Weiler, K. W., Nature **300**, 155 (1982).

Wilson, A. S., Samarasinha, N. H., and Hogg, D. E., in: *The Crab Nebular and Related Supernova Remnants*, M. Kafatos and R. B. C. Henry (eds.), Cambridge University Press, Cambridge (1985a).

Wilson, A. S., Samarasinha, N. H., and Hogg, D. E., Astrophys. J. Lett. **294**, L121 (1985b).

Wilson, W. J., Schwartz, R. P., Epstein, E. E., Johnson, W. A., Echeverry, R. D., Mori, T. T., Berry, G. G., and Dyson, H. B., Astrophys. J. **191**, 357 (1974).

Winkler, P. F. and Kirshner, R. P., Astrophys. J. **299**, 981 (1985).

Winkler, P. F., Hearn, D. R., Richardson, J. A., and Behnken, J. M., Astrophys. J. Lett. **299**, L123 (1979).

Winkler, P. F., Canizares, C. R., Clark, G. W., Markert, T. H., and Petre, R., Astrophys. J. **245**, 574 (1981).

Winkler, P. F., Canizares, C. R., and Bromley, B. C., in: IAU Symp. No. 101, *Supernova Remnants and their X-Ray Emission*, J. Danziger and P. Gorenstein (eds.), Reidel, Dordrecht (1983), p. 245.

Winkler, P. F., Tuttle, J. H., Kirshner, R. P., and Irwin, M. J., in: IAU Colloq. No. 101, *Supernova Remnants and the Interstellar Medium*, R. Roger and T. Landecker (eds.), Cambridge University Press, Cambridge (1988), p. 65.

Wisotzki, L. and Wendker, H. J., Astron. Astrophys. **221**, 311 (1989).

Wolff, M. T. and Durisen, R. H., Mon. Not. R. Astron. Soc. **224**, 701 (1987).

Woltjer, L. and Véron-Cetty, M. P., Astron. Astrophys. **172**, L7 (1987).

Wood, P. R., Astrophys. J. **339**, 1073 (1989).

Woodgate, B. E., Stockman, H., Angel, J. R., and Kirshner, R. P., Astrophys. J. Lett. **188**, L79 (1974).

Woodgate, B. E., Kirshner, R. P., and Balon, R. J., Astrophys. J. Lett. **218**, L129 (1977).

Woodgate, B. E., Lucke, R. L., and Socker, D. G., Astrophys. J. Lett. **229**, L119 (1979).

Woodward, P. R., Astrophys. J. **207**, 484 (1976).

Woosley, S. E., Astrophys. J. **330**, 218 (1988).

Woosley, S. E. and Weaver, T. A., in: *Supernovae: A Survey of Current Research*, M. J. Rees and R. J. Stoneham (eds.), Reidel, Dordrecht (1982), p. 79.

Woosley, S. E. and Weaver, T. A., Ann. Rev. Astron. Astrophys. **24**, 203 (1986).

Woosley, S. E., Pinto, P. A., Martin, P. G., and Weaver, T. A., Astrophys. J. **318**, 664 (1987).

Wootten, H. A., Astrophys. J. **216**, 440 (1977).

Wootten, H. A., Astrophys. J. **245**, 105 (1981).

Wootten, H. A., Slair, G. N., and Vanden Bout, P., Bull. Amer. Astron. Soc. **7**, 418 (1975).

Wright, E. L., Fazio, G. G., and Low, F. J., Astrophys. J. Lett. **208**, L87 (1976).

Wright, M. C. H. and Forster, J. R., Astrophys. J. **239**, 873 (1980).

Wu, Chi-Chao, Leventhal, M., Sarazin, C. L., and Gull, T. R., Astrophys. J. Lett. **269**, L5 (1983).

X

Xu, Y., Sutherland, P., McCray, R., and Ross, R. R., Astrophys. J. **327**, 197 (1988).

Y

Yakovlev, D. G. and Urpin, V. A., Pis'ma Astron. Zh. **7**, 157 (1981) [Sov. Astron. Lett. **7**, 88 (1981)].

York, H. W., Tenorio-Tagle, G., and Bodenheimer, P., in: IAU Symp. No. 101, *Supernova Remnants and their X-Ray Emission*, J. Danziger and P. Gorenstein (eds.), Reidel, Dordrecht (1983) p. 393.

Z

Zarnecki, J. C., Culhane, J. L., Toor, A., Seward, F. D., and Charles, P. A., Astrophys. J. Lett. **219**, L17 (1978).

Zealey, W. J., Elliott, K. H., and Malin, D. F., Astron. Astrophys. Suppl. Ser. **38**, 39 (1979).

Zealey, W. J., Dopita, M. A., and Malin, D. F., Mon. Not. R. Astron. Soc. **192**, 731 (1980).

Zealey, W. J., McGillivray, H. T., Malin, D. F., and Hartl, H., in: IAU Symp. No. 101, *Supernova Remnants and their X-Ray Emission*, J. Danziger and P. Gorenstein (eds.), Reidel, Dordrecht (1983), p. 267.

Zeinalov, S. K., Musaev, F. A., and Chentsov, E. L., Pis'ma Astron. Zh. **13**, 223 (1987) [Sov. Astron. Lett. **13**, 90 (1987)].

Zel'dovich, Ya. B. and Raizer, Yu. M., *Physics of Shock Waves and High-Temperature Hydrodynamic Phenomena*, Academic Press, New York (1967).

Zwicky, F., Rev. Mod. Phys. **12**, 66 (1940).

Index of Celestial Objects

Printed in the United States
By Bookmasters